Studies in Inorganic Chemistry 18

Structure and Chemistry of the Apatites and Other Calcium Orthophosphates

Studies in Inorganic Chemistry

Other titles in this series

1. **Phosphine, Arsine and Stibine Complexes of the Transition Elements**
   by C.A. McAuliffe and W. Levason

2. **Phosphorus: An Outline of its Chemistry, Biochemistry and Technology (Second Edition)**
   by D.E.C. Corbridge

3. **Solid State Chemistry 1982**
   edited by R. Metselaar, H.J.M. Heijligers and J. Schoonman

4. **Gas Hydrates**
   by E. Berecz and M. Balla-Achs

5. **Sulfur: Its Significance for Chemistry, for the Geo-, Bio-, and Cosmosphere and Technology**
   edited by A. Müller and B. Krebs

6. **Phosphorus: An Outline of its Chemistry, Biochemistry and Technology (Third Edition)**
   by D.E.C. Corbridge

7. **Inorganic High Pressure Chemistry: Kinetics and Mechanisms**
   edited by R. van Eldik

8. **Graphite Fluorides**
   by N. Watanabe, T. Nakajima and H. Touhara

9. **Selected Topics in High Temperature Chemistry: Defect Chemistry of Solids**
   edited by Ø. Johannesen and A.G. Andersen

10. **Phosphorus: An Outline of its Chemistry, Biochemistry and Technology (Fourth Edition)**
    by D.E.C. Corbridge

11. **Chemistry of the Platinum Group Metals**
    edited by F.R. Hartley

12. **Luminescence and the Solid State**
    by R.C. Ropp

13. **Transition Metal Nuclear Magnetic Resonance**
    edited by P.S. Pregosin

14. **Chemistry of Inorganic Ring Systems**
    edited by R. Steudel

15. **Inorganic Polymeric Glasses**
    by R.C. Ropp

16. **Electron Paramagnetic Resonance of d Transition Metal Compounds**
    by F.E. Mabbs and D. Collison

17. **The Chemistry of Artificial Lighting Devices. Lamps, Phosphors and Cathode Ray Tubes**
    by R.C. Ropp

Studies in Inorganic Chemistry 18

Structure and Chemistry of the Apatites and Other Calcium Orthophosphates

J.C. Elliott

Department of Child Dental Health
The London Hospital Medical College
Turner Street, London E1 2AD, U.K.

ELSEVIER
Amsterdam — London — New York — Tokyo   1994

ELSEVIER SCIENCE B.V.
Sara Burgerhartstraat 25
P.O. Box 211, 1000 AE Amsterdam, The Netherlands

Library of Congress Cataloging-in-Publication Data

Elliott, J. C. (James Cornelis), 1937-
    Structure and chemistry of the apatites and other calcium
orthophosphates / J.C. Elliott.
        p.   cm. -- (Studies in inorganic chemistry ; 18)
    Includes bibliographical references and index.
    ISBN 0-444-81582-1
    1. Apatite.   2. Calcium phosphate.   I. Title.   II. Series.
QD181.P1E44   1994
546'.39324--dc20
                                                        94-8066
                                                        CIP

ISBN 0-444-81582-1

This book is printed on acid-free paper.

Printed in The Netherlands

PREFACE

The apatites and related calcium phosphates have been of considerable interest to biologists, mineralogists, and inorganic and industrial chemists for many years. The reasons for this are clear: the apatites form the mineral component of bones and teeth, and the more acid calcium phosphates are probably also involved in mineralisation processes; many of the calcium phosphates occur in pathological calcifications; apatites and other calcium phosphates are increasingly used as biocompatible materials for bone replacement or for the coating of bone prostheses; they are widely distributed minerals, providing the world's supply of phosphorus, particularly phosphates for the production of fertilisers; and they are used as phosphors in fluorescent light tubes. The great versatility of apatites in accepting a large variety of substitutional ions complicates their study, but also presents considerable interest in its own right.

During the last thirty years or so, substantial progress has been made in the understanding of the structure and chemistry of the apatites and related calcium phosphates. Although, the basic apatite structure was known at the beginning of this period (its structure had been determined in 1930), important aspects of the structures of the other calcium orthophosphates were unclear; even the existence of one of them, octacalcium phosphate, had barely been universally accepted and little was known about the amorphous calcium phosphates. As regards the apatites, there were still heated disputes about the location of the carbonate that biological, many mineral and synthetic apatites contain. Another major puzzle was that apatites could be precipitated from solution with Ca/P molar ratios from 1.667 (that for hydroxyapatite) to about 1.5, without any apparent changes in the X-ray diffraction powder pattern.

Single crystal X-ray diffraction studies have provided much of the detail of the structures and crystal chemistry of the apatites and other calcium phosphates. However, progress in understanding the relationship between carbonate and apatites and the variable Ca/P ratios of precipitated apatites has had to rely on indirect methods because of the absence of suitable single crystals. Infrared spectroscopy, which was little used before the beginning of this period, and more recently, nuclear magnetic resonance spectroscopy have probably been the most useful.

Another area of progress has been in studies of the solution chemistry of the calcium phosphates. The aqueous phase diagram in the neutral and alkaline dilute region is now accurately known, as well as some aspects of the processes of their nucleation, crystal growth and dissolution. However, much remains to be understood, particularly how these are affected by interactions with other "impurity" ions (*e.g.* CO_3^{2-} and Mg^{2+} ions) and organic molecules. The answers to these questions lie at the heart of an understanding of mineral deposition and removal during the formation, remodelling and destruction of calcified tissues.

The aim in writing this book has been to provide an integrated account of the present knowledge of the structure of the apatites and other calcium orthophosphates, particularly in a biological context. Thus, in addition to a wide range of apatites which form the main subject of the book, the structures, crystal chemistry, preparation, solution chemistry and thermal decomposition of the monohydrogen phosphates, octacalcium phosphate, brushite and monetite, as well as the tricalcium phosphates, are also discussed in detail. References to studies of other alkaline earth phosphates, arsenates and vanadates, particularly those with an apatitic structure are included.

The arrangement of the major divisions of the material (either chapters or sections within chapters) is generally based on the type of compound, with subdivisions on specific aspects, such as structure, preparation, infrared spectra or reactions in solutions. This overall scheme is modified to emphasise the interrelationships between many of the compounds and to minimise duplication of material. The overriding principal has been to group material in the most natural way, and to provide cross-references from other sections if appropriate. This is sometimes done explicitly and sometimes by giving section references as well as literature references where other aspects of the cited work are discussed.

Section 1.1 contains a brief introduction to the calcium phosphates and a more detailed introduction to the apatites. This introduction includes references to review papers, conference proceedings and theses that deal with more specific aspects. The nonapatitic calcium orthophosphates are considered individually in other sections of Chapter 1. The three remaining chapters of the book are required to cover the much more extensive work on apatites. These three chapters are arranged in an order of increasing complexity. Chapter 2 starts with a discussion of the apatite structure based on the simplest compounds (fluor- and chlorapatite), followed by an account of the many possible lattice substitutions. The remainder of this chapter is on specific studies of fluor- and chlorapatite. Chapter 3 is devoted to hydroxyapatite and related nonstoichiometric apatites. In this chapter, apatites with oxygen in different oxidation states are discussed and consideration is given to the major complexities of precipitated nonstoichiometric apatitic calcium phosphates with Ca/P ratios substantially less than that of hydroxyapatite. As single crystals for X-ray diffraction analysis are not available for the study of these nonstoichiometric compounds, various special analytical methods have been developed, as discussed in a section in this chapter. Precipitated apatitic calcium phosphates generally have submicron dimensions and hence large surface areas; this results in reactivity in solution via exchange and adsorption of chemical species. Thus, these topics form a further significant part of Chapter 3.

Carbonate apatites are considered in the last chapter, starting with an introduction that contains a summary of work done prior to 1965. This is convenient because, until then, it was often not appreciated that many carbonate apatites from different sources contained CO_3^{2-} ions in different environments. Subsequent work is discussed under headings related to the type of carbonate apatite considered. This discussion starts with mineral apatites, then high temperature preparations, followed by synthetic and biological apatites which are formed in aqueous systems. These latter two are the most complicated as they combine the difficulties of nonstoichiometry from a variable Ca/P ratio, as discussed in Chapter 3, with the complications that arise from their carbonate content. The final section in Chapter 4 is a short summary of the properties of carbonate apatites which is followed by an Appendix of calculated X-ray diffraction patterns for the calcium orthophosphates.

J.C. Elliott

January, 1994

ACKNOWLEDGEMENTS

I have tried to give a comprehensive account of the chemistry and crystal chemistry of the apatites and related calcium phosphates and this has led me into areas in which I am by no means expert. I am therefore particularly grateful to Dr Edgard Moreno for his comments on various aspects of the physical chemistry of solutions and Dr James Yesinowski for improvements on sections dealing with nuclear magnetic resonance studies of apatites. Mr Bruce Fowler and Professor Ray Young kindly read and provided comments on the infrared and structural aspects of the apatites respectively. I am specially grateful for many detailed comments and suggestions for improvement from Dr Stephanie Dowker, particularly in the chapter on carbonate apatites. As a result of all these comments, many errors have been avoided. Dr Vitus Leung kindly calculated the $Ca(OH)_2$-H_3PO_4-H_2O phase and speciation diagrams. Daniel Morgenstein, Kamala Dawar and Alan Elliott are thanked for their substantial assistance in the preparation of camera-ready diagrams and tables, and for editorial and word processing assistance. I also thank the many colleagues who provided preprints of their papers and permission to quote from these, many of which have now appeared in print. Dr A. Sakthivel and Professor R.A. Young kindly provided a copy of their program for Rietveld analysis that was used for calculating the X-ray diffraction patterns in the Appendix.

CONTENTS

Preface   v

Acknowledgements viii

Chapter 1
General chemistry of the calcium orthophosphates   1
  1.1 Introduction   1
  1.2 Monocalcium phosphates (monohydrate and anhydrous)   9
  1.3 Octacalcium phosphate 12
    1.3.1 Introduction 12
    1.3.2 Structure 12
    1.3.3 Preparation 13
    1.3.4 Optical properties, density and habit 13
    1.3.5 Kinetics of nucleation and crystal growth 15
    1.3.6 Solubility and reactions in solution 16
    1.3.7 Thermal decomposition 18
    1.3.8 Infrared, Raman and NMR spectroscopy 20
    1.3.9 Double salts of octacalcium phosphate and dicarboxylic acids 22
  1.4 Dicalcium phosphate dihydrate, brushite 23
    1.4.1 Introduction 23
    1.4.2 Structure 24
    1.4.3 Preparation 24
    1.4.4 Optical properties, density and habit 26
    1.4.5 Kinetics of crystal growth and dissolution 26
    1.4.6 Solubility and reactions in solution 27
    1.4.7 Thermal decomposition and properties 29
    1.4.8 Infrared, Raman and NMR spectroscopy 30
  1.5 Dicalcium phosphate anhydrous, monetite 30
    1.5.1 Introduction 30
    1.5.2 Structure 31
    1.5.3 Preparation 32
    1.5.4 Optical properties, density and habit 32
    1.5.5 Solubility and reactions in solution 33
    1.5.6 Thermal decomposition and properties 33

1.5.7 Infrared, Raman and NMR spectroscopy 34
1.6 Anhydrous tricalcium phosphates and whitlockite 34
    1.6.1 Occurrence and importance 34
    1.6.2 Structures 35
    1.6.3 Optical properties, density and habit 42
    1.6.4 Preparation 43
    1.6.5 Solubility and reactions in solution 46
    1.6.6 Thermal decomposition 47
    1.6.7 Infrared and NMR spectroscopy 50
1.7 Tetracalcium phosphate 50
1.8 Amorphous calcium phosphates 53
    1.8.1 Introduction 53
    1.8.2 Preparation 53
    1.8.3 Chemical composition 54
    1.8.4 Transformation reactions 55
    1.8.5 Infrared and NMR spectroscopy 58
    1.8.6 Structure 60
    1.8.7 Thermal decomposition 61
    1.8.8 ESR of X-irradiated ACP 61

Chapter 2
Fluorapatite and chlorapatite 63
    2.1 Introduction 63
    2.2 Structures 64
        2.2.1 Fluorapatite and the apatite structure 64
        2.2.2 Structural relationships of apatite 70
        2.2.3 Chlorapatite structure 75
    2.3 Substitutions in apatites 80
        2.3.1 Hexad axis substitutions 80
        2.3.2 Substitutions for calcium ions 82
        2.3.3 Substitutions for phosphate ions 94
    2.4 Preparation of powders 95
        2.4.1 High temperature methods 95
        2.4.2 Solution methods 97
    2.5 Growth of single crystals 97
        2.5.1 Fluorapatite 97
        2.5.2 Chlorapatite 98
    2.6 Infrared and Raman spectra 99
        2.6.1 Fluorapatite 99

2.6.2 Chlorapatite 102
2.7 Other physical and chemical studies 104

Chapter 3
Hydroxyapatite and nonstoichiometric apatites 111
3.1 Introduction 111
3.2 Structure of hydroxyapatite 112
3.3 Preparation of stoichiometric hydroxyapatite powders 118
  3.3.1 Introduction 118
  3.3.2 Syntheses based on theoretical compositions 119
  3.3.3 Equilibrium syntheses in solution 121
  3.3.4 Miscellaneous methods 124
3.4 Preparation of other apatites with hydroxyl ions 125
  3.4.1 Calcium-deficient and nonstoichiometric apatites 125
  3.4.2 Calcium-rich apatites 127
  3.4.3 Apatites with oxygen in different oxidation states 127
  3.4.4 Miscellaneous preparations, including solid solutions 133
3.5 Growth of hydroxyapatite single crystals 137
3.6 Special analytical methods 139
  3.6.1 Hydroxyl ion content 139
  3.6.2 Acid phosphate content 141
  3.6.3 Miscellaneous analyses 145
3.7 Structure of calcium-deficient hydroxyapatites 148
  3.7.1 Introduction 148
  3.7.2 Surface adsorption 148
  3.7.3 Lattice substitutions 149
  3.7.4 Intercrystalline mixtures of OCP and OHAp 153
  3.7.5 Summary 154
3.8 Kinetics of nucleation and crystal growth 154
3.9 Solubility and interfacial phenomena 157
3.10 Reactions in solution 159
  3.10.1 Adsorption and surface reactions 159
  3.10.2 Reactions with fluoride ions 160
  3.10.3 Rate of dissolution 164
3.11 Infrared and Raman spectroscopy 169
  3.11.1 Introduction 169
  3.11.2 Hydroxyapatite spectrum 170
  3.11.3 High temperature OH stretching bands 173
  3.11.4 Bands from surface hydroxyl ions 175

3.11.5 Perturbations of OH bands by *c*-axis substitutions 175
3.11.6 Other "hydroxyapatites" 179
3.12 NMR spectroscopy 182
3.13 Other physical and chemical studies 186

Chapter 4
Mineral, synthetic and biological carbonate apatites 191
4.1 Introduction 191
    4.1.1 Occurrence, nomenclature and early structural work 191
    4.1.2 Growth of single crystals 196
4.2 Francolite and dahllite 199
    4.2.1 X-ray diffraction, chemical and optical studies 199
    4.2.2 Infrared and Raman spectroscopy 207
    4.2.3 Other physical and chemical studies 212
4.3 A-type carbonate apatite, $Ca_{10}(PO_4)_6CO_3$ 213
    4.3.1 Preparation 213
    4.3.2 Infrared and Raman spectroscopy 215
    4.3.3 Structure 218
    4.3.4 Other physical and chemical studies 222
4.4 Synthetic high temperature B-type carbonate apatites 223
    4.4.1 Apatites containing fluoride ions 223
    4.4.2 Fluoride-free compounds 225
    4.4.3 Sodium-containing compounds 227
4.5 Carbonate apatites from aqueous systems 229
    4.5.1 Introduction 229
    4.5.2 Assignment of IR and Raman carbonate bands 230
    4.5.3 Precipitated apatites (fluoride- and alkali-free) 234
    4.5.4 Precipitated apatites containing monovalent cations 239
    4.5.5 Precipitated apatites containing fluoride 243
    4.5.6 Reaction between alkaline phosphate solutions and calcium
          carbonate 246
    4.5.7 Thermal decomposition 248
    4.5.8 Other physical and chemical studies 254
4.6 Biological apatites 259
    4.6.1 Introduction 259
    4.6.2 X-ray diffraction studies 263
    4.6.3 Infrared and Raman spectroscopy 267
    4.6.4 Thermal decomposition 275
    4.6.5 Reactions in solution 281

4.6.6 NMR spectroscopy and other physical and chemical studies . 287
4.6.7 Calcium phosphates in biomaterials 295
4.7 Electron spin resonance of X-irradiated carbonate apatites 298
4.8 Summary 301

Appendix
Calculated X-ray diffraction patterns of the calcium orthophosphates . . . 305

References 311

Index 371

Chapter 1

GENERAL CHEMISTRY OF THE CALCIUM ORTHOPHOSPHATES

1.1 Introduction

The calcium orthophosphates are salts of the tribasic phosphoric acid, H_3PO_4, and thus can form compounds that contain $H_2PO_4^-$, HPO_4^{2-} or PO_4^{3-} ions. Those with $H_2PO_4^-$ ions only form under rather acidic conditions, and are therefore not normally found in biological systems. However, both HPO_4^{2-} and PO_4^{3-} ions occur in the mineral of bones and teeth, and in various pathological calcifications. Some calcium phosphates are hydrated, and those that belong to the basic apatitic calcium phosphate family contain OH^- ions. The system of abbreviations used throughout this book for these compounds is given in Tables 1.1 and 1.2. Calculated X-ray diffraction patterns of many of the calcium orthophosphates are given in the Appendix.

Pyrophosphates ($P_2O_7^{4-}$, dipolyphosphates) and polyphosphates, which contain P-O-P bonds, are of less biological importance than the orthophosphates, although calcium pyrophosphates occur in some pathological calcifications. Pyrophosphates are potent inhibitors of nucleation and crystal growth of CaPs in aqueous systems, and are also sometimes formed on ignition of hydrogen orthophosphates.

In addition to their biological importance, many CaPs occur as minerals [1-5], the most important of these being the apatites whose mineralogy will be briefly mentioned below.

The CaPs are all white solids (unless doped with a coloured ion): most are sparingly soluble in water, and some are very insoluble, but all dissolve in acids. The preparation and chemistry of the CaPs [6-8], ortho- and polyphosphates [9,10], and the alkaline earth ortho- and polyphosphates [11-13] have been reviewed. Structural aspects of biological CaPs have been discussed [14]. Details of the preparation and characteristics of the calcium pyrophosphates [15], and the preparation, X-ray, optical and IR characteristics of alkaline earth and other ortho- and pyrophosphates [16] have been published.

Phase diagrams show the thermodynamically stable phases, and give an indication of the likely conditions required for synthesis. However, the actual phase that forms under any given conditions is often dictated by kinetic, rather than thermodynamic, considerations. The phase diagram for the system $Ca(OH)_2-H_3PO_4-H_2O$ at normal temperatures and pressures is given in Fig. 1.1 and for the more acidic and soluble region, in Fig. 1.3 (Section 1.2). The phase diagram for the $CaO-P_2O_5-H_2O$ system has been extended to higher

Table 1.1 Abbreviations for various nonapatitic calcium phosphates.

Abbreviation	Explanation
CaP	Any calcium orthophosphate.
MCPM	Monocalcium phosphate monohydrate, $Ca(H_2PO_4)_2.H_2O$.
MCPA	Monocalcium phosphate anhydrous, $Ca(H_2PO_4)_2$.
OCP	Octacalcium phosphate, $Ca_8H_2(PO_4)_6.5H_2O$.
DCPD	Dicalcium phosphate dihydrate, $CaHPO_4.2H_2O$. Not used as abbreviation for the mineral brushite.
DCPA	Dicalcium phosphate anhydrous, $CaHPO_4$. Not used as abbreviation for the mineral monetite.
TetCP	Tetracalcium phosphate, $Ca_4(PO_4)_2O$.
β-TCP	$β-Ca_3(PO_4)_2$ without structural HPO_4^{2-} or Mg^{2+} ions.
β-TCa,MgP	$β-(Ca,Mg)_3(PO_4)_2$ without structural HPO_4^{2-} ions.
Mg whitlockite	β-TCP-like precipitates with structural HPO_4^{2-} and Mg^{2+} ions.
ACP	Amorphous calcium phosphate that gives an X-ray diffraction pattern without discernible peaks from lattice periodicities.
ACa,Mg,CO$_3$P	As above, but containing Mg^{2+} and CO_3^{2-} (and HCO_3^-) ions.

temperatures and/or pressures [17-21]. Parts of the quaternary systems formed by the addition of CaF_2 [22] or $CaCO_3$ [23] have been investigated.

The processes of nucleation and growth of CaPs have been reviewed [24-27]: kinetic factors are particularly important in this system. Another complicating factor in the solution chemistry of the CaPs is the formation of calcium phosphate ion pairs, particularly $[CaHPO_4]^0$ and $[CaH_2PO_4]^+$. If these are not taken into account, the solubility product may appear to be dependent on the pH [28]. The speciation diagram (Fig. 1.2) for the $Ca(OH)_2-H_3PO_4-H_2O$ system for the same conditions as the phase diagram (Fig. 1.1) shows how the concentration of these and other complex ions varies with pH. References to phase diagrams and solubility studies of various quaternary systems of interest to the fertilizer industry have been cited [12].

The standard enthalpies ($\Delta_f H°$), free energies of formation ($\Delta_f G°$), entropies ($S°$), specific heats (C_p) and calculated solubility constants (K_s) for a number of CaPs at 25 °C are given in Table 1.3. Values have been taken from a self-consistent compilation [39] made by the former National Bureau of Standards, Washington, DC (now the National Institutes of Standards and Technology). $\Delta_f H°$ and $\Delta_f G°$ refer to formation from the elements and the assumption that these values for H^+ ions are zero. The table includes data for the constituent

Table 1.2 Abbreviations for various apatites.

Abbreviation[a]	Explanation
OHAp	Hydroxyapatite, $Ca_{10}(PO_4)_6(OH)_2$.
s-OHAp	Stoichiometric OHAp. Used when emphasis is required that the compound has a chemical composition corresponding to the ideal, $Ca_{10}(PO_4)_6(OH)_2$.
ns-OHAp	Nonstoichiometric hydroxyapatite (CO_3-free). Indicates that the chemical analysis shows (or would show) a departure from that of s-OHAp. This may involve lattice H_2O, HPO_4^{2-}, Ca^{2+} ions and/or $2OH^-$ replaced by O^{2-}.
Ca-def OHAp	A precipitated apatitic calcium phosphate (CO_3-free) with a Ca/P molar ratio in the range 1.66 to 1.5 or less. Compounds that might contain interlayers of OCP are included.
$CaCl_2$-def ClAp	An apatite that departs from the ideal for ClAp, $Ca_{10}(PO_4)_6Cl_2$, in its calcium and chlorine contents in the molar ratio 1:2. Likewise for FAp.
BCaP	Precipitated basic apatitic calcium phosphate with an uncertain composition which may include CO_3^{2-} or HCO_3^- ions.
A-CO_3Ap	A-type carbonate apatite, ideally $Ca_{10}(PO_4)_6CO_3$.
B-CO_3Ap	B-type carbonate apatite, *i.e.* a carbonate-containing apatite in which PO_4^{3-} ions are replaced by CO_3^{2-} ions.
AB-CO_3Ap	Carbonate-containing apatite in which both the above substitutions take place.
CO_3Ap	Any carbonate-containing apatite.

[a]The above abbreviations are generalised; thus OAp, FAp and IAp are oxy-, fluor- and iodoapatite respectively. B-CO_3FAp would be FAp in which some of the PO_4^{3-} ions have been replaced by CO_3^{2-} ions, *i.e.* francolite. The abbreviations refer to the calcium and phosphate compounds unless indicated otherwise, so that $BaVO_4$OHAp is barium vanadate hydroxyapatite. Solid solutions are indicated by commas between the involved ions. For example, $Ca,BaAsO_4F,ClAp$ abbreviates $(Ca,Ba)_{10}(AsO_4)_6(F,Cl)_2$.

ions of the various CaPs so it is possible to calculate their standard free energies of solution ($\Delta_sG°$) from the sum of the standard free energies of solution of the constituent ions minus the standard free energy of formation of the salt. K_s is then given by $\Delta_sG° = -RT\ln K_s$, where R is the gas constant (8.314 J K^{-1} mol^{-1}) and T the absolute temperature. As values of K_s calculated in this way depend on the difference between two numbers of nearly the same size, both determined from thermochemical experiments and subject to experimental error, it is better to use values of K_s determined by chemical analyses of solutions in equilibrium with a pure salt, if these are available. Such values are given later (those for $Ca_{10}(PO_4)_6(OH)_2$ and $Ca_{10}(PO_4)_6F_2$ in Section 3.9, and others in this chapter). Solubilities of synthetic and biological CO_3Aps are discussed in Sections 4.5.8 and 4.6.5 respectively. Note also that K_s values for apatite can be referred to either $Ca_5(PO_4)_3X$ or $Ca_{10}(PO_4)_6X_2$, where X is a

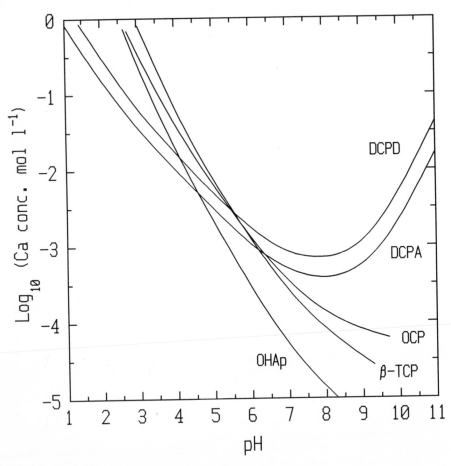

Fig. 1.1 Solubility isotherms of CaP phases in the system $Ca(OH)_2$-H_3PO_4-H_2O at 37 °C calculated with the program RAMESES [29-33]. References for the solubility products and stability constants are: DCPD [34] (Section 1.4.6); DCPA, $[CaHPO_4]^0$ and $[CaH_2PO_4]^+$ [35] (Section 1.5.5); OCP [36] (Section 1.3.6); β-TCP [37] (Section 1.6.5); OHAp [38] (Section 3.9); and H_3PO_4, $H_2PO_4^-$, HPO_4^{2-}, $[CaPO_4]^-$ and $[CaOH]^+$ as given in ref. [27].

halogen or OH$^-$ ion, and that K_s for the latter is the square of K_s for the former. Another extensive compilation of thermochemical data for phosphate minerals and complex aqueous species in equilibrium with them is to be found in ref. [40].

Apatites

Apatites have the formula $Ca_5(PO_4)_3X$, where X can be a F$^-$ ion

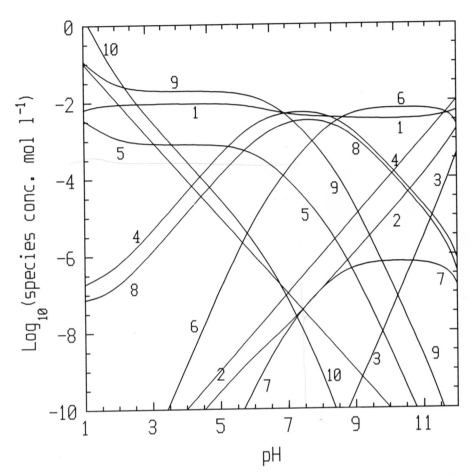

Fig. 1.2 Speciation diagram for the $Ca(OH)_2$-H_3PO_4-H_2O system at 37 °C. Key to species: (1) Ca^{2+}; (2) $[CaOH]^+$; (3) $[Ca(OH)_2]^0$; (4) $[CaHPO_4]^0$; (5) $[CaH_2PO_4]^+$; (6) $[CaPO_4]^-$; (7) PO_4^{3-}; (8) HPO_4^{2-}; (9) $H_2PO_4^-$; (10) H_3PO_4. Total calcium concentration is 10^{-2} mol l^{-1}. Other details as for Fig. 1.1.

(fluorapatite, FAp), OH⁻ ion (hydroxyapatite, OHAp) or a Cl⁻ ion (chlorapatite, ClAp) for example. Gerhard reported in 1786 [41][1] that the name apatite had been coined by AG Werner from the Greek $\alpha\pi\alpha\tau\alpha\omega$, to deceive, because it was frequently confused with other minerals such as aquamarine, crystollite (olivine), amethyst, fluorite, schorl *etc.* Often the formula is written as double that given above. The apatite structure is very tolerant of ionic substitutions,

---

[1]I am indebted to Dr A.M. Clark of the British Museum (Natural History) for this early reference.

Chemistry of the calcium phosphates

Table 1.3 Standard thermodynamic quantities and calculated solubility product constants of CaPs at 298.15 K (25 °C) [39].

Compound	$\Delta_f H°$	$\Delta_f G°$	$S°$	C_p	K_s Calc. Sol.
	kJ mol^{-1}	kJ mol^{-1}	J mol^{-1} K^{-1}	J mol^{-1} K^{-1}	Product Const.[a]
$CaHPO_4$	-1814.39	-1681.18	111.38	110.04	1.83×10^{-7}
$CaHPO_4.2H_2O$	-2403.58	-2154.58	189.45	197.07	2.59×10^{-7}
$Ca_8H_2(PO_4)_6.5H_2O$	-	-12263	-	-	1.01×10^{-94}
β-$Ca_3(PO_4)_2$	-4120.8	-3884.7	236.0	227.82	2.07×10^{-33}
α-$Ca_3(PO_4)_2$	-4109.9	-3875.5	240.91	231.58	8.46×10^{-32}
$Ca_{10}(PO_4)_6(OH)_2$	-13477	-12677	780.7	769.9	6.62×10^{-126}
$Ca_{10}(PO_4)_6F_2$	-13744	-12983	775.7	751.9	6.30×10^{-137}
H_2O	-285.830	-237.129	69.91	-	-
$Ca^{2+}(aq)$	-542.83	-553.58	-53.1	-	-
$OH^-(aq)$	-229.994	-157.244	-10.75	-148.5	-
$H_2PO_4^-(aq)$	-1296.29	-1130.28	+90.4	-	-
$HPO_4^{2-}(aq)$	-1292.14	-1089.15	-33.5	-	-
$PO_4^{3-}(aq)$	-1277.4	-1018.7	-222	-	-
$F^-(aq)$	-332.63	-278.79	-13.8	-106.7	-

[a]Based on the formula given in the right-hand column. Units have been omitted for convenience. Note particularly that generally more accurate values of the solubility product are given in other sections, see text in this section for an explanation of this.

and for example, Ca^{2+} ions can be partly or completely replaced by Ba^{2+}, Sr^{2+} or Pb^{2+} ions, and PO_4^{3-} by AsO_4^{3-} ions. Thus there are lead minerals with the apatite structure which include $Pb_5(PO_4)_3Cl$ (pyromorphite), $Pb_5(VO_4)_3Cl$ (vanadinite) and $Pb_5(AsO_4)_3Cl$ (mimetite). Solid solutions of halide and OH⁻ ions are of frequent occurrence and will be abbreviated as OH,FAp, Cl,OHAp *etc.* (Table 1.2). It is to be understood that the cation is calcium, unless indicated otherwise, *e.g.* SrOH,ClAp abbreviates strontium hydroxychlorapatite.

Coupled substitutions frequently occur in apatites. In these, one ion is replaced by another of the same sign, but different charge, and neutrality is maintained by substitutions of ions with dissimilar charges or vacancies elsewhere. The CO_3Aps are an important and complex example that have been the subject of much controversy. These CO_3Aps include the minerals francolite, dahllite and the rock-phosphates (see next paragraph), and the biological apatites. Nonstoichiometry, with vacant lattice sites, occurs in the biological apatites and frequently in synthetic apatites (both precipitated and high temperature preparations), and considerably complicates their crystal chemistry. When it is important to distinguish between nonstoichiometric and stoichiometric compositions, the above abbreviated formulae are given the prefixes ns- or s- respectively (Table 1.2).

Apatites (mostly FAp or CO_3FAp) form an important series of minerals [2,3,42-45]. They occur as a minor constituent of many igneous rocks, although a few large igneous deposits are known, *e.g.* the Kola peninsular. Apatite is also present in most metamorphic rocks, especially crystalline limestones. Less well-crystallised deposits of rather variable composition, usually referred to as rock-phosphates or phosphorite, occur in large deposits, some of which were formed by the reaction between phosphatic solutions from guano and calcareous rock, or precipitated from sea water. The rock-phosphates provide most of the world's supply of phosphorus for the fertilizer and chemical industries, and are, from a crystal chemical point of view, closely related to francolite. Dahllite is the corresponding carbonate-containing OHAp mineral. The distinction between these minerals is discussed in Section 4.1.1. A carbonate-containing OHAp-like salt forms the mineral of bones and teeth, and is a component of many pathological calcifications. Many tons of F,ClAps, doped with manganese and antimony, are produced annually for use as phosphors in fluorescent light tubes.

The basic apatite structure is hexagonal with space group $P6_3/m$ and approximate lattice parameters $a = 9.4$ and $c = 6.9$ Å with two formula units (as given above, *e.g.* $Ca_5(PO_4)_3F$) per unit cell. Typical lattice parameters, refractive indices and calculated densities are given in Table 1.4. The values often depend slightly on the mode of preparation because of frequent nonstoichiometry, so lattice parameters and refractive indices are given, if available, for the specific preparations discussed in later sections. Some lattice substitutions cause a lowering of the symmetry, so that the unit cell may be doubled (usually a, but occasionally c) and/or slightly distorted from hexagonal. These changes in symmetry often depend on the stoichiometry of the apatite.

The importance and complexity of the apatites are reflected in the fact that their preparation and structure have been the subject of numerous review papers [2,13,61-68] conference proceedings [69-74] and research theses [51,60,75-105]. Reviews that particularly relate to the structure of apatites in relation to biological mineral are to be found in refs [67,106-123].

Table 1.4 Lattice parameters, refractive indices (sodium light) and densities calculated from lattice parameters (except for francolite) for various synthetic apatites and francolite.

Apatite	a Å	c Å		E	O		Density g cm^{-3}
FAp	9.367(1)[a]	6.884(1)	[46]	1.629	1.633	[16]	3.202
ClAp[b]	9.628(5)	6.764(5)	[47]	1.670±0.001[c]		[48]	3.185
BrAp	9.761(1)	6.739(1)	[49]	1.694	1.682	[50]	3.376
OHAp[b]	9.4176[d]	6.8814[d]	-	1.644	1.650[e]	[51]	3.156
A-CO$_3$Ap[b]	9.557(3)	6.872(2)	[52]	-	-	-	3.160
SrOHAp	9.760 ±0.003	7.284 ±0.003	[53]	-	-	-	4.090
BaOHAp	10.177	7.731	[54]	-	-	-	4.735
CdOHAp	9.335(2)	6.664(3)	[55]	-	-	-	5.706
PbOHAp	9.879	7.434	[56]	-	-	-	7.072
CaVO$_4$OHAp	9.818	6.981	[57]	-	-	-	3.204
CaAsO$_4$OHAp	9.72$_8$	6.98$_1$	[58]	-	-	-	3.681
SrAsO$_4$OHAp	10.05$_0$	7.39$_5$	[58]	-	-	-	4.476
Francolite[f]	9.34 ±0.01	6.89 ±0.01	[59]	1.624 ±0.002	1.629 ±0.002	[59]	3.145 ±0.006

[a]Figures in parentheses for this and other entries are estimated standard deviations.
[b]Pseudohexagonal. For OHAp and ClAp, b is marginally greater and smaller respectively than $2a$ (Bauer and Klee, personal communication, 1991) [60]. For A-CO$_3$Ap, Ca$_{10}$(PO$_4$)$_6$CO$_3$, $b = 2a$ and $\gamma = 120.36\pm0.04$ [52].
[c]Mean value of α, β and γ; the birefringence is very low and the purest monoclinic ClAp is biaxial negative [48] (Section 2.7).
[d]Estimated error ±0.0005Å (Section 3.3.2).
[e]The crystals were monoclinic and birefringent when observed down the c-axis [51], so presumably this must be the mean of β and γ.
[f]Type specimen of francolite from Wheal Franco with 3.4 wt % CO$_2$ and 3.71 wt % F [59]; density corrected for presence of haematite. See Section 4.2.1 for further details of lattice parameter and refractive index changes with CO$_2$ content.

Our basic knowledge of the structures of the CaPs comes from single crystal X-ray diffraction (XRD) and neutron diffraction studies. However, for many of the most important apatites (precipitated, biological and many minerals), this has not yet been possible because of the absence of suitable single crystals for study. As a result, there is great interest in the development and application of new spectroscopic and diffraction methods that give information about the internal and surface structures of polycrystalline solids. The results of such investigations will be reported without any detailed experimental or theoretical background. Such information for infrared (IR), nuclear magnetic resonance (NMR), magic angle spinning (MAS) NMR, electron spin resonance (ESR), extended X-ray absorption fine structure (EXAFS) and Mössbauer spectroscopy can be found in a recent publication [124]. The application of IR to the study of minerals [125], and IR and Raman to the study of calcified tissues [126] have been reviewed. The Rietveld method of structure determination from X-ray and neutron powder diffraction patterns [127,128] has been used to a limited extent to study apatites, but its use is likely to increase.

Experimental results have been reported in their original units and abbreviations *e.g.* ppm (parts per million), psi (pounds per square inch), atm (atmospheres), *etc.*, but most of these have been converted to SI units. An exception is that unit cell parameters and atomic distances have been left in Ångström units (10 Å = 1 nm). Other abbreviations are: EM (electron microscopy), SEM (scanning electron microscopy), IR (infrared), FT IR (Fourier transform infrared), and DTA (differential thermal analysis).

1.2 Monocalcium phosphates (monohydrate and anhydrous)

Monocalcium phosphate occurs as the monohydrate (MCPM, $Ca(H_2PO_4)_2.H_2O$) and the anhydrous salt (MCPA, $Ca(H_2PO_4)_2$) in the acid and most soluble region of the $CaO-P_2O_5-H_2O$ phase diagram (Fig. 1.3). MCPM is an important constituent of superphosphate made by the acidulation of rock-phosphate with H_2SO_4, H_3PO_4 or HNO_3 whilst the anhydrous form occurs when superphosphate is made by treating rock-phosphate with concentrated H_3PO_4 at elevated temperatures [16]. MCPA also occurs as an intermediate when rock-phosphate is acidulated with hot fuming HNO_3 or superphosphates are heated to 110 to 180 °C [16].

The highest solubility in the phase diagram is at the invariant point when MCPM and $CaHPO_4$ are in equilibrium with the aqueous phase. At 25 °C, the solution has a composition 24.10 wt % P_2O_5, 5.785 wt % CaO and a density (it appears in fact to be specific gravity that was measured) of 1.3018 (25°C/25°) [129]. The solubility product constant (expressed as the pK) of MCPM is 1.1436 at 25 °C [130]. The phase diagram shows that the solubilities of the monocalcium phosphates increase with temperature [129], unlike the other, less soluble calcium phosphates (the retrograde solubility of $CaHPO_4$ can be seen in Fig. 1.3, others are discussed in later sections). The temperature at the quintuple point, *R* in Fig. 1.3, is 152 °C [131]. The positions of other quintuple points and other aspects of the acid region of the $CaO-P_2O_5-H_2O$ phase diagram have been discussed [12,131].

Structures

MCPM has a triclinic space group $P\bar{1}$ and unit cell $a = 5.6261(5)$, $b = 11.889(2)$, $c = 6.4731(8)$ Å, $\alpha = 98.633(6)$, $\beta = 118.262(6)$ and $\gamma = 83.344(6)°$ at 25 °C with two formula units per unit cell [133]. The first full structure determination [134] and subsequent refinements [133,135] showed the presence of $\cdots Ca(H_2PO_4)^+$ $Ca(H_2PO_4)^+$ $Ca(H_2PO_4)^+\cdots$ chains parallel to the c-axis that formed corrugated sheets in the (010) plane. These were similar to the sheets originally seen [136] in DCPD formed by $\cdots Ca\ HPO_4$ $Ca\ HPO_4$ $Ca\ HPO_4\cdots$ columns (Fig. 1.5, Section 1.4.2). Layers of $H_2PO_4^-$ ions and water molecules lie between the sheets of $Ca(H_2PO_4)^+$ chains [133]. Hydrogen atom positions have been approximately located [133].

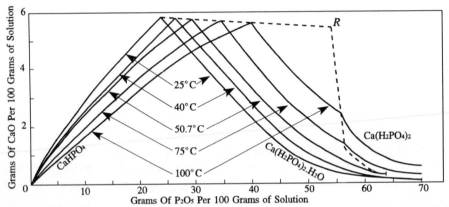

Fig. 1.3 Solubility isotherms at various temperatures in the acid region of the system $CaO-P_2O_5-H_2O$. R is the quintuple point between MCPM, MCPA, DCPA, the saturated solution and the saturated vapour. The dotted line to the left of R is the univariant curve for saturated solutions of DCPA and MCPM and to the right of R for MCPM and MCPA. Isotherms at 25, 40 and 50.7 °C from data in ref. [132] and at 75 and 100 °C from data in ref. [129]. (After Fig. 1 of Elmore and Farr [129] and Fig. 9-18 of van Wazer [12])

The structure of MCPA has been determined from X-ray and neutron diffraction data [137]. The space group is $P\bar{1}$ and the unit cell parameters are $a = 7.5577(5)$, $b = 8.2531(6)$, $c = 5.5504(3)$ Å, $\alpha = 109.87(1)$, $\beta = 93.68(1)$ and $\gamma = 109.15(1)°$ at 25 °C with two formula units per unit cell [137]. There are several ways of describing the structure; in one of these, hydrogen-bonded PO_4 groups in layers parallel to (010) are considered [137]. These layers are joined together by hydrogen bonds on one side and Ca^{2+} ions on the other side. A striking feature of the structure is the occurrence of $H_2PO_4^-$ ions held together into infinite chains by very strong, centred, O-H-O hydrogen bonds [137].

Preparation

Both the anhydrous and hydrated forms can be grown as well-formed crystals by lowering the temperature of an acidic solution with the appropriate composition, as given in Fig. 1.3. The monohydrate has been prepared by adding $Ca(OH)_2$ or CaO to concentrated H_3PO_4, followed by recrystallising the product from 50 % H_3PO_4 with slow cooling and constant stirring, washing with dry acetone, and drying under vacuum [16,138]. The anhydrous salt can be crystallised over a four day period at 130 °C from a filtered solution of 500 g of recrystallised MCPM in 2000 ml of 81.5 % H_3PO_4 at its boiling point [16,138]. The crystals are then washed free of acid with anhydrous acetone on a heated funnel, dried for 3 h at 85 °C and vacuum dried for 4 days over anhydrous magnesium perchlorate.

Habit, optical properties and density

The hydrate crystallises as triclinic (010) plates elongated in the *c*-axis direction; polysynthetic twinning with (010) as the composition plane according to the albite law is frequent [138]. The optical constants at 610 nm of the crystals prepared as given above were $\alpha = 1.496$, $\beta = 1.515$ and $\gamma = 1.529$, biaxial (negative) with $2V = 81°$ 30′ (calculated) [138]. The obtuse bisectrix, **Z**, is parallel to (010) and is inclined to the *a*-axis by 2° 40′ in acute β, whilst the **XY** plane is perpendicular to (010), with **X** inclined to (010) by 37° in obtuse α [138]. The calculated density is 2.231 g cm^{-3} [133].

MCPA crystallises as triclinic rods elongated in the *a*-axis direction and as elongated tablets with (01$\overline{1}$), and sometimes (010), as the tabular face, but without twinning [138]. The optical constants at 610 nm were $\alpha = 1.548$, $\beta = 1.572$, $\gamma = 1.602$, biaxial (positive) with $2V = 85°$ 14′ (calculated) [138]. The **XY** plane is perpendicular to (010) and essentially parallel to the *a*-axis; the optic normal, **Y**, is inclined to (010) by 15° in obtuse γ [138]. These optical constants were measured on crystals grown as described above. Optical constants at 425 nm for the anhydrous and hydrated monocalcium phosphates have also been reported [138].

Thermal decomposition and other properties

MCPM is stable in air, but loses water above 108 °C to form MCPA or calcium acid pyrophosphate [16]. MCPA alters to $CaH_2P_2O_7$ and calcium polyphosphates above 186 °C (Lehr *et al.* [16] attribute this to W.J. Howells, J. Chem. Soc. 1931,3208-12, but this seems to be an error).

In moist air, MCPA slowly alters to randomly oriented crystals of MCPM [138]. MCPA readily incorporates NH_4^+, K^+, Cl^- and Br^- ions in its lattice, with slight change in properties [139]; lattice parameters and optical properties were reported.

The structures of $CaH_2(PO_4)Br.4H_2O$ and $CaH_2(PO_4)I.4H_2O$ have been determined [140].

1.3 Octacalcium phosphate

1.3.1 Introduction

Octacalcium phosphate (synonyms: OCP, octacalcium bis(hydrogenphosphate) tetrakis(phosphate) pentahydrate, tetracalcium hydrogen triphosphate trihydrate) has the formula $Ca_8(HPO_4)_2(PO_4)_4.5H_2O$ although the water content is somewhat variable. According to Bjerrum [7], a CaP with the composition of OCP was first described by Berzelius in 1836 [141]; however, its existence was not universally accepted until the middle of the 20th century, and its importance unrecognised until the late 1950's and early 1960's. The early [7] and more recent history and revival [7,142] of OCP have been discussed.

OCP often occurs as a transient intermediate in the precipitation of the thermodynamically more stable OHAp (Section 3.8) and biological apatites (Section 4.6.1) because OCP nucleates and grows more easily than OHAp. Thus understanding the growth of OCP and its hydrolysis is potentially of considerable importance to an understanding of the processes of mineralisation in bones and teeth, and remineralisation in carious lesions. The close structural relationship between OCP and OHAp (Sections 1.3.2 and 2.2.2) has been used to explain the incorporation (via hydrolysis) of impurities, particularly CO_3^{2-}, Mg^{2+} and Na^+ ions, and hence the nonstoichiometry of precipitated apatitic CaPs (Sections 1.3.6 and 4.5.4). In addition to these roles in the formation kinetics and composition of apatites, OCP has been found as a constituent of dental calculus [143,144] and other pathological calcifications. It has also been reported in X-ray powder diffractometer studies of bone from the distal metaphysis of a two day old rabbit [145] (Table 4.8, Section 4.6.2).

The crystal chemistry of OCP up to 1981 has been reviewed [142]. Recently, it has been discovered that double salts of OCP and various dicarboxylic acids can be prepared: the preparation and properties of these are discussed in Section 1.3.9. An amorphous OCP has been reported [146] (Section 1.8.3).

1.3.2 Structure

The initially determined structure [147] of OCP has recently been refined [148]. The lattice is triclinic, space group $P\bar{1}$ and unit cell $a = 19.692(4)$, $b = 9.523(2)$, $c = 6.835(2)$ Å and $\alpha = 90.15(2)$, $\beta = 92.54(2)$ and $\gamma = 108.65(1)°$ [148]. The asymmetric unit (the largest structural unit in which none of the atoms are related by symmetry) is $Ca_8H_2(PO_4)_6.5H_2O$, with two asymmetric units per unit cell. "Apatite" layers (about 1.1 nm thick) alternating with "hydrated" or "water" layers (about 0.8 nm thick) parallel to (100) are a conspicuous feature of the structure (Fig. 1.4). The presence of the "apatite" layer explains the similarities of the lattice parameters with those of OHAp (*a* = 9.4176 and *c* = 6.8814 Å, Table 1.4). The "apatite" layer consists of alternating sheets of phosphate ions interspersed with Ca^{2+} ions; and the "water" layers consist of more widely spaced phosphate and Ca^{2+} ions with a slightly variable number of water molecules between them. Six of the Ca^{2+} ions and two of the phosphate ions are in the "apatite" layer. The other two Ca^{2+} ions and one phosphate ion are in the "water" layer. The remaining three phosphate ions lie at the junction of the "water" and "apatite" layers. The phosphate ion in the "water" layer and one at the junction between the layers are protonated. The relationships between OHAp, OCP and other structures are considered in Section 2.2.2. The apatite structure is described in Section 2.2.1 and the positions of the OH⁻ ions in OHAp in Section 3.2.

1.3.3 Preparation

Hydrolysis reactions. Very thin blades of OCP up to 250 μm long can be prepared by the slow hydrolysis of $CaHPO_4.2H_2O$ (DCPD, Section 1.4.3) at 40 °C in CH_3COONa solution (0.5 mol l⁻¹) which is renewed when the pH falls to 6.1 [149]. OCP can also be prepared by the hydrolysis of DCPD at 80 to 85 °C in dilute HNO_3 at pH 5 for 30 min, but if the reaction is prolonged, hydrolysis to OHAp occurs [16]. Hydrolysis of DCPD to OCP can also be carried out over a period of three months at room temperature by the dropwise treatment with distilled water [150]. A pH-stat has been used to maintain the pH at 6.8 during the hydrolysis of DCPD at room temperature to yield OCP with a composition of Ca 32.9 and P 18.8 wt % (theoretical 32.63 and 18.91 wt % respectively) [151]. α-TCP (Section 1.6.4) can be hydrolysed to OCP in 3 h by a CH_3COONa solution (0.5 mol l⁻¹) at 50 °C and pH 5.4±0.1 [152,153].

Precipitation reactions. In a method that relies on the retrograde solubility of OCP with temperature (Section 1.3.6), 1 mm long × 0.1 mm thick crystals have been grown from saturated CaP solutions buffered at pH 6.5 to 6.6 prepared at room temperature, then warmed slowly to 50 to 60 °C and maintained at this temperature for several hours [154]. Spherulitic growths of OCP can be obtained by the diffusion of Ca^{2+} ions into phosphate-containing gels [155]. Pure OCP has been prepared by the dropwise addition of a $(CH_3COO)_2Ca$ solution (Ca^{2+} 0.02 mol l⁻¹) to an equal volume of a sodium phosphate solution (phosphate 0.02 mol l⁻¹) (or *vice versa*) at pH 5 and 60 °C or at pH 4 and 70 to 80 °C during 3 to 4 h [156]. Doubled concentrations can also be used [156].

1.3.4 Optical properties, density and habit

OCP is biaxial (negative) with $2V$ = 50 to 55° (calculated 54°) with α = 1.576, β = 1.583 and γ = 1.585 [157]. The optic axial plane is inclined to

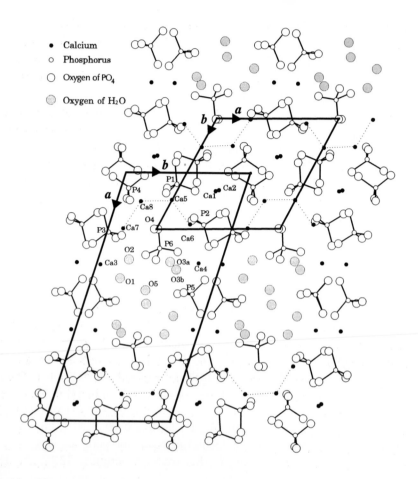

Fig. 1.4 Projection onto the (001) plane of the structure of OCP showing its relationship to the structure of apatite (Section 2.2.1). The c-axis is out of the plane of the diagram for OCP and into the plane for OHAp. Hydrogen atoms have been omitted for clarity. The oxygen atoms of the water molecules have been shaded, so the water layer, parallel to the b- and c-axes and passing through $x \approx 0.5$ of the unit cell of OCP (lower left cell), can be readily seen. The apatite layers, also parallel to the b- and c-axes, are at $x \approx 0$ and 1.0. A unit cell corresponding to that of hexagonal OHAp has been drawn for the apatite layer at $x \approx 0$ at the upper right. Note that two of the oxygen atoms on P(6) of OCP correspond to the oxygen atoms on the OHAp hexad axis (Fig. 2.4, Section 2.2.1), and the pairs Ca(5), Ca(7) and Ca(6), Ca(8) of OCP correspond to pairs of calcium atoms in the Ca(2) triangles at $z = \frac{1}{4}$ and $\frac{3}{4}$ in OHAp (Fig. 2.4). (After Fig. 2 of Mathew *et al.* [148])

(100), and its trace on this face is at 78° with respect to the *c*-axis in acute α. The acute bisectrix lies in the quadrant defined by **a**, -**b**, and -**c**. One optic axis is approximately normal to (100) [157]. The wavelength was not given for these constants, but the same values are quoted in ref. [16] for sodium light. The measured density is 2.61 g cm^{-3} [157].

OCP crystallises as thin {100} blades of triclinic pinacoidal symmetry which are elongated along the *c*-axis, and modified by the set {010} and the terminating set {011}: contact twinning with (100) as the twin plane is often observed optically [157]. Blades of OCP often grow so that they radiate from a common origin to form beautiful spherulitic growths (see Fig. 11 in ref. [8], Fig. 2.7c in ref. [118] or Fig. 7 in ref. [158]). The predominance of the {100} form can be predicted from theoretical calculations of the crystal morphology using the Hartman-Perdok theory [159].

1.3.5 Kinetics of nucleation and crystal growth

The constant composition method has been used [160] (Section 3.8) to study the growth kinetics of OCP onto seeds of OCP at pH 6 and 37 °C. Initially, a solution supersaturated with respect to OCP, β-TCP (Section 1.6.4) and OHAp was prepared (typically Ca 4.0×10^{-3}, phosphate 3.0×10^{-3} and ionic strength (from added KNO_3) 6.6×10^{-2} mol l^{-1}) and crystallisation initiated by addition of the seeds (specific area 14.6±0.2 m^2 g^{-1} and typically 0.203 g l^{-1}). The pH was monitored so that changes from the set value (pH 6) initiated the automatic addition of a $Ca(NO_3)_2$ (58.95×10^{-3} mol l^{-1}) solution and a KH_2PO_4 (44.21×10^{-3} mol l^{-1}) plus KOH (63.74×10^{-3} mol l^{-1}) solution. The two solutions were added in equal volumes to give a Ca/P molar ratio for the added reactants of 1.333 (*i.e.* that of the precipitating solid). In this way, growth conditions were kept constant. Analysis of the solid after more than 50 times the original amount of OCP had grown, gave a Ca/P molar ratio of 1.33$_6$. The rate of growth for the conditions given above was 1.70×10^{-6} mol of $Ca_8H_2(PO_4)_6.5H_2O$ min^{-1} m^{-2}, which was much higher than would have occurred for OHAp under the same conditions. At low concentrations, a rate of crystallisation proportional to the fourth power of the relative supersaturation (defined as (ion product/solubility product)$^{1/16}$ - 1) indicated a polynuclear mechanism, but at higher concentrations, the rate became proportional to the first power of the supersaturation [160].

Constant composition studies of the growth of OCP on β-TCP have been published [161]. Nucleation studies, again under constant composition conditions, of OCP on seeds of $CaHPO_4$ (DCPA, Section 1.5.3) at 60 to 80 °C, allowed the surface energy of OCP to be calculated (51 to 59 mJ m^{-2}) [162]. The presence of Mg^{2+} ions appreciably retards the rate of precipitation, probably via adsorption at active sites on the OCP crystal surfaces [163]. In solution and gel systems, it has been reported that pyrophosphate inhibited

OCP formation, and caused ACP (amorphous calcium phosphate, Section 1.8.2) to form instead, while CO_3^{2-} or citrate ions favoured the precipitation of "apatitic" calcium phosphates [155].

The oriented growth of OCP on ion selective membranes at 37 °C has been studied [164]. The system had three chambers with the central chamber formed by a cation- and an anion-selective membrane 2 mm apart. The pH of the solution in the central chamber was held constant by pH-stat controlled addition of base or acid. Calcium (5 to 48 mmol l^{-1}) and phosphate (7.2 to 28.5 mmol l^{-1}) solutions were pumped through the outer chambers formed by the cation- and anion-selective membranes, respectively. At pH 7.4 and concentrations of calcium and phosphate of 12 and 7.2 mmol l^{-1} respectively, well-crystallised ribbons of OCP grew with their *c*-axes perpendicular to the cation-selective membrane on its inner surface. The orientation was demonstrated with SEM and XRD. The effects of changes in concentration and pH on the morphology and orientation were investigated. The OCP could be hydrolysed in a week *in situ* by the addition of Ca^{2+} ions via Reaction 1.1 to give well-oriented OHAp.

The presence of CO_3^{2-}, Mg^{2+}, and F^- ions suppresses growth of OCP in the *c*-axis direction under most conditions, particularly when present simultaneously [164,165]. Growth and morphology are particularly sensitive to F^- ions [164-166] (Section 4.6.6). Both citrate and phosphocitrate influence the crystallisation of OCP, but the inhibition of crystal growth by phosphocitrate is considerably more marked [167].

1.3.6 Solubility and reactions in solution

A recent determination of the solubility of OCP gave values for pK_{sp} (referred to $Ca_4H(PO_4)_3.2.5H_2O$) of 48.3±0.2, 48.3±0.2, 48.2, 48.3, 48.4±0.1 and 48.7±0.2 at 4, 4.8, 6.0, 18, 23.5, and 37 °C respectively, so there is a very slight decrease in solubility as the temperature rises [36]. The values at 25 and 37 °C were about one unit less than reported earlier [168,169], which was ascribed mainly to the use of different ion pair formation constants [36]. Another recent determination of the pK_{sp} at 37.0 °C gave 49.04±0.03, which is about half way between the two earlier determinations [170]. One of the difficulties of these measurements is that OCP is only in metastable equilibrium with the solution, and in time, will hydrolyse to OHAp. The presence of CO_3^{2-} ions does not significantly affect the equilibrium constant because this ion is not a lattice constituent of OCP [36].

The effects of solution Ca/P ratio and degree of saturation on the dissolution kinetics of OCP at constant pH 5.75 and 25 °C have been studied [171]. The results indicated that the dissolution of OCP can apparently be controlled by two processes (probably transport and nucleation), depending on the undersaturation and Ca/P ratio. The dissolution has also been investigated [170] at constant undersaturation using constant composition methods (Section 3.8). The kinetic data were analysed in terms of recent crystal growth theories using a nonlinear least squares procedure. A rate equation was derived for a spiral dissolution, following detachment-desorption-volume diffusion mechanism at very low kink densities. The results indicated that processes at the crystal surface were much more important than volume diffusion in determining the rate of dissolution.

Hydrolysis and other reactions in solution

OCP hydrolyses in water to OHAp with contributions from both the reactions

$$Ca_8H_2(PO_4)_6.5H_2O + 2Ca^{2+} \rightarrow 2Ca_5(PO_4)_3OH + 3H_2O + 4H^+ \qquad 1.1$$

$$1.25Ca_8H_2(PO_4)_6.5H_2O \rightarrow 2Ca_5(PO_4)_3OH + 1.5H_3PO_4 + 4.25H_2O \qquad 1.2$$

depending on the availability of Ca^{2+} ions [142]. Single crystal XRD and polarised light microscopy showed [157] that pseudomorphs of the OCP were produced, whose properties progressively approached those of OHAp as the reaction proceeded. A partly hydrolysed crystal gave the XRD pattern of OHAp and OCP with the *b*- and *c*-axes collinear. Owing to the similarities between the structures of OCP and OHAp (Sections 1.3.2 and 2.2.2), it was suggested that these crystals consisted of intercrystalline mixtures of OCP and OHAp. This idea followed from an earlier suggestion [149] that precipitated apatitic CaPs with Ca/P molar ratios less than 1.667 (the theoretical value for OHAp) were composed of similar intercrystalline mixtures. The structures of these Ca-def OHAps are discussed in more detail in Section 3.7. Further evidence of the formation of intercrystalline OCP and OHAp derives from the calculation of the peak shifts in X-ray powder patterns from such structures on the assumption that individual layers do not scatter independently [172]. The calculated peak shift for d_{200}(OCP) to d_{100}(OHAp) agreed with that previously observed [173] (Section 1.8.4) for an "apatite" produced by the hydrolysis of ACP which appeared to have an anomalously large *a*-axis parameter. High resolution electron microscopy of partially and fully hydrolysed OCP also provides evidence for epitaxial intergrowths of OCP and OHAp [174]. Some partially hydrolysed crystals (collapsed OCP) showed periodic contrast variations with an approximate periodicity of 16.5 Å which were interpreted as being regions of OCP that had lost water with a resultant 2.4 Å contraction of the *a*-axis parameter: fully hydrolysed OCP also contained a few such regions surrounded by apatite [174]. Further discussion of possible types of interlayered structures can be found in refs [14,142,175]. (See also the discussion of the "central planar defect" seen in biological apatites, Section 4.6.6.) The hydrolysis of oriented crystals of OCP (grown on a cation-selective membrane) into oriented OHAp crystals has been discussed in Section 1.3.5 [164].

The hydrolysis of OCP to OHAp at different calcium and phosphate concentrations has been studied [176]. Carbonate ions slow the rate [36] and are retained in the solid [150,157]. The amount of carbonate incorporated depends markedly on which alkali carbonate is used [177] (Section 4.5.4). These results have led to the suggestion that it might be possible for part of the carbonate to be situated in some type of gross defects introduced during hydrolysis, rather than on true lattice sites [142,177,178] (Section 4.5.4). Na^+ ions can also be retained during hydrolysis [150]. Mg^{2+} ions inhibit the reaction [157]. A recent study [158] of the transformation of OCP to apatite using scanning and transmission electron microscopy and IR demonstrated the inhibiting effect of Mg^{2+}, citrate and pyrophosphate ions and facilitation by F^-, CO_3^{2-}, HPO_4^{2-}, and Ca^{2+} ions. Solution pH, ionic concentrations and OCP crystal size also influenced the rate of transformation.

Heat or F^- ions accelerate the transformation of OCP to OHAp [154]. At 40 °C and 0.01% NaF, the reaction was completed very quickly, but with 0.002% NaF, it took 48 h. Furthermore, the presence of 10 γ (10 μg) of NaF l^{-1} in a solution from which OCP might be expected to crystallise, yielded apatite directly instead [154]. EM studies of lattice images show that, in the presence of 1 ppm of F^- ions, an apatite with a "central planar defect" is precipitated, whereas without F^- ions, OCP is formed [166] (Section 4.6.6). The effect of F^- ions is not completely understood, but is probably associated with the greater stability of FAp compared with OHAp which creates a larger driving force for the precipitation of F,OHAp. Whatever the detailed mechanism, this effect is clearly important to an understanding of the biological role of F^- ions in relation to bones and teeth.

FAp is formed at 25 to 40 °C by the reaction of F^- ions at concentrations of 200 to 1000 ppm with OCP [179]. The rate limiting step in this process was found to be diffusion of F^- ions through a dense FAp surface layer.

1.3.7 Thermal decomposition

The thermal decomposition of OCP is complex. OCP with partial loss of water of hydration, collapsed OCP, DCPA ($CaHPO_4$, Section 1.5), OHAp, β-$Ca_2P_2O_7$ and tripolyphosphate are formed as the temperature increases, with these compounds persisting over different temperature ranges. The products also often depend on the duration of heating.

Heating to just above 100 °C causes a slight loss of water without any detectable change in the XRD pattern [157]. Water lost under these conditions has been likened to zeolitic, rather than true water of crystallisation [180]. However, on further heating, the d_{100} spacing decreases from 18.7 Å, until at

~180 °C, after about two-thirds of the water has been lost, the X-ray pattern (presumably powder) is that of a poorly crystallised apatite [149,157]. Samples of OCP heated at 150 °C for 4 h have a strong doublet in their XRD powder patterns comprising a d_{100} spacing at 18.6 Å, and a collapsed OCP spacing at 16.5 Å [174].

Single crystal studies [157] of OCP dried at 180 °C showed that the apatite formed had its *c*-axis and (100) face coincident with those elements in the original OCP. This observation can be understood from the close relationship between the two structures (Sections 1.3.2 and 2.2.2). Evidence was also seen for collapsed OCP layers [157].

The equation for the initial loss of water, *i.e.* before the formation of apatite, is:

$$Ca_8H_2(PO_4)_6.5H_2O \rightarrow Ca_8H_2(PO_4)_6.nH_2O + (5-n)H_2O. \qquad 1.3$$

It has been reported that the contraction of the d_{100} spacing on slight dehydration of OCP reversed after storing the sample for one year in the laboratory (MS Tung and WE Brown, personal communication, 1990). It is interesting to note that a similar reversible contraction of the *a*-axis parameter on the dehydration of a precipitated ns-OHAp has been reported [181] (Section 3.7.3).

The single crystal studies [157] of OCP dried at 180 °C also showed DCPA with its *b*-axis and (001) plane parallel to the *c*-axis and (100) plane of the original OCP, respectively. However, it was not possible to detect DCPA by IR for any thermal treatment [182], and only after heating to 220 °C by powder XRD [149]. This observation shows the greater sensitivity of single crystal techniques over IR and XRD powder methods. The reaction for the formation of DCPA and OHAp from OCP can be written:

$$Ca_8H_2(PO_4)_6.5H_2O \rightarrow Ca_5(PO_4)_3OH + 3CaHPO_4 + 4H_2O. \qquad 1.4$$

The fact that DCPA is not easily detected shows that little is formed, so Equation 1.4 can only be a minor mode of decomposition. However, this reaction appears to be significant at room temperatures because powder XRD and IR showed the presence of OHAp, and powder XRD and NMR showed the presence of DCPA, in a sample of previously well-characterised OCP that had been stored for a number of years [183]. The inference was made that some OCP had decomposed via Equation 1.4 through a solid-state disproportionation.

Between 325 and 600 °C, the major products were found to be OHAp and β-$Ca_2P_2O_7$ [182]. Chemical analyses and IR spectroscopy showed that the pyrophosphate formed in the reaction

$$Ca_8H_2(PO_4)_6.5H_2O \rightarrow Ca_5(PO_4)_3OH + 1.5Ca_2P_2O_7 + 5.5H_2O \qquad 1.5$$

increased greatly in the range 150 to 400 °C. The maximum occurred at 500 °C when about 40 mol % of the phosphorus was in this form, which compares with a theoretical value of 50 mol % for Equation 1.5. This finding indicated that at lower temperatures, Equation 1.5 was incomplete, and that at higher temperatures, pyrophosphate was removed by the reaction:

$$2Ca_5(PO_4)_3OH + Ca_2P_2O_7 \rightarrow 4Ca_3(PO_4)_2 + H_2O. \qquad 1.6$$

Between 650 and 900 °C, the products were β-$Ca_3(PO_4)_2$ and β-$Ca_2P_2O_7$ produced according to the overall equation:

$$Ca_8H_2(PO_4)_6.5H_2O \rightarrow 2Ca_3(PO_4)_2 + Ca_2P_2O_7 + 6H_2O. \qquad 1.7$$

From 700 to 900 °C, the quantity of pyrophosphate expected from Equation 1.7 was formed (33 % of the phosphorus as pyrophosphate).

The authors [182] concluded that Reaction 1.5 cannot be used under the conditions of their experiments to assay accurately the amount of OCP present in a CaP mixture because of the incompleteness of Reaction 1.5 below 600 °C, and loss of pyrophosphate via Reaction 1.6 above 600 °C. These observations are also relevant to the validity of the determination of acid phosphate in apatites by the measurement of pyrophosphate after pyrolysis (Section 3.6.2).

Paper chromatography has been used to show that the lowest detectable temperatures for the formation of pyrophosphate and tripolyphosphate in heated OCP are 150 and 350 °C respectively [180].

The water lost with heating at linear temperature rates has been followed by mass spectrometry and shows six discrete processes [184]. The largest peak, near 200 °C, was interpreted as the disappearance of the OCP lattice accompanied by a large loss of water and increase (determined by IR) in OH⁻ and HPO_4^{2-} ion concentrations; the reaction had an activation energy of 59±5 kcal mol⁻¹ (247±21 kJ mol⁻¹). The pyrophosphate concentration reached a maximum at ~500 °C. A second peak, near 700 °C, corresponded to a loss of OH⁻ ions and pyrophosphate and had an activation energy of 122±11 kcal mol⁻¹ (511±46 kJ mol⁻¹).

1.3.8 Infrared, Raman and NMR spectroscopy

IR and Raman spectroscopy

IR assignments for OCP have been made [182]. It was reported that the HPO_4^{2-} ion in OCP has bands at 865 and 910 cm⁻¹ (both characteristic of OCP) which can be used to demonstrate its presence in mixtures with OHAp. More detailed assignments based on studies of OCP and its deuterated analogue at 20 and -180 °C have been given for the range 4000 to 400 cm^{-1} [185]. The deuterated form was prepared by exposure of OCP to 97 percent D_2O vapour in a closed system at 60 °C for four days, which shows that the protons in OCP can diffuse readily at this temperature. Less detailed assignments based on spectra at 300 and 77 K were reported in the same year [186]. Subsequently [187], an earlier assignment [185,186] of the v_2 mode was corrected on the basis of more complete assignments of IR and Raman spectra.

Recently, further detailed assignments of the Raman and IR spectra of OCP from 4000 to 300 cm^{-1} have been made, which included specific assignments for bands of the two crystallographically independent acid phosphate groups [188]. Infrared spectra indicated that two polymorphs, OCP(A) and OCP(B), might exist. The principal differences between the two forms above 2000 cm^{-1} were in band positions and relative intensities, and below 2000 cm^{-1}, in the positions of corresponding bands (differences up to 14 cm^{-1}). The A form was seen in IR spectra of OCP supported in, or on, different matrixes (dry KBr or polyethylene film), but the B form was seen for the same OCP preparation only in damp KBr pellets or in oil mulls between Tl(Br,I) or KBr plates. Interconversion in KBr pellets between the A and B forms was possible by storage in a dry or damp atmosphere as appropriate; this showed that the degree of hydration was an important factor in determining the structure. It was suggested that the structures of the polymorphs differed in the arrangement of their hydrogen bonds.

At -180 °C, bands involving stretching, bending and twisting motions of hydrogen-bonded groups are considerably enhanced, and show much greater detail [185,189]. More detail can also be seen in the v_3 PO_4 [190] and v_4 PO_4 [191] bands in deconvoluted FT IR spectra of samples at room temperature.

NMR spectroscopy

The NMR MAS spectrum of OCP shows considerable complexity. In a study [183] at 500 MHz and a spinning speed of 7.8 kHz, a ^1H resonance with a chemical shift of 13 ppm from TMS (trimethyl silane) was assigned to an acidic proton of an HPO_4^{2-} ion located in the "apatite" layer of the structure. The failure to resolve clearly a resonance from the other HPO_4^{2-} ion in the "water" layer was attributed to positional disorder of this ion and overlap with resonances from structural water. An intense spinning sideband pattern centred around 5.5 ppm was assigned to resonances from relatively immobile structural water groups, with evidence for a more mobile water species (surface or structural) contributing to the centre peak intensity. Two distinct resonances at 1.1 and 1.5 ppm were observed that were unusual in that they remained sharp down to low spinning speeds: these were assigned to the protons on an isolated O(5) water molecule (atom numbering of ref. [148]) in the "water" layer which was rapidly reorienting. It was noted that the intensities of these two resonances showed great variability between samples, which was consistent with their assignment to "zeolitic" water in the structure that could be present in variable amounts, and which might be easily lost on heating (Section 1.3.7). Further NMR studies of partially thermally dehydrated OCP would clarify this assignment. All the OCP samples studied [183] showed a resonance at 0.2 ppm which is the same chemical shift as the hydroxyl proton in OHAp [183] (Section 3.12). It was suggested [183] that this indicated the presence of OH$^-$ ions in the "apatite" layer of OCP, which supported a proposal [148] that protons from the water in OCP could relocate themselves to PO_4^{3-} ions to form OH$^-$ and HPO_4^{2-} ions (see also the occurrence of this reaction at ~200 °C [184], Section 1.3.7).

^{31}P resonances (referenced to external 85 % H_3PO_4) in OCP and other CaPs have been studied [192] at 68 MHz with proton decoupling methods and MAS (1 to 2.5 kHz) at 20 and -165 °C. A resonance with a chemical shift of -0.1±0.4 ppm was assigned to phosphorus in HPO_4^{2-} ions in the "water" layer and a resonance with a shift of 3.4±0.3 ppm to phosphorus in the "apatite" layer. A just detectable "third peak" between these was assigned to "apatitic layer" PO_4^{3-} ions that were near protons of water molecules or adjacent to HPO_4^{2-} ions. Later studies [193] at 40 and 121 MHz, also with high power proton decoupling and MAS (5 kHz), reported similar assignments. A resonance with a chemical shift of -0.2 ppm was assigned to phosphorus in HPO_4^{2-} ions in the "water" layer, a resonance with a shift of 2.2 ppm to phosphorus in an ion in an environment intermediate between the "hydrated" layer and the "apatite" layer, and a resonance with a shift of 3.3 ppm to phosphorus in the "apatite" layer (shifts relative to 85% H_3PO_4).

1.3.9 Double salts of octacalcium phosphate and dicarboxylic acids

Double salts of OCP and certain dicarboxylic acids can be prepared by the hydrolysis of α-TCP (Section 1.6.4) in a solution of the sodium or ammonium salt of the acid. For example, hydrolysis for 3 h at pH 6.0±0.1 in an ammoniacal solution containing 0.025 mol l^{-1} succinate yielded

$$Ca_8(HPO_4)_{2-z}(PO_4)_4(C_4H_4O_4)_z.mH_2O \qquad\qquad 1.8$$

with $z = 1.17$ and $m = 5.5$ [153,194]. This is a general phenomenon with dicarboxylic anions, $C_nH_{2n}C_2O_4^{2-}$, so that as n increases from 1 to 6, there is an essentially linear expansion of d_{100} from 19.6 to 26.1 Å due to incorporation of dicarboxylate ions in the lattice [194]. (These values compare with 18.7 Å for d_{100} of OCP.) The b- and c-axes hardly changed as a result of the substitution; oxalate ($n = 0$) formed calcium oxalate instead of an OCP derivative [194,195]. Succinate, adipate, fumarate and citrate also form double salts, but not maleate, aspartate and glutamate [196-198]. A detailed study of the IR and Raman spectra of OCP dicarboxylic acid double salts has been made [198] and the kinetics of the formation of the OCP succinate double salt investigated [199].

Changes that take place when the OCP-succinate double salt is heated have been studied [200]: there was a gradual contraction and weakening of d_{100} until it reached 16.3 Å at 600 °C, and disappeared altogether at 650 °C; at 800 °C, the XRD pattern of TCP (polymorph not stated) could be seen in addition to apatite; and at 800 °C, the IR bands of OH⁻ ions (Section 3.11.2) and of CO_3^{2-} ions replacing OH⁻ ions (Section 4.3.2) in apatite appeared.

It has been proposed [153,194] that the dicarboxylate ion is incorporated in the lattice within the "water" layer (Fig. 1.4) by the substitution of a phosphate by a dicarboxylate ion in pillars comprising Ca(3)-HP(5)′O₄-Ca(4)′ and Ca(4)-HP(5)O₄-Ca(3)′ (atom numbering of ref. [147], see also ref. [201]). This model has been criticised [201] because of the way in which the dicarboxylate ions would have to pack into the structure; an alternative was proposed for the succinate double salt based on similarities between the unit cell dimensions and the structures of OCP and calcium succinate trihydrate. The succinate ions were still located in the "water" layer: one in effect replaced two HPO_4^{2-} ions and linked Ca(3)‴ and Ca(3)′ of OCP to form a bridge in the direction of the *a*-axis and thus produced a corresponding expansion of the *a* parameter; the other linked Ca(3) and Ca(3)′ without affecting the *a*-axis parameter. This type of model could not be applied to the malonate double salt because it was not possible to find a transformation between the lattice parameters of OCP and calcium malonate or a structural correlation [201]. Consequently, it was thought that there may be substantial structural differences between the various OCP dicarboxylate double salts.

1.4 Dicalcium phosphate dihydrate, brushite

1.4.1 Introduction

Dicalcium phosphate dihydrate (synonyms: DCPD, calcium monohydrogen phosphate dihydrate, dibasic calcium phosphate dihydrate, calcium hydrogen orthophosphate 2-hydrate, brushite) has the formula $CaHPO_4.2H_2O$. The mineral, brushite, was discovered in 1865 in phosphatic guano from Avis Island in the Caribbean, and named after the American mineralogist G. J. Brush [202]. Brushite occurs in small amounts as a component of insular and continental rock-phosphate deposits and encrustations on ancient bones [1], and in dental calculus [144] and other pathological calcifications [203]. DCPD can occur as an intermediate in the precipitation of OHAp. It has been proposed as an intermediate in apatitic mineralisation and dissolution processes (bone mineralisation [120]; dissolution of enamel in acids [204] (and refs cited in this paper); and during topical fluoride treatment of teeth [205]). Prior DCPD formation on enamel has been shown to result in enhanced F⁻ ion uptake from topical solutions [206] (Section 4.6.5). DCPD in bone [145] (Table 4.8, Section 4.6.2) and fracture callus [207] has been reported on the basis of powder XRD studies. However in later work, DCPD could not be detected in embryonic chick or bovine bone by XRD [208,209] (Section 4.6.2). On the other hand, DCPD-like NMR resonances from very young bone have been seen [209] (Section 4.6.6).

1.4.2 Structure

The original structure determination [210] was refined in a centrosymmetric space group, but this was inconsistent with the subsequent finding of piezoelectricity in DCPD which signifies a noncentrosymmetric structure. The absence of a centre of symmetry has been confirmed [211] by refinement of the original measurements without this symmetry element. DCPD is monoclinic, space group *Ia* [211] with lattice parameters $a = 5.812\pm0.002$, $b = 15.180\pm0.003$, $c = 6.239\pm0.002$ Å and $\beta = 116° 25\pm2'$ [210]. There are four formula units per unit cell with an asymmetric unit $CaHPO_4.2H_2O$, so that none of the five hydrogen atoms are related by symmetry. Their positions were initially investigated by single crystal proton magnetic resonance [212] and subsequently located by neutron diffraction [213]. There was no evidence for disorder in the hydrogen atom positions and the acidic hydrogen atom was bound to an oxygen atom with a conspicuously long P-O bond (1.608(6) compared with 1.506(5) to 1.532(5) Å for the other P-O bonds). The structure contains columns, parallel to the short diagonal of the (010) face of the unit cell, of alternating Ca^{2+} and HPO_4^{2-} ions. These columns are joined together to form corrugated sheets (Fig. 1.5) which provide a prominent feature of the structure [210]. The sheets have a composition of $CaHPO_4$, are normal to the *b*-axis, and are linked together by water molecules. Similar sheets are to be found in monocalcium phosphate monohydrate, $Ca(H_2PO_4)_2.H_2O$, [134] (Section 1.2). The structure is closely related to that of gypsum, $CaSO_4.2H_2O$ [214,215], and a calcium phosphate-sulphate hydrate, $Ca_2HPO_4SO_4.4H_2O$ [216].

1.4.3 Preparation

DCPD has been directly precipitated by the simultaneous addition, during 2.5 to 3 h with mechanical stirring, of two solutions (1 l containing 90 g $Na_2HPO_4.2H_2O$ and 5 g KH_2PO_4, and 1 l containing 110 g $CaCl_2.6H_2O$) to a third solution (0.5 l containing 10 g KH_2PO_4) at 25 °C and pH 4 to 5 (maintained by addition of KH_2PO_4) [217]. The yield was about 83 g and the weight loss on ignition at 900 °C within ½ wt % of the theoretical for condensation to $Ca_2P_2O_7$ (Equation 1.10, Section 1.4.7). Other double decomposition methods have been published [218,219]. Neutralisation of dilute H_3PO_4 by the addition of calcium sucrate (prepared from $CaCO_3$ and sucrose) below 10 °C led to the precipitation of DCPD which gave a weight loss on

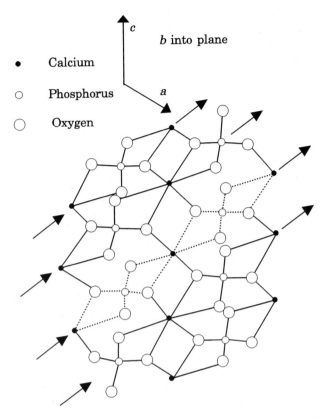

Fig. 1.5 View down the *b*-axis of one of the corrugated sheets of composition CaHPO$_4$ that occur in DCPD. There are four ···Ca HPO$_4$ Ca HPO$_4$ Ca HPO$_4$··· chains with directions marked by arrows with the bonds in one dotted for emphasis. The sheets are parallel to (100) and are linked together by water molecules. For clarity, these, and the protons on the phosphate ions, are not shown. (After Fig. 9 of Young and Brown [14])

ignition at 900 °C within 0.2 wt % of the theoretical [220]. DCPD can also be prepared by the ammoniation of a saturated solution of MCPM and DCPD at 20 to 25 °C [221] or at 10 °C [222]: chemical analyses of the product were within 0.1 percent of the theoretical [28,222]. Statistical analysis has been used to show an invariant composition and to predict yield and physical properties for the precipitation of DCPD by the addition of a CaCl$_2$ solution (3.0 mol l^{-1} raised to pH ~10 with NH$_4$OH) to an NH$_4$H$_2$PO$_4$ solution (1 mol l^{-1}, adjusted with H$_3$PO$_4$ or NH$_4$OH to a pH in the range 2.0 to 5.0) at 30 °C [223].

Large crystals of DCPD (1 × 2 × 0.5 cm^3) can be grown in four months at 25 °C by the interdiffusion through a nitric acid (pH 3) bridge of solutions of

$NH_4H_2PO_4$ (100 g l^{-1}) and $Ca(NO_3)_2.4H_2O$ (100 g l^{-1}) also at pH 3 [224]: similar crystal growth is possible at pH 4 [225]. DCPD can also be grown by the diffusion of Ca^{2+} ions into silica gel containing HPO_4^{2-} ions at pH 4 to 8 [225]. Large crystals grow (personal communication by MA Aia in ref. [13]) by the slow hydrolysis of urea in a quiescent solution of CaO in H_3PO_4, as described for DCPA (Section 1.5.3).

1.4.4 Optical properties, density and habit

Optical properties of brushite have been published for sodium light [5,16]. It is biaxial (positive) with $2V = 86°$ (calculated 85.5°) and $\alpha = 1.540$, $\beta = 1.545$, and $\gamma = 1.551$. The optic axial plane is perpendicular to (010), Z parallel to **b,** and the angle between **X** and **c** is 30° in obtuse β. The calculated density is 2.318 g cm^{-3} [210].

The habit is usually platy or prismatic, tabular on (010) *i.e.* parallel to the corrugated sheets described in Section 1.4.2; (010) and (001) are perfect cleavage planes [16].

1.4.5 Kinetics of crystal growth and dissolution

The growth rate of DCPD is proportional to the concentration product $[Ca^{2+}][HPO_4^{2-}]$ over a wide range of Ca/P ratios which suggested a predominantly surface reaction controlled process [218]. The rate is also proportional to the square of the amount to be precipitated before equilibrium is reached [226]. Constant composition studies (Section 3.8) have shown that crystal growth is pH dependent and is predominantly in two-dimensions [227]. The crystallisation rate at 37 °C and pH 6 (solution composition in mol l^{-1}: Ca 5.20×10^{-3}; phosphate 5.20×10^{-3}; KNO_3 background electrolyte 0.1) was 3.32 $\times 10^{-4}$ mol DCPD min^{-1} m^{-2} [228]. This rate was much higher than for OCP (~2 $\times 10^{-6}$ mol of $Ca_8H_2(PO_4)_6.5H_2O$ min^{-1} m^{-2}, [160], Section 1.3.5) and OHAp (2.7×10^{-7} mol of $Ca_5(PO_4)_3OH$ min^{-1} m^{-2}, [24] Section 3.8) under comparable conditions of composition and temperature [24]. (Note that these comparable conditions do not give similar thermodynamic driving forces for crystal growth because the solubility products differ.) A spiral growth mechanism with rate constants that could be explained in terms of surface tension and frequencies related to partial dehydration of cations has been proposed to explain the measured constant composition kinetics of growth and dissolution of DCPD [219,229]. Constant composition methods have been used to show that, on DCPD seeds, lower pH and higher temperatures favour the growth of DCPD rather than OCP in the 60 to 80 °C range; at 70 °C the reaction order was two, but in contrast, at 80 °C, where the rate of DCPD growth was much higher, it was four [162].

DCPD grows on OHAp seeds after an induction period [230] with the same kinetics as determined earlier [226] for the growth of pure DCPD (see above).

In systems in which CaPs are precipitated on OHAp seeds [231], the presence of F⁻ ions increases the quantity of the apatitic phase, and ethylidinediphosphonic acid encourages the formation of DCPD: this was explained by competition between the nucleation of DCPD and crystal growth on sites already present on OHAp seeds. DCPD nucleates and grows on OCP seeds after well-defined induction periods at higher supersaturations [232].

Poly-L-glutamic acid has an accelerating effect on the growth of DCPD, but polyacrylic acid shows no such effect [233]. This difference was explained on the basis of a better match for poly-L-glutamic acid compared to acrylic acid between the separation of the negatively charged -COO⁻ groups and the positions of the positively charged Ca^{2+} ions in DCPD [233]. Unlike the situation for OCP and OHAp, Mg^{2+} ions have no effect on the constant composition growth of DCPD [163], but Sr^{2+} ions cause a small decrease in growth rate accompanied by uptake of Sr^{2+} ions by the lattice which results in a small, but significant, expansion of the *a*- and *c*-axes [234]. Detailed studies have been made of the influence of isolated α_{s1}-, β- and κ-casein on the precipitation of DCPD and OCP at pH 6.1 and 6.15 under constant pH conditions [235]. The individual α_{s1}- and β-caseins induced extensive formation of DCPD at the cost of OCP, but less DCPD was formed in the presence of κ-casein.

It has been reported that the dissolution of DCPD is diffusion controlled [236], but assumptions about certain diffusion coefficients in this work have been questioned [219]. Dissolution in H_3PO_4 in the pH range 1.39 to 4 has also been reported to be diffusion controlled [224]. Constant composition methods have been used to study the kinetics of dissolution in terms of the same spiral model as was used for growth studies (see above) [219,229,237]. In such an investigation at 37 °C, it was found that the rate of dissolution was sensitive to solution hydrodynamics at high dissolution driving forces and under mild agitation which suggested that diffusion processes in the liquid provided a substantial resistance to dissolution [237]. However, at low undersaturation, the dissolution rate was insensitive to the stirring speed which showed that surface processes were rate determining. There was evidence from a detailed analysis of the results that surface diffusion appeared to be involved in transporting dissolved ions into the solution, although this process was not necessarily rate determining [237].

1.4.6 Solubility and reactions in solution

In 1966, it was found that the solubility product of DCPD appeared to be a parabolic function of pH with a minimum near pH 5, but this pH dependence could be removed if the existence of the ion pairs $[CaHPO_4]^0$ and $[CaH_2PO_4]^+$, with appropriate stability constants, were postulated [28]. In subsequent experiments and data analysis [34], values for K_s, the solubility product, (based on the activity product $(Ca^{2+})(HPO_4^{2-})$ and with dimensions $mol^2\ l^{-2}$), and stability constants for $[CaHPO_4]^0$ (K_x l mol^{-1}) and $[CaH_2PO_4]^+$ (K_y l mol^{-1}) as a function of the temperature T (K), were reported:

$$\log_e K_s = -8403.5/T + 41.863 - 0.09678T$$
$$\log_e K_x = 51090/T - 341.14 + 0.5880T \qquad\qquad 1.9$$
$$\log_e K_y = 19373/T - 122.81 + 0.1994T.$$

These give values of K_s, K_x, and K_y at 37 °C of 2.386×10^{-7}, 385.65, 4.469 respectively (units as above). The first of the Equations 1.9 shows that the solubility falls with an increase in temperature. The value of the solubility product from this equation compared well with the experimental value obtained in the system $Ca(OH)_2$-H_3PO_4-$NaCl$-H_2O at 25 °C [238]. The DCPD/OHAp singular point is at pH 4.3 at 25 °C [239], and at a pH between 5.132 and 5.281 at 37.5 °C [28]. The DCPD/OCP singular point is at pH 6.4 at 25 °C, and at pH 5.0 at 37 °C [168].

The quintuple point between DCPD, DCPA, and a phase given the formula $Ca_3(PO_4)_2.xH_2O$ occurs at 36 °C, or a little higher, [131], so DCPD should not occur in the phase diagram above this temperature.

Hydrolysis and other reaction in solution

It has been known for a long time that the ultimate product of DCPD hydrolysis is OHAp, but unlike OCP, the crystals formed are not pseudomorphs of the starting material. The kinetics of this reaction and formation of intermediates have been studied [239-243].

Hydrolysis to OCP occurs in the pH range 6.2 to 7.4 if the temperature is within 25 to 37 °C, but at higher pH values and/or temperatures, apatitic compounds are formed [239]. The reasons for this [239] are that at pH values above the DCPD/OHAp singular point (see Fig. 1.1), hydrolysis to OHAp should occur because it is the stable form, but this does not happen because the rate of growth of OHAp is too slow (except at high pH and/or temperatures). However at pH values above the DCPD/OCP singular point, OCP has a lower solubility than DCPD and grows much faster than OHAp, so that hydrolysis to OCP takes place. The kinetics of the transformation to OCP have been studied using a pH-stat technique and Mg^{2+} ions have been shown to markedly inhibit the reaction [243,244]. Hydrolysis of DCPD in ultrafiltered serum has been investigated [245].

DCPA is formed on heating DCPD at 60 to ~90 °C in water [242]. It was found that the time for dehydration decreased with a fall in the relative amount of DCPD in the water and an increase in its specific surface area.

The presence of F$^-$ ions during hydrolysis has a profound effect on the course of the reaction, the final product invariably containing apatitic products

[76,205,239,240,246,247]. The reaction is speeded up and FAp (or F,OHAp) is the predominant phase for dilute F$^-$ ion solutions (less than 1000 ppm), and CaF$_2$ for more concentrated solutions [205]; however the presence of F$^-$ ions can also promote the formation of OHAp (presumably containing very little F$^-$) [239]. DCPA can be formed as an intermediate under some conditions of hydrolysis in the presence [240,246] or absence [239,240] of F$^-$ ions.

Whitlockite can be formed in the presence of Mg^{2+} ions (Section 1.6.4). In one study, 0.2 g of DCPD per 100 ml of a solution containing Mg^{2+} ions were stirred at 37 °C for 24 days [248]. OCP plus OHAp, OCP plus OHAp (and on one occasion a trace of whitlockite), and largely whitlockite plus some OHAp were formed for magnesium concentrations of 0, 10^{-4} and 10^{-3} mol l^{-1} respectively.

1.4.7 Thermal decomposition and properties

DCPD loses its water of crystallisation slowly at ambient temperatures [5], but the ease of dehydration depends on the extent of crystal imperfections or physical damage [16,249]. DCPA is formed when DCPD is heated at 180 °C; further heating to between 320 and 340 °C yields γ-Ca$_2$P$_2$O$_7$ from the condensation

$$2HPO_4^{2-} \rightarrow P_2O_7^{4-} + H_2O, \qquad\qquad 1.10$$

and at 700 °C, the β form is obtained which changes at 1200 °C to α-Ca$_2$P$_2$O$_7$ [250]. The dehydration of DCPD to DCPA is catalysed by water, and in humid air proceeds via a single reaction at 135 °C; in the absence of moisture, the reaction takes place through several steps, including the formation of an amorphous phase which persists up to 530 °C [251]. Thermal gravimetric analysis, differential thermal analysis, IR and X-ray investigations of the various steps of the dehydration and condensation processes have been made [252] and changes in surface area, pore size distribution and volume measured [253]. The kinetics of the conversion of DCPD to pyrophosphate have been studied under vacuum at constant heating rates [254]. Paper chromatography has been used to show that the lowest temperatures for the detectable formation of pyrophosphate and tripolyphosphate are 110 and 200 °C respectively [180].

The heat capacity of DCPD has been measured from 10 to 310 K from which were calculated a standard entropy, $S°$, of 47.10 cal mol^{-1} K^{-1} (197.2 J mol^{-1} K^{-1}) and standard enthalpy ($H° - H°_0$) of 7490 cal mol^{-1} (3135.9 kJ mol^{-1}) at 298.15 K [222]. The standard entropy value can be compared with the self consistent compilation in Table 1.3.

1.4.8 Infrared, Raman and NMR spectroscopy

The IR spectrum of DCPD has been investigated from 4000 to 200 cm^{-1} at 20 and -180 °C [255] and spectra of DCPD and its deuterated analogue from 4000 to 300 cm^{-1} [186,256]. Detailed assignments were made and both sets of investigators reported that there was evidence from the IR spectra of two crystallographically different water molecules which was consistent with the structure (Section 1.4.2). There was also clear evidence of factor-group splitting of major bands which was consistent with the absence of a centre of symmetry [255]. This absence was supported by the coincidence in frequency of the Raman and IR bands within experimental error, as expected for a noncentrosymmetric structure [187,257]. Raman (polarised and nonpolarised) and IR investigations [187,257] resulted in some additional assignments and corrections to earlier [255,256] work. Recently, deconvoluted FT IR spectra of the v_3 PO$_4$ [190] and v_4 PO$_4$ [191] regions have been published.

The ^{31}P NMR MAS (~2 kHz) spectrum (including ^{1}H decoupling) at 68 MHz of DCPD has been reported [192]. A single resonance with an isotropic chemical shift of 1.7±0.3 ppm from 85% H$_3$PO$_4$ was observed, which is consistent with only one crystallographic type of phosphorus atom in the unit cell. The same resonance at 40 MHz with a chemical shift of 1.3 ppm with MAS at 5 kHz has been reported [193].

^{1}H shifts of 6.4 and 10.4 ppm from TMS have been assigned to structural water and acidic protons respectively in studies at 200 and 500 MHz with MAS (up to 8 kHz) [183].

1.5 Dicalcium phosphate anhydrous, monetite
1.5.1 Introduction

Dicalcium phosphate anhydrous (synonyms: DCPA, anhydrous dicalcium phosphate, dicalcium phosphate, calcium hydrogen orthophosphate, monetite) has the formula, CaHPO$_4$. The mineral, monetite, was first described in 1882 in rock-phosphate deposits from the islands of Moneta (from which the name derives) and Mona in the West Indies [258]. DCPA does not appear to occur in dental calculus or other pathological calcifications, nor has it been found in normal calcifications. However, DCPA has been reported in XRD studies of fracture callus (reviewed in ref. [207]) and possibly in bone [145]. Since it is less soluble than DCPD under all conditions of normal temperature and pressure (Section 1.5.5 and Fig. 1.1), it might be expected to be of more frequent occurrence. However its slow rate of crystal growth relative to DCPD [14] might explain why DCPD forms instead of DCPA, even though it is less stable. This observation again emphasises the importance of kinetic factors in determining which phase is formed in the Ca(OH)$_2$-H$_3$PO$_4$-H$_2$O system. It has been suggested [14] that DCPD precipitates more readily than the less soluble anhydrous salt because the hydrated ions in solution would be more readily incorporated into a hydrated crystal and also that a hydrated structure has a lower surface energy at the nucleation stage, and would therefore be favoured thermodynamically over anhydrous crystal nuclei. DCPA is also not very easy to detect by XRD methods, so may occur more frequently than presently thought [14].

1.5.2 Structure

The room temperature form of DCPA is triclinic, space group $P\bar{1}$ [259,260] with lattice parameters a = 6.910(1), b = 6.627(2), c = 6.998(2) Å, α = 96.34(2), β = 103.82(2) and γ = 88.33(2)° at 25 °C [261]. There is a low temperature form with space group $P1$ and lattice parameters at 145 K of a = 6.916(2), b = 6.619(2), c = 6.946(3) Å, α = 96.18(3), β = 103.82(3) and γ = 88.34(3)° [260]. The best estimate [260] of the phase transition temperature comes from heat capacity measurements [262] and is between 270 and 290 K; the exact temperature seems to be related to the presence of impurities in the sample.

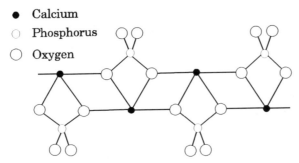

● Calcium
○ Phosphorus
○ Oxygen

Fig. 1.6 Double chain of Ca^{2+} and phosphate ions that forms the basis of the DCPA structure. (After MacLennan and Beevers [263])

There are four formula units per unit cell of DCPA. Double chains of ···Ca PO_4 Ca PO_4 Ca PO_4 $CaPO_4$··· (Fig. 1.6), extending along the a-axis, are linked together in the b-axis direction by Ca-O bonds to form distorted sheets of atoms approximately in the (001) plane: the centre of symmetry produces another sheet below this in the c-axis direction [263]. Thus the structure consists of a three-dimensional network of phosphate tetrahedra held together by Ca^{2+} ions in their interstices. In the high temperature form, there are two crystallographically independent phosphate ions and three independent hydrogen atoms [260]. One hydrogen atom is located on a centre of symmetry, another is in a general position, and the third is in two-fold disorder about a centre of symmetry. In the low temperature form without a centre of symmetry [260], the hydrogen atoms that were in two-fold disorder become ordered, so that one of the two previously disordered sites becomes fully occupied and the other completely vacant; this changes the hydrogen bonding predominantly in the *c*-axis direction, which explains why it is the *c*-axis parameter that shows the greatest change through the phase transition.

1.5.3 Preparation

The preparation of pure, stoichiometric DCPA is not entirely straightforward because the conditions of synthesis have to be carefully controlled. DCPA can be synthesised by the same method described earlier [217] for DCPD (Section 1.4.3) by the simultaneous addition of the same phosphate and calcium solutions to the same phosphate solution, but at 100 °C instead of room temperature. The resultant precipitate was washed with dilute H_3PO_4 at pH 3, and then absolute ethanol. Literature on the synthesis of DCPA has been reviewed and detailed conditions for precipitation of pure DCPA by double decomposition from solutions of $CaCl_2$ and $(NH_4)_2HPO_4$ plus H_3PO_4 have been investigated [264]. DCPA has also been prepared by the dehydration of 400 g of DCPD in 4 l of dilute H_3PO_4 (0.07 mol l^{-1}) by boiling for 72 h [35]. After filtration, the product was washed with small volumes of water until the wash solution had a pH of 6, then washed with acetone and dried *in vacuo* (3 h at 80 °C, then 60 h at 50 °C). Polarised light microscopy showed the absence of DCPD, and a particle size of about 20 μm. The chemical analyses agreed to within 0.1 wt % of the theoretical [35].

Single crystals have been grown by the slow interdiffusion, through a barrier of HNO_3 at pH 3, of 20 wt % $Ca(NO_3)_2.4H_2O$ adjusted to pH 3 with HNO_3 and 20 wt % $NH_4H_2PO_4$ solutions at 83 °C for 43 days; the crystals were washed with hot water and acetone, and dried at 105 °C for 72 h [138]. Crystals up to 1 cm have been grown (MA Aia, personal communication cited in ref. [13]) from a solution of 0.1 to 0.2 mol of CaO dissolved in 0.5 to 2.0 mol of H_3PO_4 (presumably in a total volume of 1 l of water [16]) that contained 0.5 to 2.0 mol of urea. The urea was allowed to hydrolyse at 65 °C for 12 days, without stirring, to give a final pH between 2.5 and 6.0.

1.5.4 Optical properties, density and habit

Optical properties have been reported [138]. At 610 nm, and 25 °C, $\alpha = 1.588$, $\beta = 1.616$, $\gamma = 1.640$, biaxial (negative) with a calculated $2V$ of 84° 4' (values were also given for 425 nm). The acute bisectrix, **X,** is nearly perpendicular to (11$\overline{1}$) and is inclined at 52° to the normal to (010). The trace of the optic axial plane onto the (11$\overline{1}$) plane is inclined to the trace of the (010) plane by 32° in obtuse γ. The extinction angles on the (010) plane are 15° with the (101) trace and 38° with the (01$\overline{1}$) trace. Because of the temperature at which these measurements were made, they almost certainly relate to the centrosymmetric form; optical constants for the low temperature modification do not appear to have been reported. It is noteworthy that the birefringence is relatively high, so that the salt is easily detected under the polarised light microscope, even when present in very small quantities. The calculated density is 2.89 g cm^{-3} [138].

DCPA crystallises as triclinic {010} tablets elongated in the *a*-axis direction, but prismatic and rod habits also occur; the crystals are brittle with a "hackly fracture", without twinning [138].

1.5.5 Solubility and reactions in solution

Like DCPD (Section 1.4.6), the solubility product of DCPA was found to depend on pH unless account was taken of the formation of the ion pairs [CaHPO$_4$]^0 and [CaH$_2$PO$_4$]$^+$ [35]. The solubility products and their standard errors (both × 10^7) were 1.97±0.03, 1.64±0.01, 1.26±0.02 and 0.92±0.02 mol^2 l^{-2} at 5, 15, 25 and 37 °C respectively. At the same respective temperatures, association (stability) constants and standard errors for [CaHPO$_4$]^0 were 240±30, 190±20, 380±50 and 390±90 l mol^{-1}; and for [CaH$_2$PO$_4$]$^+$ were 5±1, 10±1, 10±1 and 11±1 l mol^{-1} (no multiplier). These association constants compared favourably with one set of previous measurements [34] (Section 1.4.6), but were less than those from other potentiometric measurements [265] by more than would have been expected. The above measurements show that DCPA is less soluble than DCPD (Section 1.4.6 and Fig. 1.1) and that the solubility decreases with an increase in temperature (like DCPD, Section 1.4.6).

Hydrolysis and other reactions in solution. DCPA, like the other acid phosphates, hydrolyses to OHAp in excess water. Recently, the hydrolysis of DCPA into ns-OHAp has been investigated in detail under controlled solution conditions (1 to 7.5 mmol DCPA per 250 ml solution, 25 to 100 mmol l^{-1} CaCl$_2$, 37 °C, constant pH in the range 6.35 to 9.0) to identify factors which influenced the nucleation, growth and crystal morphology of the apatitic phase [266]. It was found that generally, hydrolysis times and the size of the ns-OHAp crystals at completion of conversion were inversely related to pH, except for a sharp fall in size below pH 6.5. Crystal morphology was also very dependent on pH, with the most equidimensional crystals formed at pH 7.0 [266]. FAp is formed in the presence of F$^-$ ions [246]. DCPA reacts in hot carbonate solutions to yield an apatite with a reduced *a*-axis parameter due to CO$_3^{2-}$ ions substituting in the lattice (Section 4.5.3), and often with the simultaneous formation of calcite and aragonite [79]; an amorphous product is formed if its CO$_3$ content exceeds about 15 wt % [267,268]. Whitlockite is formed in the presence of Mg^{2+} ions [269] (Section 1.6.4).

1.5.6 Thermal decomposition and properties

The thermal decomposition of DCPA has already been considered as it is a thermal decomposition intermediate of DCPD (Section 1.4.7). Paper chromatography showed that pyrophosphate formation starts at 300 °C and is essentially complete at 350 °C; tripolyphosphate was not detected [180].

The heat capacity of DCPA has been measured from 10 to 310 K from which were calculated a standard entropy, $S°$, of 26.62 cal mol^{-1} K^{-1} (111.5 J mol^{-1} K^{-1}) and standard enthalpy ($H° - H°_0$) of 4455 cal mol^{-1} (1865.2 kJ mol^{-1}) at 298.15 K [262]. These results can be compared with the self consistent compilation in Table 1.3.

1.5.7 Infrared, Raman and NMR spectroscopy

Assignments for the IR spectrum at 300 and 77 K from 4000 to 300 cm^{-1} of DCPA and its deuterated analogue have been reported [186,256], and from 4000 to 200 cm^{-1} for DCPA [187,270]. The Raman spectrum has been studied [187,270]. The IR spectrum is conspicuously more complex at low, compared to room temperatures [186,256]. This led to the suggestion that this difference might originate from a change in crystal structure (anonymous referee in ref. [186]). In view of the now known phase transition at 270 to 290 K (see above), this is the most likely explanation, although detailed assignments for the low temperature, noncentrosymmetric form, do not appear to have been published.

The ^{31}P NMR MAS (~2 kHz) spectrum at 68 MHz with proton decoupling has been reported [192]. Two partially resolved peaks with intense sidebands were observed with isotropic chemical shifts of 0.0±0.4 and -1.5±0.4 ppm from 85% H$_3$PO$_4$ which corresponded to at least two different types of phosphate group, as expected from the structure (Section 1.5.2). Similar chemical shifts of -0.2 and -1.5 ppm have been seen [193].

^{1}H shifts of 13.6 and 16.2 from TMS have been assigned to three acidic proton environments (unresolved) and to a more strongly hydrogen-bonded acidic proton, respectively, in studies at 200 and 500 MHz with MAS (up to 8 kHz) [183].

1.6 Anhydrous tricalcium phosphates and whitlockite
1.6.1 Occurrence and importance

The adjective anhydrous is used to distinguish these compounds from hydrated apatitic precipitates that have a similar Ca/P molar ratio of ~1.5 (Section 3.4.1). Synonyms are tricalcium diorthophosphate and TCP; the nomenclature for whitlockite is discussed below.

According to a review [271] on the anhydrous tricalcium phosphates and whitlockite up to 1956, α-Ca$_3$(PO$_4$)$_2$ (α-TCP) and β-Ca$_3$(PO$_4$)$_2$ (β-TCP) were first reported in 1932 by Trömel [272]. The β-TCP to α-TCP transition temperature is 1125 °C [273] (Section 1.6.6). A new high temperature phase, $\overline{\alpha}$-Ca$_3$(PO$_4$)$_2$ (super-α-TCP, $\overline{\alpha}$-TCP), stable above 1430 °C and unable to survive quenching to room temperature, was discovered in 1959 [274] (Section

1.6.6). The occurrence of these phases in the high temperature $CaO-P_2O_5$ phase diagram is discussed in Section 1.6.6. A high pressure polymorph is also known [275,276] (Section 1.6.2).

Whitlockite was first described in 1941 as a hydrothermal product from a granite pegmatite in the Palermo Quarry, North Groton, New Hampshire, and named after the mineralogist Herbert P. Whitlock [277]. Other occurrences in insular phosphate deposits are known [1], some of which contain carbonate (*e.g.* "martinite").

The mineral was reported with a supposed formula of $Ca_3(PO_4)_2$ and an XRD pattern similar to β-TCP [1,277]. As a result, the terms whitlockite and β-TCP have been used interchangeably, but since there are subtle structural differences between these substances which have now been elucidated (Section 1.6.2), this practice is unsatisfactory. The term whitlockite should be reserved for the mineral or similar synthetic materials in which Mg^{2+} and HPO_4^{2-} ions play a structural role, and β-TCP for the low temperature polymorph of $Ca_3(PO_4)_2$, in which Mg^{2+} and HPO_4^{2-} ions are absent. This nomenclature will be followed in this book (Table 1.1). The term β-TCa,MgP will be used to describe substances in which HPO_4^{2-} ions are absent, and which have the β-TCP structure in which some of the Ca^{2+} ions are replaced (not necessarily randomly) by Mg^{2+} ions. The presumption will be made that β-TCP-like substances, made in aqueous systems in the presence of Mg^{2+} ions, contain structural Mg^{2+} and HPO_4^{2-} ions, even though this may not have been proved; thus they will be called whitlockites, or more specifically magnesium whitlockites, and not β-TCa,MgP or β-TCP, irrespective of the nomenclature used by the author(s). The precedence of the name whitlockite and alternatives has been discussed [278].

Whitlockite occurs in various pathological calcifications [248] and as a major constituent of human dental calculus [271]. It has been identified in carious dentine by electron diffraction [269,279]. Its occurrence in some crevices of tufts and lamellae of old, caries-free, human enamel has been reported [280]. An amorphous whitlockite precursor is said to occur in *Animalia*, and $Ca_3Mg_3(PO_4)_4$ in *Annelida* [281]. Whitlockite forms the mineral phase in the tooth plates of *Chimaera phantasma* [282].

Most of the following sections concern β-TCP and its derivatives, but α-TCP is of interest because of its structure (Section 1.6.2) and its use for the synthesis of double salts of OCP with dicarboxylic acids (Sections 1.3.9 and 1.6.5).

1.6.2 Structures

Structure of α-TCP

α-TCP crystallises in the monoclinic space group $P2_1/a$ with lattice parameters $a = 12.887(2)$, $b = 27.280(4)$, $c = 15.219(2)$ Å and $\beta = 126.20(1)°$

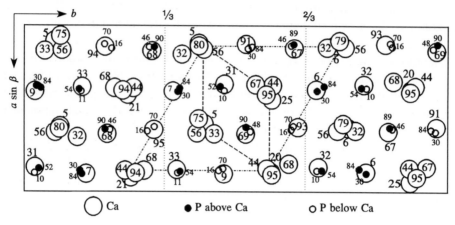

Fig. 1.7 Projection of the α-TCP structure onto the (001) plane. Only the calcium and phosphorus atoms are shown. The dashed and dashed-dotted lines outline a glaserite, $K_3Na(SO_4)_2$, and apatite unit cell, respectively. (After Fig. 3 of Mathew *et al.* [283] and Fig. 4.11 of Monma and Nagai [284])

with 24 formula units per unit cell [283]. There is a prominent approximate subcell with a *b*-axis parameter of *b*/3 (9.09 Å) that contains eight formula units which corresponds to the unit cell reported earlier for α-$Ca_3(PO_4)_2$ [285]. The structure comprises columns of Ca^{2+} and PO_4^{3-} ions parallel to the *c*-axis [283] arranged as shown in Fig. 1.7. The pseudocell translation of **b**/3 can be readily seen. There are cation columns, ···Ca Ca Ca Ca···, and cation-anion columns, ···PO_4 Ca PO_4 □ PO_4 Ca PO_4 □ PO_4 Ca PO_4···, where □ is a vacancy. Analogous cation-anion columns occur as the A columns in glaserite, $K_3Na(SO_4)_2$, except that the vacancy is occupied by a K^+ ion (Fig. 2.6a, Section 2.2.2). These columns are arranged to form a pseudohexagonal pattern (Fig. 1.7): each cation column is surrounded by six cation-anion columns, and each cation-anion column is surrounded by alternating cation-cation and cation-anion columns. A similar arrangement is seen in glaserite (Fig. 2.6a, Section 2.2.2), hence these two structures are closely related. The part of the structure of α-TCP that corresponds to a unit cell of glaserite is outlined with dashed lines in Fig. 1.7. (Note that this unit cell is translated from the cell in Fig. 2.6a in the direction of its long diagonal by half the diagonal length.) α-TCP is also related to the structure of apatite, as must be the case because of the relation between the glaserite and apatite structures (Fig. 2.6, Section 2.2.2). The section of the structure of α-TCP that corresponds to a conventional unit cell of apatite is marked with dashed-dotted lines in Fig. 1.7. The correspondence can also be seen from a consideration of the dimensions of the unit cell of α-TCP with those of apatite ($a \approx 9.4$ and $c \approx 6.8$ Å, Table 1.4). The α-TCP approximate subcell *b*-axis parameter of 9.09 Å corresponds to the apatite *a*-axis parameter, whilst half the *c*-axis parameter of α-TCP (7.6 Å) corresponds to the *c*-axis parameter of apatite. The apatite structure (Section 2.2.1 and Fig. 2.6b, Section 2.2.2) can thus be derived from α-TCP by replacing the cation-cation columns at the corners of the apatite cell in Fig. 1.7 by anion columns (F^- ions in FAp and OH^- ions in OHAp). The remaining cation columns in α-TCP become the columnar Ca(1) ions in apatite, whilst the PO_4^{3-} and Ca^{2+} ions that form the cation-anion columns in α-TCP have very approximately the same positions as the PO_4^{3-} and Ca(2) ions in apatite.

Each formula unit occupies 180 Å^3 in α-TCP compared with 168 Å^3 in the β form [283]. α-TCP is therefore a "looser" structure than β-TCP and has a higher internal energy; this is consistent with the α modification being the high temperature form [283] and with it having a higher reactivity in water than β-TCP (Section 1.6.5).

The structure of $\overline{\alpha}$-TCP, stable above 1430 °C, does not seem to have been determined.

Structure of β-TCP

β-TCP has the rhombohedral space group *R3c* with unit cell *a* = 10.439(1), *c* = 37.375(6) Å (hexagonal setting) with 21 formula units per hexagonal unit cell [286]. The structure of β-TCP has been determined and described in terms of a distortion from the $Ba_3(VO_4)_2$ structure [286] (see also the descriptions of related structures: synthetic [287] and mineral [278] whitlockites; and β-TCa,MgP [288]). The β-TCP structure is similar to $Ba_3(VO_4)_2$, but has three fewer formula units per hexagonal unit cell (Table 1.5). The $Ba_3(VO_4)_2$ structure [289] has layers perpendicular to the *c*-axis comprising two planes of intermeshed VO_4^{3-} ions with the tetrahedra oriented so that three of the oxygen atoms from each tetrahedron form the external surfaces of the layer. There is a centred plane of Ba(1) ions between them that are twelve-fold coordinated by oxygen atoms from the VO_4^{3-} ions. The layers are stacked together with another plane of Ba(2) ions between them, but this plane is displaced from the centre of the opposing faces of the layers. The Ba(2) ions are only ten-fold coordinated by oxygen atoms.

An alternative description of the structure is possible which is more useful for explaining the relationships between $Ba_3(VO_4)_2$ (or the isostructural $Ba_3(PO_4)_2$), β-TCP and whitlockite [278,286]. In this description, emphasis is given to columns of ions of the form $\cdots VO_4$ Ba Ba Ba $VO_4 \cdots$ that can be identified running parallel to the *c*-axis (Fig. 1.8a with P replaced by V and the Ca by Ba). The central Ba^{2+} ion in this motif (Ca(3) in the Figure) corresponds to the centred twelve-fold coordinated Ba(1) ions referred to above, and the ions on either side (Ca(1) and Ca(2) in the Figure) are the ten-fold coordinated Ba(2) ions. All the atoms in the motif are collinear, except for three oxygen atoms from each VO_4 tetrahedron. Each column is surrounded by six others

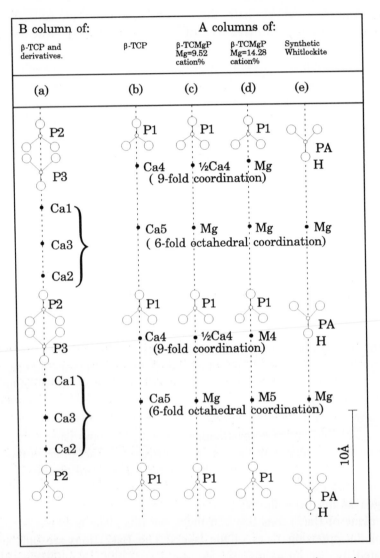

Fig. 1.8 Configuration of A and B columns (length shown is *c*-axis parameter of the hexagonal unit cell plus an additional PO$_4$ to show continuity of the columns). (a) depicts all the columns in Ba$_3$(VO$_4$)$_2$ and B columns in β-TCP and related compounds. (b) to (e) depict A columns in: (b) β-TCP; (c) β-TCa,MgP (Ca$_{18}$Mg$_2$(Ca,□)(PO$_4$)$_{14}$), all the Mg^{2+} ions have been put in the Ca(5) sites, but structure analysis shows that some of them are in the Ca(4) sites; (d) β-TCa,MgP (Ca$_{18}$Mg$_2$(Mg,□)(PO$_4$)$_{14}$); and (e) whitlockite (Ca$_{18}$Mg$_2$H$_2$(PO$_4$)$_{14}$). Note the inversion of the PO$_4$ tetrahedra that takes place in the A columns of whitlockite in (e).

(Fig. 1.9 with A = B), but with alternating displacements from each other in the *c*-axis direction to provide the oxygen coordination of the Ba^{2+} ions described above.

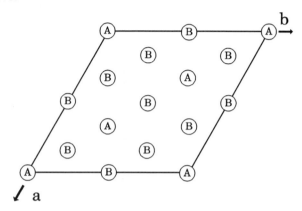

Fig. 1.9 Arrangement of A and B columns in the hexagonal unit cell of β-TCP and its derivatives. The *c*-axis is out of the plane of the diagram.

The 12 identical columns per equivalent hexagonal cell of $Ba_3(VO_4)_2$ become three A and nine B columns in β-TCP and its derivatives (Figs 1.8 and 1.9). The B columns are essentially unaltered, so that all the major changes take place in the A columns (Figs 1.8b to 1.8e). In β-TCP, the number of oxygen atoms coordinated to the cations must be reduced from that in $Ba_3(VO_4)_2$ because of the smaller size of the Ca^{2+} ion compared to the Ba^{2+} ion. This reduction causes the reorientation of some of the oxygen tetrahedra so that there is insufficient space for all the tetrahedra that were in $Ba_3(VO_4)_2$. The result is that the group of ions, $8 \times Ba_3(VO_4)_2$, is replaced by $7 \times Ca_3(PO_4)_2$, thus reducing the number of formula units per hexagonal cell by three (Table 1.5). This reduction takes place via the loss of half the ions from each A column (equal to $3Ca + 2PO_4$ per column per hexagonal *c*-axis length, Fig. 1.8b). The two lost PO_4 tetrahedra become new cation sites (designated Ca(4) sites) that are nine-fold coordinated by oxygen atoms. The particular significance of this site will be seen shortly. The three original cation sites are lost with the formation of a new site, Ca(5), which has six-fold octahedral coordination by oxygen. The structure determination [286] showed that there was only one Ca^{2+} ion per pair of Ca(4) sites in a column, that is to say, these Ca^{2+} ions are in two-fold disorder. Thus the new cation sites contain only three Ca^{2+} ions per column instead of the original six, so that charge balance is maintained. The two-fold disorder in one of the cation sites allows the space group, which has a glide plane parallel to the *c*-axis that generates an even number of lattice sites per unit cell, to be reconciled with the hexagonal unit cell contents of $21 \times Ca_3(PO_4)_2$, which contains an odd number of Ca^{2+} ions. This difficulty had been the subject of much previous discussion, summarised in ref. [278].

Table 1.5 Comparison of ionic contents of barium vanadate, β-TCP and its derivatives in equivalent volumes (one third of the hexagonal unit cell of β-TCP).

Compound	Ionic contents[a]			
	Site:	Ca(5)		Ca(4)
1. Barium vanadate, $Ba_3(VO_4)_2$	Ba_{24}			$(VO_4)_{16}$
2. β-TCP, $Ca_3(PO_4)_2$	Ca_{18}	Ca_2	$(Ca,\square)$	$(PO_4)_{14}$
3. β-TCa,MgP Mg 9.52 cation %	Ca_{18}	Mg_2	$(Ca,\square)$	$(PO_4)_{14}$
4. β-TCa,MgP Mg 14.28 cation %	Ca_{18}	Mg_2	$(Mg,\square)$	$(PO_4)_{14}$
5. Mg whitlockite[b]	Ca_{18}	Mg_2	H_2	$(PO_4)_{14}$

[a]$\square$ indicates an unoccupied site.
[b]This is the end-member formula. Mg whitlockites are solid solutions of Formulae 3 and 5. Mineral whitlockite also has Fe^{2+} and Mn^{2+} ions in the Ca(5) site.

There is evidence from fluorescence studies of tin-activated β-TCP of sluggish phase transitions at -40 and +35 °C that could be confirmed by calorimetry [290]. It is possible that at least the lower temperature transition involves the loss of the two-fold disorder of Ca(4) with a lowering of the lattice symmetry.

The high pressure form of $Ca_3(PO_4)_2$ has the $Ba_3(VO_4)_2$ structure [275].

Structure of β-TCa,MgP

It is possible to replace up to about 15% of the Ca^{2+} by Mg^{2+} ions without essentially changing the β-TCP structure [248,291,292] (although values from 7.3 to 22.2% have been reported [291]). The structures of a number of crystals of β-TCa,MgP with different magnesium contents (preparation in Section 1.6.4) have been determined [288]. The Ca^{2+} ions are not randomly replaced by Mg^{2+} ions: if there is a low total magnesium content (the crystals studied had 4.8 % of the cation sites filled with Mg^{2+} ions), the Mg^{2+} ions replace Ca^{2+} ions in Ca(4) and Ca(5) sites essentially randomly, but any further addition of Mg^{2+} ions is accommodated by the further random replacement of Ca^{2+} ions only in Ca(5) sites (in this case the crystal studied had 11.6 % of the cation sites filled with Mg^{2+} ions) [288]. This substitution is illustrated in Fig. 1.8c in which all the Mg^{2+} ions have been put into the Ca(5) sites to give an idealised composition $Ca_{19}Mg_2(PO_4)_{14}$ (Table 1.5, Formula 3). It should be recalled that the two sites, Ca(4) and Ca(5), that are occupied by Mg^{2+} ions do not occur in

$Ba_3(VO_4)_2$, but are generated in response to the replacement of Ba^{2+} by the smaller Ca^{2+} ion: it is perhaps not surprising that it is these sites that are favoured by the even smaller Mg^{2+} ion. Furthermore, the Ca(5) site that is most readily occupied, has octahedral coordination which is uncomfortably constrained for a Ca^{2+} ion, but very suitable for the smaller Mg^{2+} ion.

There is an unusual change in lattice parameters with the extent of magnesium substitution in the lattice [248,291,292]. The *a*-axis parameter decreases from 10.437 to 10.308 Å as the magnesium content increases to a maximum of about 15 % of the cations (with a slight change in slope at 10 %), but the *c*-axis decreases from 37.42 to a minimum of 37.07 Å at about 10 %, then it increases to 37.30 Å for a 15 % cation replacement [248]. The changes in slope at 10 % have been explained [291] on the basis of the structural work discussed above. As the number of Mg^{2+} ions in the lattice increases, they replace essentially only Ca(5) ions with octahedral coordination until the theoretical maximum of $100 \times 6/63 = 9.52$ % for this site (Table 1.5, Formula 3 and Fig. 1.8c) is reached (the multiplicity of the site is 6, its occupancy is 1, and there are 63 cation sites in the hexagonal unit cell). Note that the initial random replacement of Ca^{2+} ions in both Ca(4) and Ca(5) sites by small amounts of Mg^{2+} ions has been ignored in this argument. Thereafter, Mg^{2+} ions replace Ca^{2+} ions in the Ca(4) sites (Table 1.5, Formula 4 and Fig. 1.8d) which can accommodate a further of $100 \times 3/63 = 4.76$ % (the multiplicity is 6 and the occupancy 0.5), so that the total magnesium content that can be accommodated in the Ca(5) and Ca(4) sites together is $9.52 + 4.76 = 14.28$ % [292], which is close to the maximum experimental value of 15 % cited above for the fraction of cations that can be replaced by magnesium without substantial structural changes. The structural work on which these explanations are based did not extend to magnesium above 11.6 %, so the substantial occupation of Ca(4) sites needs to be confirmed.

Preparations can be made with a magnesium content greater than about 14% of the total cations, but in these cases, $Ca_7Mg_9(Ca,Mg)_2(PO_4)_{12}$ is formed in addition to a β-TCa,MgP ([292] and refs therein). The β-TCa,MgP that was synthesised had about a 10 atomic percent of calcium replaced by magnesium (*i.e.* only about 2/3 of the maximum possible) and this decreased with very slow cooling rates to room temperature (10°C h^{-1}) [292]. The structure of $Ca_7Mg_9(Ca,Mg)_2(PO_4)_{12}$ has been determined [293]; it seems to have a close structural relationship with stanfieldite, $Ca_4(Mg,Fe)_5(PO_4)_6$ (found with some mineral deposits of whitlockite [293]) and with α-TCP and $\overline{\alpha}$-TCP [283,293].

ESR studies of β-TCP doped with Cu^{2+} indicated that isolated Cu^{2+} ions were located on Ca(3) sites in the B columns (Fig. 1.8a) [276]. In the high pressure polymorph, which has the $Ba_3(VO_4)_2$ structure, ESR showed that the Cu^{2+} ions were in Ba(1) sites (these are equivalent to Ca(3) in β-TCP). Thus Cu^{2+} and Mg^{2+} occupy different sites [276].

Structure of whitlockite

The structure of a synthetic Mg whitlockite, $Ca_{18}Mg_2H_2(PO_4)_{14}$, has been determined [287] and is very closely related to the structure of β-TCP and β-TCa,MgP. The space group is $R3c$ with unit cell a = 10.350(5), c = 37.085(11) Å (hexagonal setting) with three formula units per hexagonal unit cell. The Ca(5) sites (Fig. 1.8e), with nearly octahedral coordination, accommodate all the Mg^{2+} ions, so the Ca(4) sites are not occupied by cations. However, the PO_4 tetrahedron that lies immediately above the Ca(4) site on the same triad axis (Fig. 1.8e), becomes inverted. This means that instead of having its base facing the Ca(4) site, its apex faces this site. This inversion allows the acidic proton that is attached to the oxygen atom located at its apex to bond weakly to an oxygen atom in an adjacent PO_4 group. This change takes place at all the Ca(4) sites, irrespective of whether the sites are occupied in β-TCP by Ca^{2+} ions or are vacant. This means that the Ca(4) site (or at least one nearby) accommodates two hydrogen atoms per formula unit in whitlockite, whereas it accommodates a single Ca^{2+} ion in two-fold disorder in β-TCP. Thus the cation disorder found in β-TCP does not occur in whitlockite. However, if the hydrogen atom is displaced from the triad axis as implied above, it must be statistically disordered about this axis, or the space group must have a lower symmetry.

Mineral whitlockite has been shown to be a solid solution of $Ca_{18}(Mg,Fe)_2(Ca,\square)(PO_4)_{14}$ and $Ca_{18}(Mg,Fe)_2H_2(PO_4)_{14}$ [278]. The first compound will be recognised as a β-TCa,Mg,FeP (Fig. 1.8c and Table 1.5, Formula 3) with all the octahedral Ca(5) sites occupied randomly by Mg^{2+} and Fe^{2+} ions, (half of the Ca(4) sites are unoccupied and indicated by $\square$). The second compound, without any unoccupied Ca(4) sites, is like Mg whitlockite (Fig. 1.8e and Table 1.5, Formula 5), but again with the Ca(5) sites randomly occupied by Mg^{2+} and Fe^{2+} ions. ESR studies have shown that Mn^{2+} ions in the mineral probably also occupy the Ca(5) sites [278]. This position has been confirmed in a single crystal XRD structure determination of a hydrothermally grown synthetic manganese whitlockite, $Ca_{18}Mn_2H_2(PO_4)_{14}$ [294].

It has been suggested [1] that the carbonate associated with some mineral whitlockites (*e.g.* "martinite") is present as a lattice substitution by analogy with francolite (Section 4.2). This idea is supported by the fact that the refractive indices of "martinite" are lower than those of whitlockite (next section), just as those of francolite are when compared with FAp (Table 1.4, Section 1.1). If this suggestion is true, the HPO_4^{2-} and CO_3^{2-} ion substitutions of whitlockite would parallel these two substitutions thought to occur in precipitated apatites (Section 3.7.3 and Sections 4.5.3 to 4.5.5, respectively).

1.6.3 Optical properties, density and habit

The density of α-TCP, calculated from the lattice parameters given in the previous section, is 2.863 g cm^{-3} [283]. The experimentally determined density, 2.814±0.01 g cm^{-3}, is somewhat less [285]. α-TCP is biaxial (positive) with α = 1.588, β = 1.588$_3$, γ = 1.591 and 2V = 35° (sodium light, orientation of indicatrix not known) [16].

β-TCP is uniaxial negative with O = 1.622 and E = 1.620 (sodium light, error ±0.001) [295]. The whitlockite from the Palermo quarry has slightly higher indices with O = 1.629 and E = 1.626 [1], but carbonate-containing mineral whitlockites (*e.g.* "martinite") have significantly lower indices (O = 1.607 and E = 1.604) [1]. The higher refractive indices of β-TCP, compared with those of α-TCP, are consistent with the smaller volume per formula unit in β-TCP than α-TCP (Section 1.6.2). The density (not clear if calculated or theoretical) of a synthetic whitlockite, $Ca_{18}Mg_2H_2(PO_4)_{14}$, is 3.033 g cm^{-3} [287], and the calculated density of β-TCP is 3.067 g cm^{-3} [286].

Crystal habit. β-TCP crystallises as euhedral rhombohedra modified by {11$\overline{2}$0} and {0001} forms [16]. It is this rhombohedral habit that makes whitlockite so easily recognisable in electron microscope studies of carious dentine, the so called "caries crystals" [269,279] and in dental calculus [144].

1.6.4 Preparation
Powders prepared at high temperatures

The CaO-P_2O_5 phase diagram, giving the stability regions of the TCP polymorphs, is shown in Figs 1.10 and 1.11 (Section 1.6.6). β-TCP can be prepared by heating an intimate mixture of the calculated quantity of DCPA and $CaCO_3$ [37]. Precautions have to be taken to ensure the correct stoichiometry, complete reaction, and absence of the high temperature polymorph (α-TCP). After heating for at least a day at 1000 °C, the product was examined under the polarising microscope and by XRD, and the composition adjusted by adding DCPA or $Ca(OH)_2$ solution, depending on the identity of any impurity phase [37]. It was then reground, and the process repeated until the Ca/P molar ratio was 1.50. Finally, the product was converted to α-TCP by heating to between 1150 and 1200 °C to remove inhomogeneities. The α-TCP was then reconverted to β-TCP by prolonged heating at 800 to 950 °C [37]. Less elaborate preparations of β-TCP (and β-TCa,MgP) have been reported many times. In these, an intimate mixture of the calculated quantities of $CaCO_3$ (plus $MgCO_3$ if required) and DCPA or DCPD (the latter is not so good because of its uncertain water content) are heated at about 1000 °C for several hours. MgO instead of $MgCO_3$ can be used in such preparations [296,297].

Another approach has been used to prepare pure β-TCP [298] and β-TCa,MgPs [292] that avoids the difficulty of grinding together the reactants into an intimate mixture and the possibility of incomplete reaction. An ACP with a Ca/P or (Ca + Mg)/P molar ratio of 1.5 is precipitated at high pH

(Section 1.8.2) and can then be heated at 900 °C for 2 h with the confident expectation that the reaction will go to completion.

α-TCP can be made from β-TCP by aging above 1200 °C [16] or by heating at 1400 °C, well above the transition temperature, as in the preparation of single crystals (see later).

Powders prepared in solution

β-TCP does not form in aqueous systems under normal laboratory conditions (*i.e.* up to 100 °C and at atmospheric pressures). However, whitlockites can be precipitated in such systems with the following molar % substitution of Ca^{2+} ions in solution: Mg^{2+} 0.1 to 1; Mn^{2+} about 1; and Fe^{2+} about 10 (Al^{3+}, Fe^{3+}, Ni^{2+}, Co^{2+}, Cu^{2+}, Zn^{2+}, Cd^{2+} and Ba^{2+} were not effective) [299]. It has been shown from X-ray and ESR work (Section 1.6.2) that Mg^{2+}, Fe^{2+} and Mn^{2+} ions can occupy the Ca(5) sites. By contrast, at least in β-TCP, Cu^{2+} ions occupy Ca(3) sites (Section 1.6.2). The substitution of an ion somewhat smaller than the Ca^{2+} ion into the six-fold coordinated Ca(5) site probably increases the stability of the lattice because Ca^{2+} ions usually have a higher coordination number. The substantially reduced solubility (Section 1.6.5) and raised β to α transition temperature (Section 1.6.6) of β-TCa,MgP compared with the magnesium-free compound is further evidence of this increased stability. It has been suggested that Mg^{2+} ions must comprise at least ~5 % of the cations in whitlockite for it to be sufficiently stable to precipitate in aqueous systems [248].

Magnesium whitlockites have been prepared by the dropwise addition of 250 ml of a solution containing Ca^{2+} and Mg^{2+} ions (total Ca + Mg = 10 mmol l^{-1}) into 750 ml of a stirred phosphate solution (10 mmol l^{-1}) at various temperatures, followed by a 4 h digestion, filtration and drying [269]. Magnesium whitlockite, uncontaminated by other phases, can be obtained by this method from solutions with a Ca/Mg molar ratio of 0.25 and a starting pH of 5 (RZ LeGeros, personal communication, 1990). The slow simultaneous addition of a mixed $CaCl_2$ plus $MgCl_2$ solution and a Na_2HPO_4 solution to a large volume of water at 100 °C for various fixed pH values from 5.5 to 9 (concentrations and volumes not given; pH controlled by the addition of ethylenediamine) has also been used [248]. This latter procedure has the advantage that the conditions of precipitation will be more uniform during the course of the preparation, but both procedures suffer from the formation of apatite, amorphous products and variations in the crystallinity of the whitlockite to an extent that depends on the exact experimental conditions. The most important determinant of the type of CaP obtained is the solution Mg/P molar ratio; a ratio of 0.05 favours the precipitation of whitlockite, and this is better crystallised if formed at 100 °C [269]. Acidic conditions (pH 5 to 6) also favour a better crystallised whitlockite [248,269], a finding that might be related to the fact that HPO_4^{2-} ions must be present in order to be incorporated into the whitlockite lattice. Whitlockite has also been prepared at 37 °C by the dropwise addition of 500 ml of an alkaline solution of Na_2HPO_4 to 500 ml of an alkaline solution of a mixture of calcium and magnesium acetates in the stoichiometric amounts required to form 10 g of $Ca_{20-n}Mg_nH_2(PO_4)_{14}$, where $1.4<n<3.8$ [300]. The hydrolysis of DCPD [248,269,301] (Section 1.4.6) and DCPA [269] (Section 1.5.5) in the presence of Mg^{2+} ions has also been used for whitlockite preparation. The constant composition method (Section 3.8) has been used to study the crystallisation of whitlockite on whitlockite seeds at 37 °C and pH 6.00 from metastable calcium-magnesium phosphate solutions [302]. The formation of whitlockite was confirmed. Whitlockite has occasionally been found during the hydrothermal formation of bone implants in phosphate solutions from coral skeletal calcite that contained magnesium [303] (Section 4.5.6).

Contractions of the *a*- and *c*-axis parameters with increasing magnesium content occur in synthetic whitlockite, but these are less than the contractions observed between β-TCP and β-TCa,MgP for the same magnesium content [248]. The magnesium content in synthetic whitlockites is generally insufficient to take the *c*-axis parameter contraction beyond the point when, by analogy with β-TCa,MgP (Section 1.6.2), it would start to expand.

The maximum reported magnesium content incorporated into the lattice for a synthetic whitlockite preparation is about 12 % of the total cations [248]. This compound was made by the simultaneous addition of a mixed $(Ca,Mg)Cl_2$ solution with a Ca/Mg molar ratio of 5 and a Na_2HPO_4 solution to a large volume of water at 100 °C and pH 5.5 to 6 as described above. Thus it seems possible to fill all the Ca(5), and probably some of the Ca(4) sites, by Mg^{2+} ions if appropriate conditions are used (the Ca(5) sites comprise 10 % of the total cation sites in the Mg whitlockite end-member, Formula 5, Table 1.5). Interestingly, it was also reported [248] that the axial parameters of some samples of whitlockite in pathological calcifications from tracheal and disc cartilage indicated an even greater magnesium incorporation.

Single crystals

Single crystals of α-TCP have been prepared by heating stoichiometric amounts of finely ground DCPA and $CaCO_3$ with 1 wt % cornstarch and a few drops of water at 1400 °C for two days (*i.e.* above the β-TCP to α-TCP transition at 1125 °C, Section 1.6.6) [283]. For the preparation of single crystals of β-TCP, α-TCP (prepared as above) was heated below the transition for seven days at 1100±10 °C [286]. These crystals were used in the structure determinations of α- and β-TCP discussed in Section 1.6.2.

Single crystals of β-TCP and two β-TCa,MgPs with different magnesium contents have been made by grinding alcoholic slurries of $CaCO_3$ (0.58, 1.79

and 2.57 g), DCPA (prepared by heating 2, 7 and 10 g of DCPD at 80 °C for two days) and MgO (0, 0.1 and 0.45 g) respectively [288]. The slurries were dried, heated to 1000 °C for 5 h, reground, pelletised with cornstarch and a little water, and then heated in platinum foil. The magnesium-free sample was heated at 1400 °C for 24 h, and then at 1100 °C for 162 h. The other two samples were heated at 1350 °C for 24 and 5 h respectively. The two samples of β-TCa,MgP were used in the structure analysis described in Section 1.6.2.

1.6.5 Solubility and reactions in solution
Solubility

As mentioned in the previous section, β-TCP (*i.e.* without additional Mg^{2+} or other divalent ions) does not form in aqueous systems. Nevertheless, it is possible to calculate its solubility product from analyses of a solution in equilibrium with β-TCP synthesised at high temperatures. The results of such studies [37] showed that the solubility product, K (mol^5 l^{-5}), at a temperature T (Kelvin), was given by:

$$\log_{10} K = \frac{-45723.26}{T} + 287.4536 - 0.546763T \qquad 1.11$$

which gave $K = 1.201$(standard error 0.056) $\times$ 10^{-29} mol^5 l^{-5} at 25 °C. The results showed that β-TCP is always more soluble than OHAp, but above pH ~ 6, less soluble than the other CaPs (Fig. 1.1) and, like these, the solubility decreases with an increase in temperature. The calculated pH values at 25 °C for the DCPD/β-TCP and DCPA/β-TCP singular points were 5.88 and 6.36 respectively [37]. A more recent study [297] gave $(1.14\pm0.10) \times 10^{-30}$ mol^5 l^{-5} for the solubility product at 25 °C. The difference between this figure and that given above [37] could be ascribed only partly to the use of different dissociation constants [297].

β-TCa,MgP [297] and whitlockite [248,300] have a solubility remarkably lower than β-TCP, which is a reflection of the increased stability of the lattice when Ca(5) is replaced by a smaller cation. For β-TCa,MgP in the pH range 5.4 to 6.0, the product,

$$\text{(activity of } Ca^{2+})^3 \times \text{(activity of } PO_4^{3-})^2 \qquad 1.12$$

was reduced by a factor between 900 and 3700 for a magnesium content in the solid between 8.3 and 15.3 molar % of the total cations respectively, so that the solubility of these compounds was less than that of OHAp [297]. (Note that a comparison based on ionic products that ignore lattice Mg^{2+} ions, and therefore does not reflect the stoichiometry of the solid phase at equilibrium, is open to question.) There was evidence that the equilibrating solid was a magnesium whitlockite rather than the original β-TCa,MgP phase, which indicated that the former compound was even more stable than the latter [297]. The solubility of magnesium whitlockite has been little studied [248], but the stabilising effect of magnesium would seem to explain the occurrence of whitlockite in dental calculus and various pathological calcifications.

Hydrolysis and other reactions in solution

α-TCP is fairly reactive in aqueous systems and can be hydrolysed to mixtures of DCPA, DCPD, OCP, and Ca-def OHAps in varying proportions depending upon the conditions. α-TCP is hydrolysed to OCP in sodium acetate solution [152,153], but if the ammonium or sodium salts of certain dicarboxylic acids are used instead, the dicarboxylate anion can replace some of the HPO_4^{2-} ions in the OCP lattice (Section 1.3.9). The properties of a Ca-def OHAp prepared by the hydrolysis of α-TCP have been reported [304]. Under certain circumstances, the reaction mixture sets solid [304-307].

1.6.6 Thermal decomposition

The high temperature phase diagram for the system $2CaO.P_2O_2$-CaO is shown in Fig 1.10.

β-TCP is stable to 1125 °C, but above this, and up to 1430 °C, α-TCP becomes the stable phase, then $\overline{\alpha}$-TCP forms between 1430 °C and the melting point, 1756 °C [273] (1120, 1470 and 1800 °C respectively have also been reported for these transition temperatures [308]).

In later studies [309], the presence of solid solutions shown in Fig. 1.10 could not be confirmed. It was suggested [309] that the most likely source of error arose through the loss of P_2O_5 from the system during observation of the melting behaviour on the high-temperature microscope stage that had been used. Thus the reported solid solution limits probably represented the extent of the P_2O_5 loss from the system [309]. The further suggestion was made that the temperature maximum in the α-TCP ($\overline{\alpha}$-TCP intended?) solid solution region was suspect, and that 1777 °C could be the congruent melting point of stoichiometric TCP. As a result of these considerations, a new diagram for the CaO-P_2O_5 system (Fig. 1.11) was proposed [309]. This was based on earlier phase diagrams in the literature, analogy with the SrO-P_2O_5 system, and some limited experimental data.

The phase diagrams in Figs 1.10 and 1.11 are only applicable to dry systems. In the presence of even very small amounts of water vapour (*e.g.* 73.6 mm Hg, 9.81 kPa, at 1575 K, Section 1.7), TetCP (C_4P in the phase diagrams) will react with the water to form OHAp according to Equation 1.15. Furthermore, a mixed oxyhydroxyapatite, O,OHAp, rather than pure OHAp, is often likely to form, again dependent on the water vapour pressure and temperature (Equation 3.4, Section 3.4.3). Pure OAp ($C_{3.33}P$ in the notation of

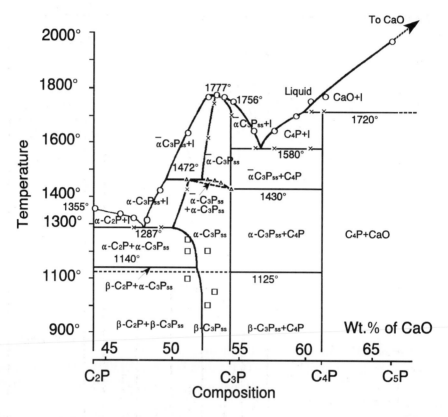

Fig. 1.10 High-temperature phase diagram for the system $2CaO.P_2O_5$. $\circ$ liquidus temperature, $\triangle$ temperature of α-TCP $\to$ $\overline{\alpha}$-TCP transition by optical examination, $\times$ temperature of initial liquid formation. $\square$ points at which determinations were made with a focusing X-ray Guinier camera on quenched specimens. ---- broken lines indicate inferred boundary curves. C = CaO, P = P_2O_5, l = liquid, ss = solid solution. Temperatures according to the International scale of 1927. (From Welch and Gutt [273], with permission)

the phase diagrams), even though it is theoretically part of the system studied, does not appear. Its existence (at least with a small, but detectable OH⁻ ion content) is now well-established (Section 3.4.3). It may be absent because it genuinely does not exist as an equilibrium phase under the conditions studied, or it might not appear for kinetic reasons, or it (or OH,OAp) occurred experimentally, but was undetected or mistaken for OHAp. As proposed earlier by the authors of Fig. 1.11 [309], the $CaO-P_2O_5$ system should be carefully reinvestigated; this should also be extended to the $CaO-P_2O_5-H_2O$ system at very low, carefully controlled, partial pressures of water.

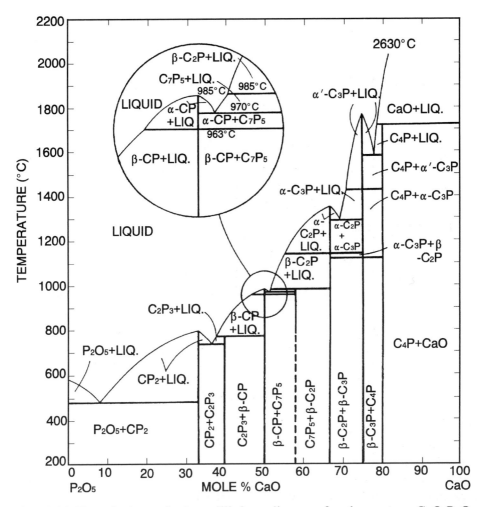

Fig. 1.11 Tentatively revised equilibrium diagram for the system CaO-P$_2$O$_5$ with melting points and inversion temperatures taken from the literature: C = CaO, P = P$_2$O$_5$ (Fig. 3 from Kreidler and Hummel [309], with permission)

The 1/30 molar replacement of Ca^{2+} by Mg^{2+} ions to form β-TCa,MgP raises the upper temperature limit of the stability range of the β polymorph to at least 1450 °C [310]. If β-TCP is heated at 950 °C and 40 kbar (4000 MPa), there is a phase change to the Ba$_3$(VO$_4$)$_3$ structure (Section 1.6.2) which is stable at room temperature if the product is quenched [275]. α-TCP can sometimes be formed at temperatures well below the β to α transition temperature, *e.g.* by heating ACP at ~530 °C (Section 1.8.7).

The lattice of whitlockite contracts when heated at 600 °C [248], so losing at least part of the expansion [248] (Section 1.6.4) over that of β-TCa,MgP with the same magnesium content. After heating at 900 °C, β-Ca$_2$P$_2$O$_7$ could be detected by XRD, which was interpreted as indicating the presence of acid phosphate in the original material [248]. Thermogravimetric analysis showed weight losses at 400 and 750 °C, ascribed to the formation of P$_2$O$_7^{4-}$ and PO$_4^{3-}$ ions, respectively, from HPO$_4^{2-}$ ions in the lattice (HPO$_4^{2-}$ ions could be seen in the IR spectrum of the unheated material) [269] (following section).

1.6.7 Infrared and NMR spectroscopy

The IR spectra of α-TCP and β-TCP show distinct differences [16]. These are that v_4 PO$_4$ is a single broad band at ~600 cm^{-1} in α-TCP, but this is split into bands at 550 and 616 cm^{-1} in β-TCP; there are also weak, but distinct, bands at 460 and 740 cm^{-1} in α-TCP that are absent in β-TCP (frequencies measured from published spectra). Some of the structure in the bands from 900 to 1200 cm^{-1} is lost as magnesium replaces calcium in β-TCP [291]. β-TCP and synthetic whitlockite show spectral differences, including the presence of a P-(OH) stretching band at 850 cm^{-1} in the latter [269]. Many differences between Mg whitlockite and β-TCP are also seen in the v_3 PO$_4$ [190] and v_4 PO$_4$ [191] spectral regions in deconvoluted FT IR spectra.

The ^{31}P NMR MAS (5 kHz) spectrum at 40 MHz of β-TCP shows chemical shifts of 0.1, 1.2 and 4.2 ppm from external 85% H$_3$PO$_4$ [193].

1.7 Tetracalcium phosphate

Tetracalcium phosphate (synonyms: TetCP, tetracalcium diphosphate monoxide, hilgenstockite) has the formula Ca$_4$(PO$_4$)$_2$O. The name hilgenstockite derives from its discovery by G. Hilgenstock in 1883 [311] in Thomas slag from blast furnaces [132]. Its major economic importance stems from the fact that it is formed by the reaction between phosphorus, oxygen and lime in the manufacture of iron, and through this reaction, it has a significant role in controlling the properties of the metal [312]. TetCP occurs under anhydrous conditions in the CaO-P$_2$O$_5$ phase diagram (Figs 1.10 and 1.11), but there is no convincing evidence [13] of its formation in aqueous solutions at normal temperatures. TetCP, mixed with either DCPD or DCPA, is used in a proposed bone cement which sets to OHAp in the presence of water (Reaction 1.13). The driving force for this reaction is that OHAp is less soluble than the reactants at physiological pH [313-315].

$$Ca_4(PO_4)_2O + CaHPO_4 \rightarrow Ca_5(PO_4)_3OH \qquad\qquad 1.13$$

Structure

TetCP is monoclinic with space group $P2_1$ and unit cell parameters $a = 7.023(1)$, $b = 11.986(4)$, $c = 9.473(2)$ Å and $\beta = 90.90(1)°$ at 25 °C [316]. There are four formula units per cell and the calculated density is 3.051 g cm^{-3} [316]. The lattice parameters and X-ray intensities showed that the structure is pseudo-orthorhombic [312]. There is also a close correspondence between the lattice parameters of TetCP and OHAp (Table 1.6) which indicated [312] that they had a close structural relationship, as was confirmed by the full structure determination [316].

Table 1.6 Correspondence between lattice of TetCP [316] and OHAp (Table 1.4).

	TetCP		OHAp
a	7.023(1)Å	c	6.881Å
b	11.986(4)	$d(100) \times 3/2$	12.233
c	9.473(2)	a	9.417
α	90.00°	γ - 30°	90.00°
β	90.90	β	90.00
γ	90.00	α	90.00

The Ca^{2+} and PO_4^{3-} ions in the structure are located in four sheets (adjacent sheets are related by symmetry) with their planes perpendicular to the b-axis [316]. Within a unit cell, each sheet is made up of one cation column, $\cdots$Ca Ca Ca Ca$\cdots$, and two cation-anion columns, $\cdots PO_4$ Ca PO_4 Ca $PO_4\cdots$, that run in the a-axis direction. Two adjacent sheets together form a layer with a structure closely related to that of apatite. This layer corresponds to the central part of the "apatite" layer seen in OCP (Fig. 1.4). Thus the cation and cation-anion columns that are parallel to the a-axis in TetCP are parallel to the apatite c-axis (Table 1.6). This direction is into the plane of Fig. 1.4. The columns in TetCP correspond to the $\cdots$Ca1 Ca2$\cdots$ and $\cdots$P1 Ca5$\cdots$ (OCP numbering) columns, respectively, seen end-on in the "apatite" layer in Fig. 1.4. The corresponding atoms in the FAp structure (Section 2.2.1) are the columnar Ca(1) atoms (Fig. 2.2) and the $\cdots PO_4$ Ca(2) PO_4 Ca(2) $PO_4\cdots$ sequence of ions. The O^{2-} ions in TetCP are near the edges of the "apatite" layer, but the corresponding OH$^-$ ions in OHAp, are outside the layer at the corners of the OHAp unit cell (Fig. 1.4) [316].

The isostructural compounds $Me_4(XO_4)_2O$, where Me is Ca, Sr or Ba and X is P, V or As (all combinations except Sr,V and Ba,V) have been synthesised and their lattice parameters measured [317].

Preparation

Several preparations have been published [16,284,312,317-320]. These are solid state reactions at high temperatures, usually between equimolar quantities of DCPA and $CaCO_3$ (Reaction 1.14). This reaction should be carried out in a dry atmosphere or vacuum, or with rapid cooling (to prevent uptake of water and formation of OHAp, Reaction 1.15). The borderline at which the reverse of Reaction 1.15 occurs is at 1575 ± 10 and 1648 ± 10 K for water vapour partial pressures of 73.6 and 760 mm Hg (9.81 and 101.3 kPa), respectively [19,21].

$$2CaHPO_4 + 2CaCO_3 \rightarrow Ca_4(PO_4)_2O + 2CO_2 + H_2O \qquad 1.14$$

$$3Ca_4(PO_4)_2O + H_2O \rightarrow Ca_{10}(PO_4)_6(OH)_2 + 2CaO \qquad 1.15$$

In one study, pure (from an XRD point of view) TetCP was synthesised by heating homogenised equimolar quantities of DCPA and $CaCO_3$ in air, or in a flow of purified nitrogen [319]. Those samples heated in air were subsequently cooled over silica gel; the others were cooled in nitrogen. Pure TetCP could be synthesised in air by heating for 6 h at 1300 °C, followed by 4 h at 1400 °C. Uncontaminated TetCP was also formed if the mixture was heated in air at 1200 °C for 2.5 h, followed by 23 h at 1300 °C in nitrogen. In another preparation, stoichiometric amounts of $Ca_2P_2O_7$ and $CaCO_3$ were heated at 1300 °C for 6 h, and then quenched to room temperature [284,318].

The single crystal ($\sim0.33 \times \sim0.11$ mm^2) of TetCP that was used for the structure determination described above was selected from a sample prepared by heating an equimolar mixture of DCPA and $CaCO_3$ in a platinum foil envelope at 1500 °C for 24 h in vacuum [312].

Optical properties and habit

TetCP has optical constants (sodium light) as follows: biaxial (positive), α = 1.643, β = 1.645 and γ = 1.650 and calculated $2V$ of 66 ° [16]. It occurs as microscopically small pseudo-orthorhombic prisms of simple form, but is frequently granular and anhedral; polysynthetic twinning on (010) and (001) is common [16]. This twinning is described in more detail in ref. [312].

Other properties

TetCP reacts readily with water vapour at temperatures between 1100 and 1420 °C (Equation 1.15) [272,284,321-323]. Reaction temperatures at two water vapour pressures were given above. The reaction has been studied by IR, TGA and XRD [284,318]. TetCP melts at 1720 °C [273] (Fig. 1.10).

TetCP is more basic than OHAp and has a higher catalytic activity and selectivity for alcohol decomposition [284]. It was found to be stable in water at room temperature for four weeks, but at 80 °C, hydrolysis to OHAp was complete in 80 h [324].

1.8 Amorphous calcium phosphates

1.8.1 Introduction

Amorphous calcium phosphate (ACP) shares with OCP the fact that there were early reports of its occurrence in the literature [9] but it was not until the mid 1960's that its formation and structure were investigated in detail. ACP is of interest because it often occurs as a transient phase during the formation of CaPs in aqueous systems, particularly ns-OHAps (Section 1.8.4). It was also reported [325] that bone contains ACP as a separate phase whose concentration is higher in younger animals than older ones, but later investigations indicated that ACP is not present in bone [326] (Sections 4.6.1 and 4.6.2). However ACa,CO_3Ps have been found in the calcarious corpuscles of tapeworms, in the tips of lobster claws and in various pathological tissues [268]. Some of the literature on the structure and properties of ACP up to 1974 (a period in which most of the basic chemical work was done) has been reviewed [327]. Detailed studies have been made of the thermal decomposition and solid state conversion of ACP into apatitic salts [92,328,329], ACa,MgP [92,328] and ACP containing $P_2O_7^{4-}$, CO_3^{2-} or F^- ions [92]. Other papers and more recent work up to 1986 have been cited in ref. [330]. The formation of a second kind of ACP (ACP2), in addition to the usual form (ACP1), has been proposed to explain the variation of pH during the transformation of ACP into crystalline CaP phases [331] (Section 1.8.4). The synthesis of an amorphous OCP has been reported [146] (Section 1.8.3).

1.8.2 Preparation

ACP has been prepared at 25 °C by the rapid addition, with stirring, of a solution of $(NH_4)_2HPO_4$ (0.25 mol l^{-1}) to a solution of $Ca(NO_3)_2.4H_2O$ (0.75 mol l^{-1}) to achieve a final phosphate concentration of 0.15 mol l^{-1} and a Ca/P molar ratio of 1.71 [332]. All solutions were adjusted to pH 10 with concentrated NH_4OH prior to mixing, and the reaction carried out in a closed system to reduce CO_2 contamination. The initial solid phase, formed immediately on mixing, was separated and washed with ammoniated water at pH 10, then freeze-dried. The Ca/P molar ratio of the solid was very reproducible at 1.52. Similar preparations, also at high pH, have been reported [92,298,328,329].

ACP has also been prepared under conditions different from those given above, particularly at lower pH values and/or with more dilute solutions, but usually with rapid mixing and separation of the solid phase (generally shortly after initial solid formation), followed by freeze-drying [92,298,326,328,329,333-338]. Rapid precipitation from alkaline alcoholic solutions has also been used [146] (Section 1.8.3).

The induction time before the initial formation of ACP is highly sensitive to solution composition. It is reduced by the following factors: higher initial

[Ca] × [P] products or Ca/P molar ratios, higher temperatures or pH values, or by a lower dielectric constant or the presence of $P_2O_7^{4-}$, F^-, Mg^{2+} or CO_3^{2-} ions or collagen [333]. The time is increased by citrate, poly-L-glutamic acid or polyacrylate [333].

ACPs with CO_3^{2-}, Mg^{2+} ions etc. ACPs can also be prepared with significant and even substantial amounts of other ions in addition to Ca^{2+} and phosphate ions. Formation of ACP by the methods discussed above in the presence of large amounts of carbonate leads to its incorporation in the solid in amounts that increase with increasing pH [339] and initial solution CO_3/P ratio [92,339]. The addition of Mg^{2+} or citrate ions to the precipitating system enables ACP to be prepared under more acid conditions than would otherwise be possible [340,341]. Thus an ACP can be prepared [341] at pH 6 and ambient temperature by adding a phosphate solution (KH_2PO_4 48, K_2HPO_4 22 and KNO_3 81 mmol l^{-1}) to an equal volume of a calcium solution ($Ca(NO_3)_2$ 32, KNO_3 81 and potassium dihydrogen citrate 6 mmol l^{-1}). The pH was then raised to 6, and maintained there during precipitation by the addition of KOH solution. The ACP was separated and freeze-dried. It was suggested that ACP could form under these more acid conditions because the added ions inhibited the growth of crystalline phases [341]. In studies of ACa,MgPs prepared at pH 11.2 and with a (Ca + Mg)/P molar ratio of 1.5, no segregation of Mg^{2+} ions between solid and solution was found for a wide magnesium concentration range [92]. ACP has also been synthesised at 37 and 100 °C by the dropwise addition, with stirring, of 250 ml of a calcium acetate solution (0.02 mol l^{-1}) to 750 ml of a phosphate solution (0.016 mol l^{-1}) containing pyrophosphate or CO_3^{2-} ions [268]. In some preparations, Ca^{2+} ions were replaced by Mg^{2+}, Ba^{2+} or iron ions on a mole for mole basis [268]. At 37 °C and pH in the range 7 to 9, the order of efficacy in promoting the formation of ACP was $Mg^{2+} + P_2O_7^{4-} > P_2O_7^{4-} > Mg^{2+} + CO_3^{2-} > Mg^{2+} > CO_3^{2-}$ [268]. In contrast to magnesium, pyrophosphate is preferentially incorporated in the solid phase: the ratio (pyrophosphate)/(pyrophosphate + orthophosphate) is up to 10 % higher in the precipitated ACP compared with this ratio in the solution [92]. The formation of ACa,CO_3P from DCPA [267,268] has already been noted (Section 1.5.5). The presence of Mg^{2+}, pyrophosphate or CO_3^{2-} ions greatly increases the stability of ACP against conversion to apatitic salts (Section 1.8.4). The precipitation of CO_3Aps is discussed in Section 4.5.1.

1.8.3 Chemical composition

The early studies of the Ca/P molar ratio of ACPs reported values close to 1.50 [332,333,342]. However, in a detailed study of the composition of ACP, it was shown [335] that ACP formed in the pH range 6.6 to 10.6 varied in Ca/P molar ratio from 1.18 to 1.50, but if the precipitates were washed before freeze-drying, the Ca/P molar ratios were 1.50±0.03 (except for the very lowest part of the pH range). Washing also considerably reduced the percentage of the total phosphorus present as acid phosphate from 33 to 18 % at pH 6.6, and from 10 to 5 % at pH 10.6 (determined from the pyrophosphate formed on pyrolysis, Section 3.6.2). Neither the increase in Ca/P ratio nor the reduction in acid phosphate on washing could be attributed to trapped supernatant ions. It was concluded that ACPs should be recognised as a class of salts having variable chemical, but identical glass-like properties [335]. On the other hand, others have reported that the departure from a Ca/P molar ratio of 1.5 can be attributed to surface adsorbed soluble phases that can be washed away, or to occluded Ca^{2+} ions [343].

ACPs with Ca/P molar ratios of 1.43±0.01, 1.5±0.003 and 1.53±0.01 have been prepared by rapid precipitation at pH 8.9, 9.9 and 11.7 with formulae $Ca_{8.58}(HPO_4)_{1.11}(PO_4)_{4.89}(OH)_{0.27}$, $Ca_{9.00}(HPO_4)_{0.16}(PO_4)_{5.84}(OH)_{0.16}$ and $Ca_{9.21}(HPO_4)_{0.08}(PO_4)_{5.92}(OH)_{0.50}$, respectively (analyses based on pyrophosphate formation on heating, Section 3.6.2) [92,328]. About half the total phosphorus in an ACa,MgP prepared at pH 7 [340], and three quarters in a magnesium-free ACP prepared at pH 6.5 [341] were protonated. The determinations were based on an oxalate precipitation/titration method (Section 3.6.2) with the assumption that OH⁻ ions were absent from the solid, so the acid phosphate content is a minimum value. The chemical analyses of these ACPs gave a (Ca + Mg)/P molar ratio of ~1.2 [341].

An ACP corresponding to an amorphous OCP with a Ca/P molar ratio of 1.33 has been reported [146]. This was prepared at 37 °C by rapid precipitation from an alcoholic phosphate solution $((NH_4)_2HPO_4$ 30 mmol, H_2O 250 ml, C_2H_5OH 295 ml, NH_3 solution density 0.92 g cm^{-3} 45 ml) by a calcium solution $(Ca(NO_3)_2$ 30 mmol, H_2O 100 ml, C_2H_5OH 100 ml). Following precipitation, washing with an alcoholic solution $(H_2O$ 180 ml, NH_3 solution density 0.92 g cm^{-3} 30 ml, C_2H_5OH 210 ml), the solid was freeze dried. Chemical analyses gave Ca 31.8±0.1, P 18.5±0.1, C≤0.1 and N 0.3±0.1 wt %. These results, together with the determination of pyrophosphate on heating, gave a formula $Ca_8(PO_4)_4(HPO_4)_2$ (no H_2O given in formula).

1.8.4 Transformation reactions

Transformation in the absence of an aqueous phase

An important distinction between the transformation of ACP in the absence and presence of an aqueous phase is that in the former situation, there can be no change in the Ca/P ratio. Thus moist ACP cake (synthesis and formulae given in previous section) converts to a ns-OHAp at room temperature in about 24 h (much faster at higher temperatures) [92,328,329]; even vacuum dried samples slowly convert to apatite over a period of months at ambient temperature [344].

The transformation process at 20 °C of ACP with a Ca/P molar ratio of 1.5±0.003 prepared at pH 9.9 (formula in previous section) has been investigated in detail [92,329]. There was a continuous hydrolysis reaction

$$PO_4^{3-} + H_2O \rightarrow HPO_4^{2-} + OH^- \qquad\qquad 1.16$$

to give a compound with a general formula

$$Ca_9(HPO_4)_x(PO_4)_{6-x}(OH)_x \qquad\qquad 1.17$$

where x increased with time and $0 \leq x \leq 1$. This compound was amorphous for the first 11 h, then, during the next ~3 h at which time the hydrolysis was half complete ($x \approx 0.5$), it changed to a crystalline apatitic single phase. The hydrolysis was essentially complete ($x = 1$) after a total time of ~24 h. The formation of HPO_4^{2-} and OH^- ions was confirmed qualitatively by IR. A product with maximal hydrolysis (250 h at 70 °C in an atmosphere saturated with water) had a formula, $Ca_{9.00}(HPO_4)_{0.96}(PO_4)_{5.04}(OH)_{0.96}$, as determined from pyrophosphate formation on heating (Section 3.6.2) and the requirement of charge balance [92,328].

The hydrolysis is endothermic, and the more rapid crystalline conversion exothermic, with ΔH values based on $Ca_9(PO_4)_6$ of 12.0±2.0 and -5.9±1.0 kJ mol^{-1}, respectively [92,329]. The activation energy for the hydrolysis reaction is 105.3±5.8 kJ mol^{-1} [92].

^{31}P NMR MAS studies of the conversion process at 16.5 °C showed a gradual decrease in hydrated ACP and the formation of two new PO_4^{3-}-containing components [345] (see also Sections 1.8.5 and 3.12). One was similar to poorly crystalline OHAp and the other a protonated PO_4^{3-} ion, probably an HPO_4^{2-} ion, in a DCPD-like environment [345]. There was a marked decrease in the line-width as the conversion proceeded.

The effect of Mg^{2+}, $P_2O_7^{4-}$, CO_3^{2-} and F$^-$ ions in the solid on the hydrolysis reaction at 20 °C of the ACP discussed above has been studied [92]. With 4.7 mol % $Mg^{2+}/(Mg^{2+} + Ca^{2+})$, 5.3 mol % P(pyro)/P(total) or ~5 wt % CO_3^{2-}, the time at which the amorphous to crystalline phase change took place increased to 80, 73 or ~45 h, respectively (above these levels, no crystalline phase was seen). On the other hand, up to at least 3 wt % F$^-$ ion did not significantly change the time, but the thermal effects were different; above 0.3 wt % F$^-$ ion, the long time-scale endothermic peak associated with the hydrolysis reaction was replaced by a long time-scale exothermic peak in addition to the shorter exothermic peak due to the amorphous to crystalline conversion. This observation was not understood. Experiments were also made at 70 °C.

Transformation in the presence of an aqueous phase

In water, the usual product after a few hours when ACP is in alkaline or slightly alkaline solutions is OHAp or ns-OHAp. DCPD [346,347] and OCP [348] can also form from ACP (in one case in the presence of Mg^{2+} ions [347]). In one study [173] of ACP matured at a constant pH of 7.4, the first formed crystals gave an apatitic XRD pattern with an exceptionally large *a*-axis parameter (10.5 Å), but in the electron microscope, their appearance was closer to OCP than apatite; this suggested that conversion proceeded via an OCP-like crystalline phase, followed by hydrolysis into apatite. (The increased *a*-axis parameter has been explained by partially coherent X-ray scattering from intercrystalline layers of OCP and OHAp [172], Section 1.3.6.) Solubility studies also indicated that an OCP-like material was an intermediate phase of the transformation [349]. Further, more detailed investigation of the hydrolysis, again at a constant pH of 7.4, in a solution "saturated" with ACP showed that conversion involved two processes [350]. The first consumed acid with the formation of an acidic intermediate with the solubility properties of OCP, and in the second process, the intermediate converted to an apatite with the consumption of base. It also seemed that direct hydrolysis of ACP to OHAp was possible [350].

Study of the pH changes during the precipitation and transformation of ACP to crystalline phases has been explained by the initial formation of an amorphous phase, ACP1, followed by formation and growth of another amorphous phase, ACP2 [331,351,352]. The apparent solubilities of ACP1, ACP2 and OCP were determined from 303 to 315 K, from which ΔG^0, ΔH^0 and ΔS^0 for the dissolution of the three phases were calculated.

ACP is more stable when made in solutions at high pH, ionic strength, rich in Ca^{2+} ions relative to HPO_4^{2-} ions, or low in initial supersaturation or temperature [353]. It is also more stable in the presence of poly-L-glutamate, polyacrylate, phosvitin, casein, large amounts of protein-polysaccharides, or on the addition of small amounts of Mg^{2+}, $P_2O_7^{4-}$, CO_3^{2-} or F^- ions [353]. The presence of sufficient Mg^{2+} and/or $P_2O_7^{4-}$ ions can increase the stability of some ACP samples so that they remain amorphous in hot solution for many hours [268]. The influence of pyrophosphate ions on the transformation has been investigated in detail [354], as well as chemical changes that take place during hydrolysis of ACP in the presence of CO_3^{2-} ions [355]. ACa,MgP can transform to DCPD at pH 7 if allowed to remain in the solution in which it was formed [340]. Addition of dentine phosphoprotein has no specific effect on the rate of transformation, but if it is added to the solution before ACP precipitates, apatite forms directly [356]. There is an increase in the induction time for the ACP transformation at 25 °C in the presence of Mg^{2+}, Sr^{2+}, Zn^{2+}, $P_2O_7^{4-}$ and $P_3O_{10}^{5-}$ ions, and combinations of these [357]. Synergistic effects were observed for combinations of Mg^{2+} or Sr^{2+} ions and $P_2O_7^{4-}$ or $P_3O_{10}^{5-}$ ions.

Heats of solution in phosphoric acid of ACP, previously heated at temperatures up to 700 °C, have been reported [358]. Up to 500 °C, no change was observed, but the heat of solution fell after heating above this temperature.

Recently, calorimetric heat power measurements, pH and Ca^{2+} ion activity measurements as a function of time have been made in a study of the effect of Mg^{2+} ions on the precipitation of CaPs at 30 °C [352]. The results could be understood if Mg^{2+} ions mainly inhibited the formation of ACP2 (see above) and the growth of OCP. Magnesium did not appear to be incorporated significantly in the solid phases or affect the solubilities. Dehydration of Ca^{2+} and Mg^{2+} ions appeared to contribute significantly to the changes in enthalpy during the precipitation process.

Kinetics of conversion to OHAp

If c is the fraction of ACP that has been converted to OHAp, it has been shown that at time t,

$$\frac{dc}{dt} = k_1 + k_2 c \qquad\qquad 1.18$$

where k_1 and k_2 are constants with $k_2 >> k_1$ [359]. k_2 is a constant associated with the autocatalytic conversion and k_1 a constant to take into account all other nucleating processes. k_1 will depend on the initial number of seed crystals. In more detailed kinetic studies [336], it was found that k_2 decreases with increasing pH at constant temperature, and that the activation energy for the conversion process was 16.4 kcal mol^{-1} (68.7 kJ mol^{-1}), showing that it was very temperature dependent. The kinetics of the transformation have been studied in the presence of different magnesium concentrations [360]. With a Mg/Ca molar ratio in the range 0.004 to 0.04, the induction period increased with increasing magnesium concentrations. If the Mg/Ca molar ratio exceeded 0.2, no conversion took place. However the rate constant for the first order transformation was independent of magnesium concentration. It was shown that Mg^{2+} ions affected the transformation by reducing the ACP solubility [360].

1.8.5 Infrared and NMR spectroscopy

The IR spectra of ACPs have broad, featureless phosphate absorption bands (Fig. 3a in ref. [335], Fig. 7D in ref. [268] and Fig. 6 in ref. [340]). This character is as to be expected for an amorphous material because the ions are not subjected to the regular distortions that they would experience if they were in a crystal lattice. Thus, the spectrum is similar to that from phosphate ions in solution [361] (Table 2.2, Section 2.6.1), including the absence of the symmetric stretch at 938 cm^{-1} which is IR inactive in the undistorted ion. Nevertheless, some slight structure in the phosphate bands between 1250 and

890 cm^{-1} can be seen by careful examination of FT IR spectra [340]. The strong ν_3 and ν_4 PO$_4$ bands (Table 2.2) at 1085 and 555 cm^{-1}, respectively, in ACP prepared at pH 7 move to slightly lower frequencies in ACP prepared in alkaline conditions [340].

ACP prepared in approximately neutral conditions has a medium intensity band at 865 cm^{-1}, due to the P-(OH) stretch of the HPO$_4^{2-}$ ion [335,340,341], which reduces in intensity if the ACP is made in slightly more alkaline conditions [341]. A similar band occurs in ACa,MgPs [268,347]. There do not appear to be any reports of deuteration studies of this band in ACP. Spectra of ACP, after heating at 225 and 450 °C, show the gradual disappearance of the P-(OH) stretching band [335]. Attempts have been made to correlate the area under the band at 890 cm^{-1} with the acid phosphate content determined by the oxalate precipitation/titration method [340,341] (Section 3.6.2). The intensities for samples prepared at pH 6.0 and 6.5 were in accord with the chemical determinations, but the integrated intensity for an ACP prepared at pH 7.0 was about twice that expected [341] (see Section 3.6.2 for further discussion).

IR spectroscopy has been used to follow quantitatively the transformation of ACP into OHAp in water. The study was based on the observation that the splitting of the ν_4 P-O antisymmetric bending mode at 550 to 600 cm^{-1} increases as the "percent crystallinity" increases [362].

Although the PO$_4^{3-}$ ion in ACa,CO$_3$P does not show any marked loss of degeneracy from a lowering of its site symmetry, ν_3, the doubly degenerate asymmetric stretch of the undistorted CO$_3^{2-}$ ion near 1500 cm^{-1} (Section 4.2.2), is split into two components [268,335]. These components nearly merge into one weaker band after vacuum drying at 225 °C for 24 h (Fig. 3b in ref. [335]). It is noticeable from published spectra [268] that the relative intensity of the two components of ν_3 is not constant. The loss of degeneracy and variable intensity suggests that the CO$_3^{2-}$ ions are subjected to some form of ordered, but slightly variable distortion. Alternatively, a variable fraction of the ions might be protonated, which would cause a loss of degeneracy and/or a shift of frequency; this could also explain the changes observed on heating (see above), because HCO$_3^-$ ions are thermally unstable.

The ^{31}P NMR spectrum at 68.4 MHz and MAS (up to 1.35 kHz) of ACP with a Ca/P molar ratio of 1.5±0.02 (presumably prepared at pH 7.4 [343]) had a chemical shift of +1.7 ppm from 85% H$_3$PO$_4$, which was similar to that of OHAp [363]. Without proton enhancement via cross polarisation, the sidebands in ACP were more intense than those in OHAp, but weaker than those in DCPD and DCPA. With proton enhancement, there was no change in the sideband intensities of ACP from -120 to 25 °C (except a minor enhancement at very short cross polarisation contact times, 0.1 ms); however, OCP, DCPD, DCPA and mixtures of the latter two with OHAp showed large differential enhancement of the sidebands [363]. From these studies and other evidence, it was concluded that ACP was neither a poorly crystallised OHAp, nor a mixture of OHAp with CaP phases containing HPO_4^{2-} ions, but rather a unique, well-defined compound, although the minor enhancement of the sidebands indicated some minor heterogeneity. The strength of the ACP sidebands was attributed to a "characteristic structural distortion of unprotonated phosphate and *not* to a mixture of protonated and unprotonated phosphates" [363]. There is clearly some inconsistency between this conclusion and the observation of the 865 cm^{-1} IR band due to the P-(OH) stretch of HPO_4^{2-} ion discussed above. In other rather similar NMR studies to those discussed above, it was concluded that ACP contains PO_4^{3-} ions that are slightly distorted and/or weakly bound to protons [364].

1.8.6 Structure

Electron microscopy of ACP usually shows spherical particles with diameters in the range 200 to 1200 Å without structure, provided the electron beam current is kept as low as practicable [365]. The particles are smaller if prepared under conditions of high supersaturation and/or pH; for a given pH, higher temperatures give larger particles [343].

There is the possibility that ACP has an apatitic structure, but with a crystal size so small that it appears to be amorphous. Calculations have been made [366] of the XRD pattern of OHAp with very few unit cells which were compared with the observed pattern for ACP. As the match was poor, it was thought that ACP was structurally distinct from OHAp, a conclusion in agreement with later ^{31}P NMR studies [363] (previous section). Subsequently, the proposal was made that the basic structural unit of ACP is a 9.5 Å diameter, roughly spherical, cluster of ions comprising $Ca_9(PO_4)_6$ [367]. This idea was mainly based on the determination of the radial distribution function calculated from XRD data which showed peaks out to about 10 Å that did not depend significantly on the water content or composition. Others [326], have obtained similar radial distribution functions. By comparison with the radial distribution function for OHAp and its known structure, peaks from ion pairs were identified as follows: P-O (1.5 Å), O-O and Ca-O (2.4 Å), and predominantly Ca-Ca and P-P (3.4 and 4.1 Å) [367]. Peaks at larger distances could not be uniquely identified. It was suggested that the larger particles of ACP seen in the electron microscope consisted of roughly spherical $Ca_9(PO_4)_6$ clusters aggregated randomly with their inter-cluster spaces filled with water [367].

EXAFS spectroscopic studies on the high energy side of the Ca K absorption edge have shown that the spectra from ACPs produced in basic [338] and acidic [340,341] conditions are very similar. The spectra from both these preparations could be interpreted in terms of a model [368] developed for the EXAFS spectrum of OHAp. Only three shells of atoms surrounding a Ca^{2+}

ion out to a radius of 3.08 Å were required to fit the experimental results [338], instead of the eight shells needed for OHAp. The radii of these shells, which corresponded to the two oxygen and one phosphorus shell nearest to a Ca^{2+} ion, differed from their values in OHAp by less than 0.04 Å. However, their Debye-Waller temperature factors were considerably greater which indicated greater disorder. Calculations of the EXAFS spectrum of acidic ACP based on DCPD have been made [369].

There is IR evidence (P-(OH) stretch at 865 cm^{-1}) that at least some of the PO_4^{3-} ions in ACPs made under neutral conditions are protonated [335,340,341] (previous section). Some ^{31}P NMR studies also indicate that a fraction of the PO_4^{3-} ions might be weakly bound to protons [364] (previous section).

1.8.7 Thermal decomposition

ACP heated to ~530 °C, so that volatiles can escape, remains amorphous, but further heating above 650 °C produces crystalline α-TCP and/or β-TCP [370,371]. Thus α-TCP can be produced well below 1125 °C, the reported β- to α-TCP transition temperature [274] (Section 1.6.6). Short term heating (30 min) at 900 °C yields α-$Ca_2P_2O_7$, β-$Ca_2P_2O_7$ and β-TCP [335], so that again a high temperature polymorph can be produced below the β- to α-$Ca_2P_2O_7$ transition point (1413 K [372]). The kinetics of the formation of α-TCP from ACP have been studied [373].

The fraction of the phosphorus as pyrophosphate, as a function of the temperature of heating, has been studied [92,328] (Section 3.6.2). For Ca/P molar ratios of 1.43, 1.5 and 1.53, maximum formation occurred at 650, 350 and 400 °C respectively. There was a substantial amount of pyrophosphate for a Ca/P molar ratio of 1.43, indicating a high acid phosphate content, but much less for the others (see formulae for these ACPs in Section 1.8.3). The presence of magnesium in the solid lowers the temperature of maximum pyrophosphate formation, for example, to 450 °C for a molar ratio of Mg/(Mg + Ca) of 0.16 [92,328].

ACPs prepared in the presence of sufficient Mg^{2+}, CO_3^{2-} or $P_2O_7^{4-}$ ions remain amorphous after heating to 400 °C, but at 600 °C, a variety of compounds (calcite, CO_3Ap, β-$Ca_3(PO_4)_2$ and γ-$Ca_2P_2O_7$) are formed, depending on the type of ACP heated [268]. Heating ACa,CO_3P with a low carbonate content at 500 to 600 °C yields well-crystallised AB-CO_3Ap [92] (Section 4.5.3).

1.8.8 ESR of X-irradiated ACP

X-irradiation of ACP at 20 °C produces a radical that looks like, but is not as well-defined as, the radical in irradiated OHAp with g = 2.0023 [374]. Decay studies showed there were at least two types of radicals which disappeared very quickly (<20% remained after 40 h) and were independent of each other [374]. The radical yield at low doses was rather high, and saturation was reached rather quickly. It has been reported that the PO_4^{2-} radical is the dominant species in irradiated ACP [375]. The ESR spectrum of ACP, heated at 900 °C and then irradiated, has been investigated [375].

Chapter 2

FLUORAPATITE AND CHLORAPATITE

2.1 Introduction

General references to the preparation, properties and structures of apatites, including FAp and ClAp, have been given (Section 1.1). FAp and ClAp are sometimes referred to as halophosphates, and FAp as calcium fluorophosphate, particularly in the fluorescent lamp industry. These compounds (and solid solutions of them) are very efficient phosphors when doped with antimony and manganese [78,376,377] and are used extensively in fluorescent light tubes. FAp and various silicate oxyapatites are efficient laser materials when doped with Nd^{3+} or Ho^{3+} ions [378-380] (Section 2.7).

FAp and the related $CO_3OH,FAps$, francolite, are important minerals (Section 1.1). Gemstone deposits of FAp occur in Cerro de Marcado, Durango, Mexico [381]. The reported formula, based on chemical analyses, was

$$Ca_{9.83}Na_{0.08}Sr_{0.01}RE_{0.09}$$

$$(PO_4)_{5.87}(SO_4)_{0.05}(CO_3)_{0.01}(SiO_4)_{0.06}(AsO_4)_{0.01} \qquad 2.1$$

$$F_{1.90}Cl_{0.12}(OH)_{0.01}$$

where RE abbreviates rare-earths [381]. The refractive indices were O = 1.6362±0.0005 and E = 1.6326±0.0005 (wavelength not stated, but presumably sodium light), lattice parameters a = 9.391±0.001 and c = 6.878±0.002 Å, and density 3.216±0.002 g cm^{-3} at 22 °C [381]. FAp occurs as hexagonal prisms, usually elongated along c, with the form $\{10\overline{1}0\}$, sometimes modified by $\{10\overline{1}1\}$ and $\{11\overline{2}1\}$; it has imperfect $(10\overline{1}0)$ and (0001) cleavages [16]. Various more complex forms and habits of mineral apatites have been described and illustrated (see pages 7 and 8 in ref. [2]). FAp is used to define point 5 on the Mohs hardness scale, but ClAp is somewhat softer. ClAp is a rare mineral and is sometimes sufficiently stoichiometric to be in the monoclinic form [382] (Section 2.2.3).

Biological interest in FAp, and more particularly solid solutions of FAp and OHAp, arises from the reduction in the incidence of dental caries that occurs when water supplies are fluoridated, and the use of NaF to treat some bone mineral disorders. The enameloid of sharks is essentially pure FAp [383,384] based on lattice parameter measurements (Section 4.6.2) and chemical analyses. However, because it contains some carbonate with IR absorption bands very similar to the carbonate in mineral francolite (Section 4.2.2), it is better regarded as a form of francolite [385]. Bone (including fossil bone) can also accumulate large amounts of F⁻ ion, and levels in bones from cows that grazed near a brickworks of up to 2.0 wt %, and with an *a*-axis parameter of 9.350 Å, have been reported [386] (Section 4.6.2). ClAp, substituted- and ns-ClAps are of interest as model compounds for studying the crystal chemistry of apatites because single crystals are easily grown. Biological apatites (Section 4.6.3) and, as mentioned above, apatitic phosphors, contain Cl⁻ ions.

FAp has a melting point of 1644 °C [387]. It is fairly stable thermally, but coarse grained FAp can be converted to OHAp when heated in steam at 1360 °C for 48 h [388] and partial conversion is possible at 1000 °C [389]. Slight loss of CaF_2 from FAp has been reported at 1000 °C [389] (Section 2.4.1). FAp decomposes with the loss of fluorine, probably as POF_3, when heated in vacuum above 1600 °C [390]. ClAp is much less stable and is easily converted to OHAp when heated in steam above 800 °C [75] (Section 3.3.4) and can also lose up to 23 % of its $CaCl_2$ whilst still remaining apatitic when heated *in vacuo* at 900 to 1200 °C for 16 h (the loss is partially reversible) [391] (Section 2.4.1). The effects of SiO_2 and water on the decomposition of FAp and ClAp up to 1300 °C have been investigated [392,393].

2.2 Structures

2.2.1 Fluorapatite and the apatite structure

The FAp structure will be considered in detail as a model for other apatites, and as a starting point to describe the numerous apatitic substitutions that are possible. FAp is hexagonal with space group $P6_3/m$ and lattice parameters $a = 9.367(1)$ and c (the hexagonal axis) $= 6.884(1)$ Å [46] with one formula unit of $Ca_{10}(PO_4)_6F_2$ per unit cell. The possibility of a low temperature monoclinic form is discussed at the end of this section.

The structure of FAp, the first apatite structure to be determined, was published in 1930 [394,395]. The only essential difference between these two reports was in the position of the F⁻ ion (0,0,¼ in ref. [395] and 0,0,½ in ref. [394]). Agreement was soon reached on 0,0,¼ [396] which was later confirmed [397]. The FAp structure has been refined from diffractometer single-crystal X-ray data [46].

Description of the structure

Many descriptions of the apatite structure have been given, for example, in refs [14,111,397]. The following is based on that of Beevers and McIntyre [397]. The $P6_3/m$ space group has three kinds of vertical symmetry elements (Fig. 2.1): (1) six-fold screw axes passing through the corners of the unit cells marked by dashed lines in the figure. These symmetry elements are equivalent to a three-fold rotation axis with a superimposed two-fold screw axis; (2) three-

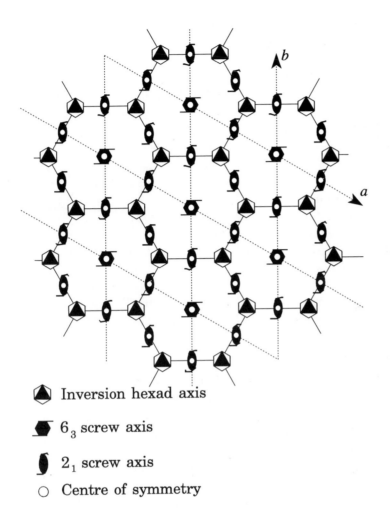

◭　Inversion hexad axis

◭　6_3 screw axis

◖　2_1 screw axis

○　Centre of symmetry

Fig. 2.1 Vertical symmetry elements of the space group $P6_3/m$. The dashed lines indicate the apatite unit cells with the c-axis out of the plane of the diagram. There are also horizontal mirror planes at $z = ¼$ and $¾$, and numerous centres of symmetry. (After Fig. 1 of Beevers and McIntyre [397])

fold rotation axes passing through $⅔,⅓,0$ and $⅓,⅔,0$; and (3) two-fold screw axes passing through the midpoints of the cell edges and its centre. There are also mirror planes perpendicular to the c-axis at $z = ¼$ and $¾$, and numerous centres of symmetry.

There are columns of Ca^{2+} ions spaced by one-half of the c-axis parameter along the three-fold axes at $⅔,⅓,0$ and $⅓,⅔,0$ which account for two-fifths of the

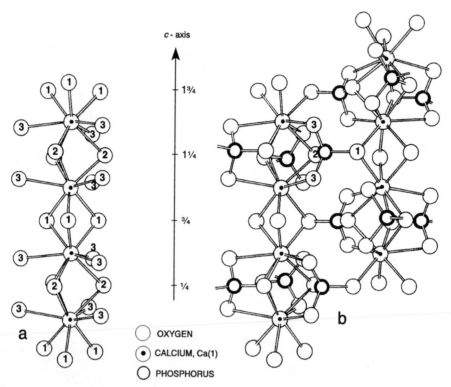

Fig. 2.2 (a) Oxygen coordination of columnar Ca(1) ions in apatite. (b) linking of columns via PO$_4$ tetrahedra. The oxygen atoms in (a) and in one tetrahedron in (b) have been numbered, and positions of the horizontal mirror planes at ¼, ¾ *etc.* marked on the *c*-axis. (After Fig. 3 of Beevers and McIntyre [397])

Ca^{2+} ions in the structure. These ions are given the designation Ca(1), Ca(I) or Ca$_I$. The site they occupy is often called the columnar site, and corresponds to the Wyckoff *f* position [398] with multiplicity 4 and point-group symmetry 3 (C_3 in the Schönflies system used by spectroscopists). Each of these Ca^{2+} ions is connected to its neighbouring Ca^{2+} ions above and below by three shared oxygen atoms that lie in the mirror plane (Fig. 2.2a); on one side, there are three O(1) at 2.397(1) Å, and on the other side, three O(2) atoms at 2.453(1) Å. Each Ca(1) ion is also coordinated by three further oxygen atoms, O(3), at a greater distance (2.801(1) Å) at approximately the same *z* parameter as the Ca^{2+} ion (this and previous Ca(1)-O bond lengths are from ref. [399], which were based on ref. [46]). Thus, the columnar Ca^{2+} ions are nine-fold coordinated by oxygen atoms. These columns of Ca^{2+} ions and their coordinating oxygen atoms are linked together (Fig. 2.2b) by PO$_4$ tetrahedra in

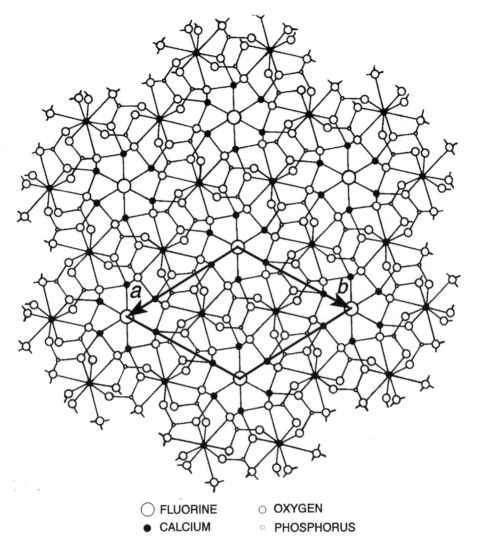

○ FLUORINE   ○ OXYGEN

● CALCIUM   ○ PHOSPHORUS

Fig. 2.3 Projection of the apatite structure on to the basal plane (0001). The c-axis is out of the plane of the paper and the *a*- and *b*-axes are as marked on the outline of a unit cell. (After Fig. 4 of Beevers and McIntyre [397])

which three oxygen atoms (two O(3) and either an O(1) or O(2)) come from one column, and the fourth (O(2) or O(1), respectively), from the adjacent column. The result is a three-dimensional network of PO_4 tetrahedra with enmeshed columnar Ca^{2+} ions and with channels passing through it (Fig. 2.3 without the F^- and their adjacent Ca^{2+} ions). The axes of these channels coincide with the six-fold screw axes and the corners of the unit cells, and one

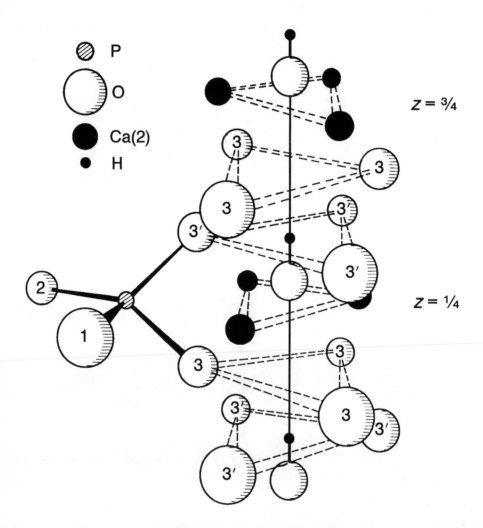

Fig. 2.4 Arrangement of the ions around the vertical axis at the corners of the apatite unit cell in Fig. 2.3. This axis is sometimes called the hexad or *c*-axis channel. The F⁻ ion is at the centre of the triangle of Ca(2) ions, but the OH⁻ (illustrated) and Cl⁻ ions are above this position. (Fig. 4 of Elliott and Mackie [400], adapted from Fig. 3 of Young [401])

forms the *c*-axis. The directions of the *a*- and *b*-axes are as marked in Fig. 2.3. The remaining ions, F⁻ and their adjacent Ca(2) ions in Fig. 2.3 that are required to complete the structure, are located in the channels. These channels have six "caves" per *c*-axis repeat of the unit cell, centred on the mirror planes at $z = \frac{1}{4}$ and $\frac{3}{4}$, into which the Ca(2) ions can fit. These ions form two triangles of Ca^{2+} ions rotated by 60° from each other about the *c*-axis (Fig. 2.4) at whose centres the F⁻ ions are located. Thus the F⁻ ions are three-fold coordinated by Ca^{2+} ions, with all four ions lying on a mirror plane. The Ca(2) ions have been called the triangular Ca^{2+} ions. The Wyckoff designation of the site is *h* with a multiplicity of 6 and point-group symmetry *m* ($C_s \equiv C_{1h} \equiv C_{1v}$ in the Schönflies system). The Ca(2) ions are seven-fold coordinated, six oxygen atoms (O(1) at 2.814(1), O(2) at 2.384(1), two O(3) at 2.344 and two O(3) at 2.398 Å) and an F⁻ ion at 2.231(1) Å (lengths from ref. [399] which were based on ref. [46]). The six atoms, Ca(2), O(1) and four O(3), lie nearly in a plane, with axial bonds, Ca-O(2) and Ca-F, nearly perpendicular to the plane.

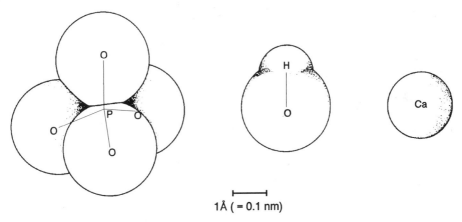

1Å (= 0.1 nm)

Fig. 2.5 Sizes of the ions in apatite. The F⁻ ion is ~3 percent smaller than the OH⁻ ion. (Fig. 1 of Elliott [111])

The above description reveals the coordination of the ions, but does not show very clearly how the ions are packed together, nor how the space in the structure is filled. The relative sizes of the ions (Fig. 2.5) show that most of the space will be taken up by the oxygen atoms of the PO_4^{3-} tetrahedra. If these ions are taken as spheres, they are approximately hexagonally close-packed, with channels through the structure parallel to the hexagonal axis [111]. Two-thirds of these channels are filled with the columns of Ca(1) ions which are in six-fold (octahedral) coordination by the PO_4^{3-} spheres. The remaining one-third of the channels are filled with the columns of F⁻ ions, with each one surrounded by three Ca(2) ions, as described above. The other octahedral site at 0,0,½ in the approximately close-packed spheres is vacant in FAp, but in ClAp, the larger Cl⁻ ion is located close to it at 0,0,0.44 (Section 2.2.3). This description of the apatite structure, based as it is on hexagonal close packing, shows that at one level, it is very simple. This simplicity is probably the basis of the remarkable versatility of apatites to accept a wide range of F⁻ (Section 2.3.1) and Ca^{2+} ion (Section 2.3.2) substitutions because these can take place within the interstices between the PO_4^{3-} ions without any great disturbance, and also the reason why the PO_4^{3-} ions themselves can be replaced by a large variety of polyatomic ions (Section 2.3.3).

A possible phase transition to a noncentrosymmetric monoclinic space group $P2/b$ ($P2_1/b$ was also mentioned) at ~133 K has been reported on the basis of ESR studies of CrO_4^{3-}-doped FAp and single crystal XRD studies of FAp [402]. This transition was thought to arise from ordered displacements of the F⁻ ions from the positions at 0,0,¼ [402]. A phase change involving small structural changes from the hexagonal cell was apparently confirmed by further ESR studies at 4.2 K [403]. Ordered displacements of Cl⁻ (Section 2.2.3) or OH⁻ (Section 3.2) ions lead to a lower symmetry and doubled b-axis parameter in ClAp and OHAp. Above the proposed transition for FAp, it might be expected that the F⁻ ion would be in two-fold disorder about 0,0,¼, but this was not reported in the refined structure [46]. It is noticeable however, that the β_{33} thermal parameter for the F⁻ ion in FAp (0.0124(4)) is slightly higher than that for the O(H)⁻ ion when in two-fold disorder in OHAp (0.0102(8)) [404], even though all the other thermal parameters in FAp are smaller than in OHAp. Furthermore, although BaFAp also has the space group $P6_3/m$, the F⁻ ions are in two-fold disorder with z displacements of ±0.24 Å about 0,0,¼ [405]. Also, $CaAsO_4FAp$ and $CaVO_4FAp$ have pseudohexagonal lattices [85], which are characteristic of apatites with ordered displacement of ions from $z = $ ¼ and ¾ in the c-axis channels (*e.g.* ClAp and OHAp as mentioned above). However, in contrast to manganese-doped monoclinic ClAp at 78 K, manganese-doped FAp at 78 K did not exhibit splitting of the 8600 cm⁻¹ line in the fluorescent emission spectrum into three components [406] (Section 2.2.3); in monoclinic ClAp, this splitting arises from the three nonequivalent manganese sites [406]. If FAp has a phase transition below 133 K, it too might have been expected to show this splitting. Clearly, further investigation is desirable to confirm, or otherwise, the existence of a low temperature phase transition in FAp.

2.2.2 Structural relationships of apatite

From a biological point of view, the most important relationship is between OHAp and OCP because of the role of OCP as an intermediate in the precipitation of OHAp (Section 3.8), the possible presence of intercrystalline mixtures of OCP and OHAp in Ca-def OHAps (Section 3.7.4) and because the "central dark line" seen in EM studies of developing enamel apatite crystals has been ascribed to the presence of a thin layer of OCP (Section 4.6.6). The structural basis for these phenomena, and for the orientational relationship found between OCP and OHAp when OCP is partially hydrolysed (Section

1.3.6) or dehydrated (Section 1.3.7), is that the OCP structure (Section 1.3.2) contains "apatite" layers about 1.1 nm thick parallel to its (100) plane which correspond to the (010) plane of OHAp (Fig. 1.4). This relationship has also been predicted by computer investigations of epitaxial relations between OHAp and OCP using metric criteria [407].

A slightly thinner, and slightly more distorted "apatite" layer is also found in $Ca_4(PO_4)_2O$ (TetCP) [316] (Section 1.7). Thus OHAp has the distinction of having close structural relationships with CaPs both more basic (TetCP) and more acidic (OCP) than itself. The structural relationship between OCP, OHAp and samuelsonite (a complex basic calcium aluminium phosphate) has been discussed [408]. They all have columns of composition $Ca_4(PO_4)_{12}$ as underlying structural units, which are fused in OCP to generate sheets of composition $Ca_4(PO_4)_8$ and, in apatite, a framework of composition $Ca_4(PO_4)_6$.

Although the structures of apatite and β-TCP (or whitlockite) are not closely related, the environment of the columnar Ca(1) ions in apatite (six-fold coordination by the closest oxygen atoms) has some similarity with that of the Ca(5) ions in β-TCP. (These are the Ca(5) ions in the A columns in Fig. 1.8b, and are six-fold coordinated by oxygen [278]).

Formal relationships exist between apatite, the glaserite-type structures [62,409] (glaserite itself is $K_3Na(SO_4)_2$) and α-TCP [283]. These similarities are most marked when the structures are viewed down the *c*-axis (Fig. 1.7, Section 1.6.2, and Fig. 2.6). The relationship between α-TCP and apatite was discussed in Section 1.6.2.

The glaserite structure contains three types of columns running parallel to the *c*-axis (Fig. 2.6a): columns of alternating SO_4^{2-} and K^+ ions (marked A); columns of alternating Na^+ and K^+ ions (marked B); and columns of K^+ ions (marked C). The apatite structure (Fig. 2.6b) can be derived from glaserite by replacing the B columns of -K-Na-K-Na- by -F-F-F-F- columns; the K^+ ions in the C columns by Ca^{2+} ions to form the apatite Ca(1) columns; and SO_4^{2-} by PO_4^{3-} and K^+ by Ca^{2+} ions in the A columns. The charge reversal in the B columns causes the movement of the cations in the A columns towards the F^- ions to give the three-fold triangular Ca(2) coordination of the F^- ions in apatite [409].

This description of the glaserite structure can also be used to show its relationship with the structure of silico-carnotite, $Ca_5(PO_4)_2SiO_4$, and hence also to apatite [409]. $K_3Ca_2(SO_4)_3F$ is a very distorted analogue of FAp [410]; and nasonite, $Pb_6Ca_4(Si_2O_7)_3Cl_2$, has an apatite-like structure [411].

There is an interesting relationship [412] (Fig. 2.7) between the structures of apatite and aragonite, the orthorhombic form of $CaCO_3$ with space group *Pcmn* and a = 7.961, b = 4.958 and c = 5.739 Å. The orthorhombic cell may be described in terms of a pseudohexagonal one of twice the volume with a = 9.378, c = 5.739 Å and γ = 117.0° [412]. In aragonite, there are columns of

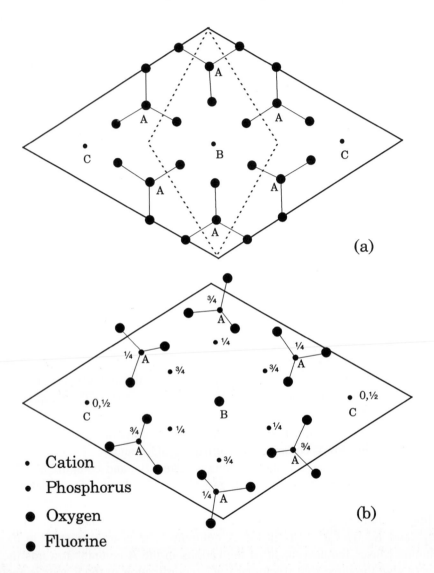

Fig. 2.6 Relationship between the structures of (a) glaserite, $K_3Na(SO_4)_2$, and (b) apatite projected in the direction of the c-axis. The unit cell of glaserite is indicated by the broken line in (a) and the conventional origin of the apatite unit cell is at the centre of the cell in (b). In (a), A is a cation-anion column, and B and C are cation-cation columns; in (b), A is a cation-anion column, B an anion-anion column and C is a cation-cation column. (After Fig. 8 of Dickens and Brown [409])

Ca^{2+} ions parallel to the c-axis that form six-membered rings in projection at whose centres the CO_3^{2-} ions are situated (Fig. 2.7a). The same six-membered

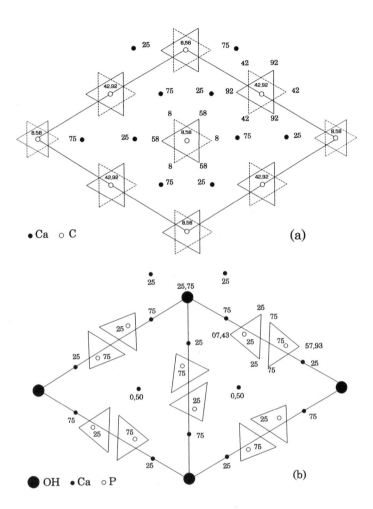

Fig. 2.7 Relationship between (a) orthorhombic aragonite, $CaCO_3$, and (b) apatite structures. The pseudohexagonal cell is shown for aragonite and the hexagonal cell for apatite with its origin coinciding with the columns of OH^- ions. Both cells are projected along the *c*-axis with CO_3^{2-} ions in aragonite and PO_4^{3-} ions in apatite indicated by triangles. *z* parameters of atoms are multiplied by 100. (After Figs 1 and 2 of Eysel and Roy [412])

rings in projection, formed by the Ca(2) triangles and centred about the OH^- ion columns, occur in apatite (Fig. 2.7b). However, the rings in apatite are rotated by 30° with respect to those in aragonite, so that a four-membered ring in projection (comprising two Ca(1) and two Ca(2) atoms) occurs in the middle of the apatite cell, instead of the six-membered ring in the middle of the aragonite cell. This four-membered ring contains two PO_4^{3-} ions per unit cell instead of the two CO_3^{2-} ions per unit cell in aragonite. The structural correspondence can be described [412] by the formulae for the unit cells of aragonite and apatite:

$$Ca_6 \quad Ca_2 \quad \square_2 \quad (CO_3)_6 \quad (CO_3)_2 \quad = 8CaCO_3$$
$$Ca(2)_6 \quad Ca(1)_2 \quad Ca(1)_2 \quad (PO_4)_6 \quad (OH)_2 \quad = Ca_{10}(PO_4)_6(OH)_2$$

2.2

In aragonite, the first six Ca^{2+} ions form a complete six-membered ring in the centre of the cell (Fig. 2.7a), and the remaining two Ca^{2+} ions are members of different six-membered rings in adjacent cells at the left and right. In apatite, the six Ca(2) atoms contribute half the atoms to two different six-membered rings in adjacent cells at the top and bottom of the cell (Fig. 2.7b). Two of the Ca(2) atoms are shared with the central four-membered ring in apatite which is completed with the addition of the first of the two Ca(1) atoms. The remaining two Ca(1) atoms are missing in aragonite, indicated by $\square_2$. The structural relation between aragonite and apatite is emphasised [412] by the fact that both anionic sites in apatite can be replaced by CO_3^{2-} ions (Section 4.1). This relation also explains the alignment found between the aragonite and apatite lattices when the former is transformed to the latter by alkaline phosphate solutions [412] (Section 4.5.6).

A rather surprising structural relationship exists between apatite and a group of metallic compounds known as Nowotny phases [413,414]. Nowotny phases are hexagonal with similar c/a ratios as apatites, but have a single metal atom in place of the tetrahedral PO_4^{3-} ion. An example is Mo_5Si_3C, in which molybdenum occupies Ca^{2+} ion sites, silicon occupies PO_4^{3-} ion sites, and carbon is located on the c-axis at $z = 0$. There are also many binary metallic compounds of structure type $D8_8$, for example Mn_5Si_3, that have structures analogous to apatite, but with vacant c-axis sites [414].

The normal apatite structure, with space group $P6_3/m$, can be viewed as a distortion from a higher symmetry structure with space group $P6_3/mcm$ in which O(1) and O(2) become equivalent [415]. Although no apatite has yet been found with this higher symmetry, it might occur as a high temperature modification [415].

Epitaxial relations between OHAp and other calcium orthophosphates have been investigated theoretically [407]. The only close relationships found were between OHAp and OCP (discussed at the beginning of this section and in Section 1.3.2) and TetCP (discussed in Section 1.7).

2.2.3 Chlorapatite structure

Monoclinic stoichiometric ClAp

The single crystal XRD pattern of stoichiometric ClAp has weak reflections that show that the *b*-axis parameter is doubled giving a monoclinic space group $P2_1/b$ [416]. These reflections can also be seen in powder X-ray diffractometer patterns (Fig. 2.8) with the strongest lines having about 1% of the intensity of the strongest line in the hexagonal pattern. The monoclinic structure is very similar to the hexagonal one described above, the most significant difference being an ordered arrangement of the Cl^- ions above and below $z = \frac{1}{2}$ on the pseudohexagonal axis (Fig. 2.9). This ordering causes the mirror planes at $z = \frac{1}{4}$ to become glide planes and the *b*-axis parameter to be doubled [47]. Thus the lattice parameters are $a = 9.628(5)$ Å, $b = 2a$, $c = 6.764(5)$ Å, $\gamma = 120°$ [47]. Recent synchrotron radiation measurements based on the observed splitting of the hexagonal 300 reflection have shown that $2a = 19.292$ and $b = 19.282$ Å (error in difference ~0.001 Å), with practically no deviation from $\gamma = 120°$; thus *b* is very slightly less than $2a$ (M Bauer and WE Klee, personal communication, 1991 and ref. [60]). The opposite is found in monoclinic OHAp, Section 3.2.

The Cl^- ion is located at 0.0016, 0.2493, 0.4439 (parameters for $P2_1/b$) and the "ordering information" from one Cl^- ion column to another over a distance of 6.7 Å is propagated via tilts of the PO_4^{3-} tetrahedra and ~0.05 Å shifts of Ca(1) [47]. Clearly, the monoclinic structure will not develop unless the Cl^- ion columns are essentially intact. The loss of Cl^- ions required to prevent the formation of the monoclinic structure is probably somewhat less than 7 % if calcium is simultaneously lost to preserve charge balance [391], between 16 and 36 % if it is through the replacement of Cl^- by F^- ions [406], and between 28 and 37 % for the replacement of Cl^- by OH^- ions [417]. Clearly the simultaneous loss of Ca^{2+} and Cl^- ions is the most disruptive of these changes.

Monoclinic to hexagonal thermal phase transition

A transition to the hexagonal space group $P6_3/m$ between 185 and 210 °C, the exact temperature depending on the crystal studied, has been reported [406]. The criteria for determining the transition temperature were the disappearance of the birefringence looking down the *c*-axis and the loss of the doubled unit cell as determined by XRD precession photographs. Heating a crystal so it lost $CaCl_2$ lowered the temperature to 174 °C; on further $CaCl_2$ loss, the crystals were always hexagonal (even down to 6 K). Addition of F^- ions raised the temperature to 310 °C (for 16 % Cl^- replaced by F^- ions), after which the crystals were always hexagonal (even at 6 K) [406]. In studies of the optical properties through the transition, transition temperatures of 211 to 216 °C were reported [48].

A detailed study of the transition has been made, based on dielectric

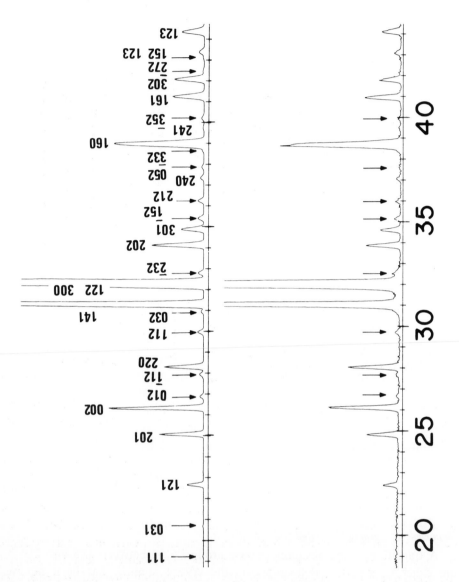

Fig. 2.8 Calculated (left) and observed (right) XRD powder patterns of monoclinic ClAp (CuKα). The "monoclinic" lines are marked by arrows. The ClAp was made by General Electric Co, Cleveland, Ohio, by disproportionation of Ca_2PO_4Cl to ClAp and $CaCl_2$ at 1200 °C [391] (Section 2.4.1). The vertical scale is the scattering angle, 2θ. (Fig. 7 of Elliott and Mackie [400])

properties and XRD [418]. On heating from 25 to 300 °C, the Cl^- ion displacement from $z = \frac{1}{2}$ decreased from 0.4 to 0.3 Å, and the root mean

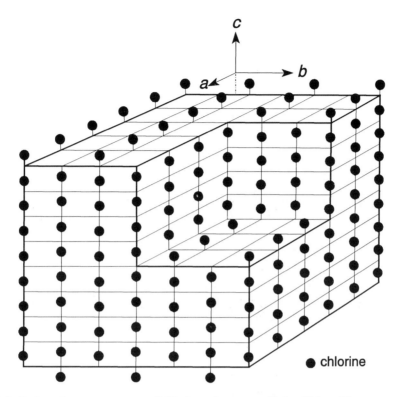

Fig. 2.9 Ordered arrangement of Cl⁻ ions in monoclinic ClAp. The axes show the size and origin of the monoclinic unit cell. Note that the origin is shifted ¼ **b** from the pseudohexagonal axis.

square (rms) thermal vibrational amplitude increased to nearly 0.2 Å, but this was still 0.1 Å short of allowing the Cl⁻ ion, on average, to reach and cross over the position at $z = \frac{1}{2}$. It was concluded that the transition near 200 °C was not a true monoclinic to hexagonal phase change. However, there was evidence for a structural change because: (1) the a-axis parameter and Cl⁻ ion rms thermal vibrational amplitude plots against temperature changed slope (but see below); (2) the Cl⁻ ion abruptly changed its position (0.025 Å shift towards $z = \frac{1}{2}$); and (3) the X-ray intensities of the reflections indicating a monoclinic cell rapidly became very much weaker (but could still be seen at 400 °C) and broader (the other reflections did not show these changes). These results (and dielectric studies, see later) were thought to show that, near 200 °C, the domain size, within which the original room temperature inter-column monoclinic ordering existed, became very much smaller. It was proposed that above ~200 °C, they persisted as microdomains, possibly no larger than 60 Å, and that a true monoclinic to hexagonal transition, when the thermal energy of the Cl⁻ ion would be sufficient to carry it over the energy barrier at $z = \frac{1}{2}$, should occur at 400 to 500 °C. However, in a further investigation of the same crystal (JC Elliott and PE Mackie, unpublished results, 1977), it was found that the additional weak reflections could still be seen at 580 °C, and furthermore, in a structure determination at 544 °C (a = 9.668 and c = 6.823 Å), the Cl⁻ ion was 0.33 (σ = 0.02) Å from z = ½ with an rms thermal vibrational amplitude of 0.170 (σ = 0.07) Å, so the proposed "true transition", if it occurs, seemed to be above this temperature (but see comments below on temperature measurements).

Recently, further aspects of the transition have been studied in detail (M Bauer and WE Klee, personal communication, 1990 and ref. [60]). High precision optical measurements showed that the birefringence observed down the c-axis persisted up to ~320 °C (the limit depended on the thermal history of the sample), which indicated that the domain size must be larger than the 60 Å mentioned above. The a-axis parameter was measured from 20 to 850 °C; little or no evidence was found for the change in slope reported earlier [418] (see (1) above). The suggestion was made that the higher temperatures in the earlier work might have been slightly lower than thought. Differential thermal analysis showed a weak endothermic peak ($\Delta H \approx 0.3$ cal g⁻¹, 1.25 J g⁻¹) near 320 to 340 °C, indicating that the phase transition is at least partly first order (M Bauer and WE Klee, personal communication, 1990 and ref. [60]). No other peaks were observed up to 500 °C.

Dielectric measurements have been made on ClAp [60,418,419]. ClAp shows hysteresis loops characteristic of a ferroelectric and, above ~200 °C, conduction effects [418,419].

The birefringence observed when looking down the c-axis has been measured for various externally applied fields in this direction (M Bauer and WE Klee, personal communication, 1990 and ref. [60]). The birefringence fell as the field increased, and at 5000 V mm⁻¹, was zero. A structure determination for a field of 1300 V mm⁻¹ indicated that ~3% of the Cl⁻ ions had shifted from below 0,0,½ to above this position. It was suggested that the uniaxial optical behaviour of ClAp at higher fields was due to all the Cl⁻ ions shifting, so the space group became $P6_3$.

Electric dipole relaxations in ClAp have been studied with the fractional polarisational mode of the thermally stimulated currents method [96,420]. Compensation temperatures of 202±2, 202±2, 420±4 and 644±4 °C with relaxation times 1.3×10^{-7}, 3.2×10^{-6}, 8.82×10^{-6} and 2.3×10^{-4} s respectively, were observed. It was proposed that the 202 °C compensation was associated with loss of ordering between Cl⁻ ion columns, and that possibly the 420 °C compensation occurred at a temperature at which the Cl⁻ ions individually had sufficient thermal energy to maintain the hexagonal form dynamically. However, in view of the structure determination at 544 °C reported above, it is more likely that the 644 °C compensation corresponds to this.

In summary, the thermal studies and more detailed X-ray and optical work show that the thermal transition of ClAp takes place at ~320 °C, rather than ~200 °C. However, details of the changes involved, and the structure above the transition, are not yet clear. There are also similar phase transitions in OHAp (Section 3.2) and A-CO_3Ap (Section 4.3.3).

$CaCl_2$-deficient ClAp

A comparison of the calculated and measured densities of melt-grown $CaCl_2$-def ClAp shows that there is a loss of $CaCl_2$ from the unit cell [391] (Section 2.5.2). The structure of a similar melt-grown $CaCl_2$-def ClAp has been reported [421]. The $CaCl_2$ deficiency was 10 % from the lattice parameters (a = 9.615, c = 6.771, σ = 0.002 Å) and ~15% from the occupancy of the c-axis channel sites by Cl⁻ ions. Additional satellite atoms close to atoms in their usual positions were found for Cl (displaced 0.867 Å along the c-axis towards the mirror plane), Ca(2) (displaced 0.385 Å towards the c-axis), O(3) (displaced 0.369 Å in a direction making an angle of 78.7° to the P-O(3) bond) and two for Ca(1) (displaced 0.316 and 0.313 Å in opposite directions along the Ca(1) column). The satellite(s) of Cl, Ca(2), and Ca(1) had occupancies of ~7, ~24, and ~1.4 % respectively of the main atom and O(3) atom and its satellite had similar occupancies (JC Elliott, RA Young, PE Mackie and E Dykes, unpublished results, 1976). It was proposed that the satellites of the Cl, Ca(2) and O(3) atoms arose from atoms adjacent to Cl⁻ ion vacancies, and the two Ca(1) satellites from atoms adjacent to Ca(1) vacancies [421]. Thus, charge compensation for a Cl⁻ ion deficiency involved loss of Ca(1) and Ca(2) atoms, because both had satellites. There was evidence that the Cl vacancies and Cl⁻ ions were segregated along the c-axis because the Cl⁻ ion satellite occupancy was lower than expected for isolated vacancies. Satellite O(3) atoms can also be seen in structure refinements of hexagonal Holly Springs OHAp because only half of the O(3) atoms are coordinated by the hydrogen atom of the hydroxyl ion (Section 3.2).

The short-range ordering of Cl⁻ ions found in the above structural work agrees with the conclusions reached from measurement of the fluorescent emission spectrum at 78 K of FAp and ClAp doped with small amounts of MnO_4^{3-} ions [406]. In monoclinic ClAp, but not FAp, the fluorescent line near 8600 cm⁻¹ was split into three components with 5 cm⁻¹ separations which were attributed to Mn^{5+} ions in P^{5+} sites that were equivalent in the hexagonal space group, but became three nonequivalent sites in the monoclinic structure. In manganese-doped $CaCl_2$-def ClAp, even though it was hexagonal, three closely spaced components were still seen (at 6 K), which were attributed to small coherent regions in which the Cl⁻ ions were ordered.

2.3 Substitutions in apatites

2.3.1 Hexad axis substitutions

Vacancies, neutral molecules and monovalent ions

Several lead (Section 2.3.2) and strontium [422,423] apatites have been prepared with completely or partially empty F⁻ ion sites. FAp single crystals with up to a 10% deficiency of CaF_2 can be grown [387,424] (Section 2.5.1), and also powders [16] (Section 2.4.1), which presumably have a substantial number of vacant F⁻ ion sites. The structure determination of a ~15 % $CaCl_2$-def ClAp showed the movement towards 0,0,¼ of Cl⁻ ions adjacent to vacancies and vacancy clustering [421] (previous section). OH⁻ site vacancies are also thought to occur in precipitated Ca-def OHAps (Section 3.7.3). Both precipitated Ca-def OHAps (Section 3.7.3) and precipitated CO_3-Aps (Sections 4.5.3 to 4.5.5) are thought to have *c*-axis vacancies.

Several authors (Table 3.2, Section 3.7.3) have proposed that water can be located on the hexad axis, and oxygen-rich apatites are known that contain molecular oxygen, again probably on the hexad axis [85,87,425,426] (Section 3.4.3). Molecular oxygen, argon or carbon dioxide appear to be trapped in the apatite lattice when finely ground enamel or bone powder is subjected to low temperature ashing at 80 °C and 1 torr (130 Pa) in the respective excited gas molecules for 3 h [427]. The evidence for this came from a small increase in the *a*-axis dimension which increased in the sequence oxygen, argon and carbon dioxide. For example, the *a*-axis dimension of enamel increased from 9.443±0.002 to 9.466±0.002 Å. An increase in the *a*-axis (not always the same) was also observed if the gas was not in an excited state; the specimens reverted to nearly their original axial dimension if kept at 10^{-5} torr (1.3 mPa) for several days. The molecular oxygen, argon or carbon dioxide was most likely located in the *c*-axis channels [427]. The ESR spectra of samples ashed in excited oxygen had an asymmetrical first derivative absorption line; for an OHAp sample, the *g*-tensor was $g_1 = 2.002$, $g_2 = 2.010$ and $g_3 = 2.017$. This signal was attributed to O_3^- paramagnetic radicals [427].

When present in low concentrations, either F⁻ or OH⁻ ions in OH,FAp appear to be randomly distributed; their positions have been studied by IR (Section 3.11.5) and NMR spectroscopy (Section 3.12). X-ray and neutron single crystal studies of Holly Springs OHAp showed an 8 atom % replacement of OH⁻ by F⁻ ions at 0,0,¼ [404] (Section 3.2). The ordering of Cl⁻ or OH⁻ ions in pure ClAp and OHAp that leads to the monoclinic structures (Sections 2.2.3 and 3.2, respectively) is destroyed by the replacement of these ions by small amounts (~10 %) of other ions. F⁻-Cl⁻ ion positional interactions have been studied by single crystal XRD of ClAp with ~14 and ~34 mole % replacement of Cl⁻ by F⁻ ions [428]. Some of the substituting F⁻ ions were shifted along the *c*-axis by about ~0.6 Å from their normal positions at *z* = ¼, and likewise, some of the Cl⁻ ions were displaced by >0.2 Å further away from their normal positions at $z = 0.44$ in a direction away from $0,0,\frac{1}{2}$. Single crystal XRD studies [429] of six crystals of F,Cl,OHAp have shown that the z parameters of the Cl⁻ and OH⁻ ions are primarily sensitive to, and depend roughly linearly on, the fractional amount of Cl⁻ ions present. As the chlorine content decreases from that in pure ClAp, the Cl⁻ ions move away from their normal position at $z = 0.44$, so that, for a very low chlorine content, the z parameter is ~0.36. Simultaneously, the associated Ca(2) triangle expands to maintain a normal Ca-Cl distance. These studies of apatites with different ions located on the c-axis show that there can be a strong interaction between the ions, particularly if Cl⁻ ions are involved. The direction of movement of the Cl⁻ ion is always towards $0,0,\frac{1}{4}$ when the Cl⁻ ions in ClAp are replaced by vacancies, OH⁻ or F⁻ ions.

Single crystals of BrAp can be synthesised [50,75,80]. An XRD structure determination showed that BrAp has the space group $P6_3/m$, $a = 9.761(1)$, $c = 6.739(1)$ Å with most of the Br⁻ ions at $0,0,0$ [49]. Apatites with I⁻ ions on the hexad axis are also known [430-433], including an impure IAp [80]. A feature of apatites with Br⁻ or I⁻ (also often Cl⁻) ions on the hexad axis is that the c-axis parameter is apparently less than twice the conventional ionic diameter (Table 2.1). This situation for BrAp has been discussed in detail [49]. A particularly interesting occurrence arises in CdPO₄ClAp, CdVO₄BrAp, CdAsO₄BrAp, CdPO₄BrAp and CdVO₄IAp which have an increasing halide deficiency from the "chlorapatite" (15 %) to the "iodoapatite" (27 %), as the radius of the halide ion increases [431]. A model has been proposed for the halide ion positions as a result of detailed single crystal XRD studies [432]. In this, the ions occurred in chains with modal positions at $0,0,\frac{1}{4}$ or $0,0,\frac{3}{4}$. The "oversize" (diameter $>\frac{1}{2}c$) halide ions in succeeding half-cells are forced to deviate further and further away from this modal position until the steric strain is relieved by a vacancy, after which a new chain starts. The number of atoms in a chain was about three for CdVO₄IAp, to about eight for CdAsO₄BrAp.

Cyanate (NCO⁻) ions can occur in low concentrations on the OHAp hexad axis [94,434,435] (Section 4.5.7). Apatites, $Sr_{9.402}Na_{0.209}(PO_4)_6B_{0.996}O_2$ (trigonal space group $P\bar{3}$, $a = 9.734(4)$ $c = 7.279(2)$ Å, B at $0,0,\frac{1}{2}$ and O at $0,0,0.3278(13)$) [436] and $Ca_{9.64}\{(PO_4)_{5.73}(BO_3)_{0.27}\}\{BO_2)_{0.73}O_{0.27}\}$ (space group $P\bar{3}$, $a = 9.456(1)$ $c = 6.905(1)$ Å, B at $0,0,\frac{1}{2}$ and O at $0,0,0.3150(12)$) [437] (Section 4.1.2), have been prepared that have the linear BO_2^- ion coincident with the c-axis (Sections 4.1.2 and 4.5.7). The structure of $Ba_5(ReO_5)_3O_2$ (space group $P6_3cm$, $a = 10.912(2)$ $c = 7.774(3)$ Å) with the O_2^- ion at $0,0,0.673$ has been reported [438].

"Nitrated" apatites are thought to have nitrate (NO₃⁻) and nitrite (NO₂⁻) ions in their c-axis channels (also NO₂²⁻ ions, see following paragraph) [439]. They can be prepared by heating A-CO₃Ap in dry nitrogen monoxide at 900 °C for 48 h and have IR absorption bands at 1380 and 1270 cm⁻¹, assigned to nitrate and nitrite ions respectively [439]. The *c*-axis location was based on a comparison of the lattice parameters (a = 9.462 and c = 6.873 Å) with those of OHAp (a = 9.421 and c = 6.882 Å) and with A-CO$_3$Ap (a = 9.56 and c = 6.86 Å). This comparison showed that the "nitrated" apatite had an expanded *a*-axis and slightly reduced *c*-axis by comparison with OHAp, a characteristic typical of *c*-axis substitutions.

Divalent ions

Many substitutions are known in which there is a partial or total replacement of 2F⁻ or 2OH⁻ ions by one or more divalent ion(s), for example O²⁻ [62,380,440-448] (see Section 3.4.3 for refs for OAp), O$_2$²⁻ (peroxide) [85,87,449] (Section 3.4.3), S²⁻ [85,450-452], NCN²⁻ (cyanamide) [94,434,435,453,454] (Section 4.5.7), CO$_3$²⁻ (see Section 4.3) or SO$_4$²⁻ [455] ions. Generally, detailed structures of these are not known. However, PbOAp has a doubled *c*-axis parameter (a = 9.84, c = 2 × 7.43 Å) and 007 reflection (based on the doubled cell) which shows the loss of the 6$_3$ screw axis [62]. These results indicated an ordered arrangement of the O²⁻ ions and vacancies along the *c*-axis [62]. Various defect centres involving O²⁻ ions and halogen ion vacancies have been investigated by optical and ESR spectroscopy, before and after X- and UV-irradiation, in ClAp [91,456,457] and FAp [424]. ESR studies at liquid nitrogen temperature of the "nitrated" apatites discussed in the previous paragraph were thought to show that the nonlinear NO$_2$²⁻ ions are present in the *c*-axis channels with the O-O direction parallel to the *c*-axis [439]. The ESR spectrum was much broader above this temperature, and was interpreted as showing free rotation of the ion about the *c*-axis. The variety of species with oxygen and nitrogen in different oxidation states in the *c*-axis channels is comparable to that found with oxygen alone (Section 3.4.3).

2.3.2 Substitutions for calcium ions
Introduction

Possible substitutions for the Ca²⁺ ions are numerous and complex. There are two cation sites, Ca(1) and Ca(2), so there may be an unequal distribution of substituting cations between them. If the substituting cation does not have a +2 charge, the charge compensation mechanism might have an overriding influence on the distribution between the two sites. If there is no difference in charge, the distribution is more likely to be determined by the relative sizes of the substituting and host cations in relation to the two sites, although polarisability and ability to form partial covalent bonds may be important in particular cases. There is also the possibility that the distribution might change with the extent of substitution, temperature and thermal treatment (*e.g.* quenching from high temperatures), and in some situations, with the method of preparation (probably because of incorporation of small amounts of impurities or change of valency). Furthermore, the substituting ion might replace a subset of the Ca(1) or Ca(2) ions with a lowering of the space-group symmetry, so there may be thermal order to disorder phase transitions.

Many investigations of the Ca(1) and Ca(2) site distributions have been based on XRD studies (both powder and single crystal), but the difference in symmetry between the sites (Section 2.2.1) enables electron spin resonance, optical fluorescence and other spectroscopic methods to be used in some cases (some of these methods are discussed in ref. [124]). IR spectroscopy can also give information in some instances when there is a change in symmetry or covalent bond formation.

The coordination number of the Ca(1) site is nine, whereas for Ca(2), it is only seven. This difference has been used to suggest that, purely on the basis of size, the Ca(1) site should be occupied preferentially by ions larger than Ca^{2+} [448,458]. If geometrical considerations alone are of overriding importance, the replacement of F^- by a Cl^- ion (or another ion) near Ca(2) might also be expected to change the distribution between the two sites because of their differences in position and size. Considerations of local charge compensation, via Pauling's electrostatic valency rule [459], have been used to explain and predict the distribution between the two sites, particularly in relation to luminescence properties of rare-earth silicate oxyapatites and calcium halophosphates [460]. Pauling's rule states that the total electrostatic bond strength of the bonds from the cations surrounding an anion should equal the magnitude of the charge on the anion. This rule is effectively obeyed for O(1), O(2) and O(3). However, the total electrostatic bond strength from trivalent cations in Ca(2) sites towards an ion in the F^- site is 9/7, and for divalent cations, only 6/7 [460]. Thus, it was proposed that it would be very unfavourable to have cations with a large radius and/or lower charge in Ca(2) sites with a divalent anion, say an O^{2-} ion, in the F^- site, but less unfavourable for monovalent anions [460]. Other factors thought to be important in determining the distribution between the sites include the electrostatic energy between, and packing of, cations replacing Ca(1) along the three-fold rotation axis [458], and the formation of partial covalent bonds (see subsequent discussion on Pb^{2+} ions in this section).

The structure fields for the total replacement of ions in the FAp and ClAp lattices have been investigated [461]. The conclusion was reached that for FAp, the conditions were: 0.29 Å$\leq R_p \leq$0.60 Å and 0.95 Å$\leq R_c \leq$1.35 Å; and for ClAp, 0.29 Å$\leq R_p \leq$ 0.60 Å and 0.80 Å$\leq R_c \leq$ 1.35 Å, where R_p and R_c were the radii of the ions (Table 2.1) occupying phosphorus and calcium positions respectively. The FAp lattice tended to be distorted from hexagonal when the calcium sites were occupied by small cations, whereas no such distortions occurred in ClAp, but limited substitution of ions that fell outside these ranges was possible [461].

Table 2.1 Effective ionic radii (Å) for some of the ions that partly or completely substitute in apatites[a]. Data from ref. [462].

Mg^{2+}	Co^{2+}	Sb^{3+}	Fe^{2+}	Mn^{2+}	Y^{3+}	Eu^{3+}	Cd^{2+}
0.720	0.745	0.76	0.780	0.830	0.900	0.947	0.95
Nd^{3+}	Ca^{2+}	Na^+	La^{3+}	Sr^{2+}	Pb^{2+}	Ba^{2+}	K^+
0.983	1.00	1.02	1.032	1.18	1.19	1.35	1.38
F^-	OH^-	O^{2-}	Cl^-	S^{2-}	Br^-	I^-	-
1.33	1.37	1.40	1.81	1.84	1.96	2.20	
C^{4+}	P^{5+}	Si^{4+}	Mn^{5+}	As^{5+}	Cr^{5+}	V^{5+}	Ge^{4+}
-0.08	0.17	0.26	0.33	0.335	0.345	0.355	0.390

[a]Radii are for six-fold coordination, except the bottom row, which is three-fold for carbon and four-fold for the others. Transition metal radii are for high spin states.

Vacancies, mono- and divalent ions

The CaF_2-def FAp single crystals [387,424] mentioned earlier (Section 2.5.1) in connection with hexad axis vacancies presumably also have vacant calcium sites, but the distribution of these between Ca(1) and Ca(2) is not known. However, in $CaCl_2$-def ClAp, XRD studies have shown that both calcium sites have vacancies [421] (Section 2.2.3).

There are many examples of rare-earth apatites that contain monovalent ions or vacancies in place of Ca^{2+} ions, but these will be discussed later under the discussion of trivalent substitutions. K^+ and Na^+ ions, and probably small amounts of NH_4^+ ions, replace Ca^{2+} ions in the apatite lattice in precipitated CO_3Aps (Section 4.5.4). $Sr_{9.402}Na_{0.209}(PO_4)_6B_{0.996}O_2$ has an apatite-like structure with space group $P\overline{3}$ in which only Sr^{2+} ions occupy Ca(2) sites, but strontium alternates with M atoms in Ca(1) sites along the calcium columns, where M is randomly strontium or sodium [436] (Section 2.3.1). In $Pb_8Na_2(PO_4)_6$, Na^+ ions also occupy Ca(1) sites [62]. Single crystal X-ray studies of $Pb_8K_2(PO_4)_6$ have shown that the two K^+ ions per unit cell randomly replace Pb^{2+} ions in the four Ca(1) sites and only Pb^{2+} ions occupy the Ca(2) sites; charge balance is maintained by the empty F^- ion sites [458]. In $Pb_4Na_6(SO_4)_6Cl_2$, caracolite [463], and $Pb_4Na_6(BeF_4)_6F_2$, a fluoroberyllate [464], the four Pb^{2+} ions per unit cell randomly occupy the six Ca(2) sites. In the system Pb,CaOHAp (Section 3.4.4 for synthesis), a break in the variation of the c-axis parameter with the lead content at $Pb_6Ca_4(PO_4)_6(OH)_2$, and a consideration of powder XRD intensities, led to the conclusion that Pb^{2+} ions are located at Ca(2) sites, and Ca^{2+} ions at Ca(1) sites [465]. This result has been confirmed by further powder XRD work [466]. The apatite, $Pb_9(PO_4)_6$, has all hexad axis sites vacant, all six Ca(2) sites occupied by Pb^{2+} ions, and the four Ca(1) sites statistically occupied by only $3Pb^{2+}$ ions, so leaving vacancies in these sites [467]. These results show that in lead apatites, the Pb^{2+} ions have a strong preference for Ca(2) sites (but see below), and Na^+ and K^+ ions and cation vacancies for the Ca(1) sites. The preference of the Pb^{2+} ion for the Ca(2) site has been explained by its ability to form partial covalent bonds [465]. This idea is supported [458] by the short Pb(2)-O(2) distance in $Pb_8K_2(PO_4)_6$ which lies within the range for covalent bonds (2.30 Å in tetrahedral PbO as cited in ref. [458]). The bond lengths for the lead apatite (with those for CaFAp from Section 2.2.1 in brackets) are Pb(2)-O(1) 2.673 (2.814), Pb(2)-O(2) 2.238 (2.384), Pb(2)-O(3) 2.558 (2.344), and Pb(2)-O(3) 2.661 (2.398) Å. The Pb(2)-O(2) distance is conspicuously shorter than the other lead bond lengths, unlike the calcium compound. It has been suggested [458] that the presence of lone pairs of electrons from the Pb^{2+} ion in the vicinity of the hexad axis could explain why the axis sites are vacant. The above deductions on the position of the Pb^{2+} ion were based on mineral and synthetic samples prepared at elevated temperatures. However, Rietveld analysis of X-ray powder patterns from Pb^{2+} ion exchanged powders of OHAp, ClAp and FAp in acid solutions, indicated that Pb^{2+} ions had no preference for Ca(1) or Ca(2) sites [468]. As the lead site occupancies of these samples were rather high (0.52 to 0.86), this result might not apply at lower levels of replacement.

Relative intensities of powder XRD lines showed a preferential occupation of Ca(2) sites by Sr^{2+} ions in Ca,SrOHAp [469], and by Ba^{2+} ions in Ca,BaFAp and Sr,BaFAp [470]. The degree of ordering of the Sr^{2+} ions in the Ca,Sr system decreased, almost linearly, with an increase in strontium content. On the other hand, the preference of Ba^{2+} ions for the Ca(2) sites increased as the barium content increased. This preference was more marked in the Ca,Ba than the Sr,Ba system, and was stronger than the preference of Sr^{2+} ions for the Ca(2) sites in the Ca,Sr system. It was noted [470] that these observations showed that the degree of preferential occupation of the Ca(2) sites by the larger cation increased with an increase in the difference in radii of the two cations (given as 0.16, 0.20 and 0.36 Å for the pairs Ca^{2+}-Sr^{2+}, Sr^{2+}-Ba^{2+} and Ca^{2+}-Ba^{2+}, respectively). The degree of ordering was independent of the substitution of F^- by OH^- ions, and did not change with thermal treatment (annealing or quenching from high temperatures) [469,470]. The independence of ordering with the temperature of quenching was discussed in terms of the opposing thermodynamic driving forces that might produce segregation of the ions [469]. Ordering of cations in the Ca,SrOHAp system has been confirmed by intensity measurements of powder XRD patterns, but no breakdown in Vegard's law could be detected [471]. XRD studies have shown that Ba^{2+} ions have a preference for Ca(2) sites in Ba,CaClAps, a result that was consistent with Raman scattering studies [472] (Section 2.6.2). The relative discriminations of Pb^{2+}, Sr^{2+} and Ba^{2+} ions against Ca^{2+} ions during the precipitation of OHAp are discussed in Section 3.4.4.

Single crystal X-ray studies of Ca,SrClAp showed that at 2.3 and 7.3 % strontium (expressed as percentages of the total occupancies of the Ca sites), all the Sr^{2+} ions were located in Ca(2) sites, but at 48 % strontium, ~68% of the Ca(2) and ~19% of the Ca(1) sites were occupied by Sr^{2+} ions [473]. The Cl⁻ ion position depended on the strontium content, shifting from its position in ClAp (0,0,0.44), to its position in SrClAp (0,0,½) [474] at, or before, 48% of the calcium was replaced by strontium. The lattice parameter changes obeyed Vegard's law. It will be seen that the behaviour of Ca,SrClAp and that of Ca,SrOHAp discussed earlier are the same, both with respect to the initial occupation of Ca(2) sites by Sr^{2+} ions, and the decrease in ordering with an increase in strontium content.

The study of Mg^{2+} ion substitutions is difficult because of the lack of suitable spectroscopic methods, *e.g.* ESR or Mössbauer spectroscopy (^{25}Mg NMR MAS spectroscopy might be feasible, but appears not to have been attempted, JP Yesinowski, personal communication, 1992), the X-ray scattering of magnesium is small, and no single crystals containing sufficient magnesium are available (although it is stated in ref. [380] that single crystals of MgFAp were grown by the Czochralski method, but not investigated). Evidence for magnesium substitution is usually based on small lattice parameter changes, but these could be caused by other minor changes in stoichiometry. The Mg^{2+} ion is rather small (Table 2.1), and as a consequence, it is near, or at the limit of acceptance by the apatite lattice. It occurs in whitlockite and β-TCa,MgP, where it occupies a six-fold coordinated site (Section 1.6.2), but apatite has only seven- and nine-fold coordinated cation sites. However, the nine-fold coordinated Ca(1) site can be occupied by a Mn^{2+} ion, with a concomitant change of the oxygen coordination towards six-fold coordination [475] (see later), and also by an Fe^{2+} ion (see later). As both Mn^{2+} and Fe^{2+} ions occupy the same site as Mg^{2+} in whitlockite, it is reasonable to expect the same in apatite, thus Ca(1) is the most likely site for Mg^{2+} ions in apatite. Apatitic rare-earth oxyapatites with the general formula $Mg_2RE_8(SiO_4)_6O_2$ have been prepared and lattice parameters measured [440]; this formula also suggests that it is the Ca(1) sites that are occupied by Mg^{2+} and RE^{3+} ions, possibly alternating with each other. The relationship between Mg^{2+} ions and apatites has been reviewed [43,118,476].

It has been reported [461] that Mg^{2+} ions can occupy 7 to 9 % of the calcium sites in FAp, based on a small lattice parameter change. Subsequently this was attributed to nonstoichiometry due to the high temperature synthesis used [476]. In another study, a series of FAps with magnesium contents up to 2.43 mmol g⁻¹ were made by precipitation at 80 °C and pH 7.4 [477]. The *a*- and *c*-axes and the crystallinity decreased with an increase in magnesium content. The lattice parameters were *a* = 9.398 and *c* = 6.90 Å for a preparation without magnesium with composition Ca 9.80, P 5.90, and F 1.84 mmol g^{-1}. These parameters compared with $a = 9.384$ and $c = 6.873$ Å for a composition Ca 9.02, Mg 0.457, P 5.57 and F = 1.76 mmol g^{-1}. The precipitate was essentially amorphous for a magnesium content greater than 1.8 mmol g^{-1}. It was suggested that Mg^{2+} ions may substitute into the apatite lattice to a limited extent because a and c decreased with an increase in magnesium content [477]. The appropriateness of the observed change in lattice parameters attributed to Mg^{2+} ions can be gauged from a consideration of other apatites. When Fe^{2+} ions substitute in FAp, both a and c decrease (for 15 mol % replacement of calcium by iron, $a = 9.318$ and $c = 6.833$ Å) [478], and for the larger Sr^{2+} ions in OHAp, a and c increase [53] (Table 1.4). The changes in unit cell volume per Ca^{2+} ion replaced by Mg^{2+}, Fe^{2+} or Sr^{2+} ions can be calculated from the lattice parameters assuming Vegard's law and are: -7.9, -5.3, and +7.20 Å^3 respectively. These values are consistent with the ionic radii (Table 2.1), given the small lattice changes involved, and strengthen the belief that a limited substitution of Ca^{2+} by Mg^{2+} ions in FAp is possible under the preparation conditions used.

In sedimentary apatites (impure forms of CO$_3$FAp, Section 1.1), a clear correlation between the magnesium and carbonate contents has been demonstrated [42,43] (Table 4.3, Section 4.2.1). Thus, $y = 0.463 \times z/(6 - z)$ with $R^2 = 0.928$, where y and z were the moles of magnesium and carbonate respectively per unit cell and R is the correlation coefficient [43]. This relationship implies that without CO$_3^{2-}$ ions in the lattice, Mg^{2+} ion substitution in FAp does not occur. However, for samples with $z \approx 0$, y ranged from 0.01 up to 0.07; the maximum y was ~0.17 for $z/(6 - z) \approx 0.2$. It was suggested that Mg^{2+} and CO$_3^{2-}$ ion substitutions were linked to enable the structure to physically compensate one substitution for the other [42,43]. The presence of CO$_3^{2-}$ (and F$^-$) ions also increases the uptake of Mg^{2+} ions in OHAp synthesised in aqueous systems [479]. The partial replacement of Ca^{2+} by Mg^{2+} ions has been proposed as one of the factors to explain the short c-axis parameter found in shark dentine (c = 6.84±0.01 Å) [383] (Section 4.6.2).

The total replacement of Ca^{2+} by Mg^{2+} ions in a OHAp precipitated in solution at high pH has been reported [480,481]. Systematic changes in the IR spectrum and lattice parameters (from $a = 9.429$, $c = 6.884$ to $a = 9.298$, $c = 6.886$ Å) were published, as the Ca^{2+} ions were completely replaced by Mg^{2+} ions [481]. An attempt to confirm this work was unsuccessful [476].

The presence of Mg^{2+} ions during the precipitation of apatites leads to lower crystallinity, but the precipitate only contains a limited amount of magnesium (< 1 %), [118,269,479]. ^{28}Mg exchange from OHAp precipitated under physiological conditions in the presence of Mg^{2+} ions and aged for 21 days, showed that nearly 90% of the magnesium was located in readily exchangeable (surface) positions [482]. Magnesium-containing OHAps have been prepared by precipitation and hydrolysis methods that suggested that a limited (up to 0.3 wt %) replacement of Ca^{2+} by Mg^{2+} ions in OHAp was possible [479]. The lattice parameters were $a = 9.438$, $c = 6.885$ Å in the absence of magnesium, $a = 9.423$, $c = 6.876$ Å for Mg 0.14 wt %, and $a = 9.417$, $c = 6.867$ Å for Mg 0.27 wt %. As with FAp (above), both axes decrease with an increase in magnesium content. The reduction in unit cell volume per Ca^{2+} ion replaced by a Mg^{2+} ion is about 30 Å^3, which is very much larger than expected on the basis of the change in volume of the ions. This result suggests that other changes are also occurring (perhaps a change in lattice water). In summary, it seems that in both OHAp and FAp, a limited replacement of Ca^{2+} by Mg^{2+} ions is possible, probably in Ca(1) sites.

The solubility limits for cobalt replacing calcium in FAp and ClAp are 15 and 25 mol % respectively, but less than 10 mol % in the strontium analogues [483]. In ClAp with ~20 % replacement of Ca^{2+} by Co^{2+} ions, single crystal XRD studies indicated that all the Co^{2+} ions were located at Ca(2) sites, as the authors expected [484]. Most of the chlorine, Cl(1), was at 0,0,0.4370(11) with an occupancy of 0.79(1). As the Co^{2+} ion is smaller than the Ca^{2+} ion (Table 2.1), most of the remaining chlorine, Cl(2), was shifted along the c-axis towards the Co^{2+} ion to maintain a reasonable Co-Cl(2) bond distance (2.50 Å). Cl(2) was at $(0,0,¼)$ with an occupancy of 0.10(2). The sum of the occupancies indicated an 11 % deficiency in chlorine, but there was evidence for further Cl$^-$ ions in disordered positions, so the deficiency was probably less. There was no indication of a doubled b-axis parameter as in stoichiometric ClAp (Section 2.2.3), so the presence of the cobalt (and possibly a slight chlorine-deficiency) prevented the ordering of the Cl$^-$ ions [484].

Mössbauer spectroscopy has been used to determine the location of $^{57}Fe^{2+}$ ions in substituted FAp [478]. At low levels (<1 % replacement of Ca^{2+} by Fe^{2+}), there was essentially a random distribution between the Ca(1) and Ca(2) sites, but at higher levels towards the solubility limit of Fe^{2+} ions in apatite (~15 % replacement of Ca^{2+} by Fe^{2+}), most of the Fe^{2+} ions went to Ca(1) sites. Quenching from 700 °C did not change the distribution. Chemical shift measurements showed that the effective charge on Fe^{2+} ions in Ca(2) sites was about 0.15 electrons larger than for Fe^{2+} ions in Ca(1) sites. It was thought that, because the Fe^{2+} ion was more electronegative than the Ca^{2+} ion, Fe^{2+} ions preferentially entered sites where its covalent bonding was stronger, *i.e.* into the Ca(1) site. (Note that this seems to be inconsistent with ideas for lead apatites, see earlier discussion.) It is interesting to note that it is the Ca(1) site in apatite that has a similarity with the Ca(5) site in whitlockite [278] (Section 2.2.2) which is the site occupied by Fe^{2+} ions in whitlockite [278] (Section 1.6.2).

The substitution of manganese in apatites has been very extensively investigated because of its commercial importance as a constituent (up to 0.5

wt %) of halophosphate phosphors in fluorescent light tubes (Sb^{3+}, as an activator, is another important constituent, see trivalent substitutions, this section). The situation revealed is one of considerable complexity. Flux-grown ClAp has been doped with $MnCl_2$ (0.031 and 0.0033 wt %) and, because chemical analyses indicated a valency state of +5, it was thought more likely that manganese was present as MnO_4^{3-} replacing PO_4^{3-} ions [485]. SrClAp doped with MnO_4^{3-} ions has also been studied [486]. On the other hand, it has been shown that Mn^{2+} can replace Ca^{2+} ions in FAp [487-489] (and papers cited in these refs). Single crystal XRD shows that $Mn_5(PO_4)_3Cl_{0.9}(OH)_{0.1}$ is isostructural with FAp [399].

With approximately 10^{-4} to 10^{-1} of the calcium atoms replaced by manganese atoms, ESR and optical studies of single crystals of melt-grown FAp of variable stoichiometry have been particularly informative [488,490-492]. Mn^{2+} ions in Ca(1) and Ca(2) sites (designated Mn(1) and Mn(2) respectively) can be identified in FAp grown from melts with excess CaF_2, so that it is stoichiometric [488]. At low concentrations (Mn/Ca = 10^{-4} in the crystal), there was very little Mn(2), but at a higher ratio (0.3), about a quarter of the manganese was present as Mn(2). If the manganese-doped FAp was grown from stoichiometric melts, so that there was up to a 10% deficiency of CaF_2 in the crystal, the signal from Mn(2) was very weak, and was replaced by a modified signal, identified as a Mn^{2+} ion in a modified Ca(2) site, designated Mn(2m) [492]. As the Mn/Ca ratio in the melt was increased from 10^{-3} to 10^{-1}, the Mn(2m) signal decreased, and that from Mn(2) increased (the Mn(1) signal also increased, and was always greater than the Mn(2) signal). It was proposed [492] that there was a defect (called X) at a concentration of about 10^{18} cm^{-3} in FAp grown from a stoichiometric melt that was associated with a Mn^{2+} ion to form Mn(2m). As the manganese concentration increased, this association continued until there were no more X defects available, after which Mn(2) increased. The X defect was thought to be the $(VOV)^+$ defect that had been identified earlier [424] in ns-FAp. This defect was a complex of a linear arrangement of an O^{2-} ion flanked by an F$^-$ ion vacancy on either side (other similar types of complex defects probably occurred). Thus Mn(2m) would be an Mn^{2+} ion in an Mn(2) site adjacent to one of the F$^-$ ion vacancies of the $(VOV)^+$ complex. The behaviour of manganese-doped melt-grown SrFAp was rather similar, but not identical, to that described for CaFAp, there being an even greater preference for Mn(1) over Mn(2) [491]. It was suggested that the Ca(2) site was "too large" for the easy accommodation of a Mn^{2+} ion, and that this was even more so in SrFAp with its increased lattice parameters.

It is difficult to make ESR measurements at higher manganese concentrations because of line broadening, but the preference of Mn(1) over Mn(2) at higher concentrations has also been seen from IR [97] and neutron powder Rietveld analysis [475]. It was concluded that Mn(1) occurs preferentially (or exclusively) in FAp containing up to one Mn^{2+} ion per unit cell prepared by precipitation, followed by calcination at 1173 K [97]. This idea was deduced from the presence of an additional manganese perturbed v_1 PO_4 IR band about 10 cm^{-1} lower than the unperturbed v_1 PO_4 band at 962 cm^{-1} (Table 2.2). This perturbed band increased in intensity with an increase in manganese content. The origin of this perturbed band is explained below.

It has been shown [475] from Rietveld structure refinements of neutron powder diffraction data that Mn^{2+} ions go essentially to a subset of the Mn(1) sites, Mn(1a), at ⅓,⅔,z ($z \approx 0$) with a lowering of the space-group symmetry to $P6_3$, or perhaps $P3$. The material investigated was similarly prepared to that used for the IR studies, and contained 0.448 Mn^{2+} ions per unit cell. The structure determination showed that one Mn(1a) interacted with all six PO_4^{3-} ions in a unit cell, rotating them slightly. This rotation brought the six closest oxygen atoms (three O(1) and three O(2), Section 2.2.1 and Fig. 2.2) even closer to the Mn^{2+} ion, and moved the three more distant O(3) atoms further away. The rotation of only a fraction of the PO_4^{3-} tetrahedra explained the simultaneous presence of an unperturbed and perturbed v_1 band, and the reluctance of the structure to accept more than one Mn^{2+} ion per unit cell (this would require some counter rotations which are not possible) [475]. Interestingly, the structural changes that occur when calcium is replaced by manganese bring the environment for the Mn^{2+} ion closer to the octahedral coordination of Mg^{2+} and Mn^{2+} ions that is found for these ions in whitlockite (Section 1.6.2).

The breakdown of Vegard's Law for $Eu_{5-x}Ba_x(PO_4)_3F$ shows there is a segregation of Eu^{2+} and Ba^{2+} ions, but structural details are unknown [493].

Several cadmium apatites, some of which do not have the usual $P6_3/m$ apatite space group, have been prepared and investigated by XRD or IR [55,430,431,494-496].

Trivalent ions

The rare-earths (REs, lanthanides) and yttrium have radii that are close to that of the Ca^{2+} ion (Table 2.1), and are therefore readily accepted by the apatite lattice, and indeed are often associated with mineral apatites [497] (see, for example, Formula 2.1 for Durango FAp and Formula 4.10 for Magnet Cove francolite). Mechanisms of charge balance for trivalent rare-earth ions include replacing an F^- ion by one or more divalent ion(s) (O^{2-} [62,380,440-448] or S^{2-} [451,452]), or replacement of a Ca^{2+} ion by a monovalent ion (Li^+ [441]; Na^+ [405,441,498-501]; Ag^+ [502]) or by a vacancy [441,443,498], or replacement of PO_4^{3-} by SiO_4^{4-} ions [380,440-442,446-448,483,503,504]. Combinations of these also occur. The distribution of rare-earth ions between Ca(1) and Ca(2) sites is likely to depend on the charge compensation mechanism. Single crystals of some Ca and SrOAps with substantial amounts of rare-earths have been grown [380] (Section 2.5.1).

In $(Ca,RE)_{10}((P,Si)O_4)_6F_2$, rare-earth atoms preferentially occupy Ca(2) sites, but this segregation decreases to zero, as the silicon content increases, until all the phosphorus is replaced; it also decreases as the temperature is raised to 1100 °C (investigated by high temperature XRD) [505]. In $Sm_8Cr_2Si_6N_2O_{24}$ (space group $P6_3$, a = 9.469(5) and c = 6.890(4) Å), Cr^{3+} and Sm^{3+} ions randomly occupy the two different Ca(2) sites without Cr^{3+} ions in Ca(1) sites [446].

Single crystal XRD studies of FAp doped with 1.4 wt % neodymium showed that the Nd^{3+} ions are located exclusively at Ca(2) sites if the dopant vehicle is Nd_2O_3, but at Ca(1) and Ca(2) in approximately equal atomic fractions if doped with NdF_3 [506,507]. These results correlated well with earlier conclusions [379] that, when Nd_2O_3 was used, Nd^{3+} ions were located at Ca(2) sites because none of the lines in the fluorescence or absorption spectra were 100 % polarised (many would have been for the Ca(1) site). Optical absorption and ESR studies [508] of CaF_2-def neodymium-doped FAp prepared from stoichiometric melts (and therefore probably with O^{2-} ions in place of some F^- ions) showed that Nd^{3+} ions occupied Ca(2) sites, with local change compensation from an adjacent O^{2-} ion. This type of substitution was not seen in a crystal grown with a CaO deficiency, and therefore with insufficient O^{2-} ions in the lattice for this charge compensation mechanism; instead, most of the Nd^{3+} ions occupied Ca(1) sites. In all, five different types of neodymium substitutions in FAp grown under different conditions could be identified [508].

The location of the Sb^{3+} ion in commercial halophosphate phosphors is of considerable interest because of its role as an activator to absorb UV radiation produced by the mercury discharge in fluorescent light tubes (this is reradiated as visible energy by Mn^{2+} ions whose location in the apatite structure was discussed earlier). The Sb^{3+} ion, unlike Mn^{2+}, is not paramagnetic, so its position in the apatite lattice cannot be investigated by ESR. As a result, its position has been the subject of some considerable uncertainty and Ca(1), Ca(2) and P sites have been discussed [509,510]. Replacement of Ca(1) by Sb^{3+} ions was the first site to be considered, however, no direct evidence for this substitution has been found [510]. Ca(2) sites were later proposed; optical emission studies of O^{2-}- and Na^+-compensated antimony-doped ClAp and similarly doped FAp led to the suggestion that it is very likely that, in the O^{2-}-compensated apatite, the Sb^{3+} ion is located in this site with charge balance maintained by an adjacent O^{2-} ion [377]. Part of the evidence for this was that the emission spectrum depended little on whether the halide ions were F^- or Cl^-. It was noted that this interpretation of the emission results was consistent with IR studies (personal communication from J Paynter and FI Ewing in ref. [377])

of a band at ~685 cm^{-1}. This band, assigned to a Sb-O vibration from antimony chemically bonded to oxygen, was present in O^{2-}-compensated apatites, but absent for Na^{+}-compensation. More Sb^{3+} ions could be incorporated in Na^{+}-compensated apatites (formula of limiting compound, Ca$_{9.8}$Na$_{0.1}$Sb$_{0.1}$(PO$_4$)$_6$F$_2$, note that charges do not balance), but the location of the Sb^{3+} ion in this case was less certain, but thought to be also in Ca(2) sites.

Rietveld powder XRD refinements of FAp with 2.2 wt % Sb, were consistent with an Sb^{3+} ion replacing Ca(2) [127,511]. The charge compensation mechanism for this sample was not established, but IR and Mössbauer spectroscopy and luminescence studies [511] showed that the Sb^{3+} ions were essentially in the same sites as for the smaller amounts used as an activator in commercial phosphors. Thus, it seemed most probable that the Sb^{3+} ion charge was compensated by O^{2-} ions, so the Sb^{3+} ion location from the X-ray studies appeared to be the same as that deduced from the optical emission studies cited above. Rietveld analysis of another FAp that contained 3.1 wt % Sb and which differed in properties from commercial phosphors, showed no evidence of Sb^{3+} ions replacing Ca(2), but rather occupying sites at ⅓,⅔,¼ and ⅔,⅓,¼ [511].

The evidence for the popular view for the location of Sb^{3+} in Ca(2) sites discussed in the previous two paragraphs is in direct conflict with two recent studies using ^{121}Sb Mössbauer [509] and NMR MAS spectroscopy of ^{19}F [510,512] and ^{31}P [510]. The Mössbauer spectra of ^{121}Sb in antimony-doped FAp and Cl,FAp were well-defined and characteristic of Sb^{3+} compounds, and allowed the local electronic environment of the antimony nucleus to be investigated via determination of the nuclear quadrupole interaction parameters and isomer shift data [509]. The experimental quadrupole parameters were compared with calculated values for models of Sb^{3+} in a Ca(1) or Ca(2) site, and SbO$_3$$^{3-}$ ions replacing PO$_4$$^{3-}$ ions. The experimental values for the Ca(1) site were quite inconsistent with an Sb^{3+} ion in this position and, in particular, indicated that the ion was in a site that did not have three-fold symmetry. For the Ca(2) model, the calculated quadrupole coupling constant was an order of magnitude smaller than the experimental value, although the asymmetry parameter was a better fit than for the Ca(1) model. Variants of the Ca(2) model did not improve the fit of the quadrupole coupling constant. Comparison of calculated and experimental values for the PO$_4$$^{3-}$ site model showed much better agreement, particularly for the quadrupole coupling constant. The fit of the quadrupole interaction parameters was made nearly perfect if an F$^-$ ion was added to form a fourth ligand for the Sb^{3+} ion, and the Sb-O and Sb-F distances were reduced from 2.10 Å, the P-O distance in the PO$_4$$^{3-}$ ion, to 1.88 Å as found for Sb-O in a number of inorganic compounds. Thus, it was concluded that Sb^{3+} ions substituted in FAp and F,ClAp as SbO$_3$$^{3-}F^-$ [509].

The ^{31}P NMR MAS (6.15±0.1 kHz) studies [510] at 161.76 MHz of FAps containing 0.0 to 3.0 wt % of antimony showed a strong ^{31}P resonance in all the samples with an isotropic chemical shift of 2.8 ppm downfield from 85 % H_3PO_4, typical of undoped FAp; no additional peaks attributable to the substitution of Sb^{3+} for Ca^{2+} ions could be seen (an additional peak was seen if small amounts of Sr^{2+} replaced Ca^{2+} ions). These observations seemed to rule out the possibility of Sb^{3+} ions in either Ca(1) or Ca(2) sites as both these, and the phosphorus atoms, have common coordinating oxygen atoms. On the other hand, substitution of PO_4^{3-} by SbO_3^{3-} ions would not significantly perturb adjacent PO_4^{3-} ions. ^{19}F NMR MAS (8.25±0.1 kHz) at 376.2 MHz of the samples showed, in addition to the main resonance 64.0 ppm downfield from C_6F_6 typical of F$^-$ ions in FAp, a shoulder at 65.6 ppm and a sharp peak at 68.6 ppm [510,512]. Irrespective of the particular specimen, the spin-lattice relaxation times (T_1) of these antimony related peaks were always the same as the main peak and varied from 129 to 378 s. This observation indicated that both the additional 65.6 ppm shoulder and 68.6 ppm peak originated from antimony perturbed F$^-$ ions in an apatitic lattice. Quantitative studies showed that these features were associated with one and two Sb^{3+} ions per unit cell, respectively. Spin diffusion measurements showed the presence of cross-relaxation of the 68.6 ppm peak to the main peak at 64 ppm, but not to the shoulder at 65.6 ppm, which meant that each Sb^{3+} ion perturbed F$^-$ ions in at least two different chains. Although the ^{19}F NMR results did not explicitly rule out the Ca(1) position for the Sb^{3+} ion, there were arguments against this based on the ^{19}F NMR results and, particularly, the Mössbauer work [509] cited above. However, if an SbO_3^{3-} ion replaced a PO_4^{3-} ion, a model could be devised for the location of the perturbed F$^-$ ions that allowed the NMR peaks to be assigned and which was consistent with the other NMR results. Thus the NMR work provided detailed support for this substitution. However, no consideration was given to the suggestion based on the detailed interpretation of the Mössbauer results [509] (see above) that an F$^-$ ion provided a fourth ligand.

The locations of rare-earth ions in a number of oxyapatites and dioxyapatites have been studied. Attempts have been made to determine these from discontinuities in plots of the cell parameters as a function of the radius of the rare-earth ion [441,511]. Considerations of local charge compensation suggest [460] (see also the introduction to this section) that the rare-earth ion should preferentially replace Ca(2) because it is adjacent to an O^{2-} ion. This substitution takes place in a number of apatites. For example, single crystal XRD studies of $Ca_2La_8(SiO_4)_6O_2$ showed that the La^{3+} ions occupied all the Ca(2) sites, whilst the Ca(1) sites were randomly occupied by La^{3+} and Ca^{2+} ions [448]; the O^{2-} ion was at 0,0,¼, and a short La^{3+}-O^{2-} distance of 2.305 Å (0.16 Å less than the sum of the ionic radii) indicated a very strong bond. A short Eu^{3+}-O^{2-} bond has also been reported in $Ca_8Eu_2(PO_4)_6O_2$ (RA Young cited in ref. [443]). This strong RE-O^{2-} bond is also responsible for the IR [444,445] and Raman [443] bands located in the range 650 to 530 cm^{-1} seen in various rare-earth oxyapatites (recall that a similar IR band was reported above for O^{2-}-compensated Sb^{3+} halophosphates). The same local charge compensation mechanism found in rare-earth oxyapatites seems to be responsible for the exclusive substitution of Eu^{3+} ions in Ca(2) sites in a sulphoapatite, $Ca_{10-x}Eu_x(PO_4)_6S_{1+x/2}$, which has S^{2-} ions distributed statistically between the positions 0,0,0.47 and 0,0,0.97 [451], and probably also in the corresponding Sr,Eu sulphoapatite [452].

Conclusions

It does not seem possible to state general rules that will predict the distribution of substituting ions between the Ca(1) and Ca(2) sites. Clearly there is, in general, little to choose between them from an energetic point of view; note particularly how the distribution sometimes changes with the concentration of substituting ion and, in one case, with temperature. In the absence of specific effects, there seems to be a tendency for smaller and/or monovalent ions to go to Ca(1) sites, and therefore the apparent difference in coordination between the two sites is not a good guide (in this context, see the comment above about the change in the Ca(1) site when occupied by a Mn^{2+} ion). Sometimes, there is an ordered substitution of the Ca(1) ions which results in a lowering of the space-group symmetry. For restricted classes of substitution, *e.g.* the alkaline earths, it seems possible to formulate general rules based on the sizes of the ions. In some specific cases, there seems to be clear reasons why one site is favoured over another, for example, the tendency of lead to form partial covalent bonds in Ca(2) sites, and the requirement for local charge compensation that can occur for trivalent ions in Ca(2) sites if divalent ions are available on the hexad axis. Finally, it is clear that even if Vegard's law is accurately obeyed, there can still be significant segregation of substituting ions between the two sites.

2.3.3 Substitutions for phosphate ions

Apatites have been synthesised with partial or total replacement of PO_4^{3-} ions by VO_4^{3-} [57,442,461,494,495], AsO_4^{3-} [442,461,494,495,513,514], MnO_4^{3-} [485], CrO_4^{3-} [402,422,515-519], GeO_4^{4-} [422,423], SiO_4^{4-} [380,422,423,440-442,448,461,503,520], SiO_3N^{5-} [446,447,504,521-524], SO_4^{2-} [85,422,455,461,463,525-527]; SeO_4^{2-} [525] and ReO_5^{3-} ions [433,438,528-530]. Mineral examples of some of these substitutions occur (see Table 5.5 in ref. [2]), of which ellestadite, $Ca_{10}(SiO_4)_3(SO_4)_3(OH,F,Cl)_2$, was one of the earliest examples to be described [531]. CO_3^{2-} ions can partially replace PO_4^{3-} ions in mineral apatites (Section 4.2), synthetic apatites made at high temperatures

(Section 4.4), apatites formed in aqueous systems (Section 4.5) and in biological apatites (Section 4.6). As CO_3^{2-} ions can also be located on the hexad axis (Section 4.3.3), factors that determine the distribution of CO_3^{2-} ions between the two anion sites are of interest (ref. [532], Section 4.1.2; and ref. [533], Section 4.5.3). It has been suggested that CO_2 can occupy the PO_4^{3-} ion site in heated dental enamel [534] (Sections 4.5.7 and 4.6.4). Apatite precipitated in the presence of SO_3^{2-} ions have been found to contain 10 wt % SO_3^{2-} ion which was thought to replace PO_4^{3-} ions in the lattice [85]. As discussed in the previous section, there is recent good evidence that Sb^{3+} can occur in the apatite lattice as SbO_3^{3-} ions substitution for PO_4^{3-} ions, rather than as Sb^{3+} ions in Ca(2) sites [509,510,512].

2.4 Preparation of powders
2.4.1 High temperature methods
Fluorapatite

FAp can be prepared by heating an intimate stoichiometric mixture of CaF_2 and TCP at 900 °C for several hours [75,76]. A better crystallised product is obtained if the reaction is carried out at 1370 °C for 30 min in a current of dry N_2 with CaF_2 upstream to reduce the volatilisation of fluorine (as CaF_2) [16]; slight loss of CaF_2 from FAp is otherwise possible [389]. Refractive indices for sodium light were O = 1.633 and E = 1.629 (uniaxial negative) with lattice parameters $a = 9.367$ and $c = 6.882$ Å [16]. If the CaF_2 upstream to the sample was omitted, the FAp was low in F with O ≈ 1.627 and E ≈ 1.623 (sodium light) [16]. This example shows that refractive index measurements are a good indicator of nonstoichiometry.

FAp has also been prepared by heating, at 900 °C, a mixture of $Ca_2P_2O_7$ and CaF_2 with the evolution of POF_3 gas, but the product had a slight excess of fluorine (~5 wt % rather than the theoretical 3.7 wt %), and a slight deficiency of calcium [76]. The a-axis parameter was slightly larger (≤ 0.01 Å), and density slightly less, than for s-FAp; this was explained by the presence of PO_3F^{2-} ions and charge compensating Ca^{2+} ion vacancies in the lattice [76]. FAp with an even greater excess of fluorine (~10 wt %) has been reported on heating the calculated quantities of $Ca_3(PO_4)_2$ and CaF_2 with the required composition in a sealed platinum tube at 1040 to 1110 °C for periods of seven to nine days [535]. XRD surprisingly showed no evidence of changed lattice parameters or the presence of CaF_2, and single crystal Weissenberg XRD photographs showed no evidence for a change in space group. It was suggested that the excess CaF_2 was located along the c-axis. An apatitic compound of formula $Ca_7(PO_4)_4F_2$ with lattice parameters only slightly different from FAp and melting point 1710 °C (*i.e.* about 100 °C higher than FAp) has been reported [536]. (Fluorine appeared to be lost on thermal treatment of $Ca_7(PO_4)_4F_2$ and FAp, but the analyses on which this observation was based were subsequently found to be in error, J Berak, personal communication, 1974.) Although these reports of preparations of FAp with such a large excess of fluorine do not seem to have been confirmed, OHAp with an apparent large excess of $Ca(OH)_2$ has recently been reported [537] (Section 3.4.2). Furthermore, an excess of fluoride is often found in mineral [42] (Section 4.2.1) and in some synthetic precipitated CO_3FAps [99,538] (Section 4.5.5, Formulae 4.30 and 4.31).

Chlorapatite

s-ClAp with 10 to 60 μm crystals can be prepared (R Hickock, personal communication in ref. 391) by annealing chlorspodiosite, Ca_2PO_4Cl, in argon at 1200 °C for several hours, followed by rapid cooling. The chlorspodiosite was prepared by heating TetCP (synthesised by a solid state reaction at 1100 °C between DCPA and $CaCO_3$, Section 1.7) in anhydrous HCl gas at 800 °C [391]. Chlorspodiosite melts incongruently at 1040 °C to yield ClAp and $CaCl_2$ [539]; $CaCl_2$ acts as a flux to produce extremely well-formed, stoichiometric crystals, free of inclusions and extraneous phases [391]. Rapid cooling is required to prevent reconversion to chlorspodiosite. The chemical analysis was chlorine 6.81 and calcium 38.5 wt % (theoretical 6.81 and 38.5 wt % respectively). The ClAp synthesised by this method is monoclinic (Fig. 2.8) with pseudohexagonal lattice parameters $a = 9.634$ and $c = 6.763$ Å (derived from Equations 2.3 with $x = 0$ [391]). ClAp can also be made by heating $3Ca_3(PO_4)_2 + CaCl_2$ or $OHAp + CaCl_2$ at 800 °C and Cl,FAps by heating ClAp and FAp in the required proportions at 800 °C [75].

CaCl$_2$-def ClAp, $Ca_{10-x}(PO_4)_6Cl_{2-2x}$, with a maximum $x = 0.23$ can be prepared [391] by heating s-ClAp (*e.g.* the material described in the previous paragraph) *in vacuo* at 900 to 1200 °C for 16 h. The lattice parameters (mean deviation ±0.002Å) were given by the equations:

$$a = 9.634 - 0.1934x \text{ Å}$$
$$c = 6.763 + 0.0802x \text{ Å}.$$

2.3

It was reported that the limit for x for the structure to remain monoclinic was probably somewhat less than 0.07. After heating for 32 h at 1000 °C, the small birefringence in the plane perpendicular to the c-axis (typical of the monoclinic structure) had disappeared, and the birefringence in the plane containing the c-axis had increased ten-fold to -0.0030 [391] (see also optical properties of ClAp in Section 2.7).

The preparation of solid solutions of ClAp, FAp and OHAp is discussed in Section 3.4.4.

2.4.2 Solution methods

Agitated aqueous suspensions of any CaP and CaF_2 or NaF lead to the formation of FAp or F,OHAp [76]. Acicular to well-formed hexagonal prisms of FAp have been prepared by the equilibration of H_3PO_4, α-TCP, MCPM and CaF_2 at 50 °C [540]. The reaction between OHAp and F⁻ ions in solution, with the formation of FAp, F,OHAp and CaF_2, is discussed in Section 3.10.2. In contrast to the extensive investigations of ns-OHAps (Section 3.7.3), there appear to be no detailed studies on the stoichiometry of FAps from aqueous systems, except for CO_3FAps (Section 4.5.5).

It has not been possible to synthesise ClAp in aqueous systems under normal temperatures and pressures because of the strong discrimination against Cl⁻ ions by OH⁻ ions; however, Cl,OHAps can be prepared (Section 3.4.4).

2.5 Growth of single crystals

2.5.1 Fluorapatite

Hexagonal FAp prisms, up to 2 mm long and 0.3 mm across, can be grown by slow cooling of a FAp melt (mp ~1650 °C) accompanied by seeding at the appropriate temperature [541]. Much larger crystals (1 cm diameter by 5 cm long) have been grown by the Kyropoulos method in which an 80% platinum-20% rhodium rod was pulled at 10 mm h⁻¹ from a melt of FAp in an induction heated iridium crucible [541].

FAp crystals up to 30 cm long, and for shorter lengths up to 1.9 cm wide, have been grown from the melt by the Czochralski method [378,387]. The crystals were grown [378] in an argon atmosphere from a seed crystal pulled from an induction heated iridium crucible at rates of 3 to 5 mm h⁻¹ and rotations of 30 to 100 revolutions per minute. The melts used had compositions at, or near, stoichiometry; crystals could be grown along [001] or [100], depending on the orientation of the seed [378]. Chemical analysis of crystals grown from a stoichiometric melt appeared to have a 5% deficiency of CaF_2 [387]. These crystals had $a = 9.3697\pm0.0004$ and $c = 6.8834\pm0.0008$ Å, which gave a calculated density of 3.200 g cm⁻³ for a stoichiometric compound or 3.187 g cm⁻³ assuming a 5 % deficiency of CaF_2. The latter agreed with the measured density of 3.189 ± 0.003 g cm⁻³, thus confirming the 5 % deficiency of CaF_2 in the unit cell. The refractive indices for sodium light were O = 1.637 and E = 1.631. Note that these values for O and E are significantly higher than those reported earlier for ns- and s-FAps [16] (Section 2.4.1) and for the flux-grown FAp discussed below. The Czochralski method has also been used to grow large crystals of SrFAp, $Ca_8La_2(PO_4)_6O_2$, $Ca_2Y_8(SiO_4)_6O_2$, $Ca_2La_8(SiO_4)_6O_2$ and $Sr_2La_8(SiO_4)_6O_2$ [380].

The difficulty of growing s-FAp has been solved by growing the apatite from a flux of CaF_2 [406]. 45 g of CaF_2 and 55 g of FAp were heated in a platinum crucible which was covered by platinum gauze and an inverted platinum crucible to form a closed system. The crucible was heated at about 50 °C above the liquidus temperature (1325 °C) for 10 to 20 h, and then cycled up and down some 20 °C below this temperature so that the smallest crystallites dissolved, leaving only a few larger ones to act as seeds. The melt was then cooled at a linear rate of 2 to 4 °C h^{-1} from 1375 to 1220 °C. The FAp crystals were then separated by inverting the furnace so that the molten CaF_2 drained through the gauze. After cooling, the remaining CaF_2 was dissolved in 20 % aqueous boiling $Al(NO_3)_3.9H_2O$. Hexagonal prisms, a few mm in size, were grown with F, Ca and PO_4 contents within 0.4, 0.1 and 0.1 wt % respectively of the theoretical. The lattice parameters were $a = 9.364\pm0.001$ and $c = 6.89\pm0.05$ Å, and the birefringence was -0.0019 [406]. Refractive indices of O = 1.633 and E = 1.629 (presumably sodium light) for these crystals have been reported [46]. These values agree with those for s-FAp powders reported earlier [16] (Section 2.4.1). A similar, but less elaborate, flux method has been used [402].

FAp has been grown in a hydrothermal bomb from an apatitic CaP and CaF_2 heated at 973 K and 0.75×10^8 N m^{-2} for 72 h (variants of this charge were also used) [542].

The hydrothermal method has been used to prepare manganese-, cerium- and europium-doped FAp [542]. Manganese- [541] and neodymium- [378] doped FAp have been grown from the melt, and chromium- [402] doped crystals from the flux.

2.5.2 Chlorapatite

s-ClAp cannot be grown from the melt because of the severe loss of $CaCl_2$ at the melting point (~1925 K), however $CaCl_2$-def ClAp can be successfully grown by this method [543]. Such $CaCl_2$-def ClAp has a 17.3% deficiency of chlorine determined by chemical analysis, which gave $a = 9.601$ and $c = 6.777$ Å from Equations 2.3 [391]. The measured density at 26.7 °C was 3.14 g cm^{-3} which agreed with the calculated density (3.13_7 g cm^{-3}) assuming that Ca^{2+} and Cl^- ion vacancies were formed from the loss of $CaCl_2$ [391]. Aspects of the structure (Section 2.2.3) and lattice parameters ($a = 9.615$ and $c = 6.771$ Å, $\sigma = 0.002$ Å) have been reported [421] for melt-grown $CaCl_2$-def ClAp, almost certainly from the same source as discussed above. These measured parameters differ significantly from those calculated above for reasons that are not known.

The flux method described above for FAp has been used to grow s-ClAp with dimensions of 3 to 4 mm [406]. As a consequence of the water solubility of the $CaCl_2$ flux, a single tightly closed platinum crucible could be used instead of the inverting furnace and two crucibles, as used for FAp. A charge of 60 g $CaCl_2$ (prepared by dehydrating $CaCl_2.2H_2O$ in a stream of anhydrous HCl gas at 500 °C) and 23.5 g ClAp was used. This mixture was kept at 50 °C above the liquidus temperature (1230 °C) for 10 to 20 h, and then cooled at a linear rate of 2 to 4 °C h^{-1} to 1060 °C. To prevent the formation of chlorspodiosite, Ca_2PO_4Cl, growth was terminated at this temperature and the furnace switched off. The chemical analysis was: Cl 6.72, Ca 38.5 and PO_4 54.2 wt % (theoretical 6.81, 38.5 and 54.7 wt % respectively). The crystals were monoclinic with a doubled *b*-axis parameter and pseudohexagonal parameters of *a* = 9.640 and *c* = 6.73 Å. The crystals were biaxial with mimetic twinning seen in sections cut perpendicular to the *c*-axis. However, unlike the situation for monoclinic OHAp (Section 3.5), the monoclinic single crystal regions were sometimes more than 1 mm across. The fact that flux-grown crystals are monoclinic is an indication that the chlorine content is high enough to ensure that an ordered arrangement of these ions occurs (Section 2.2.3). The purification of starting materials required for flux-growth of ClAp with a minimum of oxygen impurities has been described [91]. The refractive indices of flux-grown ClAp crystals are discussed in Section 2.7.

ClAp crystals up to 1.5×1.5×2 mm^3 have been grown hydrothermally in HCl solution at pH 1 and 50,000 psi (344.7 MPa), with the growth zone at 465 °C and nutrient ClAp crystals at 360 °C [544] (see also similar growth method for OHAp [51,545], Section 3.5). The crystals were monoclinic with mean refractive index of 1.674 (sodium light) and chlorine content 6.78 wt % (theoretical 6.81 wt %). No ESR signal could be detected after X- and γ-irradiation (contrary to the situation for $CaCl_2$-def ClAp and flux-grown crystals), indicating a higher degree of perfection [544]. Refractive index studies [48] (Section 2.7) also indicate that hydrothermally grown crystals are generally purer than flux-grown ones. Crystals (0.25 to 1 mm) of $Ca_{10}(PO_4)_6Cl_{2-x}(OH)_x$, with *x* from 0.41 to 1.1, have been grown hydrothermally by the controlled hydrolysis and subsequent transport of ClAp powder by varying the HCl concentration in the mineralising fluid from 0.1 to 0.01 mol l^{-1} [417]. The crystals were monoclinic for *x* = 0.41 and 0.56, but hexagonal for *x* ≥ 0.74 [417].

Hydrothermal growth methods have been used for doping ClAp with a variety of elements [546]. Flux-growth has been used for doping with sodium or gadolinium [91].

2.6 Infrared and Raman spectra
2.6.1 Fluorapatite

As a first approximation, it is to be expected that the IR and Raman spectra of the PO_4^{3-} ion in FAp would be similar to that of the isolated PO_4^{3-} ion of symmetry T_d. The normal modes of this ion are given in Table 2.2.

This approximation cannot be very good because many more than two IR PO_4^{3-} ion bands are observed in apatites. In the more realistic site-group approximation, the normal modes are deduced for the ion after it has been distorted by the static field of the surrounding ions in the crystal. This

Table 2.2 Normal modes of the undistorted PO_4^{3-} ion. Frequencies from the Raman spectrum in solution [547].

Mode	Symmetry	Frequency cm^{-1}	Activity
v_1 P-O symmetric stretch	A_1 (nondegenerate)	938	Raman
v_2 O-P-O bend	E (doubly degenerate)	420	Raman
v_3 P-O antisymmetric	F_2 (triply degenerate)	1017	Raman, IR
v_4 O-P-O bend	F_2 (triply degenerate)	567	Raman, IR

distortion may remove degeneracies and change the selection rules governing the activities of the normal modes, so that more bands are seen. In the case of FAp, the PO_4^{3-} ion is situated on a mirror plane, so that its site-group symmetry is C_s. As a result, all the degeneracies listed in Table 2.2 are lost, so that there are six A' (in-plane vibrations) and three A'' (out-of-plane) vibrations, all of which are Raman and IR active. The site-group approximation has been used to make IR assignments for FAp in the range 4000 to 400 cm^{-1} at 20 and -186 °C [548] with a misassignment of v_2 corrected in later work [549]. The 11 force constants for the PO_4^{3-} ion with site symmetry C_s in pyromorphite have been calculated from a normal coordinate analysis based on a Urey-Bradley force field [550].

If the PO_4^{3-} ions interact dynamically to a significant extent, their internal vibrations become coupled, so that the site-group approximation, based on static interactions, is no longer completely valid. In this situation, the normal modes and selection rules are required for the vibrations of all the atoms in the primitive unit cell taken as a single unit. These modes and selection rules can be deduced from the factor group for the structure (this comprises all the symmetry elements of the space group except lattice translations, see Chapter 4 in ref. [125]). Factor-group analysis has been applied to FAp [551-553]. For FAp, the appropriate factor group is C_{6h}. This leads to 54 internal PO_4^{3-} ion modes derived from the six PO_4^{3-} ions per unit cell, of which there are six nondegenerate and nine doubly degenerate Raman active modes, and three nondegenerate and six doubly degenerate IR active modes (there are also 15 inactive modes). Notice that the number of IR active internal modes (nine) is the same as for the site-group analysis, so that it is only possible to determine if dynamical interactions are significant in apatites with space group $P6_3/m$ using both IR and Raman spectroscopy. In addition to the internal PO_4^{3-} ion modes, factor-group analysis shows there are 69 external modes, of which 18 are PO_4^{3-} ion librational modes, and the others translational modes, so the total number of modes is 123. Factor-group analysis predicts that there will be 33 lines in the Raman spectrum from the PO_4^{3-} ion internal and external modes together, many of which will be doubly degenerate.

Polarised Raman and IR reflectance spectra of Durango FAp have been used to make a detailed factor-group analysis of the two triply degenerate PO_4^{3-} ion modes, v_3 and v_4 [554]. The splittings observed in the Raman spectra were analysed in terms of dipole-dipole interionic forces; they were in qualitative agreement with experiment. Agreement was improved if account was taken of electronic and other vibrational states for each triply degenerate mode [555]. Subsequent factor-group analyses have been applied to IR (single crystals in transmission) and Raman (single crystal) spectra [551,552], and IR (single crystal reflectance) and Raman (single crystal) spectra [553]. These studies were made on Durango FAp and used polarised radiation. The IR spectrum to 60 cm^{-1} [551] and the Raman spectrum [556] of powdered FAp have been reported.

The v_1 band at 960 cm^{-1} has partial perpendicular dichroism (0.43 in Fowey Consols francolite, [77], Section 4.2.2; and 0.786 in dental enamel (JC Elliott, unpublished results, 1963), Section 4.6.3). The dichroic ratio for v_3 at 1025 cm^{-1} with partial parallel dichroism in dental enamel is 1.45 (JC Elliott 1962, unpublished results, Section 4.6.3).

Detailed assignments for FAp [552,553] are in essential agreement and are given in Table 2.3 from a study of Durango mineral. A large number of weak overtone and combination bands (for example near 2000 to 2150 cm^{-1}), as well as bands possibly originating from OH$^-$ and CO_3^{2-} ion impurities, were seen, but are not listed in the Table [552]. In the absence of coupling between PO_4^{3-} ions, there would be no differences in the frequencies within the columns of Table 2.3; in other words, there would be only nine different frequencies corresponding to the nine normal modes of the site-group approximation. The differences that occur are a measure of the dynamic interaction between the PO_4^{3-} ions, and confirm that a factor-group analysis is required to understand the FAp spectrum.

A band at about 325 cm^{-1} in the IR spectrum has been assigned [549,557] to a Ca_3-F "v_3 type stretching" of the $2(Ca_3$-F) sublattice, and bands at 280 and 230 cm^{-1} to Ca-PO_4 lattice modes, by analogy with OHAp (Section 3.11.2). The "v_3" band in FAp is very close to that calculated (327 cm^{-1}) from the shift of "v_3" from its position in OHAp (343 cm^{-1}, Table 3.4, Section 3.11.2) if the heavier F$^-$ replaces the OH$^-$ ion [549,557]. A 336 cm^{-1} band with transition moment perpendicular to the c-axis seen in the IR reflection spectrum of FAp [558] has been assigned to "v_3", in agreement with the expected direction of the transition moment [557]. The Raman lines at 310 (E_{2g}), 304 (A_g) and 267 (E_{2g}) cm^{-1} have been assigned to factor-group components of the $2[(Ca_{II})_3$-F)] sublattice, and at 285 (E_{1g}) cm^{-1}, primarily to rotary PO_4^{3-} ion motions [559,560].

Table 2.3 Frequencies (cm^{-1}), assignments and activities of the IR and Raman active internal vibrations of the PO_4^{3-} ion in FAp [552].[a]

Mode		ν_1	ν_{2a}	ν_{2b}	ν_{3a}	ν_{3b}	ν_{3c}	ν_{4a}	ν_{4b}	ν_{4c}
Species	Activity									
A_g	R	963 vs	451 vw	-	1078 w	1051 m	-	606 w	591 m	-
E_{2g}	R	963 w	446 w	-	1059 w	1033 w	-	615 vw	580 m	-
E_{1u}	IR	962 w	460 vw	-	1090 vs	1040 vs	-	601 s	575 w	-
E_{1g}	R	-	-	429 m	-	-	1040 w	-	-	591 w
A_u	IR	-	-	470 vw	-	-	1032 vs	-	-	560 s

[a]R Raman active, IR infrared active, vw very weak, w weak, m medium, s strong, vs very strong.

A rigid-ion model of FAp has been used to calculate the 33 Raman lines predicted by factor-group analysis of FAp, which were compared to the 31 lines that were observed from single crystals of Durango FAp [561]. Atomic parameters for FAp [46] were the primary input to the model. Fixed point charges for the ions were assumed, except the charge on the PO_4^{3-} ion, which was taken as an adjustable parameter. Six other adjustable parameters were used to describe short range ionic interactions. These parameters were varied to obtain the best match between the observed and calculated spectra, which in general was very good. From the calculation, it was possible to show which two lines in the observed spectrum must be unresolved doublets to explain the observation of only 31 lines [561]. A more sophisticated potential model of the polarisable ion type has been developed to reproduce the transverse and longitudinal vibrational frequencies, elastic constants, and the static and high frequency dielectric constants of FAp [562]. Short range force constants, ionic charges and ionic polarisabilities were adjusted to give the best fit. Despite the complicated bonding properties of FAp and the large number of atoms per unit cell, all the calculated physical properties agreed satisfactorily with experimental values.

2.6.2 Chlorapatite

The IR and Raman spectra of ClAp are very similar to those of FAp, as is to be expected, except for the lattice mode region (see later). As the Cl⁻ ion is not situated exactly at $z = \frac{1}{2}$, the structure has no mirror planes, so the site symmetry of the PO_4^{3-} ion is reduced from C_s for FAp, to C_1, so there will be nine A normal modes, all IR and Raman active. Thus, the number of phosphate bands and their activities are not changed, because all degeneracies are lost and all bands are IR and Raman active for both C_1 and C_s. However, if the ClAp is sufficiently stoichiometric, the Cl⁻ ions are ordered and the space group is $P2_1/b$, so the selection rules and number of predicted bands for the factor-group approximation will be different. The same possibility of a lower symmetry space group exists in OHAp, which has been studied in greater detail (Section 3.11.2).

Lines in the polarised Raman spectrum of single crystals of hexagonal melt-grown ClAp have been assigned from 1250 to 400 cm⁻¹ [563] (results also cited in ref. [556]). Laser-Raman spectra of ClAp powder have also been studied [556] and differences noted in the lattice mode region between ClAp and FAp (OHAp was quite similar to FAp). These differences and similarities were consistent with the relative masses of the Cl⁻, F⁻ and OH⁻ ions [556]. Polarised Raman spectra of single crystals of ClAp (not stated if monoclinic) have been used to assign bands in the Raman spectra of powders of CaClAp, BaClAp and Ba,CaClAps (9 Ba + 1 Ca and 8 Ba + 2 Ca atoms per unit cell) [472]. The results were consistent with a factor group analysis and the higher mass of the barium atom compared with calcium. In particular, the absence of the symmetrical A_g mode at 80 cm⁻¹ and the degenerate modes at 102 and 145 cm⁻¹, very clearly present in BaClAp, but absent in the Ba,CaClAps, was ascribed to a preference of the Ca^{2+} ions for the Ca(1) sites. This result was consistent with XRD studies [472] (Section 2.3.2).

Assignments for the PO_4^{3-} ion IR fundamental region have been given [551]. The major IR band at ~290 cm⁻¹ has been assigned to lattice modes [557]. The far IR spectrum (to 60 cm⁻¹) of ClAp from Snarum, Norway (the present author does not know if this was sufficiently stoichiometric to be monoclinic) and FAp from Durango were compared, and differences noted in the lattice mode region [551], presumably due to the same cause noted above for the Raman spectra.

It has been reported [559,560] that combined IR and Raman data for the PO_4^{3-} ion internal and external modes in monoclinic ClAp agreed better with a factor-group symmetry C_{6h} than C_{2h} or C_6; this was consistent with there being only a slight deviation of the PO_4^{3-} tetrahedron from the symmetry it has in FAp. The major Raman spectral difference found between the PO_4^{3-} ion internal modes of hexagonal and monoclinic ClAp occurred for v_3 at ~1032 cm⁻¹. This mode was a singlet for the hexagonal form and a doublet (~4 cm⁻¹ separation) for the monoclinic form. There do not seem to have been any IR or Raman studies at temperatures above the monoclinic to hexagonal phase transition. Additional Raman external bands in the lattice mode region (<350 cm^{-1}) have been observed for monoclinic ClAp [560]. Such lattice modes in monoclinic OHAp have also been reported [564,565] (Section 3.11.2).

Assignments for the Raman spectrum of BrAp have been given [566].

2.7 Other physical and chemical studies

The CaF$_2$-FAp and CaCl$_2$-ClAp phase diagrams have been published [539] and the FAp system reinvestigated [387,536] (the diagram from ref. [536] is also published as Fig. 6025 in ref. [20]).

Thermal expansion coefficients of 9.1 × 10^{-6} and 8.5 × 10^{-6} °C^{-1} and dielectric constants at 1 kHz of 10.4±0.5 and 9.5±0.5, parallel and perpendicular to the *c*-axis respectively, and a hardness of 480 Knoop with a 100 g load have been reported for CaF$_2$-def FAp (5 wt % CaF$_2$ deficient, melt-grown) [387]. Thermal conductivities of 0.020 and 0.024 W cm^{-1} °C^{-1}, parallel and perpendicular to the *c*-axis respectively, have been determined for Nd-doped FAp (2 mol % calcium replaced by neodymium) [379]. The silicate oxyapatites are harder than FAp with, for example, a Knoop hardness of 857 for a 200 g load for Ca$_2$La$_8$(SiO$_4$)$_6$O$_2$ [380]. Values and references for other physical constants for FAp and other apatites have been given [9,380,562].

Room temperature adiabatic stiffnesses for Durango FAp have been determined by the ultrasonic pulse superposition method [567]. Full chemical analyses were given of the two crystals used whose densities were 3.2147 and 3.2140 g cm^{-3}. With right-handed orthogonal axes and x_1, x_2 and x_3 parallel to [100], [120] and [001] respectively, the elastic constants were: c_{11} = 1.505±0.001, c_{33} = 1.850±0.001, c_{44} = 0.4251±0.0004, c_{12} = 0.488±0.002, c_{13} = 0.622±0.004 and c_{66} = 0.5087±0.0008 (units of megabars or 10^{11} Pa). (Values from personal communication from HS Yoon, 1971, and differ slightly from the published values in ref. [567].) Rayleigh wave velocities and attenuation measurements have been made on single crystals of Durango FAp with an acoustic microscope [568]. The velocities agreed with those calculated from the above elastic constants to within 0.13%.

The heat content (H_T) of FAp has been measured between 298.16 and 1600 K and is given by

$$H_T - H_{298.16} = 226.04T + 14.44 \times 10^{-3}T^2 + 48.82 \times 10^5 T^{-1} - 85050 \quad 2.4$$

where the units of H_T are cal mol^{-1} K^{-1} (1 cal = 4.187 J) of Ca$_{10}$(PO$_4$)$_6$F$_2$ and T is the temperature in K [569]. Equivalent expressions for the heat capacity and entropy were derived. Literature values of ΔH^0, ΔG^0 and S^0 at 298.15 K for FAp and ClAp have been compiled [570] (see also Table 1.3, Section 1.1) and used with other published data to derive a thermodynamic model for OHAp-ClAp and OHAp-FAp solid solutions [570]. It was concluded that these were essentially ideal above 500 °C. The solubility products of FAp and F,OHAp, and the rate of dissolution of FAp are considered in Sections 3.9 and 3.10.3, respectively.

X-ray topography of mineral FAp from Mexico (presumably Durango) showed that these crystals grow mainly by deposition of $\{10\overline{1}1\}$ layers [571]. The crystals had a low density of screw dislocations aligned with the *c*-axis and mixed dislocations nearly aligned with this axis; the long prismatic habit of natural apatite crystals suggested that these dislocations promoted the growth in the *c*-axis direction [571]. Dislocations in melt-grown FAp, which are present at much higher densities, have been studied in detail by various decoration techniques [572].

NMR studies of Durango FAp with a constant magnetic field parallel to the *c*-axis showed that the ^{19}F resonance was symmetrically split into three lines with a separation between the outer satellites of 3.64±0.20 Oe (0.290±0.016 kA m^{-1}) [573]. These satellites disappeared as the magnetic field was rotated away from the *c*-axis. The ^{31}P NMR absorption line did not show a similar structure nor such a marked dependence of the line-width on orientation. A theoretical separation between the ^{19}F satellites of 3.71 Oe (0.295 kA m^{-1}) was calculated for a model in which only interactions within a linear chain of F nuclei were taken into account. NMR investigations of F,OHAp (Section 3.12) and OHAp after the reaction with F$^-$ ions (Sections 3.10.2 and 3.12) are discussed later.

The diffusion of anions and cations in FAp has been studied, but because this process is very slow, these have nearly all been carried out at elevated temperatures (>800 °C). As the ions are charged, a distinction has to be made between self-diffusion, and the situation when charge balance is maintained by the simultaneous movement of another ionic species, the nature of which will, in part, determine the behaviour of the system.

Self-diffusion coefficients of ^{45}Ca and ^{32}P from vapour deposited films have been determined in Durango FAp between 1000 and 1800 K [574]. For calcium, the diffusion coefficients were

$$D\bot = 1.28\times10^8\exp(-6.0/kT) + 7.92\times10^{-8}\exp(-1.15/kT)\ cm^2\ s^{-1} \qquad 2.5$$
$$D\| = 9.3\times10^6\exp(-5.64/kT) + 3.6\times10^{-3}\exp(-2.46/kT)\ cm^2\ s^{-1}$$

where the activation energies are in eV. These results showed that at 1420 K, $D_\| < D_\bot$, whereas at 1615 K, $D_\| > D_\bot$. The diffusion coefficients of PO_4^{3-} ions at 1700 K were $D_\| = 1.2\times10^{-12}$ and $D_\bot = 3.6\times10^{-12}$ cm^2 s^{-1}, which are at least two orders of magnitude less than the diffusion coefficient for Ca^{2+} ions. It was suggested that the diffusion of Ca^{2+} ions involved vacancy migration [574].

The diffusion of OD$^-$ ions has been studied by IR intensity measurements on a crystal of FAp heated in D$_2$O [574]. Above about 1250 K, the activation energy was 2.7 eV, and below this, ~1.0 eV. At 1300 K, $D \approx 2\times10^{-11}$ cm^2 s^{-1}.

The OD⁻ ion uptake could be enhanced by 50 % by the application of an electric field of a few volts per cm. Rather surprisingly, given the linear arrangement of OH⁻ ions, the reported OD⁻ ion diffusion coefficients were similar for directions parallel and perpendicular to the *c*-axis [575]. However, faster diffusion in the *c*-axis direction has been found for ions involved in interchange processes along the hexad axis when single crystals of FAp [389] and BrAp [576] were heated at 1000 °C in steam. ^{1}H and ^{19}F NMR studies of the temperature dependence (20 to 400 °C) of the spectrum from a single crystal of mineral OH,FAp have been used to determine the activation energy for the diffusion of F⁻ ions along the *c*-axis and to study the free rotation of the OH⁻ ions [577] (Section 3.12).

Diffusion studies of rare-earths, strontium and lead have been published for Durango FAp at 900 to 1250 °C in which the crystal was in contact with a silicate melt, pre-equilibrated with FAp and enriched with the component under study [578]. The measured coefficients were therefore not values for self-diffusion because of the possibility of simultaneous movement of other ions to maintain charge balance. The diffusion coefficients for large and small rare-earth ions were essentially the same and independent of the crystallographic direction. For samarium, $D = 2.3 \times 10^{-6} \exp(-E/RT)$ cm^2 s^{-1} where $E = 52.2 \pm 5.8$ kcal mol^{-1} (218±24 kJ mol^{-1}) and T is the absolute temperature. The diffusion coefficient for Sr^{2+} ions parallel to the *c*-axis was slightly less than the perpendicular coefficient, with a mean $D = 412 \exp(-E/RT)$ cm^2 s^{-1}, where $E = 100$ kcal mol^{-1} (419 kJ mol^{-1}). For lead, $D = 0.035 \exp(-E/RT)$ cm^2 s^{-1}, where $E = 70$ kcal mol^{-1} (293 kJ mol^{-1}). Diffusion coefficients in apatite can also be determined by fission-track annealing studies and generally give activation energies between 45 and 50 kcal mol^{-1} (188 and 209 kJ mol^{-1}) [578] (see below). Diffusion coefficients for ^{37}Ar in FAp have been published [579].

The diffusion of ^{45}Ca, ^{85}Sr and ^{32}P isotopes have been studied in polycrystalline sinter pellets of OHAp at 1000 to 1400 °C [579]. Similar values of activation enthalpies, Q, and frequency factors, D$_0$, were found for the lattice diffusion of ^{45}Ca and ^{85}Sr. These were

$$Q = 3.50 \pm 0.02 \text{ eV,}$$

$$D_0 = 41 \pm 5 \text{ cm}^2 \text{ s}^{-1} \text{ for } p_{water} < 30 \text{ Torr (4.0 kPa)}$$

and   2.6

$$Q = 3.55 \pm 0.02 \text{ eV,}$$

$$D_0 = 20 \pm 3 \text{ cm}^2 \text{ s}^{-1} \text{ for } p_{water} = 230 \text{ Torr (30.7 kPa).}$$

at two different partial pressures of water, p_{water}. It was thought that the dependence of cation diffusion on the partial pressure of water was caused by the formation of O^{2-} ions, and hence vacancies in the OH^- ion sublattice, which accelerated cation transport. The extrapolated cation diffusion coefficient at 37 °C was $10^{-(56\pm2)}$ cm^2 s^{-1}. The diffusion coefficient for ^{32}P at 1360 °C and p_{water} = 30 torr (4.0 kPa) was $1/(400\pm50)$ times the cation diffusion coefficient.

Theoretical aspects of ionic diffusion in apatites have been considered [574,580,581].

Fission-track chronothermometry

This is an important technique, developed during the last 30 years, for dating and investigating the thermal history of rocks over geological time spans [582]. It is applicable to many minerals, but apatite is particularly important because of its occurrence as a common accessory mineral and because of its relatively high uranium content (~1 to ~100 µm/g). The basis of the method is the spontaneous decay of ^{238}U by fission into two nuclear fragments with a half life of 8.2×10^{15} years. These are expelled in opposite directions leaving a trail of damage in the crystal, initially about 16 µm in total length, that can be made visible in polished sections by chemical etching [582,583]. These tracks can be studied by conventional light microscopy, and more recently [584-586] (Fig. 2.10), by confocal scanning laser microscopy.

The analysis of fission-tracks is based on the exponential decay law

$$N_D = N_p(e^{\lambda t} - 1) \qquad\qquad 2.7$$

where N_D and N_P are the number per unit volume of daughter fission-tracks and the number of original parent atoms still present, respectively. λ is the decay constant (the reciprocal of the half-life in years) and t the age of the apatite in years from when it first crystallised. N_D can be determined by direct observation and N_P, again by fission-track counting, after neutron irradiation of the sample to induce fission of ^{235}U (the $^{235}U/^{238}U$ relative abundance ratio is constant, apart for rare exceptions). An apparent complication of the method, but one that can be turned to great advantage, is that the fission-tracks are subject to fading [582]. The most important parameters that determine this are time and temperature. As a consequence, extensive studies have been made of the kinetics of the annealing process as a function of temperature (see [582] for references). These studies showed that the activation energies were characteristic of diffusion processes on an atomic scale. For a time-dependent annealing function

$$\ln t - C_1 \ln(1 - l/l_0) = C_2 + C_3/T, \qquad\qquad 2.8$$

an activation energy of 1.64 eV for track-length reduction in Durango FAp has been reported, where l/l_0 (with $l/l_0 > 0.65$) is the *reduced mean confined track*

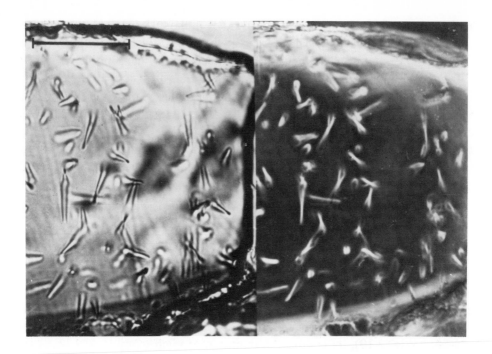

Fig. 2.10 Confocal scanning micrograph (stereo-pair) of fission-tracks in apatite from Fish Canyon Tuff. The section was etched for 20 s in nitric acid (5 mol l^{-1}) and the micrograph compiled from a stack of 32 separate images taken at 0.2 μm increments. The bar is 25 μm. (Fig. 1 of Petford and Miller [585], with permission)

length after annealing, l_0 is the mean length of unannealed induced fission-tracks (16.3 μm), t is the annealing time (s), T is the annealing temperature (K), and the constants, C_1, C_2 and C_3 are 4.47, -17.37 and 19000 respectively [587]. Later studies with a modified model [588] suggested that the activation energy depended somewhat on the degree of annealing. A direct consequence of this annealing process is that the measured age of a mineral may be less than its actual age, the difference depending on its thermal history. Thus comparison of different minerals in the same locality coupled with information from kinetic experiments, as outlined above, enables deductions to be made about the thermal history of a locality. For apatite, the *partial annealing zone*, below which the tracks are stable and above which they are unstable, is between 40 to 70 °C and 120 to 140 °C [582]. Thus it is this temperature range over which apatite can provide information about the thermal history of a geological sample.

Other studies

Thermoluminescent properties of natural and synthetic apatites before and after X-irradiation [542] and cathodoluminescence and microprobe studies of rare-earths in mineral apatites [497] have been published.

The optical properties of stoichiometric or nearly stoichiometric synthetic and mineral ClAps have been studied for sodium light [48]. The indicatrix is very nearly spherical, so small compositional changes cause the optic axial plane to change its orientation by 90°, and to change the optic sign. The purest crystals were acicular, flux-grown ClAp which were biaxial (negative), γ - β = 2.9×10^{-5}, β - α = 8.2×10^{-5}, mean refractive index = 1.6672 with the unique *c*-axis normal to the optic axial plane. Some hydrothermally grown crystals were equally pure. Mineral ClAp and most prismatic flux-grow ClAp (both monoclinic) were less pure; they were biaxial (positive) with the optic axial plane perpendicular to the unique *c*-axis with mean refractive indices 1.6672 (*i.e.* not detectably different from the acicular flux-grown crystals) and 1.6635, respectively. The prisms had γ - β = 1.6×10^{-4} and γ - α = 1.9×10^{-4}.

Pure s-ClAp is not expected to begin to absorb until the wavelength is less than ~150 nm, where intrinsic absorption begins [541]. Experimentally, s-ClAp, if heated in anhydrous HCl at 1000 °C to remove traces of OH$^-$ and O^{2-} ions, does not absorb until below 190 nm, as determined by diffuse reflectance [389]. However, after heating in wet argon for several hours at 1000 °C, then *in vacuo* at 1000 °C for 16 h so that 2OH$^-$ ions react to form an O^{2-} ion and a water molecule, a strong absorption band was seen starting at about 240 nm which, when excited with a high pressure mercury arc, gave a very weak yellow fluorescence [389].

A number of different colour centres are formed in FAp and ClAp on X-irradiation which have absorption bands in the visible and near UV regions [589]. One of these, designated the X band, which only occurred in ClAp and absorbed at 380 and 486 nm, increased in intensity after loss of more ClAp by heating [391]. This band was assigned to a hole trapped at a defect comprising adjacent Ca^{2+} and Cl$^-$ ion vacancies [391].

Apatites, for example FAp, SrFAp and various rare-earth phosphate and silicate oxyapatites, have been found to be excellent laser hosts for Nd^{3+} and Ho^{3+} ions [378-380] (crystal growth in Section 2.5.1). The high gain and efficiency of Nd-doped FAp lasers originate from the narrow, intense 1.06 µm fluorescence line and the broad absorption of Nd^{3+} ions in FAp, but the power output is limited by the relatively low thermal conductivity and elastic moduli which leads to distortion of the laser rod [380] (and refs therein). Much improved physical properties result from the use of silicate oxyapatites *e.g.* Ca$_2$La$_8$(SiO$_4$)$_6$O$_2$, which are harder and more resistant to thermal shock than FAp; this enables them to be run at higher power levels [380].

The wear characteristics, surface failure and frictional behaviour of natural FAp have been investigated in detailed studies in which a polished diamond slider of known geometry was passed across the polished basal plane of a single crystal [590-596]. The frictional behaviour depended on the sliding direction relative to the crystallographic axes; the influence of environment (air, water or dimethylformamide) was attributed to an interaction between polar water on the surface with charged species a very small distance into the substrate. The experiments showed that repeated sliding, at some contact stress greater than a threshold value, resulted in catastrophic surface damage on the second pass, even though there was little visible damage on the first pass.

Chapter 3

HYDROXYAPATITE AND NONSTOICHIOMETRIC APATITES

3.1 Introduction

General references to the preparation, properties and structure of apatites, including hydroxyapatite, $Ca_{10}(PO_4)_6(OH)_2$, have already been given (Section 1.1). Its preparation, processing, properties and applications have been recently reviewed [597]. Synonyms and abbreviations that have been given include: OHAp, HAP, HAp and pentacalcium monohydroxyorthophosphate. The alternative spelling, hydroxylapatite, is sometimes found, particularly in older literature; the origin and usage of this alternative to hydroxyapatite is discussed in ref. [598]. The abbreviations to describe the various variants of OHAp in the present work are given in Table 1.2 (Section 1.1).

The chemistry of OHAp is conspicuously more complex than FAp and ClAp, mainly because, when attempts are made to precipitate it in solution, apatitic products are formed that can have Ca/P molar ratios from 1.5 to 1.66, and sometimes even outside this range. Furthermore, precipitates usually are of submicron dimensions, so that their surface chemistry becomes extremely important. These factors complicate the preparation and study of OHAp. Apatitic precipitates (sometimes referred to as basic calcium phosphates, BCaPs) and OHAp are thermally much less stable than FAp and ClAp. Precipitated BCaPs contain variable amounts of water, and often acid phosphate and carbonate contamination, which are all lost in a complex manner on heating (Sections 3.6.2 and 4.5.7). OHAp itself loses constitutional water, that is to say its OH⁻ ions, above 800 °C (Section 3.4.3).

OHAp is a very rare mineral. Wax yellow crystals up to $6 \times 6 \times 11$ mm³ have been described from talc schists from the Old Verde Antique serpentine quarry near Holly Springs, Cherokee County, Georgia, USA. Refractive indices (sodium light) of O = 1.651±0.001 and E = 1.644±0.001 [598], and a F⁻ ion content of 0.16 [598] and 0.28±0.01 [599] wt % have been reported. At least some of these crystals contain 0.36 wt % chlorine, which was thought to explain [600] the slight expansion of the a-axis parameter ($a = 9.429 \pm 0.002$, $c = 6.883 \pm 0.002$ Å) over that of pure OHAp ($a = 9.4176$, $c = 6.8814$ Å, Table 1.4). OHAp crystals up to 3 cm long by 1.5 cm wide have been described from a serpentine-talc quarry at Kemmleten, near Hospenthal (Uri Canton, Switzerland) [601]. Refractive indices (sodium light) of O = 1.6452 and E = 1.6413 were reported.

The main interest in OHAp in the present context is as a model for the inorganic component of bones and teeth, although precipitated CO_3Aps (Section 4.5) are probably better prototypes. There is also considerable academic and commercial interest in the development of apatitic (and β-TCP) bioceramics and apatite-loaded polymers for bone replacement, and OHAp coatings on metallic joint prostheses (Section 4.6.7). OHAp occurs in ammoniated superphosphates, and in calcarious soils as a reaction product of MCPM or brushite. BCaPs with about 2 to 12 wt % iron or copper catalyse dehydrogenation and dehydration of alcohols [602,603]. OHAp is used in column chromatography for protein separation [604-607].

3.2 Structure of hydroxyapatite

s-OHAp is monoclinic with space group $P2_1/b$ [608] with lattice parameters $a = 9.4214(8)$, $b = 2a$, $c = 6.8814(7)$ Å, $\gamma = 120°$ [609]. Recent synchrotron radiation measurements have shown that b is marginally greater than $2a$ (M Bauer and WE Klee, personal communication, 1990 and ref. [60]).

In the first structure determination of OHAp (a hydrothermally grown crystal), the structure was refined in the space group $P6_3/m$ with the hydroxyl oxygen atom at $z = ¼$, by analogy with FAp [610]. Later neutron diffraction studies of Holly Springs OHAp and further study of the earlier X-ray measurements showed [599] that the hydroxyl oxygen and hydrogen atoms were located at 0,0,0.201 and 0,0,0.062 respectively, and that no OH ions straddled the mirror plane at $z = ¼$ within experimental error; this is illustrated in Fig. 2.4 (Section 2.2.1), except the OH⁻ ions are in the positions related by the mirror plane at $z = ¼$. The positions of the OH⁻ ions indicated that there was at least short range ordering of these ions into columns, ···OH OH OH OH···, otherwise the hydrogen atoms would be too close. Two models were postulated to reconcile this ordering with the mirror plane of the space group $P6_3/m$ deduced from diffraction results: (1) a "disordered column" model in which the OH⁻ ion orientation was reversed at various places within a column; and (2) an "ordered column" model in which all OH⁻ ions in a given column were oriented the same way, but the choice of direction was random [599]. These models could not be distinguished from the diffraction data. It was proposed that the F⁻ ion impurity provided a point for column reversal in Holly Springs OHAp via the sequence ···OH OH F HO HO···. Furthermore, synthetic OHAp might follow either the "ordered column" model, or alternatively, a "disordered column" model with an occasional OH⁻ ion astride a pseudo-mirror plane at $z = ¼$, or an occasional pair of OH⁻ ions pointing generally towards each other from adjacent pseudo-mirror planes, but making a substantial angle with the c-axis so that the hydrogen atoms did not approach each other too closely [599]. Neutron diffraction results on the position of the hydrogen atom [599] confirmed the orientation of the O-H bond determined by polarised IR studies [611] (Section 3.11.2). They also showed that there was a large rocking motion of the hydrogen atom perpendicular to the *c*-axis about the oxygen atom, which was later confirmed by identification of a librational IR mode at 630 cm^{-1} (Section 3.11.2). ESR measurements on monoclinic OHAp, after X-irradiation, have suggested that, at 92 K, instead of having a large rocking motion, the OH$^-$ ions were tilted at 6 to 7° to the *c*-axis with six orientations about the axis [612] (Section 3.13). A precise XRD study of Holly Springs OHAp has confirmed that the hydroxyl oxygen is on the *c*-axis with a *z* parameter in the range 0.1950(7) to 0.1960(6) (there was evidence of variability between crystals); direct refinement of the degree of substitution showed that 8 % of the OH$^-$ ions were replaced by F$^-$ ions at 0,0,¼ [404].

An XRD study [609] of a synthetic OHAp single crystal (made by heating ClAp in steam at 1300 °C for two weeks [613], Section 3.5) showed that the space group was $P2_1/b$. This lowering of the symmetry from hexagonal to monoclinic can also be seen in the XRD powder pattern of some preparations of OHAp (Fig. 3.1). The explanation given [609] for the monoclinic space group of OHAp was that the crystal was sufficiently pure so that there were few reversal points within the OH$^-$ ion columns with the result that the crystal could not adopt a "disordered column" structure, as described above for Holly Springs OHAp. Instead, there must be some arrangement of "ordered columns". The structure determination confirmed that the OH$^-$ ions were ordered within each column, so that they all had the same direction of displacement from *z* = ¼. Furthermore, all the columns pointed in the same direction within a plane parallel to the *a*- and *c*-axes, but the direction of the displacements alternated between adjacent planes (Fig. 3.2). The perpendicular separation between the planes was b_{hex}cos30°. This ordered arrangement of OH$^-$ ions changed the mirror planes in $P6_3/m$ into *b* glide planes in $P2_1/b$, and doubled the *b*-axis parameter. The similarity with the ordering of Cl$^-$ ions in s-ClAp (Fig. 2.9) is very evident.

The distortions of the monoclinic from the hexagonal structure in OHAp are very similar to those in ClAp, with the O'(3) triangle that is enlarged by a Cl$^-$ ion near its centre in ClAp [47] corresponding to the O'(3) triangle that is enlarged by the hydroxyl hydrogen atom near its centre in OHAp [609]. Further analysis (JC Elliott, unpublished results, 1976) of the diffraction data for Holly Springs OHAp showed the presence of two O(3) atoms forming two slightly different sized triangles about the *c*-axis (O(3a) and O(3b) in Table 3.1). This result can be correlated with the structure of monoclinic OHAp in which two different sized O(3) triangles occur with a hydrogen atom at the centre of the larger one [609]. As the O(3a) atoms form the larger triangle (Table 3.1), they must have a hydrogen atom at their centre and correspond to the O'(3) atoms of the monoclinic structure. An analogous situation with two different sized O(3) triangles occurs in the structure of hexagonal CaCl$_2$-def ClAp (Section 2.2.3).

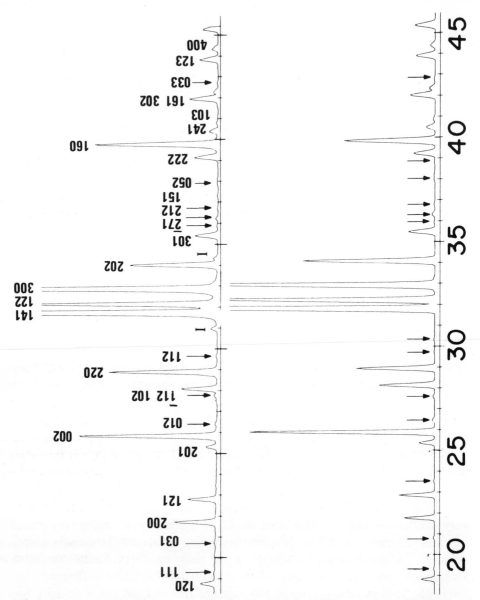

Fig. 3.1 Observed (left) and calculated (right) XRD patterns (CuKα) of monoclinic OHAp made by heating monoclinic ClAp powder in steam at 1000 °C. "Monoclinic" lines are arrowed and the I lines are β-TCP impurity. Vertical scale is the scattering angle in degrees 2θ.(Fig. 6 of Elliott and Mackie [400])

There do not seem to have been any studies of the fraction of OH⁻ ions that can be replaced by Cl⁻ ions and the structure remain monoclinic. However,

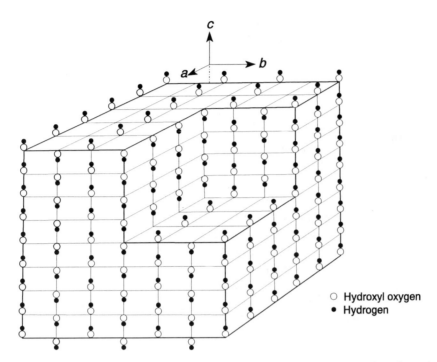

Fig. 3.2 Ordered sheets of OH⁻ ions in monoclinic OHAp. Note that the origin of the axes is displaced by ¼ **b** from the pseudohexagonal axis. (Fig. 1 of Elliott *et al.* [609])

Holly Springs OHAp with F⁻ ions in 8 % of the OH⁻ sites [404] and a synthetic F,OHAp with F⁻ in 10 % of the OH⁻ ion sites [564,565] are hexagonal; this indicates a greater tendency of F,OHAp to become hexagonal compared with F,ClAp, which birefringence measurements [406] (Section 2.2.3) showed is still monoclinic with F⁻ ions in 16 % of the Cl⁻ sites.

Electron microscope images at a resolution of 2.0 to 2.5 Å of OHAp along ⟨010⟩ and ⟨001⟩ have been reported [614]. Good agreement between computed and experimental images along ⟨010⟩ over a range of crystal thicknesses, defocus conditions and apertures were obtained. Along ⟨001⟩, image contrast was very sensitive to experimental conditions, and a good match with the computed images was difficult to obtain.

The structures of other OHAps have been published: $Sr_5(PO_4)_3OH$, O(H) at $z = 0.1476(69)$, $a = 9.745(1)$, $c = 7.265(1)$ Å [615]; $Ca_5(VO_4)_3OH$, O(H) at $z = ¼$, apparently was not tried off the mirror plane, but B_{33} was very large (11.1 Å²), $a = 9.818$, $c = 6.981$ Å [57]; $Cd_5(PO_4)_3OH$, O(H) at $z = 0.1880$, $a = 9.335(2)$, $c = 6.664(3)$ Å [55]; and $Ca_5(CrO_4)_3OH$, O(H) at ¼ but apparently not tried off the mirror plane ($B_{33} = 2.28$ Å²), $a = 9.683$, $c = 7.01$ Å [515].

Table 3.1 Positional parameters and occupancies of O(3) for three Holly Springs OHAp crystals compared with parameters in monoclinic OHAp[a].

Crystal	Occupancy	x	y	z
O(3) for Holly Springs OHAp with one oxygen atom in refinement				
X-23-4	10000[b]	3437(2)	2579(1)	702(2)
X-23-6	10000[b]	3434(2)	2579(2)	704(2)
X-23-10	10000[b]	3438(2)	2581(2)	704(2)
O(3a) and O(3b) for Holly Springs OHAp with two oxygen atoms in refinement (1st and 2nd data sets respectively)				
X-23-4	4691	3559(36)	2683(38)	625(32)
X-23-6	3598[c]	3545(13)	2650(13)	650(19)
X-23-10	5000[c]	3519(30)	2577(39)	662(46)
X-23-4	5874	3337(115)	2500(66)	762(61)
X-23-6	6632	3358(6)	2537(7)	740(9)
X-23-10	5519	3345(27)	2575(33)	753(43)
Parameters for two of the six independent oxygen and hydrogen atoms in monoclinic OHAp				
	O(3a)	3280(4)	2492(4)	821(4)
	O'(3a)	3578(3)	2654(2)	4404(3)
	H	0	0	4392

[a]Values and standard deviations $\times 10^4$. Standard deviations are not reported for occupancies because of high correlation between values for O(3a) and O(3b). Data for O(3) from ref. [404], and O(3a) and O(3b) from further analysis [Elliott, 1976 unpublished results] of X-ray data used for O(3) results [404]. Data for O(3a) and O'(3a) in the monoclinic structure (transformed to hexagonal cell) from ref. [609] and data for H from ref. [599].
[b]Theoretical value for O(3) and O(3a) + O(3b).
[c]Not varied during refinement.

None of these were reported to be monoclinic.

Monoclinic to hexagonal thermal phase transition

Polarised light studies of single crystals of monoclinic OHAp and ODAp, previously found to be monoclinic by X-ray precession photographs, showed that the birefringence observed down the c-axis disappeared reversibly, without hysteresis, at 211.5(5) °C [616]. This finding was ascribed to a monoclinic to hexagonal phase transition [616] (but see comments at the end of this section on recent polarised light measurements).

The fractional polarisation version of the thermally stimulated currents method applied to ClAp (Section 2.2.3) has also been used to study the thermal transition in OHAp [96,617,618]. Compensation temperatures of 211.5, 211.5, 356 and 620 °C were found with relaxation times of 7.0×10^{-4}, 7.7×10^{-4}, 2.1

$\times$ 10^{-4} and 4.75 $\times$ 10^{-5} s respectively [618]. By analogy with ClAp (Section 2.2.3), the first two compensations were associated with cooperative reorientations of lengths of strings of OH$^-$ ion dipoles at the quasi-statically stabilised monoclinic to hexagonal phase transition. In this process, reorientation would have to await a statistical fluctuation of the right size and direction so that an OH$^-$ ion at the end of a string was momentarily reversed, which then set off progressive reorientations along the string. The compensation at 356 °C was assigned to the onset of dynamic stabilisation of the hexagonal form, in which every OH$^-$ ion had sufficient energy to be able to reorient continuously. The last compensation was not identified.

If 16 % of the OH$^-$ ions in OHAp were replaced by F$^-$ ions, five compensations occurred, three at essentially the same temperature (209 to 216 °C), but with different relaxation times (the others were at 383 and 684 °C) [619]. These three were ascribed to three different OH$^-$ ion chain lengths between interrupting F$^-$ ions. When 25 % of the OH$^-$ ions were replaced by F$^-$ ions, the OH$^-$ ion chains were too short and too bound at the ends to show cooperative phenomena of the sort associated with the monoclinic to hexagonal phase transition, although compensations at 900 and 1037 °C were seen. For Ca$_5$(PO$_4$)$_3$Cl$_{0.5}$(OH)$_{0.5}$, the absence of compensation below 800 °C was ascribed to short chain lengths and direct OH$^-$-Cl$^-$ interactions; FAp showed no compensation below 1200 °C [619]. The activation energies for diffusion of F$^-$ ions and the onset of dynamic disorder of OH$^-$ ions about the mirror plane have been deduced from ^{1}H and ^{19}F NMR studies of OH,FAp (20 % F$^-$ replaced by OH$^-$) over the temperature range 20 to 400 °C [577] (Section 3.12).

No changes in the Raman spectra of powders could be detected on going through the monoclinic to hexagonal phase transition at 211 °C because of spectral line broadening [564,565].

Recently, high precision optical measurements have shown that a weak birefringence in OHAp persists beyond 211 °C, up to ~370 °C (M Bauer and WE Klee, personal communication, 1990 and ref. [60]). There is also recent evidence [60] (Section 2.2.3) that the corresponding monoclinic to hexagonal transition in ClAp occurs at ~320 °C, rather than near 200 °C, as previously thought [406]. Thus, it might be that the true transition in OHAp is also at a higher temperature. Aspects of the corresponding thermal transition involving the CO$_3^{2-}$ ions in A-CO$_3$Aps (Section 4.3.3), and the as yet not fully explained presence of an additional OH IR band seen in hot OHAp (Section 3.11.3) may also be relevant. Clearly, the behaviour of the OH$^-$ ions on raising the temperature through the thermal transition is rather complex and not yet well understood.

3.3 Preparation of stoichiometric hydroxyapatite powders
3.3.1 Introduction

The large number of different synthetic routes and published preparations demonstrates the difficulty of making s-OHAp. The variability that is found in such products has been reviewed [620].

Most synthetic methods eventually have to rely on the assumption that s-OHAp is the equilibrium phase under the preparation conditions used. Although this may be approximately true, there is no thermodynamic reason why an apatitic equilibrium phase should be exactly stoichiometric. The stoichiometry has to be determined experimentally by accurate chemical analyses and preferably also from the unit cell contents calculated from lattice parameter and density measurements (Section 3.6.3). What evidence there is, suggests that the equilibrium phase is not generally exactly stoichiometric. In studies of the phase diagram at 825 °C and 1000 bar (100 MPa), it was found that excess calcium reduced the a-axis parameter slightly, and excess water had the opposite effect [17]. At 2000 bar (200 MPa) and between 300 and 600 °C, slight changes in the a-axis parameter (range 9.4151(7) to 9.4225(5) Å) have been found, depending on temperature and composition [18]. For OHAp formed at the singular point with DCPA or β-$Ca_2P_2O_7$, the a-axis parameter increased with temperature, but at the singular point with $Ca(OH)_2$, it decreased with temperature. A broad existence region at high temperatures has been found for the water content (present as OH^- ions) of OHAp, but the region was limited for changes in CaO/P_2O_5 ratio [621] (Section 3.4.3). On the other hand, the synthesis of an OHAp with excess $Ca(OH)_2$ at 1000 °C has been reported [537] (Section 3.4.2).

There do not seem to have been any precise investigations at near normal temperatures and pressures of the stoichiometry of the OHAp at equilibrium in the $Ca(OH)_2$-H_3PO_4-H_2O system, but there is no reason why the generally small deviations in stoichiometry found at higher temperatures and pressures should not also occur under milder conditions. The fact that OHAps prepared in boiling water by the prolonged hydrolysis of DCPD have a slightly enlarged a-axis parameter and are hexagonal rather than monoclinic [620] suggests that there is indeed a small variable compositional range at equilibrium, caused by a limited solubility of $Ca(OH)_2$, H_2O and/or H_3PO_4 in OHAp under these conditions. It seems then that, at equilibrium, the precise composition may depend significantly on temperature (and pressure). In addition to these thermodynamic reasons why the equilibrium phase may not be stoichiometric, there will be many instances when deviations (sometimes quite large) are caused by kinetic factors. As a result of these considerations, there can be no precise boundary between the preparation of s-OHAp (this section) and Ca-def OHAps in aqueous systems (Section 3.4.1).

One method that has been used to prepare s-OHAp is to use accurately stoichiometric quantities for the synthetic reaction, and attempt to ensure formation of a single phase without unwanted gain or loss of chemical species (Section 3.3.2). Another approach is to use the aqueous system $Ca(OH)_2$-H_3PO_4-H_2O at a composition in which OHAp (assumed to be stoichiometric) is the only stable phase, and to attempt to ensure that the system is at equilibrium (Section 3.3.3). The calcium and phosphorus can, for example, come from the controlled addition of suitable solutions, or from a more acidic CaP that hydrolyses to OHAp; in some systems, calcium is released from a bound form into a phosphate solution.

Yet another technique is to synthesise a ns-OHAp under nonequilibrium conditions, and to process this to give a stoichiometric composition. For example, solid state reactions have been used in which samples of nearly stoichiometric compositions are heated to a high temperature so that s-OHAp is formed as the major phase, together with a minor phase (CaO or β-TCP, depending on whether the Ca/P molar ratio is greater or less than 1.667 respectively) in an amount dependent on the departure of the overall composition from s-OHAp (Section 3.3.4). The minor phase is then extracted from the product. A further method is to rely on the easier synthesis of a related stoichiometric apatite (s-ClAp or s-BrAp) that can be converted to s-OHAp (Section 3.3.4).

3.3.2 Syntheses based on theoretical compositions

This method can be used for solid state high temperature reactions or reactions that involve precipitation. The reactants can be analysed for calcium and phosphorus, or chosen to have a well-defined composition. Analytical grade $CaCO_3$, dried at up to 500 °C, can be used without analysis, and dissolved in acid if a solution is required. Alternatively, it can be heated to constant weight at 1000 °C to provide CaO, which can be dissolved in CO_2-free water to give $Ca(OH)_2$ (great care is required to prevent absorption of CO_2). A source of phosphorus of known composition is more difficult. Analytical grade H_3PO_4 can be titrated against standard NaOH, or its composition deduced from its density [622]. DCPA or DCPD can also be used as starting materials if their compositions are known, or their purity can be established (DCPD is less favoured because of its slightly uncertain water content). For solid state reactions, it is preferable to use the better defined $Ca_2P_2O_7$ which is prepared by heating DCPA (Section 1.5.6) or DCPD (Section 1.4.7).

For quantitative reactions in solution, the reactants must be H_3PO_4 and $Ca(OH)_2$, or salts of these with ions that are unlikely to be incorporated in the apatite lattice. Thus, nitrates should be used in preference to chlorides, and ammonium salts instead of sodium or potassium salts. (It has alternatively been claimed that K^+, NH_4^+, and NO_3^- ions are not incorporated into crystalline apatites [623], but there is evidence at least for the limited incorporation of K^+ ions [181], Section 3.4.1, and very limited incorporation of NH_4^+ ions [94,434,435,624,625], Section 4.5.4.) The pH should likewise be controlled by use of NH_4OH or NH_3 gas, rather than NaOH or KOH. An advantage of using ammonium and nitrate salts is that NH_4NO_3 can readily be removed from the separated precipitate by decomposition into gaseous products on heating. Care must be taken to exclude atmospheric CO_2 during precipitation, particularly under alkaline conditions, as CO_2 is very easily incorporated into apatites. Although this can be removed afterwards by heating in steam at 900 °C, PO_4^{3-} ions displaced by CO_3^{2-} ions from the apatite might be lost whilst filtering the precipitate.

Owing to the very low solubility of apatites, the separated precipitate will contain essentially all the original calcium and phosphorus, so it must have the correct Ca/P ratio. If precipitation has been carried out at a high pH, so that there is essentially no acid phosphate in the solid, charge balance dictates that the solid also has the correct proportion of OH^- ions. However, if precipitation is carried out near neutral pH, in principle, there could be acid phosphate present (Equation 3.14, in a direction right to left), in which case the proportion of OH^- ions will be higher than the stoichiometric value. Regardless of the pH of formation, the solid will contain a large amount of water. If the solid is now heated, there is the possibility of removing excess water, volatilisation of NH_4NO_3 if present, transformation of HPO_4^{2-} to PO_4^{3-} ions via a pyrophosphate intermediate, and removal of any small amounts of carbonate contamination. However, there is also the possibility of the formation of O^{2-} ions from the reaction

$$2OH^- \rightarrow O^{2-} + H_2O \qquad\qquad 3.1$$

with the result that the product does not have the stoichiometric OH^- ion content. Studies of the thermal decomposition of OHAp [19,21,621] (Section 3.4.3) indicate that this will occur only to a minimal extent at 900 °C in steam at atmospheric pressure (0.1 MPa). At this temperature, HPO_4^{2-} ions will be already transformed to PO_4^{3-} ions, and any carbonate should be lost, given sufficient time.

In a quantitative solution preparation of OHAp [626], 300 ml of a solution containing 0.33 mol of $Ca(NO_3)_2.4H_2O$ were brought to pH 12 with CO_2-free ammonia solution and diluted to 600 ml. 500 ml of a solution containing 0.2 mol of $(NH_4)_2HPO_4$, similarly brought to pH 12 and diluted to 800 ml, were added slowly with vigorous stirring. The reaction mixture was boiled for 10 min to improve filtration properties. The precipitate was allowed to settle, the supernatant poured off, then it was filtered. Excess water was removed by heating to 900 °C. No lattice parameters or chemical analyses were reported. However, OHAp prepared by a similar method [627], in which gravimetric analyses of the starting materials and product were made [77], gave CaO 55.68, P_2O_5 42.32 and H_2O (by subtraction) 2.00 wt % (theoretical 55.79, 42.39 and 1.80 wt % respectively). The CO_2 content was <0.1 wt %. The lattice parameters were $a = 9.422\pm0.003$ and $c = 6.883\pm0.003$ Å [416].

In solid state quantitative reactions, the main problem is to ensure complete reaction to a single phase. It is of little importance what starting materials are used, except insofar as this affects the accuracy of the composition determination. Starting materials that have been used include $CaCO_3$, CaO, $Ca(OH)_2$, DCPD, DCPA and $Ca_2P_2O_7$. To minimise formation of O^{2-} ions, the final high temperature heating should be carried out in steam.

The following scheme has been used to prepare s-OHAp by quantitative solid state reactions [628]. DCPA was heated at 1000 °C in air for 12 h to give $Ca_2P_2O_7$ (Section 1.4.7). Weighed amounts of $Ca_2P_2O_7$ and $CaCO_3$ in the mole ratio 3:4 were ground in acetone in a steel ball-mill to a 10 μm particle size, pelleted to ensure completion of subsequent reactions, and dried at 200 °C. These were then heated under vacuum at 1100 °C for 20 h in a platinum crucible to remove CO_2:

$$3Ca_2P_2O_7 + 4CaCO_3 \rightarrow 10CaO.3P_2O_5 + 4CO_2. \qquad 3.2$$

The weight loss was 15.11 ± 0.00 wt % (theoretical 15.14). The pellets were then heated at 900 °C for 24 h in a platinum crucible in a stream of steam at 1.5 atm (0.15 MPa). The water weight gain was 1.801 ± 0.003 wt % (theoretical 1.826) which indicated an OH⁻ ion content of 98.6 ± 0.2 % of the theoretical. The phosphorus content of the OHAp was 18.50 wt % (theoretical 18.49). A dissolution and titration method of analysis gave the OH⁻ ion content to within ±5% of theoretical [623] (Section 3.6.1). XRD showed a single phase of well-crystallised monoclinic OHAp with pseudohexagonal lattice parameters $a = 9.4182(5)$ and $c = 6.8814(4)$ Å (BO Fowler, personal communication, 1991). Two samples prepared by a similar, but not identical method, were also monoclinic with $a = 9.4170(3)$ and $c = 6.8796(3)$ Å for a single phase sample, and $a = 9.4175(2)$ and $c = 6.8802(2)$ Å for a sample that contained some $Ca(OH)_2$ [620]. Quantitative IR of the OH stretching band indicated a 17 % OH⁻ ion deficiency in the first sample, and 1 % in the second [620]. The three pseudohexagonal lattice parameter determinations agree well, and their means ($a = 9.4176$, $c = 6.8804$ Å, estimated error in both ±0.0005 Å) probably give the best estimate of the lattice parameters for monoclinic OHAp.

3.3.3 Equilibrium syntheses in solution

This method assumes that s-OHAp is the equilibrium phase under the conditions used (see Section 3.3.1) and that the system is in equilibrium. Two approaches can be used. In the first, the composition must be controlled so that the solution is supersaturated only with respect to OHAp, and not to any other CaP. This condition requires the use of constant composition methods (Section 3.8) which can generally only be used for small quantities.

The second approach is to assume that s-OHAp is the most stable phase in solution (Fig. 1.1) at a pH above the OHAp/DCPD singular point (pH 4.3 at 25 °C, Section 1.4.6). If this assumption is correct, any CaP left in an aqueous solution for long enough should transform to s-OHAp, provided the pH is not allowed to fall below the singular point. In practice, it is usual to hydrolyse DCPD or DCPA, rather than a Ca-def OHAp, boil to speed the reaction, and to periodically replace the water to remove H^+ ions. This type of preparation seems often to lead to a ns-OHAp or Ca-def OHAp with a slightly expanded *a*-axis parameter [620], and for this reason is discussed in the section on these compounds (Section 3.4.1).

An alternative to hydrolysis is to directly precipitate OHAp above the OHAp/DCPD singular point. This reaction is again usually done in boiling water to promote equilibrium. An early example was the precipitation of OHAp by the addition of a saturated solution of $Ca(OH)_2$ to a dilute H_3PO_4 solution (14 g l^{-1}) until alkaline to phenolphthalein (pH >9) [75]. To avoid DCPA precipitation, neutralisation was initially carried out in the cold, the mixture boiled and further $Ca(OH)_2$ added until the red colour change was stable for 1 h. The precipitate (separated by filtration and dried at 80 °C) had a CaO/P_2O_5 mol ratio of 3.5 to 3.6 (theoretical 3.33) and an XRD pattern of OHAp after heating at 800 °C.

In most of the subsequent preparations of OHAp by reaction between $Ca(OH)_2$ and H_3PO_4, acid has been added to alkali to avoid formation of metastable phases, and a slurry instead of a solution of $Ca(OH)_2$ used to increase the yield. Of two such almost identical preparations [38,629], the more recent will be described in detail. A boiling aqueous suspension of $Ca(OH)_2$ (194 g of CaO, made by calcining $CaCO_3$ at 1000 °C for 24 h, then added to 7 l of freshly boiled distilled water) was titrated with H_3PO_4 (0.5 mol l^{-1} recrystallised $2H_3PO_4.H_2O$, standardised by titration) [38]. These were reacted in a 15 l vessel equipped with a reflux condenser, stirrer and ports for introduction of N_2 and acid, which was added at 1 ml per min until a Ca/P molar ratio of 1.70 was reached. After this, boiling and stirring was continued for a further two days, the solid allowed to settle, and the supernatant siphoned off. 5 l of freshly boiled distilled water were then added, and the mixture boiled and stirred for two days to remove excess $Ca(OH)_2$. The supernatant was then syphoned off, and the process repeated four times, the last using dilute H_3PO_4 (0.001 mol l^{-1}). The product was then dried at 110 °C in a stream of dry N_2. The mean refractive index was 1.636, and IR indicated that it was essentially free of carbonate. Chemical analyses gave Ca 39.25 and P 18.13 wt % (theoretical 39.89 and 18.50 wt % respectively), hence the Ca/P molar ratio was 1.67 (standard error 0.042). The BET surface area was 16.7 m^2 g^{-1} [38]. Lattice parameters of a = 9.4174(2) c = 6.8853(2) Å, 90% of theoretical OH content determined by IR, and a full c-axis channel from Rietveld structure analysis have been reported for this preparation [620]. These lattice parameters agree well with those given in the previous section. The solubility product is given by Equation 3.23 in Section 3.9.

OHAp has been prepared [630] by adding equal volumes of CO_2-free $Ca(OH)_2$ and MCPA solutions at a constant rate to rapidly stirred distilled-deionised water at 100 °C with CO_2-free N_2 flushed through the system. The precipitate was then filtered under CO_2-free N_2, washed with acetone or distilled-deionised water, and dried at 105 °C. Well-defined prism-shaped crystals from 0.7 to 7 μm long and 0.04 to 0.14 μm wide [631] were formed, depending on the rate of addition of the solutions [630]. Products with surface areas of 9.1 and 26.6 m^2 g^{-1} contained Ca 39.4 and P 18.2 wt % and Ca 38.8 and P 18.1 wt % respectively (theoretical Ca 39.89 and P 18.50 wt %) [630]. A preparation with a surface area of 20.0±0.4 (standard error, SE) m^2 g^{-1} had lattice parameters a = 9.422±0.001 (SE) and c = 6.883±0.003 (SE) [632] and a = 9.422±0.002 (standard deviation, σ) and c = 6.887±0.002 (σ) [631]. A similar process to the above has been described for large scale production of OHAp [633].

OHAp can also be prepared by methods similar to those given above in which precipitation takes place in the presence of ions additional to those in the $Ca(OH)_2$-H_3PO_4-H_2O system [634,635]. 2.5 l of $(NH_4)_2HPO_4$ solution (7.25 × 10^{-2} mol l^{-1}) and 2.5 l of $Ca(NO_3)_2.4H_2O$ (13.33 × 10^{-2} mol l^{-1}) both at pH 8.5 to 9 (adjusted with NH_3 gas) were pumped at 100 ml h^{-1} into a 22 l flask containing 9 l of CH_3COONH_4 (1 mol l^{-1}) [634]. The reaction mixture was vigorously boiled during addition of the solutions and for 3 h thereafter, and the pH was maintained at 8.5 to 9 (measured at room temperature) with NH_3. After the solid had settled, the supernatant was removed and replaced by 9 l of boiled distilled water and the mixture boiled for 2 h. This process was repeated a few times to remove soluble salts. Extensive precautions were taken against ingress of CO_2. The precipitate was filtered and dried at 150 °C for 48 h, then at 1000 °C in air. Chemical analyses were: P 18.22±0.27 and Ca 39.63±0.99 wt % (theoretical 18.50 and 39.89 wt % respectively) and 0.04 wt % CO_2. OHAps with specific surface areas from 6.39 to 50.1 m^2 g^{-1} have been prepared by a similar precipitation process by varying the conditions of temperature and seeding, and the concentration and rates of addition of the reagents [635]. This method has also been used to prepare OH,FAps [636] (Section 3.4.4).

Preparation of OHAp at 70 °C by the addition of 2.9 l of a potassium phosphate buffer (0.0805 mol l^{-1} K$_2$HPO$_4$, 0.0195 mol l^{-1} KH$_2$PO$_4$, pH 7.4) over 30 min to 11.6 l of a solution containing 0.477 mol of Ca(NO$_3$)$_2$ and 0.318 mol of KOH, followed by stirring for one month has been described [637]. The product was analysed in several different laboratories by a number of methods including XRD, IR (Section 3.6.1), BET, chemical analysis, differential thermal analysis, NMR (Sections 3.6.1, 3.6.2 and 3.12) and transmission electron microscopy. The Ca/P molar ratio was in the range 1.631 to 1.659 and the lattice parameters were $a = 9.428$ and $c = 6.882$ Å. There was a deficiency in the OH$^-$ ion content of about 10 % (Section 3.6.1). About 2 % of the total phosphate was protonated and thought to be on the surface.

3.3.4 Miscellaneous methods

OHAp has been prepared by heating a stoichiometric mixture of minus 200-mesh MCPM and CaCO$_3$ for 3 h at 1200 °C in an atmosphere of equal volumes water and N$_2$ [16]. If the system is at equilibrium and the overall composition deviates from that of s-OHAp (assumed to be the equilibrium phase), a small amount of a second phase will be formed (Ca(OH)$_2$ for excess calcium, and TCP for excess phosphorus). The reaction mixture was then extracted twice with neutral ammonium citrate solution [638] to remove nonapatitic CaPs (this would also remove any Ca(OH)$_2$), washed thoroughly, and dried at 110 °C. Neither lattice parameters nor the chemical composition were reported.

s-OHAp can in principle be precipitated in solution under conditions that are some way from equilibrium, provided these have been chosen correctly. These conditions have been studied systematically for OHAp precipitated by slow addition, during T min, of 400 ml of (NH$_4$)$_2$HPO$_4$ solution (0.25 mol l^{-1}) containing V_1 ml of NH$_4$OH solution (density 0.92 g ml^{-1}) to 900 ml of Ca(NO$_3$)$_2$ solution (0.20 mol l^{-1}) containing V_2 ml of NH$_4$OH (density 0.92 g ml^{-1}) in a 3 l flask fitted with a Liebig condenser [537]. During time T, and for a further 15 min, the reaction mixture was boiled and stirred, after which the precipitate was separated by filtration and dried overnight at 105 °C. Although stoichiometry could have been explored as a function of many experimental variables, only three were chosen (Equation 3.3). It was found that the Ca/P molar ratio was equal to $1.688 + 0.020x_1 + 0.024x_2 - 0.019x_1x_2$ to a good approximation for a fixed time T of 180 min. The coefficients, x_1 and x_2, were determined by a least squares procedure. Conditions for the synthesis of s-OHAp yielded a product, after heating at 900 °C for 5 h, with a Ca/P molar ratio of 1.67±0.01 and lattice parameters $a = 9.420±0.002$ and $c = 6.880±0.002$ Å.

OHAp has been prepared by boiling CaSO$_4$ with strongly alkaline Na$_3$PO$_4$ (0.2 mol l^{-1}) [639]. It is also possible to prepare OHAp by heating ClAp

$$x_1 = \frac{(V_1 - 80)}{40}$$

$$x_2 = \frac{(V_2 - 100)}{60}$$

$$\text{and } x_3 = \frac{(T - 180)}{60}$$

3.3

[75,620] or BrAp [80,576] in steam at about 800 to 1000 °C. The product from ClAp is monoclinic [400,620] (Fig. 3.1); OHAp from BrAp has not been investigated, but is likely to be also monoclinic. As both ClAp (Sections 2.4.1 and 2.5.2) and BrAp [50] can be easily produced in a well-crystallised form, this approach, although little used, should be a good method for producing substantial quantities of monoclinic OHAp with micrometre or larger crystals (0.4 mm twinned monoclinic crystals of OHAp have been made from ClAp [613], Section 3.5).

Oriented crystals of OHAp can be produced by the hydrolysis of oriented preparations of OCP grown on a cation-selective membrane [164] (Section 1.3.5).

A variety of different methods have been developed for the production, sometimes on a large scale, of OHAp for use directly, or after further processing, as a bone replacement material or for the coating of metal prostheses [105,597,633,640-642] (and references therein). These biomaterials applications are discussed further in Section 4.6.7.

3.4 Preparation of other apatites with hydroxyl ions
3.4.1 Calcium-deficient and nonstoichiometric apatites

There is no difficulty in preparing Ca-def and ns-OHAps in aqueous systems, the problem lies in obtaining a single phase product with larger than 1 μm crystals, so that the bulk composition will not be dominated by that of the surface. As discussed in Section 3.3.1, there seems to be a small composition range possible for OHAp, even under equilibrium conditions, so that some of the methods described previously for s-OHAp and some hydrothermal methods [17,18] can be used to produce slightly ns-OHAps.

A Ca-def OHAp has been prepared by the simultaneous addition of $Ca(CH_3COO)_2$ and $(NH_4)_2HPO_4$ solutions (both 0.1 mol l^{-1}) at a rate equivalent to 2.5 mmol of $Ca_3(PO_4)_2$ per hour to boiling CH_3COONH_4 at pH 4.5 with constant stirring [643]. The mixture was refluxed for 16 to 20 h before filtration, and the precipitate washed with cold water and C_2H_5OH before drying. The initial precipitate was OCP, but the final product only contained apatite. The crystals were lath-shaped, up to 200 μm long and 5 μm wide, with a thickness much less than the width. The Ca/P molar ratio was about 1.56 and the lattice parameters a little different (a +0.16% and c -0.09%) from those of sintered OHAp. If these are taken as a = 9.418 and c = 6.881 Å (Table 1.4), the ns-OHAp lattice parameters were a = 9.433 and c = 6.875 Å. The diffraction pattern was unchanged after heating at 330 °C; above 450 °C, a small amount of α-$Ca_2P_2O_7$ was seen, and after heating at 600 °C, the lattice parameters changed to those of sintered OHAp. For a Ca/P molar ratio extrapolated to 1.5, the weight loss on heating indicated a water content of 2.5 to 3.5 mol per unit cell. There was a 25 % phosphorus as pyrophosphate yield on heating at 600 to 650 °C, again for a Ca/P ratio extrapolated to 1.5. The lattice contraction on heating was attributed to loss of lattice water (Section 3.7.3).

Ca-def OHAps have been prepared in the pH range 6.4 to 13.38 by slow release of Ca^{2+} ions by boiling off water from a phosphate solution containing Ca^{2+} ions chelated to ethylenediaminetetra-acetic acid (EDTA) [181]. 60 g of EDTA and 60 g of DCPA or $Ca_3(PO_4)_2$ were mixed in 1.8 l water and brought to the desired pH with alkali, usually KOH. After stirring for 6 h to allow maximal dissolution, excess solids were removed by filtration. The volume was reduced to one-sixth by boiling whilst N_2 was bubbled through, then the crystals were separated, washed, and dried at 110 °C for 12 h. The preparations all had an enlarged a-axis parameter compared to normal OHAp (up to a = 9.45 Å), and the c-axis parameter seemed to be related to the pH of formation. For pH < 7, c was slightly smaller, and in alkaline conditions, slightly larger than in normal OHAp. One sample prepared at pH 6.96 had a composition (in wt %) of: K_2O 0.68, CaO 51.47, P_2O_5 43.21, CO_2 0.04, H_2O 4.6 (by difference), which gave Ca/P and (Ca+K)/P molar ratios of 1.51 and 1.53 respectively. Loss of water on heating and resultant lattice parameter changes are discussed in Section 3.7.3.

Hydrolysis of DCPD in boiling distilled and deionised water for one month with four to six changes to reduce acid build up, produced an apatite with a Ca/P molar ratio of 1.60 (chemically determined) and a lower Ca/P ratio than s-OHAp (by Rietveld analysis) [620]. Lattice parameters were a = 9.4290(3) and c = 6.8805(3) Å.

Ca-def OHAp with crystals of micron dimensions has been synthesised by slow hydrolysis of DCPD (80 °C, 94 h, 0.1 × 2 μm² crystals) or DCPA (90 °C, 71 h, 0.25 × 10 μm² crystals) in which the acid produced was neutralised by NH_3 from the slow hydrolysis of urea [644]. 5 l of water, containing 50 g of the solid and 9 g of urea, were adjusted to an initial pH of 4.0 with HNO_3. The pH after the reaction was 6.6 to 6.8. Chemical analysis gave a Ca/P molar ratio of 1.53±0.03, XRD showed only the presence of apatite, and after heating at 1000 °C, only β-TCP. Lattice parameters and refractive indices were not given, but IR showed the presence of acid phosphate (band at 875, shoulders at 1180 and 1200 cm⁻¹), and a trace of carbonate. NMR studies were consistent with the chemical formula $Ca_9HPO_4(PO_4)_5OH$ [193] (Section 3.12).

The constant composition method (Section 3.8) has been used at 37 °C to prepare nonstoichiometric apatites with Ca/P molar ratios from 1.49 to 1.65 at pH 6.0 to 9.0 [645,646]. Growth was on OHAp seeds in solutions with a Ca/P molar ratio of 1.667 and ionic strength of 0.100 mol l⁻¹ at equilibrium with respect to OCP, but supersaturated with respect to OHAp (increasingly so at higher pH). There was little difference between the IR spectrum (including the HPO_4 band at 875 cm⁻¹) and lattice parameters (a = 9.42, c = 6.88 Å) of the product compared with the seed material.

3.4.2 Calcium-rich apatites

Precipitated apatites strongly adsorb $Ca(OH)_2$ from lime solutions [647]. The possibility of precipitated BCaPs with a Ca/P molar ratio greater than 1.667 has been reviewed [6,648]. The method discussed earlier [537] (Section 3.3.4), in which the stoichiometry was explored as a function of precipitation conditions, gave products with Ca/P molar ratios from 1.63 to 1.73. The precipitates with a Ca/P molar ratio above 1.67 were thought to be mixtures of OHAp and $Ca(OH)_2$, or an apatite adsorbing Ca^{2+} ions and an equivalent amount of negative ions.

A series of calcium-rich apatites with Ca/P molar ratios above 1.67 has been prepared by heating s-OHAp and $CaCO_3$ in air presaturated with water vapour at 1000 °C for 10 days [537]. XRD showed only an apatitic phase whose c-axis parameter remained constant at 6.884 Å, whilst the a-axis parameter decreased linearly with excess $Ca(OH)_2$ from a = 9.420 to 9.373 Å at which point the Ca/P molar ratio was 1.75. Chemical tests and IR showed the absence of a separate $Ca(OH)_2$ phase, but above a ratio of 1.75, IR indicated its presence. The thermal stability of this new apatite under various conditions was reported [537], as well as its IR spectrum (Section 3.11.6), which showed additional OH bands. The possibility has been raised that the change in lattice parameters and the extra OH IR bands (see Section 3.11.5 for a discussion of these) might be due to accumulation of F⁻ ions from the muffle furnace (BO Fowler, personal communication, 1991).

3.4.3 Apatites with oxygen in different oxidation states
Oxyapatites (OAps), with O(-II) as O²⁻ ions

The existence of oxyapatite, $Ca_{10}(PO_4)_6O$, was the subject of controversy for many years [85,86]. However, its formation now appears to be well-established, though it possibly always contains a small amount of OH⁻ ions. Perhaps for this reason, it does not occur in published high temperature phase diagrams of the $CaO-P_2O_5$ system (Section 1.6.6). The name "vœlckerite" was coined in 1912 by Rogers [649] for the (possible) mineral form of oxyapatite, after JA Vœlcker whose analyses [650] first showed an apparent halogen deficiency in some mineral apatites. The IR of OAp is discussed in Section 3.11.6.

In 1933, it was demonstrated by XRD that strongly heated OHAp could lose a substantial amount of its constitutional water, whilst the integrity of the lattice was maintained [322]. In this experiment, OHAp was heated in a vacuum for 7.5 h at 1360 °C and the change in water content measured; after this treatment, the formation of a solid solution of OHAp and OAp with the formula $Ca_{10}(PO_4)_6(OH)_{0.9}O_{0.55}$ was postulated. In 1957, the loss of constitutional water on heating OHAp in air for 4 h at 1300 °C was shown to be accompanied by a contraction of the a-axis dimension of about 0.02 Å (c was not mentioned) [651]. The loss probably starts at temperatures as low as 800 to 900 °C.

The water content of $Ca_{10}(PO_4)_6(OH)_{2-2x}O_xV_x$, where V is an OH⁻ ion vacancy, at equilibrium with water vapour of partial pressure p mm Hg ($0.133p$ kPa), has been studied as a function of temperature T (Kelvin) [621]. The reaction was discussed in terms of the equilibrium:

$$2OH^-(solid) \rightleftharpoons O^{2-}(solid) + V(solid) + H_2O(gas). \qquad 3.4$$

From the Law of Mass Action [621,652]:

$$\frac{px^2}{(2-2x)^2} = K(T). \qquad 3.5$$

The equilibrium constant, $K(T)$, was found [621] to be:

$$K(T) = 1.2 \times 10^9 exp(-\Delta H/RT) \qquad 3.6$$

where R is the gas constant and ΔH the experimentally determined enthalpy change (60 kcal mol⁻¹, 251 kJ mol⁻¹). At 1230 °C, OHAp could lose up to 75% of its constitutional water without loss of the apatite structure [621]. At higher temperatures and a low water pressure (1 mm Hg, 0.13 kPa), α-TCP and TetCP were formed (Reaction 3.7) [621]. In another study [19,21], it was found that the borderline at which Reaction 3.7 occurred was at 1598±10, 1750±10, and 1838±10 K for partial water pressures of 4.6, 73.6 and 760 mm Hg (0.613, 9.81 and 101.3 kPa) respectively.

$$Ca_{10}(PO_4)_6(OH)_2 \rightarrow Ca_4(PO_4)_2O + 2\alpha-Ca_3(PO_4)_2 + H_2O \qquad 3.7$$

Thermogravimetric analysis showed that, under a vacuum of ~10⁻⁵ mm Hg (1.3 mPa) at 1000 °C, OHAp loses about three-quarters of its constitutional water to form an O,OHAp, with lattice parameters $a = 9.402\pm0.003$ and $c = 6.888\pm0.003$ Å [85,86,653]. This compound could be rehydroxylated in a stepwise fashion by successive heatings to 400 °C in an atmosphere with a partial water pressure of 20 torr (2.7 kPa) so that the lattice parameters returned to those of OHAp in successive steps [85,86]. Extrapolation of these results back to a zero constitutional water content gave lattice parameters $a = 9.40$ and $c = 6.89$ Å. Heating OHAp at 800 °C and 0.05 mm Hg (6.6 Pa) for 20 h, with cooling to 25 °C under the same vacuum, caused a 20 % loss of OH⁻ ions [654].

In studies [652] with $p = 5$ mm Hg (0.67 kPa), temperatures between 800 and 1250 °C, and x (maximum value ~0.7) calculated from Equation 3.5, the lattice parameters (Å) were given by Equation 3.8.

$$a = 9.4197 - 0.0268x$$
$$c = 6.8805 - 0.0129x$$

3.8

Putting $x = 1$, gives $a = 9.396$ and $c = 6.8676$ Å for the lattice parameters of fully dehydroxylated OHAp. The a-axis parameter agrees with the value given in the previous paragraph, but the c-axis parameter implies that dehydroxylation causes a small reduction in c, whereas the earlier result shows a small expansion. In discussing this difference, it was suggested that the quoted errors in the earlier study that showed an expansion were sufficiently large that, in fact, these results were compatible with a small contraction in the c-axis parameter and that the accuracy of the later study resolved any doubt about the matter [652]. However, examination of the plots of the c-axis parameter against composition of the earlier (Fig. 26 of ref. [85] or Fig. 8 of ref. [86]) and later (Fig. 1 [652]) studies indicates clearly that two sets of results are not consistent. Furthermore, measurement of the lattice parameters at 850 ° C of OHAp in air ($a = 9.526\pm0.005$, $c = 6.955\pm0.005$ Å) and OAp in a vacuum of 10^{-5} torr (1.3 mPa) ($a = 9.519\pm0.005$, $c = 6.978\pm0.005$ Å), also showed an expansion of the c-axis parameter. [85,86,655]. An increase in the c- and reduction in the a-axes for dense O,OHAp blocks has also been clearly demonstrated [642]. The samples were prepared by heating polycrystalline blocks of OHAp in the temperature range 1050 to 1450 °C; there was a substantial diminution in the intensity of the OH IR stretching band and the lattice parameter changes increased as the firing temperature increased [642].

It is said that OAp decomposes to α-TCP and TetCP above 1050 °C, even when traces of water are carefully excluded [655].

Sr, Ba and PbOHAps all show substantial dehydroxylation on strong heating [54]. The lattice parameters of SrOHAp changed from $a = 9.765$, $c = 7.280$ Å to $a = 9.724$, $c = 7.263$ Å, with heating at 1600 °C, and BaOHAp changed from $a = 10.177$, $c = 7.731$ Å to $a = 10.153$, $c = 7.696$Å, with the same treatment. Thus, in all three alkaline earth OHAps, loss of constitutional water causes a contraction of the a-axis parameter (but see below).

There is a report that, when SrOHAp was dehydroxylated in vacuum, a new apatitic phase sometimes occurred [54]. After heating SrOHAp at 1275 °C and 10^{-2} mm Hg (1.3 Pa) for 110 h, the unit cell parameters were $a = 9.872$ and $c = 7.199$ Å. It is possible that the change in lattice parameters was due to the formation of SrClAp rather than a SrOAp, the Cl⁻ ions deriving from a chlororganic solvent used to clean the furnace (T Negas, personal communication, 1991). The formation of this new phase has apparently been confirmed and the absence of OH⁻ ions demonstrated by IR [85,655]. However, it did give a weight loss on heating with SrF_2, which is inconsistent with it being s-OAp. A possible explanation for this contradiction is that the later preparation of the new phase was also contaminated with Cl⁻ ions, and the weight loss due to volatilisation of $SrCl_2$ (BO Fowler, personal communication, 1991).

An alternative method of preparing OAp, first used by Wallaeys in 1954 [656], is to heat A-CO_3Ap (idealised formula, $Ca_{10}(PO_4)_6CO_3$, Section 4.3.1) in a dry atmosphere. More recently, high vacuum, dry nitrogen or helium (oxygen must be excluded) has been used [85,86,653,655]. Thermogravimetric analysis under vacuum ($\sim10^{-4}$ mm Hg, 13 mPa), showed that, between 700 and 1000 °C, the loss of CO_2 corresponded to formation of almost pure OAp. However, when the apatite was cooled, even under high vacuum (10^{-6} mm Hg, 0.13 mPa), there was always a small uptake of water to form $Ca_{10}(PO_4)_6(OH)_{0.5}O_{0.75}\square_{0.75}$, where $\square$ represents an OH⁻ ion vacancy. This apatite had lattice parameters $a = 9.402\pm0.003$ and $c = 6.888\pm0.003$ Å [85,86,653]. No OH stretch could be seen in the IR spectrum [85,86,653], but a weak Raman band occurred at 3550 cm^{-1} [85,653]. The IR inactivity and Raman activity were attributed to a centrosymmetrical, weakly hydrogen-bonded group, O-H···O. The lowering of the Raman frequency from that in OHAp due to this bonding implied an O-O distance of about 3.0 Å, which could occur if the oxygen atom of the OH⁻ ion occupied its normal position, 0.3 Å from the plane of a Ca(2) triangle, with its hydrogen atom facing an O^{2-} ion in the plane of an adjacent Ca(2) triangle [85]. Although all lines in the X-ray powder pattern of OH,OAp could be indexed with a space group $P6_3/m$ [85,86,655], which indicates disordering of O^{2-} ions between 0,0,¼ and 0,0,¾, it is possible that in s-OAp, there is an ordered structure, as in PbOAp which has a doubled c-axis parameter [62] (Section 2.3.1).

High temperature XRD has been used to study the decomposition of A-CO_3Ap (preparation, Section 4.3.1) at 10^{-5} mm Hg (1.3 mPa) between 20 to 1000 °C [85,86,655]. There was evidence of a miscibility gap at 850 °C in the solid solution between A-CO_3Ap and OAp.

It is possible to prepare $Ca_{10-x}Eu_x(PO_4)_6O_{1+x/2}\square_{1-x/2}$ with $0 \le x \le 2$ [100,444] and strontium [445] dioxyapatites, in which there are more than one O^{2-} ion per unit cell and in which Ca^{2+} or Sr^{2+} ions are partially replaced by trivalent rare-earth ions. These dioxyapatites can be hydroxylated or fluoridated [100,657]. Yttrium-substituted O,OHAps have been reported [658].

Peroxyapatites (perOAps), with O(-I) as O_2^{2-} ions

If OHAp or A-CO$_3$Ap is heated in dry oxygen (instead of nitrogen, helium or vacuum as used for preparing OAps) oxygen in an oxidation state -1, O(-I), occurs in the apatitic product [85,87,449]. The lattice parameters decreased progressively as the O(-I) content increased. For 2.6_7 wt % O(-I) (the maximum obtained), the lattice parameters were $a = 9.36_9$ and $c = 6.85_1$ Å. The theoretical maximum for perOAp, $Ca_{10}(PO_4)_6O_2$, is 3.19 wt %. PerOAp was found to be diamagnetic, so it was thought to contain one diamagnetic peroxide (O_2^{2-}) ion and vacancy on the c-axis per unit cell, rather than two paramagnetic O^- ions [85,87,449]. The IR of perOAp was distinct from OAp. In perOAp, a weak Raman line, but no IR band, was seen at 750 cm^{-1} [449] (reported at 850 cm^{-1} in refs [85,87]) which was assigned to the symmetrical O-O stretch. A sharp shoulder at 1130 cm^{-1} was assigned to a PO$_4$ vibration that occurs [85,87,659] when divalent ions are associated with vacancies in the c-axis channels. Thus, the reaction between O,OHAp and dry O$_2$ at 900 °C is [85,87,449]

$$Ca_{10}(PO_4)_6(OH)_{2-2x}O(-II)_x\square_x + \tfrac{1}{2}\,yO_2 \rightarrow$$
$$Ca_{10}(PO_4)_6(OH)_{2-2x}O(-II)_{x-y}(O(-I)_2)_y\square_x, \qquad 3.9$$

or as regards the oxygen

$$yO^{2-} + \tfrac{1}{2}yO_2 \rightarrow yO_2^{2-}. \qquad 3.10$$

Oxidation number $\qquad$ −2 $\qquad$ 0 $\qquad$ −1

If perOAp is heated in dry helium, the O_2^{2-} content falls above 400 °C, and has completely gone by 900 °C to yield O,OHAp by a reversal of Reaction 3.9 [85,87,449]. On the other hand, if the heating is done in air so that water is available, loss of O_2^{2-} starts at 200 °C, and is complete by 600 °C; O^{2-} ions do not occur in the apatite, because on formation, they immediately react with water to form OH⁻ ions, so the product is OHAp. In both cases, it seems that O$_2$ is entirely lost from the lattice.

It is possible to prepare SrperOAps by the same methods used for calcium compounds [85,87,449]. Their properties are similar, but not identical. For example, on heating the strontium compound in air, some of the molecular O$_2$ that is formed is retained by the lattice up to 350 °C, so that the change in lattice parameters with temperature of heating is irregular, and the *a*-axis parameter can become larger than that of the original SrOHAp.

Oxygen-rich apatites (O-rich Aps), with O(0) as O_2 and O(-I) as O_2^{2-}

An oxygen-rich apatite has been prepared by the hydrolysis of β-TCP in boiling 30 % H_2O_2 (presumably 30 wt %) for 12 h [425]. Needles (1.5 × 0.2 μm^2) were formed, with lattice parameters $a = 9.500 \pm 0.002$, $c = 6.875 \pm 0.002$ Å, which released about 2.1 wt % O_2 on acid dissolution. The Ca/P molar ratio was 1.58, and the excess O_2 was retained after heating at 580 °C, indicating that it was an integral part of the lattice. After heating at 775 °C, the sample decomposed to apatite and β-TCP, with a small residual (0.2 wt %) excess O_2.

Further studies have been made of O-rich Aps synthesised by the hydrolysis of β-TCP in H_2O_2 solutions of up to 110 volumes [85,87,426,660-662]. It was found that a substantial fraction of the oxygen was present as O(-I) [85,87,426]. For example, an apatite with lattice parameters $a = 9.51_4$ and $c = 6.87_5$ Å had 1.6_8 and 1.0_5 wt % of O_2 and O(-I) respectively. IR studies indicated that the OH band decreased in intensity in O-rich Aps synthesised at high concentrations of H_2O_2; in the presence of F$^-$ ions, FAp with little or no extra oxygen was formed. From these results, it was deduced that the oxygen species displaced OH$^-$ ions from the *c*-axis channels and could in turn be displaced by F$^-$ ions.

ESR and Raman studies were undertaken in an attempt to identify the oxygen species responsible for the O(-I) seen in the chemical analysis. ESR studies clearly showed the absence of O$^-$ ions [85,87] and presence of superoxide (O_2^-) ions [660,661]. However, the quantity of O_2^- ions seen by ESR was less than would have been expected from the quantities of O_2 and O(-I) determined by chemical analysis. Furthermore, if O_2^- ions were the only oxygen species present in the O-rich Ap, equal quantities of O_2 and O(-I) should have been observed because O_2^- will decompose during acid dissolution in the chemical analysis via the reaction $2O_2^- \rightarrow O_2^{2-} + O_2$. The ESR study could give no information about the peroxide ion, O_2^{2-}, which was likely to be present, because this ion is not paramagnetic; however, the peroxide ion can be detected by Raman spectroscopy. As a consequence, a study of the Raman spectrum was undertaken, using a rotating cell, so that little thermal decomposition occurred [426]. The spectrum showed a weak line at 809 cm^{-1}, which was assigned to the O_2^{2-} ion (note that this frequency differs from that given earlier for perOAp). It was therefore deduced that a significant part of the O(-I) seen in the chemical analysis originated from peroxide ions. No bands attributable to molecular O_2 were observed in the Raman spectrum. However, if a rotating cell was not used, so that decomposition occurred, the 809 cm^{-1} line was not seen, but a weak line at 1553 cm^{-1} appeared, which was assigned to molecular O_2. The comment was made that the Raman studies had to be interpreted with caution because the oxygen species were present in small amounts [426]. The conclusion from these results was that O-rich Aps contained O_2 and O_2^{2-} ions and some O_2^- ions, although further studies were required [85,87,426]. IR also showed the presence of acid phosphate [85,87,426].

As oxygen-rich Aps are heated, the excess O_2 content increases to a maximum of 1.9 wt % at between 400 to 600 °C, O_2^{2-} decreases to 0.1 wt %, and acid phosphate is lost via Equation 3.15 (Section 3.6.2) to yield $P_2O_7^{4-}$ ions and water [85,87,426]. At 700 °C, neither O_2 nor O_2^{2-} ions could be detected and the lattice parameters were $a = 9.42_1$ and $c = 6.88_5$ Å. Thus, on heating, O_2^{2-} ions disproportionate via a reversal of Equation 3.10 into O_2 and O^{2-} ions; the O^{2-} ions then react with atmospheric water to yield OH ions which could be detected by IR. O_2 could be clearly seen by Raman spectroscopy (line at 1553 cm^{-1}) in the product formed at 600 °C [426].

Oxygen-rich strontium and barium, and mixed calcium/cadmium apatites cannot be made by the method used for calcium salts, but have been prepared at 80 °C by the dropwise addition of a solution of the metal nitrate containing H_2O_2, to an ammoniacal solution of $(NH_4)_2HPO_4$, also containing H_2O_2 [426,662]. These apatites all contained variable amounts of lattice oxygen species (essentially O_2 and O_2^{2-}). This concentration was highest for the strontium compound (O_2 2.1$_9$, O_2^{2-} 1.4$_6$ wt %). On heating, the O_2^{2-} ion content fell, but unlike the calcium salt, the O_2 content also fell, even though it was being formed, because it was readily lost from the lattice. This behaviour, and that of the Ca- and SrperOAps discussed earlier, was explained by analogy with inclusion compounds, namely that the c-axis channels would not retain O_2 if they were either too small or too large [426].

The location of the O_2^- ion has been studied by ESR in calcium, strontium and barium phosphate and arsenate oxygen-rich apatites [661]. The spectra were characteristic of an anisotropic g factor with two well-resolved components which enabled g_z and g_x (= g_y) to be determined. The values obtained were inconsistent with those expected if the O-O bond was parallel to the c-axis. This finding made it unlikely that the ion was at the centre of the Ca(2) triangle, as symmetry would then require it to be parallel to the c-axis. The preferred location was at $z = \frac{1}{2}$, and considerations of the minimum energy suggested an angle of ~34° to the c-axis [661].

3.4.4 Miscellaneous preparations, including solid solutions

Preparative details for many Ba, Sr, Pb, VO_4 and AsO_4 OHAps, including solid solutions, can be found in the references cited in sections that deal with the structure of OHAp (Section 3.2), structural aspects of substitutions in apatites (Section 2.3.1, hexad axis; Section 2.3.2, Ca^{2+} ions; and Section 2.3.3, PO_4^{3-} ions) and in some of the IR sections on OHAp (Sections 3.11.3, 3.11.5 and 3.11.6).

Apatites with Sr, Ba, Pb, V, As, etc.

Preparation of OHAp from $CaSO_4$ (Section 3.3.4) can be generalised to $M_{10}(XO_4)_6(OH)_2$ (M = Sr, Pb; X = P, V, As) by boiling MSO_4 with strongly alkaline Na_3XO_4 (0.2 mol l^{-1}) [639]. Lattice parameters were given and, for lead compounds, it was suggested they contained sodium. Barium and cadmium apatites could not be prepared by this method. $Eu_5(AsO_4)_3OH$ can be prepared by boiling $EuSO_4$ under argon with a solution containing an excess of Na_2HAsO_4 made strongly basic with NaOH [493].

The high temperature solid state method of synthesis of OHAp from DCPA and $CaCO_3$ (Section 3.3.2) has been used for SrOHAp, BaOHAp and mixed Ca,SrOHAp (whole range possible) [628]. A marked discrimination against Sr^{2+} ions was found in studies of the aqueous precipitation of Ca,SrOHAps, but the whole composition range could be made [53,663]. Lattice parameters of Ca,SrOHAps, for which Vegard's law was followed, have been published [53,471]. For SrOHAp, $a = 9.758$ and $c = 7.28$ Å [471]. Only small amounts of barium could be incorporated in OHAp by precipitation, but solid state reactions at 1200 °C gave solid solutions for $0.6 > y > 1$, where y is the mole fraction of barium replacing calcium, with lattice parameters [664]

$$a = 9.412 + 0.789y$$
$$\text{and } c = 6.883 + 0.872y.$$

3.11

The parameters for BaOHAp were $a = 10.191(3)$ and $c = 7.747(3)$ Å [664]. A structural explanation for the 0.6 limit could be that, in Ca,BaOHAp containing substantial amounts of barium, Ca^{2+} ions only go to Ca(1) sites. As these comprise 40 % of the total, the minimum Ba^{2+} content would be 60 %, all in Ca(2) sites. This segregation would parallel Ca,BaFAp, for which XRD has shown that Ba^{2+} ions prefer Ca(2) sites [470] (Section 2.3.2).

The complete compositional range of Ca,PbOHAp can be synthesised [465], although it has been suggested [466] that there is a miscibility gap in the composition. Ca,PbOHAps have been made by high temperature methods followed by hydrothermal treatment [465]. The hydrothermal treatment was required to reverse the formation of Ca,PbOH,OAps at high temperatures, due to the loss of water. This thermal instability increased very markedly as the lead content increased. For the lead-rich range (Pb ≥ 50 atom %), PbO, $CaCO_3$ and $(NH_4)_2HPO_4$ were heated together at a final temperature less than 900 °C. The calcium-rich range was synthesised with some difficulty because of the loss of TCP from the crystals at 800 °C. For this range, an "active" mixture was made by precipitation between Pb^{2+} and Ca^{2+} ions in solution and a solution of PO_4^{3-} ions, NH_3 and $(NH_4)_2CO_3$, which was then heated at a maximum temperature of 780 °C for 80 h. For the hydrothermal treatment, 1 g of the apatite and 10 ml of water were heated in a silver capsule at 350 °C for

24 h in an autoclave of 18 ml capacity. There were substantial deviations from Vegard's law with a break in the dependence of the c-axis parameter on composition for 60 atom % replacement of calcium by lead [465]. The lattice parameters of PbOHAp were $a = 9.879$ and $c = 7.434$ Å [56]. XRD intensities showed that, at 60 atom % Pb composition, the Ca^{2+} and Pb^{2+} ions were segregated on Ca(1) and Ca(2) sites respectively [465] (Section 2.3.2).

There is a marked discrimination by Pb^{2+} against Ca^{2+} ions in aqueous preparations [465], and Ca^{2+} ions in already precipitated OHAp will readily exchange with Pb^{2+} ions in solution (up to 0.4 g Pb per g CaOHAp) [665]. This discrimination of the larger Pb^{2+} against the smaller Ca^{2+} ion is the reverse of that seen above for Ba^{2+} and Sr^{2+} ions in relation to Ca^{2+} ions in solution which are also both larger than Ca^{2+} ions (in fact, discrimination of Ca^{2+} ions is stronger against Ba^{2+} ions, which are larger than Sr^{2+} ions). Thus, as with the distribution of different ions between Ca(1) and Ca(2) sites (Section 2.3.2), ionic size alone is a restricted guide to behaviour. There is a very limited substitution possible of Ca^{2+} by Mg^{2+} ions in OHAp (Section 2.3.2).

Solid solutions with FAp and ClAp

OH,FAps and OH,ClAp can be prepared over the whole composition range by heating OHAp with the calculated quantity of the appropriate calcium halide at 800 °C for several hours [75]. If the CaO by-product formed in the reaction is undesirable (it converts to $Ca(OH)_2$ and/or $CaCO_3$ on standing in the atmosphere), the calculated quantity of ClAp or FAp can be used instead of the calcium halide [75]. This method can clearly be extended to solid solutions between FAp, ClAp and OHAp.

A method for precipitation of s-OHAp at 100 °C [634] (Section 3.3.3) has been adapted [636] for the synthesis of OH,FAps over the whole composition range. 1.5 l of solutions of $(NH_4)_2HPO_4$ (0.121 mol l^{-1}) and $Ca(NO_3)_2$ (0.222 mol l^{-1}) were pumped at 40 ml h^{-1} into 5 l of CH_3COONH_4 (1.18 mol l^{-1}). NH_4F was dissolved into the phosphate solution to give F^- ion concentrations of 0 to 0.04 mol l^{-1}. The pH (9 to 9.5) was maintained with NH_3 as described previously (Section 3.3.3) and CO_2 carefully excluded. The crystals (3 to 6 μm long and 2 to 3 μm thick, occasionally 10 to 15 μm long and 5 to 7 μm thick) had a parallel extinction and mean refractive index 1.636±0.002 (white light). The a- and c-axis parameters changed linearly with F^- ion content from 9.422±0.002 and 6.879±0.002 Å (OHAp) to 9.382±0.003 and 6.887±0.002 Å (FAp), respectively [636]. The compositions of OH,FAps synthesised at 80 °C in an ammonium acetate solution maintained at pH 7.4±0.1 have been investigated [666]. Calcium acetate solutions and $NH_4H_2PO_4$ plus HF solutions, with a range of Ca/P molar ratios from 0.1 to 10, were pumped in at constant rates; the F^- ion concentration was adjusted such that the Ca/F molar ratio was always five (the theoretical for FAp). The F^- ion uptake at low Ca/P ratios was strongly inhibited, so that the F$^-$ ion content of the OH,FAp precipitated for a Ca/P molar ratio of 0.1 was only about half that for a ratio of 1.67, for which the F$^-$ ion content approached the theoretical for FAp [666]. Heterogeneous F,OHAps have been made in solution by supplying the F$^-$ ions during either the first, or the second, half of the precipitation process [667]. Although the precipitates had similar F$^-$ ion contents, their characteristics suggested the crystals were either OHAp covered with FAp, or FAp covered with OHAp.

OH,ClAps can be synthesised in solution under normal conditions of temperature and pressure with only a very small chlorine content because of strong discrimination of the apatite lattice against Cl$^-$ ions in favour of OH$^-$ ions. However, addition of CaCl$_2$ solution dropwise into sodium phosphate solution containing CaCl$_2$ (up to 5 mol l^{-1}) at 70 to 100 °C, or hydrolysis of DCPA in NaCl solution (up to 5 mol l^{-1}) at 95 to 100 °C, yielded OH,ClAps with up to 3 wt % chlorine [668]. Hydrothermal methods at 400 to 425 °C and 52,500 psi (362 MPa) have been used to grow OH,ClAp single crystals up to 1 mm with a chlorine content of 3.6 to 5.5 wt % [417] (Section 2.5.2).

Apatites precipitated in the presence of organic molecules

There is evidence that some organic molecules can be incorporated into OHAp (or at least coprecipitated with it), but much less than the amounts of dicarboxylic acids that can substitute into OCP (Section 1.3.9).

PbOHAps containing glycine have been made by precipitation between Pb(NO$_3$)$_2$ and sodium or ammonium phosphate solutions in the presence of glycine, sometimes with added NH$_4$OH [669]. These apatites had slightly larger *a*- and smaller *c*-axis parameters compared with the glycine-free compound; for Pb$_{9.71}$(PO$_4$)$_6$Y$_{1.42}$Gly$_{0.34}$ (Y is a negatively charged constituent required for charge balance), the changes were +0.024 and -0.011 Å respectively. As these changes were typical of the substitution of a larger ion on the *c*-axis, it was thought that glycine was located on this axis.

It has been inferred that CH$_3$COO$^-$ ions can be trapped in the lattice of precipitated CO$_3$Aps, from the appearance of the ESR signal from the ĊH$_3$ radical after annealing at 150 °C and X-irradiation [670], and also from gas chromatographic detection of CO evolved on heating [671,672] (Section 4.5.7).

Apatitic coprecipitates of amino-2-ethylphosphate from alkaline 1:1 H$_2$O-C$_2$H$_5$OH mixtures (v/v) have been prepared by adding an alcoholic solution of Ca(NO$_3$)$_2$ to an alcoholic solution of (NH$_4$)$_2$HPO$_4$ [537]. On the basis of chemical analyses and IR, the proposal was made of a limited solid solution

$$(Ca^{2+})_8(PO_4^{3-})_{3.5+1.5x}(HPO_4^{2-})_{2.5-2.5x}(PO_4^{2-}CH_2CH_2NH_3^+)_x(OH^-)_{0.5-0.5x} \qquad 3.12$$

(with 0≤*x*≤1) between the organic phosphate and the CaP, with pooling of phosphate groups [537].

3.5 Growth of hydroxyapatite single crystals

Crystals, usually needles, with maximum lengths of about 500 μm have been produced by hydrothermal synthesis. The reactants were heated with water in a closed bomb at fixed temperatures in the range 300 to 700 °C and pressures of 1250 lbs/sq inch (8.6 MPa) to 2,000 bar (200 MPa) for several days. Reactions that have been used include: recrystallisation of OHAp in the presence of NaOH [673,674]; recrystallisation in water alone [675]; hydrolysis of an acidic CaP with water [675,676] or $Ca(OH)_2$ [677]; reaction with water of a stoichiometric mixture of CaO and β-$Ca_2P_2O_7$ [678]; reaction of $Ca(NO_3)_2$, NaOH, KH_2PO_4 and H_2O (crystals up to 2.35 mm long by 0.2 mm across were synthesised) [679]; and reaction of H_3PO_4, $Ca(NO_3)_2$ and water [680]. Hydrothermal recrystallisation of a carefully prepared s-OHAp (composition within ~0.2 wt % of theoretical) at 2 kbar (200 MPa) and temperatures of 430 to 500 °C gave prismatic crystals with forms $\{10\overline{1}0\}$ and $\{10\overline{1}1\}$ and average size 0.1 mm, but some were up to 3 mm [681]. The lattice parameters were a = 9.426±0.003 and c = 6.884±0.003 Å. Although the possibility of a monoclinic space group was not discussed, examination of the published XRD pattern (Fig. 3C in ref. [681]) suggests this to be the case, as the monoclinic 212 reflection at ~36.4° 2θ (Fig. 3.1) seems to be present. The crystals contained a small amount of carbonate (<0.1 wt %); refractive indices were not reported [681]. In hydrothermal recrystallisation experiments at 300 °C and 165 atm (16.7 MPa) CO_2 pressure, needles of OHAp up to 0.5 mm were obtained after three months [681]. These crystals quite often had the $\{0001\}$ instead of the $\{10\overline{1}1\}$ form, particularly if the CO_3 content was slightly higher than 0.1 wt % [681]. On one occasion, 3 mm long crystals were grown, but this could not be repeated [681,682]. Hydrothermal recrystallisation of OHAp between 100 to 200 °C and pressures of 0.1 to 2 MPa produced prisms whose aspect ratio increased as the temperature increased; at 200 °C, the crystals were 25 nm wide by 90 nm long [683]. Chemical analyses and IR gave the formula $Ca_{10}(PO_4)_6(OH)_{1.81\pm0.03}(CO_3)_{0.095\pm0.015}$. Na^+ ions can be incorporated into OHAp if present during hydrothermal synthesis [674].

OHAp and ODAp crystals 4 to 8 mm in the *c*-axis direction, by 0.5 mm across have been grown by cooling a flux of the apatite in molten $Ca(OH)_2$ or $Ca(OD)_2$ [51]. Gold capsules, 4.5 cm long, were half filled with 90 wt % $Ca(OH)_2$ or $Ca(OD)_2$ and 10 wt % TCP and kept at 925 °C and 10,000 psi (69 MPa) for 24 h. The temperature was then lowered to 900 °C, cooled slowly to 775 °C during three to four days, quenched to room temperature and the flux dissolved in boiling dilute acetic acid [51]. The crystals were monoclinic (*b*-axis parameter doubled in precession photographs), and sections cut perpendicular to the *c*-axis showed mimetic twinning [51]. Refractive indices at 5830 Å were O = 1.650 and E = 1.644 [545].

Flux-grown crystals up to 8 mm in length by 0.5 mm in diameter have been crystallised from an apatitic CaP and $Ca(OH)_2$ flux with 25 wt % added water [684]. The temperature was oscillated between 750 °C (14,500 psi, 100 MPa) and 880 °C (17,500 psi, 121 MPa) to reduce the number of nuclei and enhance growth of preferred ones.

The method used to grow probably the largest crystals of synthetic OHAp ($7 \times 3 \times 3$ mm³) employed the retrograde solubility of OHAp between 300 °C (45,000 psi, 310 MPa) and 670 °C (60,000 psi, 414 MPa) [51,545]. Growth, predominantly perpendicular to the *c*-axis, was on a selected OHAp seed crystal in a OHAp-saturated nutrient that was provided by another OHAp crystal (both crystals were hydrothermally grown OHAp). With the nutrient crystal at 453 °C and the seed at 470 °C, a 36 mg seed grew to 50 mg in six days. Chemical analyses of the crystals were within 0.1 wt % of theoretical [545], and refractive indices at 5830 Å were O = 1.650 and E = 1.644 [51,545]. The crystals were monoclinic and mimetically twinned [51]. Attempts to grow untwinned monoclinic OHAp below the monoclinic to hexagonal transition temperature (211 °C) were unsuccessful [51]. The hydrothermal bomb method has been used to synthesise OHAp with various cationic dopants [677].

Single crystals of OHAp have been prepared by heating 0.4 mm spheres of flux-grown ClAp in steam at 1300 °C for two weeks, so that Cl⁻ ions could exchange with OH⁻ ions [613]. After the reaction, the spheres still contained 0.4 wt % chlorine and had refractive indices (sodium light) O = 1.651 and E = 1.6445 (estimated error 0.001), and an XRD pattern superficially similar to that of Holly Springs OHAp [613]. However, careful examination showed that *b* was doubled and the space group was $P2_1/b$ with three twin orientations about the *c*-axis [608]. One of these crystals was used for determining the monoclinic structure [609] (Section 3.2). Comparison of the calculated and observed intensities showed that only about 37% of the crystal was in the monoclinic form.

The presence of the doubled *b*-axis parameter, associated with ordering of OH⁻ ions, is a sensitive indicator of correct stoichiometry (Section 3.2). For most of the preparations described above, this doubling was not looked for, but in all cases when it was, the crystals were monoclinic, but microscopically twinned. Growth of mm sized single crystals of untwinned OHAp remains an unsolved problem.

Crystals of $Pb_{10}(XO_4)_6(OH)_2$, where X is P, As or V, 0.3 mm long have been grown under hydrothermal conditions at 425 °C for 4 days under a pressure of 1000 atm (101 MPa) [685].

3.6 Special analytical methods

3.6.1 Hydroxyl ion content

If an intimate mixture of an apatite and powdered CaF_2 is heated at 800 °C, all lattice OH^- ions are replaced by F^- ions with the evolution of water, so the weight loss can be used to determine the OH^- ion content, provided this is the only source of volatiles [75]. If carbonate is present, the weight loss should be corrected for CO_2 loss or the evolved water collected and weighed. Good agreement has been obtained between this method and the change in weight from loss of water on heating OHAp or O,OHAp at 1650 °C in dry O_2 [686] (but see the comment that P_2O_5 might be lost at this higher temperature [621]).

Dissolution in anhydrous methanolic HCl (5 wt %), followed by water estimation by the Karl Fischer reagent, has been used in an attempt to determine the OH^- ion content of calcified tissues [687]. This method applied to DCPD gave a water content of 96 % of theoretical [687], but does not distinguish between water already present or formed from the reaction of carbonate with the acid [623].

Another method for OH^- ion determination, involving dissolution and titration, has been described [623,688]. About 20 mg of the BCaP was dissolved in a slight excess of standard HCl (0.1 mol l^{-1}) and diluted to 25 ml. This solution was then titrated with standard KOH solution (0.1 mol l^{-1}) at 25±0.1 °C to the previously calculated end-point for neutralisation of the first proton of H_3PO_4, taking into account acid-base and Ca-phosphate ion-pair equilibria. The moles of OH^- ion in the BCaP are given by:

$$T_{OH} + 2T_P + 2T_{CO3} = T_A - T_B + T_{HPO4} + T_{HCO3} \qquad 3.13$$

in which T_{OH}, T_P and T_{CO3} are the moles of OH^- ions, phosphorus and CO_3^{2-} ions respectively. T_A is the moles of acid used to dissolve the solid and T_B the moles of base required in the back-titration. The left-hand side is the moles of H^+ ion required to bring the system to the state at the end-point when the great majority of the phosphorus is present as $H_2PO_4^-$ ions, and the OH^- and CO_3^{2-} ions are neutralised to water and CO_2 respectively. The difference between the first two terms on the right-hand side gives the moles of H^+ ion provided externally for the neutralisation, and the sum of the last two terms, the moles provided internally from the solid. T_{OH} can be calculated if T_P and T_{CO3} are known from direct chemical analysis, T_{HPO4} from pyrophosphate formation on heating (Section 3.6.2), and the reasonable assumption is made that T_{HCO3} is negligible. The quantities in Equation 3.13 have to be known as accurately as possible, because T_{OH} is determined by the difference between quantities of comparable magnitude. Nevertheless, an analysis of errors [623] indicated that the absolute error for the OH^- ion content was about ±5%. The value for the

OH⁻ ion content per PO_4^{3-} ion for s-OHAp was equal to the theoretical within this limit.

An oxalate precipitation/titration method used to determine the HPO_4^{2-} content of ACP on the assumption that OH⁻ ions were absent [340] (Section 3.6.2), should also be applicable to BCaPs. Thus, if the HPO_4^{2-} ion content is known, it could then be used to determine the OH⁻ ion content (see discussion in the following paragraph).

The basic question being addressed in chemical estimations of the OH⁻ ion content in BCaPs that contain HPO_4^{2-} ions and possibly water molecules is the equilibrium position of the reaction:

$$OH^- + HPO_4^{2-} \rightleftharpoons H_2O + PO_4^{3-}. \hspace{2cm} 3.14$$

The total amount of negative charge can be determined by the requirement of charge balance from accurate calcium and phosphorus analyses of the solid, or by dissolving the solid and titrating the ions to some predetermined state, as in the above titration and the oxalate precipitation/titration methods (following section). If either term on the left-hand side of Equation 3.14 is known, the other can be calculated. However, chemical methods have great difficulty in determining either of these individually, because the position of the equilibrium will generally change during the estimation. The only applicable methods are those in which the kinetics are sufficiently slow for the position of the equilibrium to be essentially unchanged. This basic assumption is made in the acid phosphate estimation from pyrophosphate formed on heating (following section), but as will be seen, there are also doubts about the validity of this method.

As a result of the problems with chemical analyses, any physical method that allows direct determination of OH⁻ or HPO_4^{2-} ion contents must be very attractive. Raman spectroscopy has been investigated as a means of determining the OH⁻ ion content of F,OHAp and Cl,OHAp [556] (Section 3.11.5). Studies of synthetic apatites of known composition led to the conclusion that sensitivities of 5 % for F⁻ replacing OH⁻ ions (from the 612 to 597 cm⁻¹ intensity ratio), ~2 % for Cl⁻ replacing OH⁻ ions (from the 612 to 451 cm⁻¹ intensity ratio), and 5 % for deficiency of OH⁻ ions (from the 3573 cm⁻¹ line intensity) were indicated, although it was thought that these limits could probably be improved. Analyses of dental enamel were difficult because of large amounts of visible-light fluorescence [556]. A comparative evaluation of IR and Raman spectroscopy of the OH stretching mode for estimating the OH⁻ ion content in a wide range of BCaPs (some containing carbonate) has been made [689]. Hexagonal OHAp of known OH⁻ ion content was used as a standard; the OH⁻ content was measured from the OH band area, absorbance or intensity ratio of the OH stretch to the intensity of the v_4 PO_4 (IR) or v_2 PO_4

(Raman) bands. IR methods were reproducible to ±5 % and agreed with the measurements using the titration method (see above) to within about ±15 %. Raman methods were much less accurate [689]. F⁻ ions will interfere with the IR estimations because the OH⁻ ion molar absorption coefficient depends markedly on the F⁻ ion content [690] (Section 3.11.5). The intensity of the OH Raman line at ~3570 cm⁻¹ has been used to measure the OH content of precipitated CO₃Aps [691] (Section 4.5.3).

Lattice and adsorbed water in apatites have been measured from the heights of the IR bands at 3300 and 3430 cm⁻¹, respectively [620].

NMR can, in principle, be used for quantitative estimation of hydrogen in different environments, but solid samples can present difficulties because of line broadening. Although the OH⁻ ion and water ¹H resonances can be separated, the resonances from HPO₄²⁻ ions and water molecules are very close, so this might also present difficulties (Section 3.12). In a study [637] of a precipitated apatite (preparation, Section 3.3.3; further related NMR studies [183], Section 3.12), the ¹H resonances due to water and OH⁻ ions could be distinguished (5.6 and 0.18 ppm respectively, relative to TMS) in spectra at 200 MHz with a 7.25 kHz spinning speed. Thus, the ratio of these two species could be determined from the ratio of the areas of the respective peaks, including sidebands. Integration of the ¹H spectrum at 200 MHz without spinning, gave the total hydrogen content, so that the OH⁻ ion and water contents (3.13 and 2.82 wt % respectively) could be calculated. The NMR determination of the OH⁻ content compared well with that from IR (2.76 ±0.17 wt %) [637]. A very weak resonance from HPO₄²⁻ ions was seen at 8.6 ppm in 500 MHz spectra (see next section for the NMR determination of HPO₄²⁻ ion content).

Recently, study of the multiple-quantum NMR dynamics of the quasi-one-dimensional distribution of proton spins in OHAp has been used to detect lack of coherence within the ···OH OH OH··· chains due to OH⁻ ion deficiency [692] (Section 3.12).

The total scattering from atoms in the *c*-axis channels has been determined from Rietveld X-ray powder pattern refinements [620].

3.6.2 Acid phosphate content

Chemical methods

BCaPs that contain HPO₄²⁻ ions form pyrophosphate on heating at 400 to 700 °C by the condensation [693]:

$$2HPO_4^{2-} \rightarrow P_2O_7^{4-} + H_2O. \qquad 3.15$$

There is less formed at higher temperatures because of loss from the reaction [693]:

$$Ca_2P_2O_7 + Ca_{10}(PO_4)_6(OH)_2 \rightarrow 4Ca_3(PO_4)_2 + H_2O. \qquad 3.16$$

The formation of pyrophosphate is rather slow, and as a compromise between maximum formation via Reaction 3.15 and minimum loss through Reaction 3.16, samples were heated at 600 °C for 60 h [693]. The HPO_4^{2-} ion content can be calculated from the pyrophosphate formed, assuming only Reaction 3.15 occurs. DCPD gave 86 % of the theoretical yield of pyrophosphate which indicates how low estimations might be [693]. A fairly steep loss in weight between 660 and 710 °C for $Ca_{9.00}(HPO_4)_{0.96}(PO_4)_{5.04}(OH)_{0.96}$ has been ascribed to Reaction 3.16 [92,328,329], which shows how close the temperatures for Reactions 3.15 and 3.16 are.

Detailed studies of the thermal decomposition of OCP (Section 1.3.7) and DCPD (Section 1.4.7), show how complex the mechanisms of the above reactions are likely to be. Incomplete pyrophosphate formation and loss of pyrophosphate through reaction with OHAp (Reaction 3.16) are also problems with estimations of OCP in apatitic mixtures [182] (Section 1.3.7); in this case, it was concluded that, under the conditions used, the assay of OCP by this method was not valid.

More recently, other conditions for pyrophosphate formation have been used, for example, heating at 550 °C in vacuum for 24 h [623]. The HPO_4^{2-} ion content has also been calculated from the pyrophosphate formed at the temperature at which it was a maximum for the particular sample under study [92,328,329]. This maximum was determined by heating a BCaP or ACP at 300 °C h^{-1} to different temperatures in helium or vacuum, with immediate cooling, followed by analysis [92]. The temperature for maximal pyrophosphate formation under these conditions was ~650 °C, but much lower if magnesium was present [92,328] (Section 1.8.7).

The presence of carbonate, either from admixed $CaCO_3$ [355,694], Na_2CO_3 [355] or from rinsing the apatite in a HCO_3^- ion solution [355], greatly reduced the pyrophosphate yield. Some CO_3Aps showed IR evidence of acid phosphate, but gave no pyrophosphate on pyrolysis [355]. Added CaO, but not $CaSO_4$, also gave less pyrophosphate [355]. The mechanism proposed for the reduced yield was reaction of carbonate (or O^{2-} ions formed from it) with HPO_4^{2-} ions (or already formed pyrophosphate), to give PO_4^{3-} ions with evolution of water and CO_2 (Equation 3.17). Interference from CO_3^{2-} ions was less severe for ACPs, but it was concluded that for calcified tissues, the method should be used with great caution and definitely not as an assay for total HPO_4^{2-} ion content [355]. Others have concluded that the presence of carbonate (and F$^-$ ions) invalidates the method [92]. Despite these shortcomings, extensive use has been made of the pyrolysis method to measure the HPO_4^{2-} ion content of biological apatites containing several percent of carbonate (Section 4.6.1), really because there is no practical alternative.

$$2HPO_4^{2-} + CO_3^{2-} \rightarrow 2PO_4^{3-} + CO_2 + H_2O \qquad \qquad 3.17$$

The yield of pyrophosphate in Reaction 3.15 is usually determined from the increase in the orthophosphate content on acid hydrolysis of the pyrophosphate to orthophosphate. Typical hydrolysis conditions are 1 h at 100 °C in perchloric acid (1 mol l^{-1}) [693]. Alternatively, the pyro- and orthophosphates may be separated by anion-exchange chromatography [695,696]. The orthophosphate can be eluted with Tris-HCl buffer (containing EDTA and NaCl) at pH 6.5 to 7.0; the pyrophosphate is then eluted with the same buffer, but with addition of NaCl [695]. If ^{32}P compounds have been used, the ortho- and pyrophosphates may then be assayed directly by liquid scintillation counting [695].

An oxalate precipitation/titration method has been used to determine the total HPO_4^{2-} ion content of ACPs and other CaPs on the assumption that OH$^-$ ions were absent [340]. The CaP (about 20 mg) was stirred overnight with 20 ml of a phosphate-buffered potassium oxalate solution (NaH_2PO_4 100 mmol l^{-1} and $(COOK)_2$ 20 mmol l^{-1}), adjusted to a known pH between 6 and 7, so that insoluble calcium oxalate was formed. The solution was then titrated with standard HCl (50 mmol l^{-1}) back to its original pH. Further additions were made at 24 h intervals, until the pH was constant and no more calcium oxalate was precipitated. In the following analysis, it is assumed that m mol of HCl were added, and that there were H mol of HPO_4^{2-} ions and OH mol of OH$^-$ ions in the solid, all quantities per mole of phosphorus in the solid. The total moles of H$^+$ ions added to the buffer from the solid and the HCl is ($H - OH + m$). In the final buffer solution, after titration back to the original pH, these H$^+$ ions can be regarded as being associated only with the additional phosphate that derived from the solid; the phosphate from the original buffer and its associated H$^+$ ions can be ignored because protonation of this phosphate is unchanged as the pH is unchanged. If $[H_2PO_4^-]_{solid}$ and $[HPO_4^{2-}]_{solid}$ are the moles of these ions in solution that derived from the solid per mole of phosphorus in the solid, the H$^+$ ion in solution associated with the phosphorus from the solid is $2[H_2PO_4^-]_{solid} + [HPO_4^{2-}]_{solid}$. These can be equated with the protons added from the solid and HCl (Equation 3.18).

$$H - OH + m = 2[H_2PO_4^-]_{solid} + [HPO_4^{2-}]_{solid}. \qquad \qquad 3.18$$

In addition,

$$[HPO_4^{2-}]_{solid} + [H_2PO_4^-]_{solid} = 1,$$ 3.19

Eliminating $[H_2PO_4^-]_{solid}$ and $[HPO_4^{2-}]_{solid}$ gives Equation 3.20

$$H - OH = (2 + R)/(1 + R) - m.$$ 3.20

$$\text{where} \quad R = \frac{[HPO_4^{2-}]_{solid}}{[H_2PO_4^-]_{solid}}.$$ 3.21

The ratio, R, is known, because it is equal to $K_a/[H^+]$, where K_a is the appropriate acid dissociation constant and $[H^+]$ is the H^+ ion concentration in the buffer.

Results from the oxalate precipitation/titration method agreed well with those obtained from the theoretical composition of DCPD and DCPA, and with the charge balance for a range of chemically analysed BCaPs [340]. The method will obviously only give the HPO_4^{2-} ion content if the assumption is made that the OH⁻ ion content is zero, or if it is known from some other analysis. Alternatively, if the HPO_4^{2-} content is known, the OH⁻ ion content can be determined.

IR methods

The HPO_4^{2-} ion absorbs IR at 870 cm⁻¹ in Ca-def OHAp (Section 3.11.6). The intensity of this band has been used to estimate acid phosphate. Estimates of 5 to 10 % of the phosphorus as HPO_4^{2-} ions in BCaPs were found, even though no pyrophosphate could be detected on pyrolysis [355]. When carbonate was present, IR gave a much higher value than the pyrolysis method [355]. It has been reported that the molar extinction coefficient for this band is an order of magnitude less for HPO_4^{2-} ions thought to be on the surface of apatite crystals, compared with the interior (based on HPO_4^{2-} ion concentration determined by pyrophosphate estimation on pyrolysis) [697], and also that it depends on the state of crystallisation [92]. The area of the band for an ACa,MgP has been compared with that expected from its $H - OH$ content (with OH assumed to be zero) determined by the oxalate precipitation/titration method on the basis of calibration with DCPD and partially matured ACa,MgPs [340] (Section 1.8.5). This area was about twice as large as expected, so it was concluded that resonances in addition to the P-O(H) stretch contributed to the band in ACPs. An alternative explanation is that the acid phosphate determination was too low because the assumption of the absence of OH⁻ ions was incorrect.

The carbonate out-of-plane stretching band (Section 4.2.2) is almost coincident with the HPO_4 band at 870 cm⁻¹ and will interfere with analyses. An attempt has been made to correct for this in measurements on normal and carious enamel by deducting the carbonate contribution [697]. This contribution was calculated from the extinction values (log T_2/T_1, where T_1 and T_2 are the peak and background transmissions respectively) of the carbonate in-plane stretching band at 1410 cm⁻¹, and extinction values of the out-of-plane and in-plane modes established from measurements on samples of known carbonate content. In principle, it would be possible to reduce interference from CO_3^{2-} ions by deuterating the HPO_4^{2-} ions, but surprisingly, this is difficult and requires hydrothermal conditions [698] (Section 3.11.6). In the case of bone mineral, it has been concluded that its poor crystallinity broadens the HPO_4^{2-} ion band so much by comparison with the carbonate out-of-plane stretching band, that the HPO_4^{2-} ion band cannot be detected [699].

NMR should be applicable to the quantitative determination of acid phosphate contents of apatites, but as mentioned in the previous section, there is difficulty in resolving adsorbed water and HPO_4^{2-} ion proton resonances [637,700] (Section 3.12). NMR MAS (7.25 kHz) spectra at 500 MHz of the precipitated OHAp, whose NMR determination of total hydrogen, OH⁻ ion and water content were discussed in the previous section, had a very weak, just discernible, ¹H resonance from HPO_4^{2-} ions at 8.6 ppm [637]. This observation was consistent with the small amount of acid phosphate detected by IR. It should therefore be possible to estimate the acid phosphate content in this type of compound from the NMR spectrum. In a well-crystallised Ca-def OHAp, in which the acid phosphate content was much higher and the water content much lower, the equality of the areas of the peaks at 0.8 (OH⁻ ions) and 6 (HPO_4^{2-} ions) ppm has been used to deduce equal contents of these ions, in agreement with the proposed structural formula [700] ($x = 1$ in Formula 2, Table 3.2, Section 3.7.3).

3.6.3 Miscellaneous analyses

Carbonate determination by IR

The ratio of the extinction of the 1415 cm⁻¹ carbonate in-plane stretching mode, to that of the 575 cm⁻¹ PO_4 band, has been used to measure carbonate contents (extinction coefficients were taken as log[baseline transmittance/peak transmittance]) [701]. The system was calibrated with mixtures of $BaCO_3$ and commercial tricalcium phosphate or OHAp, and gave estimates to within ±10 wt % for carbonate contents in the range 1 to 12 wt %.

An IR method has also been developed to determine the CO_3/PO_4 molar ratio (R) in francolites [702]. In an analysis of 65 rock-phosphates, it was found that $R = (CO_2\text{-}index - 0.0678)/4.184$ where the $CO_2\text{-}index$ was defined as the area under the absorption curve for the carbonate bands from 1375 to 1550 cm^{-1} divided by the PO_4 absorption band area from 530 to 690 cm^{-1}. The correlation factor, r^2, was 0.938 and standard error of the slope ±0.136. An FT IR spectrometer was used with spectral stripping to remove contributions from silicate and carbonate minerals, and from water.

IR has been used to determine the fraction of the total carbonate that is located on the *c*-axis in dental enamel [703,704] (Section 4.6.3). The method relies on the different absorption frequencies for different carbonate substitutions (Sections 4.2.2 and 4.3.2). In one estimate [703], the *c*-axis carbonate contribution (absorption at ~1545 and 880 cm^{-1}) was subtracted by matching the sample with an apatite with only this type of carbonate substitution and of known CO_2 content placed in the reference beam. The exact frequency of the ~1545 cm^{-1} band depends on the unit cell volume [82], so it is important that the reference beam apatite should be matched with the sample in this respect. In a later estimate [704], six samples with varying known proportions of A-CO_3Ap to B-CO_3Ap were prepared and the intensity ratios of the 1540 cm^{-1} carbonate band to the 560 cm^{-1} PO_4 band plotted against the proportion of A-CO_3Ap. A linear regression line fitted to the results had a slope of 0.113 and correlation coefficient 0.990 (the spectra were deconvoluted, so the slope will depend on the details of this procedure). This calibration enabled the quantity of CO_3^{2-} replacing OH^- ions to be determined from spectra of enamel run under the same conditions (Section 4.6.3).

Ca/P ratio from XRD after ignition at ~ 900 °C

When a BCaP with molar ratio between 1.5 and 1.667 is heated at ~900 °C (prolonged heating may be required to eliminate the CO_2 if present), the only solid phases formed are essentially s-OHAp and β-TCP (Reaction 3.22), provided there is sufficient water vapour present to prevent the formation of O,OHAp.

$$BCaP \rightarrow xCa_{10}(PO_4)_6(OH)_2 + y\beta\text{-}Ca_3(PO_4)_2 + \qquad 3.22$$
$$\text{water + carbon dioxide}$$

Irrespective of the details of the structure of the BCaP and carbonate content, if x mol of OHAp and y mol of β-TCP are formed, the Ca/P molar ratio is $(10x/y + 3)/(6x/y + 2)$. x/y can be determined from XRD measurements provided the system has been calibrated with known mixtures [705,706]. The advantage of the method is that it is very sensitive. A change in the Ca/P molar ratio from 1.667 through intermediate values to 1.5 corresponds to a change in the heated product from OHAp, OHAp plus β-TCP to only β-TCP, respectively. Thus, it is very easy to detect a small departure from stoichiometry in OHAp by the presence of β-TCP after heating. The method is applicable to mg quantities if a suitable powder diffraction camera is used.

Deuterium exchange and diffusion along the c-axis

Deuterium exchange kinetics on heating 1 to 20 mg of apatite in ~1 mg D_2O at 110 °C have been followed by IR and used as an indicator of ease of diffusion along the *c*-axis [620]. It was assumed that diffusion was only in this direction and that all OH⁻ ions to a depth *D*, and no others, had exchanged. *D* could be calculated from the OD⁻ ion content determined by IR, provided the BET surface area and the length/width of the crystals were known from SEM.

Hydrogen content

The hydrogen content of OHAps has been determined using a rising temperature deuterium-exchange method [707]. Samples were evacuated at 550 °C for 5 h prior to analysis. A known quantity of deuterium was circulated over the sample and the thermal conductivity of the gas measured as the temperature was raised until equilibrium was reached (typically to 500 or 600 °C for OHAp or a Ca-def OHAp respectively). The apatite hydrogen content was then calculated from the deuterium dilution at equilibrium. Clearly this method is only applicable to thermally stable samples, but as mentioned previously, integration of the ¹H NMR spectrum at 200 MHz of a non-spinning precipitated apatite has been used for total hydrogen determination [637] (Section 3.6.1). "Active" surface hydrogen on OHAp has been determined by measuring the volume of methane evolved on reaction with methyl magnesium iodide (Grignard Reagent) in isopentyl ether [708].

Unit cell contents

The chemical analysis of an apatite gives only the relative numbers of ions in the unit cell. The absolute content is often deduced from this by assuming that the total number of PO_4^{3-} ions (or PO_4^{3-} plus CO_3^{2-} ions) is six. However, the unit cell contents can be determined without this assumption through the relation, $\rho = u\Sigma A/V$ where ρ is the density (kg m⁻³), u is the atomic mass unit (1.66056×10^{-27} kg) and ΣA is the sum of the relative atomic masses of the atoms in the unit cell of volume V (m³). The accurate determination of the density of single crystals of apatites presents no problems [387] (Section 2.5.1), but for very fine apatite powders, there is the difficulty of removing entrapped air if immersion liquids are used. Nevertheless, densities of precipitated CO_3Aps have been measured with a pycnometer filled with diethyl phthalate and removal of the air under a vacuum of 10⁻⁴ torr (0.013 Pa) [99]. Densities of degassed fragments of compacted apatites have also been determined in mixtures of di-iodomethane (density 3.315 g cm⁻³) with sufficient benzene (density 0.878 g cm⁻³) added so that the densities of the liquid and solid matched [99]. The density of the liquid was then determined in a separate experiment. Densities of apatites determined by these two methods agreed to within 2 % [99]. Densities of high temperature apatites have been measured with an accuracy of about 1 % using an automatic helium pycnometer [709]. For example, the experimental density of A-CO$_3$Ap was 3.16 g cm^{-3}, which agrees well with 3.160 g cm^{-3} (Table 1.4, Section 1.1), the density calculated from the lattice parameters.

3.7 Structure of calcium-deficient hydroxyapatites

3.7.1 Introduction

Proposals that have been made to explain the variable Ca/P molar ratios of Ca-def OHAps (less than 1.5 to greater than 1.67) whilst the apatitic diffraction pattern is maintained have been reviewed [107,111,623]. These include undetected phases, surface adsorption, lattice substitutions, and intercrystalline mixtures of OHAp and OCP. Possible undetected phases include OCP, whose diffraction pattern is rather similar to OHAp (except for an intense small angle line at ~18 Å, Figs A.3 and A.10), ACP, and small quantities of DCPD (also CaO, Ca(OH)$_2$ and CaCO$_3$ in some circumstances). Although these compounds might not be detected in small concentrations by XRD, even sparsely distributed micron sized crystals of OCP, DCPD and CaCO$_3$ (particularly calcite) can readily be detected by polarised light microscopy, a rather neglected technique in this field of study.

3.7.2 Surface adsorption

Although surface adsorption of ions used to be a popular hypothesis (see ref. [107] for citations), the present fashion is to interpret the chemical composition of an apatite solely in terms of a model based on lattice substitutions. However, if it is assumed that there are two surface phosphate ions per surface unit cell, and that the {100} form provides all the surface faces, the fraction of surface phosphate ions is ~A/1200, where A is the surface area in m^2 g^{-1}. Thus, for a surface area of 60 m^2 g^{-1}, ~5 % of the PO$_4^{3-}$ ions are on the surface and can be protonated or replaced by other ions without any restrictions imposed by the stoichiometry of the lattice (although some adjacent Ca^{2+} ions may be lost to maintain charge balance), so the potential for changing the chemical composition is quite significant. A recent exception to the interpretation of chemical analyses solely in terms of unit cell contents relates to an OHAp precipitated at 70 °C with a BET surface area of 37±1 m^2 g^{-1} and 2 % of the phosphate as HPO$_4^{2-}$ ions [637] (Section 3.3.3). The comment was made that most of the evidence indicated that the HPO$_4^{2-}$ ions were adsorbed on the crystal surfaces, which meant the majority of surface phosphate ions were protonated [637].

3.7.3 Lattice substitutions

Various models for lattice substitutions are listed in Table 3.2. Some have been extended to include precipitated CO_3Aps (Section 4.5.3). Experimental evidence for these models comes mostly from estimation of the acid phosphate content from pyrophosphate formed on heating (Section 3.6.2) and other chemical analyses, and from IR and the weight loss on heating. In one study, measurements of hydrogen content as a function of Ca/P ratio were made after heating at 550 °C [707] (Section 3.6.3).

Table 3.2 Models of lattice substitutions in Ca-def OHAps.

1. $Ca_{10-x}(HPO_4)_{2x}(PO_4)_{6-2x}(OH)_2$	$0 \leq x \leq \sim 2$	[710,711,712]
2. $Ca_{10-x}(HPO_4)_x(PO_4)_{6-x}(OH)_{2-x}$	$0 \leq x \leq 2$	[713]
3. $Ca_{10-x-y}(HPO_4)_x(PO_4)_{6-x}(OH)_{2-x-2y}$	$0 \leq x \leq 2$ and $y \leq (1 - x/2)$	[623,714]
4. $Ca_{10-x}(HPO_4)_x(PO_4)_{6-x}(OH)_{2-x}$	$0 \leq x \leq 1$	[189,698]
5. $Ca_{10-x}(HPO_4)_x(PO_4)_{6-x}(OH)_{2-x}(H_2O)_x$	$0 \leq x \leq 1$	[698,707,715]
6. $Ca_{9-x}(HPO_4)_{1+2x}(PO_4)_{5-2x}(OH)$	Ca/P molar ratio 1.4 to 1.5	[698,715,716]
7. $Ca_{9+z}(PO_4)_{5+y+z}(HPO_4)_{1-y-z}(OH)_{1-y+z}$	-	[717]
8. $Ca_{10-x+u}(PO_4)_{6-x}(HPO_4)_x(OH)_{2-x+u}$	$2-x+2u \leq 2$ and $0 \leq u \leq x/2$	[83]

Changes in XRD powder intensities have been observed in ns-PbOHAp; these were consistent with intensities calculated for a structure with missing Pb(1), rather than Pb(2) or Pb(1) plus Pb(2) atoms [710]. However, this interpretation has been questioned because of the possibility of incorporation of Na^+ ions [62]. Equivalent changes in Ca-def OHAps, which would be much smaller, appear not to have been observed. However, precipitated apatites generally have an *a*-axis parameter that is from 0.01 to 0.02 Å larger than the parameter for OHAp made at high temperatures [718] (Section 3.4.1), which indicates unequivocally that lattice substitutions do take place. The *a*-axis parameter became slightly larger, and the *c*-axis parameter perhaps slightly smaller, as the Ca/P ratio decreased, but actual values were more variable than could be explained by changes in Ca/P ratios or experimental errors; it was thought that factors besides the Ca/P ratio, particularly coprecipitated water, influenced axial parameters [718]. The possible location of this water and magnitude of the resultant lattice expansion will be considered later in this section.

When ACP with a Ca/P molar ratio of 1.5 was hydrolysed in the solid state (Equations 1.16 and 1.17, Section 1.8.4), the finally produced Ca-def OHAp had a composition $Ca_{9.00}(HPO_4)_{0.96}(PO_4)_{5.04}(OH)_{0.96}$, as determined by pyrophosphate formation on heating [92,329]. This formula also corresponds to the composition given by Formula 2 (Table 3.2) for an apatite with a composition corresponding to that of the original ACP (Ca/P molar ratio = 1.5).

A detailed study has been made of the chemical composition, including OH$^-$ ion content and pyrophosphate formation on heating, of a series of Ca-def OHAps with Ca/P molar ratios from 1.40 to 1.62 made at pH 7.00 to 10.00 [623] (Section 3.6.1). It was found that the chemical composition best fitted Formula 3, Table 3.2. As the Ca/P ratio increased, x decreased from 1.1 to 0.03, whereas y was relatively constant at ~0.5. Thus loss of Ca^{2+} ions coupled with loss of OH$^-$ ions and gain of HPO$_4^{2-}$ ions was sensitive to preparative conditions, but loss of Ca^{2+} ions compensated by loss of two OH$^-$ ions was not. Furthermore, it seemed possible to prepare apatitic CaPs at neutral or only slightly alkaline conditions that contained few or no OH$^-$ ions (these conditions appeared to correspond to those when OCP or an OCP-like intermediate occurred during transformation of ACP into an apatitic compound, see Section 1.8.4). It was recognised that the unknown contribution from the large surface area (130 to 170 m^2 g^{-1}) of the samples led to uncertainty in the deduced formula. It was also found that samples stored for ~36 months had an increased IR OH stretch intensity and up to a 25 % increase in the amount of pyrophosphate formed on heating. These two observations were attributed to the reaction H$_2$O + PO$_4^{3-}$ → OH$^-$ + HPO$_4^{2-}$ during storage, but the increase in OH intensity was slightly greater than expected from the increase in OH$^-$ ion content determined from the increased acid phosphate content. This solid state hydrolysis reaction (Equation 1.16, Section 1.8.4) is the same reaction responsible for the production of the Ca-def OHAp discussed above.

As the Ca/P ratio falls, the HPO$_4^{2-}$ ion IR band at 870 cm^{-1} increases in intensity (Sections 3.6.2 and 3.11.6), and the OH librational band at 630 cm^{-1} (Section 3.11.6) decreases in intensity. The stretching mode at 3672 cm^{-1} is weaker in Ca-def OHAps [189]. These changes are in agreement with all but the first formula in Table 3.2. Quantitative IR of precipitated OHAps showed that they appeared to have a lower OH$^-$ ion content than s-OHAp and contained considerable amounts of structural water, whilst Rietveld X-ray structure analyses showed that they had more X-ray scattering on the c-axis than s-OHAp [620]. There was no evidence that the structural water and increased X-ray scattering were correlated [620]. However, one report indicated that the integrated intensity, as opposed to the peak intensity, of the OH librational mode did not appear to change with changes in Ca/P ratio [698] (Section 3.11.6); this was attributed to the replacement of OH$^-$ ions by water.

Polarised IR spectra of dental enamel (Figs 4.7 and 4.8, Section 4.6.3) show little evidence for the orientation of the water, but much of the observed absorption might be due to water adsorbed on the apatite crystals, so this observation gives no information about structural water [77]. However, the francolite minerals (Section 4.2.2, Figs 4.2 and 4.3) have water bands that are probably due to structural water, but again with little evidence of dichroism.

^{1}H NMR spectra of a well-crystallised Ca-def OHAp with a Ca/P molar ratio of 1.5 (preparation in ref. [644], Section 3.4.1) indicated equal HPO_4^{2-} and OH^- ion contents [700], in agreement with Formula 2 with $x = 1$ (Table 3.2).

The constant composition method (Section 3.8) has been used to grow a series of Ca-def OHAps at pH 6.0 to 9.0 with Ca/P molar ratios from 1.49 to 1.65, respectively [645,646] (Section 3.4.1). The quantities of titrants added and analysis of the solid indicated that its composition followed Formula 2 (Table 3.2) with x varying from ~1 to ~0.05 as the pH increased, rather than Formula 3 [645]. Protonation of surface PO_4^{3-} ions was thought to be insignificant by comparison with interior ions. Because the solutions from which the crystals grew were not appreciably supersaturated with respect to OCP, it seemed unlikely that interlayered OCP/OHAp structures could have been formed [646]. Furthermore, it was concluded that nonstoichiometry of the precipitating apatite was not due to protonation of the expanding surface, but rather growth of bulk material which contained Ca^{2+} and OH^- ion vacancies, coupled with incorporation of HPO_4^{2-} ions. The suggestion was made that the nonstoichiometry arose for kinetic reasons, namely the slow rate of deprotonation of HPO_4^{2-} ions.

There do not seem to have been any studies of fluorescence spectra of Ca-def OHAps doped with MnO_4^{3-} ions, comparable to those of $CaCl_2$-def ClAps (these showed the presence of a local monoclinic symmetry due to short-range ordering of the Cl^- ions [406], Section 2.2.3).

Effects of water and HPO_4^{2-} ions on lattice parameters

The preparation of a potassium-bearing apatite at pH 6.96 in boiling water with an expanded a-axis parameter was described earlier (Section 3.4.1) [181]. The undried moist apatite had lattice parameters $a = 9.439$ and $c = 6.876$ Å, but after drying at 120 °C for 12 h, $a = 9.432$ and $c = 6.873$ Å (all ±0.002Å). Exposure to steam at 120 °C for 12 h resulted in lattice parameters midway between the dried and undried values. These results were interpreted in terms of a model in which the undried material contained H_3O^+ ions in calcium sites which, on drying, released water with H^+ ions remaining in the structure; this process was then reversed when the apatite was heated in steam. Water was also thought to be in OH^- positions, and responsible for the major part of the a-axis parameter increase. The presence of H_3O^+ ions in Ca-def OHAps has been criticised because this ion normally only occurs in the structures of strong acids [64].

Earlier (Section 3.4.1), the preparation of a well-crystallised Ca-def OHAp (Ca/P molar ratio of 1.56) was described which showed a slight expansion of the a-axis parameter (+0.16%) and contraction of the c-axis parameter (-0.09%)

(volume expansion +0.23%) compared with sintered OHAp [643]. This apatite lost water at 600 °C and the lattice parameters changed to those of sintered OHAp [643]. The water lost was 2.5 to 3.5 molecules per unit cell for a Ca/P molar ratio extrapolated to 1.5. Modifications of Formulae 1 or 2 (Table 3.2) were proposed for the unheated apatite; these modifications were that vacant lattice positions, either calcium sites in Formula 1 or calcium and OH$^-$ ion sites in Formula 2 were filled, or over filled, with water [643]. The figures for the water lost and parameter changes give an *a*-axis parameter increase of +0.0060 to +0.0043 Å respectively per structural water molecule in the unit cell (this includes structural water attributable to HPO_4^{2-} ions). A similar contraction in the *a*-axis parameter of ~0.18% (0.017Å) accompanied by a loss of water has been observed on heating a Ca-def OHAp, prepared by the hydrolysis of DCPA at 95 to 100 °C, from 200 to 400 °C [719]. This irreversible loss was ascribed to lattice water (possibly H_2O-for-OH$^-$ and/or HPO_4^{2-}-for-PO_4^{3-}). Reversible loss of water below 200 °C that did not result in lattice parameter changes was ascribed to adsorbed water [719]. Adsorption isotherms of water on OHAp are discussed in Section 3.10.1.

The change in lattice parameters due specifically to HPO_4^{2-} ions has been investigated in a Ca-def OHAp prepared by the hydrolysis of DCPA at 100 °C [720]. A sharp decrease in the *a/c* ratio on heating between 160 to 240 °C was attributed to loss of HPO_4^{2-} ions in the lattice because it correlated with the nearly total loss of the HPO_4 band at 875 cm^{-1} and a loss of 3 to 3½ wt % HPO_4^{2-} ion as determined by pyrophosphate formation on pyrolysis. From this, it was deduced that the increase in the *a*-axis parameter attributable to HPO_4^{2-} ions in a Ca-def OHAp was 0.0015 Å per wt % HPO_4^{2-} ion (0.0144 Å per HPO_4^{2-} ion in the unit cell).

Loss of lattice water has also been used to explain the decrease in the *a*-axis parameter seen on heating enamel below 400 °C [719,721]. A "sudden" *a*-axis parameter decrease of ~0.014 Å was seen in the range 250 to 300 °C [721] (Section 4.6.4). It was concluded that, at most, only a small part of this could be attributed to loss of HPO_4^{2-} ions because the contraction was too large [720]. Thus, it was proposed that about one to two water molecules per apatite unit cell increased the *a*-axis parameter of enamel by +0.014 Å [620,721]. From this, it was expected that the Ca-def OHAp prepared by the hydrolysis of DCPA (see above) would show a decrease in the *a*-axis parameter of 0.007 Å due to water loss in the range 250 to 300 °C, but this was not seen. This finding led to the suggestion that the form of combination of structural water in the Ca-def OHAp differed from that of water in enamel [720].

In conclusion, although attempts have been made to separate the effects of lattice expansion due to water and HPO_4^{2-} ions, these have not been entirely satisfactory, although the combined effect has been clearly established.

Structure

Direct determinations of the structures of the Ca-def OHAps are lacking, but the chemical, spectroscopic and lattice parameter studies discussed above indicate that these apatites (at least the well-crystallised ones) follow structural Formula 2 (Table 3.2) or some minor variant of it. By analogy with a proposal made for the structure of precipitated B-CO_3Aps [83,722] (Section 4.5.3), the same authors suggested that the HPO_4^{2-} ions in Ca-def OHAps are adjacent to vacant OH^- and Ca(2) ion sites [83,722]. Thus there would be x clustered vacancies per unit cell of the form $\square_{OH}, \square_{Ca(2)}, HPO_4$, in which $\square$ represents a vacancy in the indicated site, and x decreases to 0 as the composition approaches that of s-OHAp. If this proposal is correct, it is probable that the proton on the phosphate ion occupies the Ca(2) vacancy as closely as possible, as happens to the vacant Ca(4) site in whitlockite (Section 1.6.2). Again by analogy with precipitated B-CO_3Aps, the suggestion was made that some of the vacant Ca(2) and OH sites might be filled with Ca^{2+} and OH^- ions (Formula 8, Table 3.2) [722]. The vacancy cluster could also be the site of lattice water.

The fact that the solid state hydrolysis of ACP stops so precisely at a point with nearly integral numbers of ions in the formula for a unit cell (see above), suggests that there is a structural interaction between the HPO_4^{2-} ions and Ca^{2+} and OH^- ion vacancies that favours having one each of these species per unit cell. This deduction supports the idea that, in the finally produced Ca-def OHAp, these three species are grouped together to form a vacancy cluster, $\square_{OH}, \square_{Ca(2)}, HPO_4$, as discussed above for Ca-def OHAps with a more general composition.

3.7.4 Intercrystalline mixtures of OCP and OHAp

Variable amounts of OCP and OHAp in the form of intercrystalline mixtures have been proposed [149] (Section 1.3.6) to explain the variable composition of precipitated apatites (reviewed in refs [14,107,142,150,175,178]). Evidence for the formation of intercrystalline mixtures (reviewed in Section 1.3.6) includes: (a) single crystal XRD [157] and electron microscopic studies [174] of partially hydrolysed OCP; (b) the strong structural relationship of OCP and OHAp (Sections 1.3.2 and 2.2.2); and (c) the anomalously large *a*-axis parameter [173] for an apatite formed during the hydrolysis of ACP which was explained [172] by partially coherent X-ray scattering between alternating layers of OCP and OHAp.

The platy or ribbon habit of precipitated and biological apatites has also been taken as evidence that these crystals were formed via an OCP intermediate, and might therefore still contain intercrystalline layers of OCP (see reviews cited above and discussion of the "central dark-line" seen in electron micrographs of enamel crystals, Section 4.6.6). The basis of this evidence is that OCP crystallises as {100} plates (Section 1.3.4) which remain as pseudomorphs of OCP when it hydrolyses (partially or completely) to OHAp (Section 1.3.6). If OHAp were to grow directly, it would be expected to have a hexagonal habit because of its hexagonal internal structure. (It has alternatively been suggested [400] that the platy morphology could arise from monoclinic OHAp, but there is no evidence for this lower symmetry for OHAps grown in solution at room temperatures.)

Against the idea of intercrystalline mixtures of OCP and OHAp is the observation that the IR spectrum at 20 and -180 °C of a Ca-def OHAp showed no evidence of absorption bands that could be attributed to OCP in interlayered structures with OHAp [189] (Section 3.11.6). Likewise, there was no evidence from ^{31}P NMR spectra of Ca-def OHAps with Ca/P molar ratios as low as 1.33 of resonances attributable to OCP [192] (Section 3.12). This observation has been confirmed [193] in similar studies of a well-crystallised Ca-def OHAp with Ca/P molar ratio of 1.5 (preparation in ref. [644], Section 3.4.1).

Carbonate ions in sites within intercrystalline mixtures of OCP and OHAp have been proposed as a mechanism for the incorporation of these ions in precipitated CO_3Aps [150,157] (Section 1.3.6). The "H_2O layers" of OCP (Section 1.3.2) in such structures would also provide lattice sites for structural water.

3.7.5 Summary

There seems to be no doubt that Ca-def OHAps can be synthesised that are sufficiently well-crystallised that their nonstoichiometry cannot be attributed to surface effects and for which there is no evidence of intercrystalline growths of OCP and OHAp. There is direct physical and chemical evidence that these contain HPO_4^{2-} ions and are deficient in OH⁻ ions. At least some have a slightly increased *a*-axis parameter due mainly to the presence of lattice water. Chemical evidence indicates that these apatites follow a composition given by Formula 2 (Table 3.2) or a minor variant of it. It seems likely that they contain vacancy clusters of the form, $\square_{OH},\square_{Ca(2)},HPO_4$.

Intercrystalline growths of OCP and OHAp undoubtable occur, so that in many instances, particularly during the formation and growth of apatite via OCP, this type of structure probably provides the most appropriate description. As equilibrium or pseudo-equilibrium conditions are reached, the structure may often be better described by Formula 2 (Table 3.2).

For very small crystals, surface effects will make a major contribution to the composition.

3.8 Kinetics of nucleation and crystal growth

Studies of growth kinetics [25] and inhibitors of precipitation and growth (particularly Mg^{2+} ions, Al^{3+} ions, condensed phosphates, phosphonates, ATP, proteoglycans, serum and salivary proteins) [723] of OHAp have recently been reviewed. Earlier sections that dealt with the formation and growth (Section 1.3.5) and hydrolysis (Section 1.3.6) of OCP and transformation reactions of ACP (Section 1.8.4) are relevant because both these occur very often as unstable intermediates during the precipitation of OHAp. Both have been proposed as precursors of biological apatites (Section 4.6.1).

A distinction has to be made between studies of the phases formed during the spontaneous precipitation, and the seeded growth of OHAp. In spontaneous precipitation experiments at relatively high supersaturations, when the solution is unstable with respect to all the phases in the near neutral or alkaline part of the $Ca(OH)_2$-H_3PO_4-H_2O phase diagram (Fig. 1.1), ACP is normally precipitated first. This then converts to OHAp, usually via OCP (Section 1.8.4). Formation of OHAp at intermediate supersaturations can occur without prior formation of ACP, but still seems to be via other crystalline phases, such as DCPD or OCP [724]. Direct formation of OHAp at low supersaturation has been reported [725].

In experiments on the seeded growth of OHAp, well-characterised OHAp seeds were added to a supersaturated solution, and changes in pH, calcium and phosphorus concentrations measured as a function of time. These quantities allowed the rate of growth and composition of the growing CaP to be determined. Such studies showed that the driving force for crystal growth was the degree of supersaturation (defined in Equation 3.35, Section 3.10.3). Analysis of the kinetics showed that the linear growth rate was dependent on the square of the degree of supersaturation, which indicated a spiral growth mechanism [726,727].

There are problems with the studies reported in the previous paragraph. Only a small amount of solid phase can be grown and the determination of its composition from changes in the solution is subject to considerable error; furthermore, the driving force during growth varies. These difficulties can be overcome by constant composition methods in which the concentration of lattice ions is kept constant by the addition of solutions containing Ca^{2+}, phosphate, and OH^- ions from mechanically coupled burettes arranged so that they are added in the same ratio as in the precipitating solid [728]. The trigger for the addition of these solutions comes from a deviation of the Ca^{2+} or H^+ ion concentrations in the reacting solution, as monitored by specific ion electrodes, from the initial values. The rate of crystal growth can be accurately calculated from the rate of addition of these solutions, even at very low levels. Using this method, more than three times the original seed weight of well-crystallised OHAp could be grown at low supersaturation and pH 6.00 to 8.50 with a linear rate of growth proportional to the relative supersaturation to the power ~1.3 [728]. A typical growth rate was 2.7×10^{-7} mol min^{-1} m^{-2} of $Ca_5(PO_4)_3OH$ at 37 °C and pH 6.00 with concentrations of calcium, phosphorus and KNO_3

background electrolyte of 5.05×10^{-3}, 3.03×10^{-3} and 0.1 mol l^{-1} respectively [24], which is much lower than the rates for OCP (Section 1.3.5) and DCPD (Section 1.4.5) under comparable conditions. In studies at 37 °C, growth of the seeds was predominantly in the c-axis direction, so that their increase in length was proportional to the extent of precipitation [729]. However, if growth at pH 8.5 was allowed to continue, so that five or six times the original seed weight had grown, a striking change in morphology occurred, as plate-like crystals began to grow. IR and XRD confirmed these plates to be OHAp, but with slightly changed lattice parameters ($a = 9.423 \pm 0.002$, $c = 6.762 \pm 0.004$ Å) compared with the initially formed crystals ($a = 9.418 \pm 0.002$, $c = 6.880 \pm 0.002$ Å). The changes in morphology and lattice parameters were ascribed to Cl$^-$ ions because plates did not grow in their absence. Thinner, plate-like crystals, were formed at lower pH. The relationship between crystallographic axes and external morphology of the crystals was not reported, nor whether XRD showed any of the preparations to be monoclinic.

The constant composition method can also be used to study the growth of Ca-def OHAps in the pH range 6.0 to 9.0 with Ca/P molar ratios of 1.49 to 1.65 [645,646] (Sections 3.4.1 and 3.7.3).

The effect of Sr^{2+} [730], Li$^+$ [731] and pyrophosphate, citrate and Mg^{2+} [732] ions on the crystal growth of OHAp on OHAp seeds has been investigated by constant composition methods. Significant amounts of Sr^{2+} were incorporated in the lattice, but the lattice Sr/Ca ratio was much lower than in the solution. Increased levels of Li$^+$ ions (up to ~40 ppm) reduced significantly the rate of crystallisation; it was thought that small quantities of Li$^+$ ions entered the lattice because the a- and c-axis parameters were slightly reduced [731]. At higher pH values, pyrophosphate and citrate ions showed an increase, and Mg^{2+} ions a decrease, in effectiveness as inhibitors [732].

Constant composition methods have also been used to study the growth of OHAp on seeds of DCPD, CaF$_2$ and calcite [733]. At low supersaturation and after an induction period that depended on the supersaturation, OHAp grew without formation of precursors and with the same kinetics as on OHAp seeds.

Recently, the constant composition method has been extended to studies of phase transformations and crystallisation of mixed phases [734]. In this dual constant composition system, two potentiostats and two electrode sets are used to maintain simultaneously two processes, so they both occur at constant driving force. The kinetics of concurrent growth of OCP and dissolution or growth of DCPD were investigated [734]. H$^+$ and Ca^{2+} ion electrodes were used to control the addition of titrants to maintain constant driving forces for the growth of OCP and the growth (or dissolution) of DCPD respectively.

Formation of BCaPs is profoundly affected by F$^-$ ions in a complex manner [735-738]. Studies have been made of the OH$^-$ ion uptake and F$^-$ ion concentration changes with time at constant pH (8.50) in a series of initially moderately supersaturated CaP solutions with initial F⁻ ion concentrations from 10^{-6} to 10^{-3} mol l⁻¹ [738]. At low F⁻ ion concentrations ($\leq 10^{-5}$ mol l⁻¹), the ACP to crystalline apatite conversion time was extended so apatite formation was retarded, but above this concentration, F⁻ ions accelerated formation. At even higher concentrations ($\geq 5.9 \times 10^{-4}$ mol l⁻¹), apatite formation was again retarded. A number of explanations were given for these observations: at low F⁻ ion concentrations, F⁻ ions might be adsorbed onto an OCP-like precursor phase and block its growth; or there might be an increase in interfacial tension of the growing apatite as a result of adsorbed F⁻ ions or from a higher interfacial tension of FAp compared with OHAp. Enhancement of growth was ascribed to an increasing supersaturation of the growing F,OHAp with an increased solution F⁻ ion concentration. Precipitation of CaF_2 was thought to be responsible for the inhibition found at the highest F⁻ ion concentrations.

3.9 Solubility and interfacial phenomena

Solubility and interfacial properties of OHAp have been reviewed [739-745]. Although there has been considerably controversy about the solubility behaviour of OHAp, it is now generally accepted that s-OHAp has a well-defined solubility with $pK_s \approx 57.5$ (based on $Ca_5(PO_4)_3OH$) [744]. This value was based on a tabulation of literature values of the pK_s of OHAp, mostly determined at 25 °C [744]. Determination [38] of K_s at 5, 15, 25 and 37 °C in the pH range 3.7 to 6.7 yielded the relation:

$$\log K_s = -8219.41/T - 1.6657 - 0.098215T \qquad 3.23$$

where T is in Kelvin and K_s in (mol l⁻¹)⁹ (see Section 3.3.3 for the preparation of the OHAp). This relation showed that there was a maximum in K_s at about 16 °C. Thus, above this point, the solubility decreases with increasing temperature like DCPD (Section 1.4.6), DCPA (Section 1.5.5) and β-TCP (Section 1.6.5). Values of K_s (mol l⁻¹)⁹ at 25 °C of 2.91 (95% confidence range 0.51 to 7.89) $\times 10^{-58}$ [746] and 3.2($\pm$0.3) $\times 10^{-58}$ [747] have been reported. These are in good agreement, but are slightly larger than 3.04 (standard error 0.25) $\times 10^{-59}$ (mol l⁻¹)⁹ calculated [38] from Equation 3.23. The reasons for this and other observed slight variations in K_s which were said to be greater than experimental error have been discussed [747]. It was proposed that variability between samples and experimental methods could only partially account for the slight variability in K_s, and that reactions occurring at the surface of the solid were mainly responsible [747].

In equilibrium studies in the pH range 4 to 8 of OHAp partially dehydrated by heating at 1000 °C, it was found that the solutions were supersaturated with respect to s-OHAp and that there was a tendency for the ion activity product to increase with pH [748]. Thermodynamic functions for the dissolution of

OHAp have been reported [38]. The equilibrium solubility behaviour of well-crystallised Ca-def OHAps does not appear to have been investigated, but the effect of CO_3^{2-} ions in the lattice has been studied in dental enamel [749,750] (Section 4.6.5) and synthetic $CO_3F,OHAp$ [751] (Section 4.5.8).

Solubility of fluorhydroxyapatites. Solubility products of $Ca_5(PO_4)_3(OH)_{1-x}F_x$ for $0 \leq x \leq 1$ precipitated at pH 9 to 9.5 in boiling solutions have been determined (preparation in Section 3.4.4) [636,752]. The samples were equilibrated for 30 days at 37 °C in phosphoric acid solutions to give a final pH from 3.6 to 6.3 and final F^- ion concentrations up to 5.4×10^{-4} mol l^{-1}. The K_s were calculated from the expression, $(Ca)^5(PO_4)^3(OH)^{1-x}(F)^x$, where the parentheses denote activities. For $x = 0$ (OHAp) and $x = 1$ (FAp), they were $7.36 \pm 0.93 \times 10^{-60}$ and $3.19 \pm 0.14 \times 10^{-61}$ (mol l^{-1})^9 respectively. Another recent determination of K_s for FAp from constant composition dissolution studies gave $9.0 \pm 1.0 \times 10^{-61}$ (mol l^{-1})^9 at pH 4.5 and 37°C [753]. Thus, the K_s of FAp is significantly lower than that of OHAp. There was a minimum in K_s of about 6.5×10^{-63} (mol l^{-1})^9 for $x = 0.56$ [636,752]. This minimum was attributed to the maximal OH directional disorder that occurred when there were equal numbers of OH^- and F^- ions; the increased stability of F,OHAps over OHAp was ascribed to this disorder together with OH$\cdots$F hydrogen bond formation [636,752]. A potential complication was that the equilibrating solution often had a lower Ca/F ratio than the original solid. This meant that a fluoride-rich solid, thought to be FAp, had formed on the surface of the equilibrating apatite. In fact, many of the solutions were saturated, or supersaturated with respect to FAp. However, in these quaternary systems, there will still be one degree of freedom, even when the solution is in equilibrium with two CaPs. Solubility products of thermally synthesised F,OHAps have been reported [754] which differ from the above in some details. The structure and formation of fluoride-rich phases during the reaction of F^- ions with OHAp is discussed further in Section 3.10.2.

Surface charges. Ionic solids in contact with solutions can become charged from the preferential dissolution or adsorption of lattice ions, or from the formation of complexes between lattice ions and species in solution. For a solid containing three lattice ions such as OHAp, its solubility can be described by a surface plotted as a three-dimensional graph whose axes are the concentrations of the three ions (note that, unlike the ternary system in Fig. 1.1, the three concentrations are independent variables because other salts are added, so a surface rather than a line describes the solubility). In general, the solid will be charged, but there will be points of zero charge (pzc) when the charges from surface anions and cations are equal. This extra condition means that the locus of the pzc's is given by a line lying in the solubility surface [739]. A set of equipotential lines exists on the solubility surface, of which the line of zero charge (lzc) is a special case. This concept has been confirmed experimentally for several apatites and three independent pzc's were determined for both CaOHAp and SrOHAp. A theoretical lzc has been calculated, assuming nearly equal affinity for the crystal surface of all lattice ions in solution, and comparison made with experimental results [740,744]. Literature values of the pzc in the absence of added calcium or phosphate salts (the isoelectric point) have been tabulated [744]. These were in the pH range 6.5 to 8.5. Zeta potentials of OHAp have been measured as a function of pH, KNO_3, $Ca(NO_3)_2$, K_2HPO_4 and KF concentrations [745] and electrophoretic mobilities in the presence of surfactants (oleic acid, oleates, dodecyl sulphonates, tridecanoic acid, tridecanoates), amines (dodecylammonium and dodecyltrimethylammonium chlorides) [755] and as a function of F⁻ ion concentration in a pretreating solution [756].

Exothermic heats of immersion (in units of $J\ m^{-2}$) of OHAp degassed at temperatures from 20 to 500 °C have been reported in water (0.4 to 0.8), 2-propanol (0.35 to ~0.5) and hexane (0.1) [708]. Exothermic heats of immersion of OHAp in water and various ionic solutions [757], and of OHAp, SrOHAp, BaOHAp and FAp in water, methanol, ethanol and n-butanol [741], have also been measured. Heats of reaction increased with ionic concentration; a substantially larger heat output for FPO_3^{2-} ions, compared with F⁻ ions, was observed which was associated with a different reaction mechanism for these ions [757]. Lowering of the dielectric constant caused a substantial decrease in the heat of immersion, consistent with the polar nature of the apatite surface; a change in the apatite cation made a greater difference than a change in the anion [741].

3.10 Reactions in solution

3.10.1 Adsorption and surface reactions

Aspects of OHAp surface chemistry have been reviewed [758]. The kinetics of $^{32}PO_4^{3-}$ and $^{45}Ca^{2+}$ ion exchange with OHAp have been interpreted in terms of a rapid exchange with the hydration shell (1 or 2 min), an intermediate rate associated with surface ion exchange (30 min), and a very slow process of incorporation of the ions into the crystals, primarily due to recrystallisation [759]. The binding between cements based on polyacrylate acid and OHAp has been shown to involve displacement of Ca^{2+} and PO_4^{3-} ions by ionised carboxyl groups of the polyacrylic acid [760]. Photo-oxidation of tetracycline adsorbed on OHAp has been studied [761]. The bright yellow fluorescence of the adsorbed molecule gradually disappeared on exposure to a mercury vapour lamp with concomitant formation of a red-purple oxidised form of tetracycline. The proposal was made that the same phenomenon was responsible for the light-induced change of colour of tetracycline stained teeth [761].

Adsorption of binary mixtures of proteins, peptides and organic acids from saliva (and analogues), onto OHAp has been studied [762]. The results did not follow Langmuir adsorption isotherms, but a thermodynamic model was developed which adequately predicted the results. This model was based on isotherms of single solutes adsorbing onto OHAp and used activity coefficients to describe the nonideal behaviour of the molecules in the adsorbed state [762]. The adsorption process could not be explained on the basis of simple electrostatic interactions and the main driving force for adsorption appeared to be entropic, because the process was endothermic [763]. In a later study, FT IR was used to investigate changes in macromolecule conformation on adsorption, and the binary thermodynamic model was extended to ternary systems [763]. The selective adsorption of enamel proteins onto F,OHAps with various degrees of F⁻ ion substitution has been investigated [764]. Most components, except for a 20 kdalton fraction, were adsorbed, and this increased with the degree of fluoridation. The preferential adsorption of fluorescence-labelled dentine and bone acidic proteins onto the (100) face of OHAp crystals has been observed [765]. Initial rate constants for adsorption and desorption on OHAp of *Streptococcus sanguis* between 18 and 37 °C have been studied in relation to bacterial adhesion mechanisms [766].

The water vapour adsorption isotherms at 18.5 °C of OHAp and FAp, heated at 500 °C, have been reported [741]. Adsorption per unit area was greater for OHAp than FAp. Isotherms of OHAp heated at 300 °C were lower than for unheated OHAp. The isotherm after a second heating of a heated sample was always the same as the first isotherm after heating, irrespective of the temperature of the second treatment; the difference between the unheated and first heated sample measurements was ascribed to the loss of chemisorbed water [741].

3.10.2 Reactions with fluoride ions

The reactions between BCaPs and F⁻ ions in solution are complex and not fully understood. Proposed reaction mechanisms have been reviewed [756]. Other sections of relevance are: 3.8 (effect of F⁻ ions in solution during formation of BCaPs); 3.9 (solubility of FAp and OH,FAps, and studies of electrophoretic properties of OHAp in the presence of F⁻ ions); 3.10.3 (effect on the rate of dissolution of OHAp); 3.12 (^{19}F NMR studies) and 4.6.5 (reactions with dental enamel).

The most important factor that controls the course of the reaction is the F⁻ ion concentration. Even at low F⁻ ion concentrations, K_s for FAp may be exceeded, so FAp (or F,OHAp) might be formed by exchange of OH⁻ ions in OHAp (Equation 3.24) or by dissolution of OHAp (Equation 3.25) followed by direct precipitation of FAp (Equation 3.26) or F,OHAp (Equation 3.27). (In these equations, the protonation of the PO_4^{3-} ions has been ignored for simplicity; in practice they will be protonated to an extent depending on the pH which will change the balance of H⁺ and OH⁻ ions in the equations.) Equation

3.24 will be very slow at ambient temperatures, and probably limited to a depth of a few unit cells, except for geological time periods.

$$Ca_{10}(PO_4)_6(OH)_2 + 2F^- \rightarrow Ca_{10}(PO_4)_6F_2 + 2OH^- \qquad 3.24$$

$$Ca_{10}(PO_4)_6(OH)_2 \rightarrow 10Ca^{2+} + 6PO_4^{3-} + 2OH^- \qquad 3.25$$

$$10Ca^{2+} + 6PO_4^{3-} + 2F^- \rightarrow Ca_{10}(PO_4)_6F_2 \qquad 3.26$$

$$10Ca^{2+} + 6PO_4^{3-} + xF^- + (2 - x)OH^- \rightarrow Ca_{10}(PO_4)_6F_x(OH)_{2-x} \qquad 3.27$$

At higher F⁻ ion concentrations (Table 3.3), CaF_2 might also be precipitated (Equation 3.28).

$$Ca_{10}(PO_4)_6(OH)_2 + 20F^- \rightarrow 10CaF_2 + 6PO_4^{3-} + 2OH^- \qquad 3.28$$

Other factors that influence the course of the reaction between F⁻ ions and OHAp include Ca^{2+} and phosphate ion concentrations, pH, surface area of the solid and its Ca/P ratio, and the temperature.

Table 3.3 F⁻ ion concentrations for CaF_2 ($K_s \approx 3.6 \times 10^{-11}$ (mol l⁻¹)³) to be precipitated from solutions saturated with respect to OHAp [756].

pH	Ca^{2+} conc. in equilibrium with OHAp (mol l⁻¹)	F⁻ conc. for CaF_2 precipitation (mol l⁻¹)
5.0	10^{-4}	6×10^{-4}
7.5	10^{-6}	6×10^{-3}
10.0	10^{-7}	1.9×10^{-2}

Adsorption of F⁻ ions by BCaPs has been studied as a function of F⁻ ion concentration, pH, Ca/P ratio and specific surface area [756]. Adsorption equilibrium was attained in ~1 h and the uptake increased with an increase in solution F⁻ ion concentration, increasing markedly when the K_s of CaF_2 was exceeded. The uptake decreased with an increase in pH, which was attributed to competition with OH⁻ ions. Products of the reaction were studied by X-ray photon spectroscopy [756,767]. The calcium and phosphorus binding energies at low F⁻ ion adsorption densities ($<2.0 \times 10^{-10}$ mol cm⁻²) were the same as those in OHAp, but at higher densities, binding energies decreased slightly and then became constant for a monolayer coverage of F⁻ ions. As the binding energies in FAp and CaF_2 are higher than in OHAp, the calcium and phosphorus were less tightly bound than in either of these, so neither could have been formed. Although the F^- ion binding energy did not change with absorption density, the F^- ion signal increased markedly at a F^- ion adsorption density of 2.0 to 3.2 × 10^{-10} mol cm^{-2}, and then remained constant. It was postulated [767] that F^- ions were adsorbed at discrete sites, possibly OH$^-$ ion sites, until there was an F^- ion coverage of about 3.2 × 10^{-10} mol cm^{-2}. Then, at higher densities, the OHAp lattice broke down to form an interfacial layer of more loosely bound ions containing Ca^{2+}, OH$^-$, PO_4^{3-} and F^- ions which acted as a precursor to nucleation of FAp or, at higher F^- ion concentrations, CaF_2. The constant F^- ion signal at higher coverages was thought to indicate that some F^- ions could penetrate the lattice and exchange with OH$^-$ ions. Owing to their greater depth, they would not contribute to the signal.

Detailed information about the chemical environment of F^- ions after reaction with OHAp can also be obtained from ^{19}F NMR MAS studies [768,769] (see Section 3.12 for details of ^{19}F NMR spectroscopy of FAp, F,OHAp, CaF_2 powders and single crystal studies of F,OHAps). In one study, OHAp with a surface area of 61.6 m^2 g^{-1} that was exposed to F^- ions (final concentration 9.7 mmol l^{-1}) for several days, had an F^- ion uptake corresponding to 1.1 F^- ions per surface OH$^-$ ion site and an NMR spectrum corresponding to $Ca_{10}(PO_4)_6(OH)_{2-2x}F_{2x}$ with 0.4<x<0.8 [768]. Thus, it seemed that a surface layer of this composition had been formed. After six month's storage in the NMR rotor at ambient temperature and humidity, the NMR spectrum corresponded to FAp, so that the F^- ion environment along the c-axis had changed from OH$^-$ to F^- ions as nearest neighbours. It was thought that this occurred by surface diffusion of ions, rather than by a bulk-phase transformation. At higher F^- ion concentrations (final concentration 155.3 mmol l^{-1}), samples aged for several months after isolation gave an NMR spectrum of FAp superimposed on a broad spectrum characteristic of CaF_2. The proposal was made that NMR could be used for quantifying the relative amounts of CaF_2 and FAp formed in the reaction between OHAp and F^- ions [768].

In a combined ^{19}F NMR MAS spectroscopic and chemical study, the formation of FAp, F,OHAp and CaF_2, with increased amounts of CaF_2 at higher F^- ion concentrations and lower pH, was confirmed [770] (Section 3.12). NMR showed that F,OHAp or FAp was always formed, irrespective of conditions, even though CaF_2 might also be formed as an additional product. Under conditions when ion exchange predominated, and NMR ruled out the formation of CaF_2, Equation 3.24 did not represent the reaction exactly because too little base was released and the phosphate in solution increased. Exchange of F^- ions with water rather than OH$^-$ ions, and desorption of HPO_4^{2-} ions from the OHAp surface were given as possible explanations for these observations. Powdered dental enamel had a similar uptake of F^- ion per unit area as synthetic OHAp, however, the ^{19}F NMR MAS spectrum showed significant differences, mainly a new sharp peak at 44 ppm shift with respect to C_6F_6 [770]. A sample of synthetic CO_3Ap similarly treated with F$^-$ ions did not show this peak, so the carbonate in enamel was ruled out as its cause. It was suggested that surface Mg^{2+} ions might be responsible. Although CO_3^{2-} ions seemed to have been excluded, it is worth noting that another environment would be provided by [CO_3^{2-},F$^-$] replacing PO_4^{3-} ions as has been proposed [538,771,772] (Sections 4.2.1 and 4.5.5), but there are objections to having two ions with negative charges close to each other [45] (Section 4.2.1). Recently, as a result of fast MAS (16 kHz) ^{19}F NMR studies, it has been proposed that the narrow 44 ppm resonance comes from nonspecifically adsorbed F$^-$ ions [769] (Section 3.12). This form of adsorbed F$^-$ ions, hydrogen bonded onto surface HPO_4^{2-} ions, had earlier been proposed as the initial location of the F$^-$ ions which then replaced OH$^-$ ions in surface sites of the apatite crystals [773].

It should be possible to study the formation of OH,FAps from the perturbation of the IR OH stretching band (Section 3.11.5), but little use seems to have been made of this approach. However, the formation of CaF_2 or CaF_2-like material in the reaction between dental enamel and F$^-$ ions has been observed through observation of the CaF_2 lattice mode at 322 cm^{-1} in Raman microprobe experiments [774] (Section 4.6.5).

Adsorption of F$^-$ ions onto OHAp has been studied under conditions when the K_s of FAp was not exceeded, so there was no precipitation of FAp [775]. Rate constants for adsorption and desorption of F$^-$ ions were determined for pH values from 5.8 to 6.75. As OHAp is more soluble than FAp, the OHAp was dissolving, so Langmuir adsorption constants for F$^-$ ion adsorption could be determined from the kinetics of dissolution. These values were always larger than constants determined from equilibrium studies. The inhibitory effect of F$^-$ ions on the rate of dissolution increased with the time of exposure to the ion.

The transformation of CaF_2 into FAp in phosphate solutions between 25 and 75 °C in the pH range 6.5 to 8.5 has been investigated [776]. The process for the formation of FAp involved the dissolution of CaF_2 to form Ca^{2+} and F$^-$ ions which then reacted with HPO_4^{2-} ions to give FAp. There was no evidence for the epitaxial growth of FAp on CaF_2, even though their crystal structures suggested this might be possible.

The reaction between SnF_2 and OHAp can result in the formation of $Sn_3PO_4F_3$ [777,778]; this affects the electrophoretic mobility of the apatite [779]. $Sn_2(OH)PO_4$ is formed if the reaction is carried out at reflux temperature [780].

Ca-def OHAps react with solutions of sodium monofluorophosphate with an uptake of FPO_3^{2-} ions [781]. Orthophosphate is liberated from the solid and addition of phosphate to the monofluorophosphate solution inhibits subsequent uptake of fluorine. The mechanism proposed was that FPO_3^{2-} replaced HPO_4^{2-}

ions, and in this way, the uptake of fluorine was related to the HPO_4^{2-} content, *i.e.* its Ca/P ratio [781]. Pyrophosphate, a common contaminant of sodium monofluorophosphate, inhibits the reaction between Ca-def OHAps and sodium monofluorophosphate [782].

3.10.3 Rate of dissolution

A clear distinction must be made between *solubility*, which is measured under equilibrium conditions and is a thermodynamic constant for a solid at a given temperature (and pressure), and *rate of dissolution*, which is measured under nonequilibrium conditions and depends on temperature, solution composition (particularly the extent to which this reflects departure from equilibrium), concentration of surface adsorbed species if present in the system, size and surface characteristics of the crystals, and possibly, the rate of stirring. Although there now seems to be general agreement that OHAp behaves as a normal ionic solid with a well-defined solubility product (Section 3.9), various different models have been proposed during the last few years to explain the observed rates of dissolution data. The rate of dissolution of CO_3Aps and enamel will also be included in this section as their behaviour is closely related to that of OHAp. However, certain aspects are left to Section 4.5.8 (synthetic CO_3Aps), Section 4.6.5 (dental enamel) and Section 4.6.6 (EM studies).

Studies on the rate of dissolution of apatites have been reviewed [25,229,783-785]. Rotating disc [786-793], stirred suspensions [794-796], constant pH [784,785,797-801] and constant composition [753,802,803] methods have been used to study OHAp, CO_3Ap, dental enamel and FAp dissolution. Electron microscopic investigations have also been made [682,804-807] (Section 4.6.6).

In many respects, there is a close relationship between studies of crystal growth and dissolution, so that constant composition methods (Section 3.8) are equally applicable [25]. Dissolution can be controlled by processes on the surface of the crystal or by the transport of ions away from the surface into the bulk solution. When controlled by surface processes, the rate of dissolution is independent of crystal size and rate of stirring, and the concentration of the ions at the crystal/liquid interface is effectively the same as in the bulk solution. The rate of dissolution per unit area of crystal surface, J (mol m^{-2}), can be interpreted in terms of an empirical kinetic equation

$$\frac{JM}{\rho} = k(1 - S)^n \qquad\qquad 3.29$$

where M is the molecular weight, ρ the density, k a rate constant and S the relative saturation or degree of saturation ($S = C/C_s$ where C is the concentration of dissolved OHAp and C_s its solubility) [25]. (This equation was originally developed for nonpolar solvents.) The effective order of the reaction, *n*, gives information about the likely dissolution mechanism.

On the other hand, if the surface processes are sufficiently rapid, the crystals will be surrounded by a thin layer of almost saturated liquid, and dissolution will be controlled by the rate at which ions can diffuse away from this into the bulk solution. In this case

$$\frac{JM}{\rho} = \frac{DV_m C_s(1 - S)}{\delta} \qquad 3.30$$

where *D* is the mean diffusion coefficient of the lattice ions, V_m the molar volume of the crystals, and δ the thickness of the diffusion layer at the crystal surface [25]. The value of δ can be estimated [25]. Equation 3.30 shows that diffusion controlled dissolution reactions are first order.

Experiments at constant pH over the range 6.6 to 7.2 showed that the rate of dissolution of dispersed 50 to 200 nm OHAp crystals is a factor of almost 10^4 less than the rate calculated from Equation 3.30, which meant that dissolution must be controlled by surface reactions [797]. The results were interpreted in terms of a general equation

$$J = k_{pH} m_0 g(C) F(m/m_0) \qquad 3.31$$

where k_{pH} is a rate constant dependent on pH, m_0 the mass of crystals at time zero and *m* the mass at a later time. $g(C)$ represents the influence of the concentration of dissolved OHAp on the rate, and depends on the mechanisms of dissolution. The empirical function $F(m/m_0)$ took into account the changes in the crystals during dissolution (surface area, roughness, shape etc). The rate constant k_{pH} was found to be equal to $k[H^+]^{0.57}$, where *k* was a constant. Specific functions for *g* and *F* were chosen with respective parameters, *p* and *q*, as exponents, so that

$$J = km_0[H^+]^{0.57}(1 - C/C_s)^p (m/m_0)^q \qquad 3.32$$

The results showed that *q* was constant at 0.6 ±0.2, but *p*, the effective order of the reaction, varied between 2.9 and 4.6 depending on the pH and m/m_0. The high value for *p* and dependence of rate constant on pH suggested a polynuclear mechanism in which H^+ ions played an important role [797,798,808]. It was proposed that H^+ ions acted as a catalyst for the exchange of phosphate between crystal surface and solution. Some criticisms of this work have been made [794].

In many studies, the undersaturation decreases with time. This difficulty can be overcome by constant composition methods (Section 3.8) which allow rates (including initial rates) to be measured as a function of time at constant undersaturation. In such a study comparing fine powders of OHAp, a precipitated CO_3Ap and enamel [802], reaction kinetics were interpreted in terms of the equation:

$$J = K(1 - (C/C_s)^{1/9})^d. \qquad\qquad 3.33$$

Here the C_s and C are the solubility product and corresponding activity product respectively, K is the rate constant and d the order of the reaction. Rates of dissolution were determined per unit area of the crystals. For all samples, the rate decreased markedly as dissolution progressed, until it appeared to have completely stopped after 60 to 80 % dissolution, depending on the undersaturation. This observation was ascribed mainly to changes in the properties of the dissolving surface. However, in experiments in which fresh OHAp crystals were added during dissolution, the rate of dissolution became higher than the starting value. The interpretation for this was that micro amounts of impurity inhibitors of dissolution were adsorbed on the crystals, with the consequence that the second batch of crystals dissolved into a purer system. Enamel and the CO_3Ap did not show this behaviour [802]. The order of the reaction, d, for OHAp was about five for up to 30 % dissolution in agreement with previous work [798], but after further dissolution, it increased sharply. The order for the dissolution of the CO_3Ap was smaller and its rate of dissolution higher than for OHAp. The rate of dissolution of enamel, initially very high, fell rapidly with the extent of dissolution; the order of the reaction increased to a constant value after about 30 % dissolution. These differences from the synthetic material were thought to be due to release of impurities on dissolution, particularly F^- ions, from enamel. It was concluded that CO_3Ap was a better model than OHAp for enamel dissolution [802].

The constant composition method has been used to compare the dissolution mechanisms of OHAp (pH 5) and FAp (pH 4.5, 5.0, 5.5 and 6.0) at 37 °C [753]. The effective order of the dissolution reaction for OHAp was ~6, in agreement with earlier work, which was attributed to a polynucleation dissolution mechanism, as discussed above. This finding contrasted with the results for the dissolution of FAp, which gave an effective reaction order of 2. This value indicated that dissolution of FAp was surface controlled, probably following a spiral dislocation mechanism [753].

The rates of dissolution of OHAp powders (calculated from the pH change and initial conditions) and corresponding degree of saturation have been determined for dissolution in acetic, lactic and dilute phosphoric acid for initial pH values from 4 to 6 [795]. The rate of dissolution was found to be independent of the rate of stirring, confirming that surface, rather than diffusion processes, were rate determining under the conditions studied. The results were interpreted in terms of the rate of dissolution R (expressed as moles of calcium appearing per litre of solution per unit time per unit area of the crystals) given by:

$$R = K(1 - DS)^m (\sum_i a_i)^n \qquad 3.34$$

where K is a rate constant, a_i the activity of the ith undissociated acid, and DS the degree of saturation defined by:

$$(DS)^9 = (Ca^{2+})^5 (PO_4^{3-})^3 (OH^-)/K_s \qquad 3.35$$

where the brackets on the right-hand side denote activities. The exponents, m and n, reflect the contribution of DS and the sum of the undissociated acid activities, respectively, to the overall order of the dissolution reaction. K, m and n were taken as adjustable parameters that were determined from the experimental data by a nonlinear regression analysis. The resulting equation described the results well, although less so for phosphoric acid. This poorer fit was ascribed to the rate of dissolution in phosphoric acid being at least one order of magnitude less than in the organic acids at the same pH. The establishment of Equation 3.34 confirmed that the rate of OHAp dissolution was not solely a function of the thermodynamic driving force, as reflected in DS, but also depended on the sum of the activities of the undissociated acids. It was pointed out that Equation 3.34 was applicable to a variety of chemical systems, whereas Equation 3.32, in which $[H^+]$ was explicitly considered, was only useful in systems with similar solution properties, for example, when the DS and acid activities changed little as the dissolution proceeded. However, m, the order of the reaction with respect to DS (Equation 3.34) was about 3, comparable to earlier determinations (see above).

The rate of dissolution of OHAp or enamel in rotating disc dissolution experiments is controlled by diffusion at low speeds and by kinetic factors at high speeds. Results from such studies have been used as a basis for a model for the dissolution of enamel and OHAp in dilute acetic acid that involved dissolution from two sites with different apparent solubilities [787-789]. Studies of the dissolution of stirred suspensions of OHAp apparently supported this model [794]. However, it has been claimed [798] that agreement with the two site model was an artifact due to the mathematical curve fitting procedure that was used.

The kinetics of dissolution of enamel [785,800] and OHAp [801] have been studied at constant pH from 3.7 to ~6.9 by continuously recording the H^+ ion uptake and Ca^{2+} ion release. The rate of dissolution of the 165 to 200 μm powders that were used was dependent on the rate of stirring at low speeds, and independent at high speeds. It was proposed that at low speeds, the rate limiting process was diffusion through the Nernst layer adjacent to the crystals, and at high speeds, by diffusion through a calcium- and/or phosphate-rich "by-product" layer.

Adsorption of various chemical species onto OHAp and their inhibitory effect on dissolution have been studied as follows: phosphoproteins [809]; methylene diphosphonate, Sn^{2+} ions and partly peptized collagen [810]; ATP, ADP, and AMP [811]; F^- [775]; and Al^{3+}, Cr^{3+} and Fe^{3+} ions [812]. Constant composition methods have been used to study the effects of methanehydroxydiphosphonate and polyacrylic acid on the kinetics of dissolution [802].

The reduced rate of acid dissolution of OHAp in the presence of F^- ions has been known for a long time, but some of the effects observed are not simple to explain [775] (see also Section 3.10.2 for the reaction between F^- ions and OHAp). In studies at constant pH, the ratio of the rates of dissolution in the presence and absence of F^- ions has been determined, whilst other parameters remained the same [775]. For a constant degree of saturation, inhibition increased, as the pH decreased. At constant pH, the inhibitory effect increased with an increase in the degree of saturation (*i.e.* if the undersaturation lessened). The inhibitory effects increased with the length of exposure of the OHAp to F^- ions; possible causes of these effects were discussed.

Dissolution of single crystals of OHAp is highly anisotropic, being very fast along the *c*-axis with the formation of a central cavity, and very much less in a direction perpendicular to the *c*-axis [682]. Similar behaviour has been seen in single crystals of CO_3Aps [813] (Section 4.6.6). Preferential acid dissolution of the centre is also seen in enamel crystals from carious lesions, shark enamel (a CO_3FAp) and other synthetic apatites [804] (and papers cited in ref. [804]). This feature has been attributed [682] to the presence of screw dislocations with the dislocation line parallel to the *c*-axis. However, at least for biological apatites, the presence of a central defect originating from a growth process via OCP, rather than a screw dislocation, has been thought to be a more likely explanation [792] (Section 4.6.6).

Permeable aggregates of OHAp that have dilute acid applied to an external surface dissolve beneath the surface, in preference to direct surface dissolution (Fig. 3.3). It has been proposed that this phenomenon, and a similar feature seen in early carious lesions in dental enamel, is a result of diffusive coupling between the incoming H^+ ions from the acid and outflowing products of dissolution; the coupling causes a build up of the products of dissolution near the surface that will retard dissolution in this region [815]. A similar suggestion has been made to explain subsurface dissolution and precipitation in systems containing calcium hydroxide and other porous solids [816]. In apatitic systems, part of the resistance of the surface to dissolution might be the result

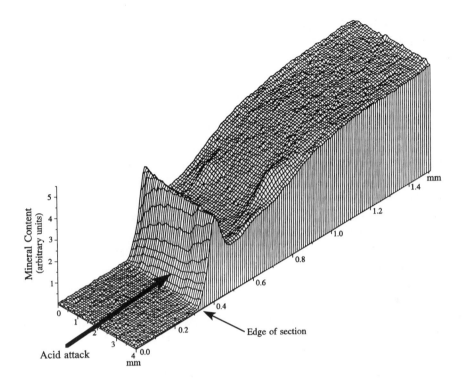

Fig. 3.3 Scanning microradiograph showing subsurface demineralisation in a 200 μm thick section of permeable OHAp aggregate. The section had been painted with acid resistant varnish, except for the surface along its edge (arrowed), and immersed in 0.1 mol l⁻¹ sodium acetate/acetic acid buffer at pH 4.4 for 140 h. The radiograph was taken by measuring the absorption of a 10 μm X-ray beam through the section as it was stepped in two dimensions. Mo target, Zr filter, scintillation counter set for MoKα radiation, 10 μm steps normal to the section edge and 100 μm steps parallel to the edge, 10 s counting time at each point. (After Fig. 5.1.1.2 of Anderson [814])

of phase changes and in dental enamel, also to a different surface composition (higher F^- and lower CO_3^{2-} ion contents), as has been discussed in a recent review of the causes of subsurface demineralisation in dental caries [817].

3.11 Infrared and Raman spectroscopy
3.11.1 Introduction

The IR and Raman spectra of OHAp are very similar to that of FAp (Section 2.6.1) except for additional OH bands. Thus, in the site-group approximation, in addition to PO_4 bands, a single IR and Raman active OH stretching mode is expected at ~3600 cm^{-1}. Librational (rotary) bands of the OH$^-$ ion are known in alkali and alkaline earth hydroxides at much lower frequencies [548], so they might also occur. There is also the potential, shared with ClAp (Section 2.6.2), for minor differences due to the local absence of a mirror plane because the OH$^-$ ion is displaced from 0,0,¼ and, for stoichiometric compositions, from an ordered monoclinic structure. There might also be small differences due to the heteronuclearity of the OH$^-$ ion.

As in the site-group approximation for ClAp (Section 2.6.2), the absence of a local mirror plane reduces the symmetry of the PO$_4^{3-}$ ion to C_1, so there are nine A normal modes (IR and Raman active). Spectra of hexagonal OHAp have been discussed in terms of the factor group approximation with space group $P6_3$ and factor-group C_6, to determine the effect of the absence of a mirror plane on the spectrum [548,557,675]. In these circumstances, there should be 9A (IR and Raman) + 9E_1 (IR and Raman) and 9E_2 (Raman) internal modes for the six PO$_4^{3-}$ ions.

3.11.2 Hydroxyapatite spectrum

Assignments for IR [557] and Raman [556,675,818] spectra of OHAp have been made (Table 3.4). The OH stretch IR assignment was based on: (a) the reduction in the frequency of this band on deuteration by a factor of ~$\sqrt{2}$ in dental enamel [77]; and (b) polarised IR studies [77,611] of francolite (Figs 4.2 and 4.3) and dental enamel (Figs 4.7 and 4.8), that showed that the transition moment was parallel to the c-axis, in agreement with the orientation found by neutron diffraction studies [599] of Holly Springs OHAp. In deuterated synthetic OHAp, the OD stretch is at 2636 cm^{-1}, and the observed frequency shift of 8 cm^{-1} between the ^{16}OH and ^{18}OH stretch is in near agreement with the calculated value of 12 cm^{-1} [557].

The 630 cm^{-1} OH librational mode assignment was based on the reduction in frequency on deuteration by a factor of ~$\sqrt{2}$, and the absence of a band near this frequency in FAp [182]. This corrected an earlier misassignment to a ν_4 PO$_4$ component [77], and ruled out an alternative assignment to a Fermi resonance effect [548]. As expected, the transition moment of this band is perpendicular to the c-axis as determined by polarised IR reflectance studies of single crystals of natural OHAp [820,821], and transmission spectra of FAp containing small amounts of OH$^-$ ions [552]. For ODAp, the centre of the OD librational peak is at 465 cm^{-1}, with components at 469 and 457 cm^{-1} (better resolved at -185 °C) [557]. It was suggested that the stronger than expected intensity of the 469 cm^{-1} band originated from vibrational coupling between the weak band at 474 and shoulder at 462 cm^{-1} seen in undeuterated OHAp (ν_2 PO$_4$, Table 3.4) with the OD librational mode; or alternatively, from the loss of degeneracy of the doubly degenerate librational mode, because the D atoms were positioned off the c-axis, resulting in C_1 symmetry for the ion. (In the

Table 3.4 Frequencies (cm^{-1}) and assignments for OHAp. IR at 48 °C with values within 1 to 2 cm^{-1} [557]; Raman at 295 K with values ±3 cm^{-1} [818,819]. See text for isotopic assignments.

IR[a]	Raman[a]		Assignment
3572	3572 m		OH stretch
1087	1075 s	1046 s	}
~1072 sh[a]	1061 w,sh	1039 m	} v_3 PO$_4$
1046	1053 sh	1028 s	}
~1032 sh			
	962 vs		
962		948 m	v_1 PO$_4$
	641 vw		
630		630 vw	OH libration
	615 m		
601	608 s	593 s	}
571		580 s	} v_4 PO$_4$
	447 s		
474		432 s	}
~462 sh			} v_2 PO$_4$
	376 w		lattice mode
	333 w[b]		
~355 sh			Ca$_3$-(OH) "v_3 stretch"
343			
	306 w[b]		
	289 w[b]	175 w	}
	267 w[b]	157 w	}
	237 w	153 w	}
~290	222 w	140 w	} Ca-PO$_4$ lattice modes
~275	206 w	132 w	}
~228	196 w	114 w[c]	}

[a]vs, very strong; s, strong; m, medium; w, weak; vw very weak; sh, shoulder.
[b]see also refs [559,560] and text.
[c]probably the same band reported [556] at 107 cm^{-1} which varied with sample. Another variable band was seen at 92 cm^{-1}.

monoclinic structure, Section 3.2, the oxygen and hydrogen atoms of the OH$^-$ ion are in general positions, the oxygen atom being 0.0012 Å from the 2$_1$ pseudohexagonal screw axis, so disordered displacements of this size are likely in hexagonal OHAp.) The observed frequency shift of 4 cm^{-1} between the ^{16}OH and ^{18}OH librations was in near agreement with the calculated value of 2 cm^{-1} [557]. The assignment of the two librational bands seen in the Raman spectrum was based on their disappearance on deuteration, presumed to be because they had shifted to a position obscured by v_2 PO$_4$ [818].

As in FAp (Section 2.6.1), the v_2 PO_4 IR band was originally misassigned in OHAp [182,548], and subsequently corrected [549,551,675]. Assignment of the weak IR band at 474 and shoulder at 462 cm^{-1} to components of v_2 was confirmed by the shift of the peaks by the appropriate amount for OHAp containing $P^{18}O_4^{3-}$ ions, and the absence of a shift if the calcium mass was changed [557]. Weak PO_4 overtone and combination bands at about 2000 cm^{-1} are not listed in Table 3.4, but frequencies and assignments have been published [557].

The IR band at 343 cm^{-1} and shoulder at 355 cm^{-1}, have been assigned to the doubly degenerate in-plane antisymmetric stretching mode ("v_3"), and the much weaker, poorly resolved band at 228 cm^{-1} (and other weak bands seen at -185 °C) to other modes (possibly "v_2" out-of-plane deformation motions), of the Ca_3-(OH) group and sublattice with D_{3h} symmetry [557]. These assignments were based on: (a) the observation that these were the only lattice modes that changed on deuteration (7 cm^{-1} for "v_3" and 2 cm^{-1} for "v_2"); and (b) by analogy with the undistorted CO_3^{2-} ion which has this symmetry and has v_3 at ~1415 cm^{-1} and the much weaker v_2 at ~879 cm^{-1} (Section 4.2.2). As the OH⁻ ion is much lighter than the surrounding Ca^{2+} ions, it will undergo the predominant motion. Observed isotopic shifts for the band assigned to "v_3" at -185 °C for $^{40}Ca_3$-(^{16}OD) and $^{40}Ca_3$-(^{18}OH) were 7 and 16 cm^{-1} respectively, compared with calculated values for "v_2" and "v_3" type motions of 9 and 16 cm^{-1} respectively [557]. The observed isotopic shifts for $^{44}Ca_3$-(^{16}OH) and $^{48}Ca_3$-(^{16}OH) of 2±0.5 and 4±0.5 cm^{-1} respectively were consistent with 2 and 3.7 cm^{-1}, respectively, calculated for a "v_2" type of motion. Shifts of "v_3" for isotopic cationic substitutions were not calculated, but were expected to be greater than those predicted for "v_2". Thus, these isotopic studies alone could not prove the above assignments, but were consistent with them. The Raman line corresponding to the 343 cm^{-1} IR band occurs at 329 cm^{-1} [557]. However, a frequency of 337 cm^{-1} with a shift on deuteration to 329 cm^{-1} for this Raman line has been reported [556]. The intensity of this Raman line in F,OHAp depends strongly on the F⁻ ion content [556] (Section 3.11.5). A "v_3" mode, analogous to that in OHAp, is seen in FAp at 325 cm^{-1} (Section 2.6.1). The Raman spectrum of OHAp at 50 K has been published [818].

From polarised Raman studies of single crystals of mineral OHAp and studies of isotopically enriched synthetics, bands at 329 (E_{2g}), 305 (A_g) and 270 cm^{-1} (E_{2g}), were assigned to vibrations of the $2[Ca_3$-(OH)] sublattice, and a band at 285 cm^{-1} (E_{1g}) primarily to rotary PO_4^{3-} ion motions [559,560].

Combined IR and Raman spectra for internal PO_4 and lattice modes of monoclinic powdered OHAp agreed better, overall, with the factor-group C_{6h} of FAp, than with a factor group of lower symmetry, indicating that the monoclinic form has a pseudo-C_{6h} factor group [564,565]. The same conclusion was reached from polarised Raman studies of mineral OHAp [559,560]. In detailed polarised Raman studies [822] of single crystals of monoclinic OHAp, it was also found that general features of the spectrum could be explained on the basis of the hexagonal factor-group C_6, indicating pseudohexagonal, rather than monoclinic symmetry. Thus, although the non-coincidence, except for v_1, of the Raman and IR PO_4 bands, indicates that factor-group analysis is required for full interpretation of the spectrum, the fact that the OH^- ions are displaced from $0,0,\frac{1}{4}$ does not affect the PO_4^{3-} ion coupling sufficiently to reduce the symmetry of the effective factor group of OHAp from the C_{6h} factor group of FAp [675]. However, like ClAp (Section 2.6.2), there are slight spectral differences between monoclinic and hexagonal OHAp [564,565,822]. Additional weak Raman lines in the v_3 (1061, 1054, 1043 and 1032 cm^{-1}) and v_4 (very weak shoulders at 592 and 587 cm^{-1}) PO_4 and lattice mode (260 - 160 cm^{-1}) regions, barely or not visible at room temperature, but clearer at -185 °C, were attributed [564,565] to the lower symmetry monoclinic form. They were absent in ns-OHAp, or if one in ten of the OH^- ions were replaced by F^- ions so that the ordering of the OH^- ions was disturbed and the structure remained hexagonal [565,564]. The E_1 polarised Raman spectrum of monoclinic OHAp has two low frequency bands at 65 and 68 cm^{-1}, whereas FAp has only a single line at 45 cm^{-1}, which led to the suggestion that the doublet in OHAp might originate from a small monoclinic distortion [822]. The approximately doubled Raman OH stretching intensity in monoclinic OHAp, compared to that in a slightly impure hexagonal form, has been attributed to OH^- ion ordering [689]. Increased asymmetry (toe extending towards lower frequencies) in the OH band of OHAps prepared at high temperatures has been seen [620], but this might be due to the Christiansen filter effect (BO Fowler, personal communication, 1991) due to a changing mismatch of the KBr pellet and sample refractive indices through the absorption band. A variable width of the OH band, which was much narrower in high temperature monoclinic preparations (but still variable amongst these) compared with aqueous preparations, was reported [620]. An increased line-width was associated with an increased ease of exchange of OH^- for OD^- ions [620]. Variability in intensity of OH bands has been observed in lead apatitic compounds [823] (Section 3.11.6).

3.11.3 High temperature OH stretching bands

An additional OH IR stretching band appears at elevated temperatures in s- and Ca-def OHAps [824]. Its intensity increased, and the original OH band at 3572 cm^{-1} decreased, with temperature, but the sum of their intensities was approximately constant. At 485 °C in a Ca-def OHAp, the main OH band had moved to a slightly lower frequency (3565 cm^{-1}), and the new band, which was still weaker than the main one, was at 3535 cm^{-1} (this frequency appeared to be independent of temperature, unlike the main band). The assignment of the new band to an OH stretch was confirmed by deuteration (OD stretch at 2610

cm^{-1}), and the thermal changes were fully reversible. It was concluded [824] that the additional OH band at higher temperatures arose from movement of OH⁻ ions from their room temperature positions to another location, possibly with the OH⁻ ion straddling the mirror plane at $z = \frac{1}{4}$.

The frequencies of the OH stretch and librational modes have been measured in $M_{10}(XO_4)_6(OH)_2$ with M = Ca, Sr, Ba, Cd or Pb and X = P, As or V (all combinations, except Cd,As and Cd,V which apparently do not exist) [823]. An additional OH stretch at 20 to 30 cm^{-1} below the main OH IR band was seen at *room temperatures* in all the barium and arsenic compounds, and in all the vanadium compounds except those with calcium and strontium. No additional OH librational modes were seen. These bands were thought [823] to have the same origin as the high temperature band seen in OHAp. The additional OH bands seen in these apatites also increased in intensity with temperature, whilst the main OH band decreased in intensity [825]. For example, the two bands absorbed equally at 590 °C for M = Ca and X = P, and at 50 °C for M = Ba and X = V. The temperature dependence was modelled on an equilibrium between OH⁻ ions in two different sites with concentrations *HT* and *LT* for the high and low temperature forms respectively. The van't Hoff relationship

$$\frac{d(\ln(HT/LT))}{d(1/T)} = - \frac{\Delta H}{R},$$

3.36

where T is the absolute temperature and R the gas constant, was reasonably well-obeyed. Values of ΔH ranged from 9 to 15 kJ mol⁻¹. Two models, equally preferred, were proposed for the high temperature band. One was the orientational or positional disorder of the OH⁻ ion discussed above [824]. The other supposed reversible formation of water from 2OH⁻ → H_2O + O^{2-} so that the high temperature band was assigned to ν_3 H_2O (ν_1 is much weaker and would not be seen). The definitive explanation of the origin of the "additional" OH stretching band is not clear, but it is probably intimately associated with the mechanism of the monoclinic to hexagonal thermal phase transition (Section 3.2), because both phenomena involve the effect of increased thermal motion on the OH⁻ ions. Note also that an additional OH band of uncertain origin is seen at 3544 cm^{-1} in calcium-rich OHAp at room temperature (Section 3.11.6). The temperature dependence of this band does not seem to have been investigated. Changes with temperature in the NMR spectra of mineral F,OHAp (20 % replacement of F⁻ by OH⁻ ions) have been attributed to a change from static to dynamic disorder of the OH⁻ ion above and below the mirror plane at $z = \frac{1}{4}$ over the range 120 to 360 °C [577] (Section 3.12).

3.11.4 Bands from surface hydroxyl ions

Surface areas of many OHAp preparations are very high (often greater than 60 m^2 g^{-1}), so it might be anticipated that there would be sufficient surface ions to cause spectral effects. An IR band at 3660 cm^{-1} in s- and Ca-def OHAp samples has been assigned to such OH$^-$ ions [824]. A pellet of OHAp, degassed at 440 °C, exposed to D$_2$O vapour at 275 °C and evacuated at 275 °C, substantially lost the 3660 cm^{-1} band and gained a band at 2725 cm^{-1}. Treatment with NH$_3$ gas at 60 mm Hg (8.0 kPa) and 35 °C did not affect intensities of the stretching mode of the OH$^-$ and OD$^-$ lattice ions, but did cause a substantial reduction of the 2725 cm^{-1} band with a concomitant increase of the original 3660 cm^{-1} band. There were also weak NH bands at 3385 and 3280 cm^{-1} which were attributed to adsorbed NH$_3$. On overnight evacuation at 130 °C, the NH bands disappeared, and the 3660 cm^{-1} OH band strengthened. These results showed that the deuterium of the OD$^-$ ions, responsible for the 2725 cm^{-1} band, could exchange with hydrogen in NH$_3$ to regenerate the OH band at 3660 cm^{-1}; the OD$^-$ ions within the lattice could not exchange. The conclusion was that the 2725 and 3660 cm^{-1} bands must originate from accessible ions in the surface.

3.11.5 Perturbations of OH bands by *c*-axis substitutions

So far, only additional OH bands in pure hydroxyl apatitic compounds have been discussed. However, if some of the OH$^-$ ions are replaced by F$^-$ ions, additional IR OH stretching bands from weak OH$\cdots$F hydrogen bonds appear at lower frequencies (Fig. 3.4) [549,557,600,690,823,826-832]. Additional OH librational bands at higher frequencies than the band in pure OHAp are also seen [549,831]. Their assignments have been confirmed by deuteration studies [549].

In one detailed study of F,OHAp, two additional OH stretching bands at 3545 and 3537 cm^{-1}, and three additional librational bands at about 735, 715 and 670 cm^{-1} were reported [549]. As the F$^-$ ion content increased to equal that of the OH$^-$ ions, the intensities of these three bands increased, whilst the intensity of the 3572 and 630 cm^{-1} bands decreased. The additional bands were attributed to at least two nonequivalent hydrogen bonds. The bands at 3545, 715 and 670 cm^{-1} predominated when there were more OH$^-$ than F$^-$ ions; these were assigned to the group OH$\cdots$F$\cdots$HO. The 3537 and 735 cm^{-1} bands predominated when there were fewer OH$^-$ than F$^-$ ions, and were assigned to OH$\cdots$F pairs. The band intensity changes were taken as good evidence that the OH$^-$ and F$^-$ ions were intermingled along the *c*-axis, rather than aggregated into regions of OHAp and FAp.

In a later study, different assignments and slightly different experimental results were reported [831].The shifts were correlated with *n* in the ion column

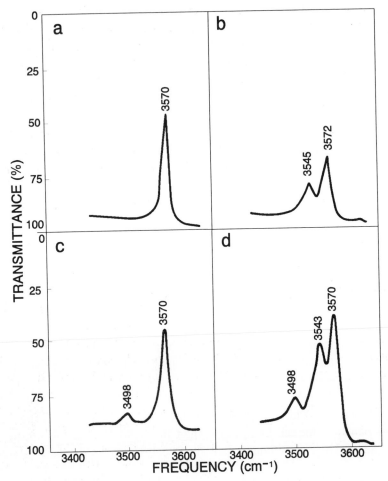

Fig. 3.4 IR spectra (bromoform mulls) of OH stretching modes in various apatites. (a) synthetic OHAp prepared by the stoichiometric reaction between $CaCO_3$ and dilute phosphoric acid followed by heating at 900 °C for 1 h. (b) F,OHAp with ~10 % of the OH$^-$ ions replaced by F$^-$ ions. (c) Cl,OHAp with a few percent OH$^-$ ions replaced by Cl$^-$ ions. (d) Holly Springs OHAp containing 0.36 wt % Cl. (Fig. 1 of Dykes and Elliott [600])

$$F\cdots HO \; (HO)_n \; HO \; OH \; (OH)_n \; OH\cdots F. \qquad 3.37$$

Bands at 3643 and 680 cm^{-1} were assigned to the central HO OH configuration, and bands at 3546 and 720 cm^{-1} to F$\cdots$HO at the ends of the column (actually OH$\cdots$F$\cdots$HO because it was implied that an F$^-$ ion always reversed the OH direction; OH$\cdots$F OH, without reversal, was not discussed). Bands at 3540 and

747 cm^{-1} were assigned to F OH$\cdots$F from isolated OH$^-$ ions in a chain of F$^-$ ions. The band at 3643 cm^{-1} has not been seen by others, and is probably due to Ca(OH)$_2$ impurity [690] which absorbs in this region. More recently, the high frequency member of the F$^-$ ion perturbed bands (at 3547 cm^{-1}) was again assigned to OH$\cdots$F$\cdots$HO, and the lower frequency member (at 3541 cm^{-1}) to F OH$\cdots$F F in which the OH$^-$ ion had at least two F$^-$ ions on its hydrogen side [690].

Two weak bands at 3650 and 3496 cm^{-1}, almost equidistant from the main band at 3573 cm^{-1}, have been assigned to combinations with lattice vibrations when only a little F$^-$ ion is present [690]. The possibility of weak bands at 3670, 3650, 3565 and 3549 cm^{-1} in F,OHAp has been discussed [830]. Spectra of OH$^-$ ions in mineral FAps have been investigated (some with polarised radiation) in the range 4000 to 3000 cm^{-1}; a number of bands were seen above 3600 cm^{-1} [829].

Correlation of IR and NMR results. Much detailed information about the interaction between OH$^-$ and F$^-$ ions can also be obtained from NMR single crystal and powder studies (Section 3.12). In particular, NMR has established nearest neighbour H$\cdots$F and F F distances for the OH OH$\cdots$F$\cdots$HO HO, OH OH$\cdots$F OH OH, and F F OH$\cdots$F F configurations. These distances, taken from Section 3.12, will be cited at appropriate times during the following discussion. Some aspects of the extent to which NMR and IR results can be correlated have been discussed [183]. Both techniques agree that F$^-$ and OH$^-$ ions are dispersed in OHAp and FAp respectively, when present as minority species. The IR results show that the high frequency band at 3548 cm^{-1} (frequency from ref. [690]) is seen first as the fluorine content in OHAp increases. This band must correspond to the weakest hydrogen bonding and hence the longest F$\cdots$H distance. The NMR results also show that, for small amounts of F$^-$ ion in OHAp, the majority configuration has the longest F$\cdots$H distance, so the IR and NMR results are consistent. Thus, it seems very probable that the high frequency band originates from the symmetrical configuration OH$\cdots$F$\cdots$HO with F at $z = \frac{1}{4}$, because NMR shows that the OH$^-$ ions have essentially their normal positions with an H$\cdots$F distance of 2.18 Å. When few OH$^-$ ions are present in FAp, the most displaced IR band at 3538 cm^{-1} (frequency from ref. [690]) is seen, which must correspond to the strongest hydrogen bonding and hence the shortest F$\cdots$H distance. Likewise, the NMR results show the shortest F$\cdots$H distance (2.035 Å) when there are few OH$^-$ ions. However, the situation as regards assigning IR bands for the configuration OH OH$\cdots$F OH OH (F$\cdots$H distance 2.12 Å) and OH OH$\cdots$F F (F$\cdots$H distance 2.035 Å) is not straightforward. It has been suggested that the NMR determination of the shorter distance is in error, and that it should be nearer 2.12 Å because structural considerations indicate that the F$\cdots$H distances for the two configurations should be rather similar [183]. As a result, the configurations were both assigned to a low frequency band at 3541 cm^{-1} [183]. On the other hand, if the NMR distances are accepted at their face value, the low frequency IR band at 3538 cm^{-1} should correspond to F OH$\cdots$F F with an H$\cdots$F distance of 2.035 Å. It is then not clear which IR band, if any, corresponds to the less frequent configuration, OH OH$\cdots$F OH OH, that is found at low F concentrations with an intermediate F$\cdots$H distance (2.12 Å). The implication [828] that, of the two bands seen in F,OHAps, the low frequency one at 3535 cm^{-1} corresponds to this configuration would seem to be incorrect because this band is not seen at low F$^-$ ion concentrations (see Fig. 2 of ref. [690]). The frequency of the IR band corresponding to an F$\cdots$H distance of 2.12 Å can be estimated from the known F$\cdots$H distances for the bands at 3548 and 3538 cm^{-1}, on the assumption that the frequency shift is linear with F$\cdots$H distance. The value obtained (3544 cm^{-1}) indicates that the band from the OH OH$\cdots$F OH OH configuration will not be clearly resolved from the 3548 cm^{-1} band, but will contribute to its width on the low frequency side. In summary, the crux of the unresolved difficulty is whether OH OH$\cdots$F OH OH absorbs at the same (low) frequency as F F OH$\cdots$F F, or the same (high) frequency as OH OH$\cdots$F$\cdots$HO HO.

The molar absorption coefficients for the OH stretching modes increase with the OH$^-$ ion content of F,OHAp, and reach a maximum with about 80 % of the F$^-$ ions replaced by OH$^-$ ions, after which it falls [690]. In the Raman spectrum of F,OHAp, a line near 3550 cm^{-1} (the exact position varies with F content) that was assigned to an OH stretching mode perturbed by an adjacent F$^-$ ion, also increases in intensity with increasing F$^-$ ion content up to at least 40 % replacement of OH$^-$ by F$^-$ ions [556]. The intensity of the 337 cm^{-1} Raman line, assigned to the Ca$_3$-(OH) "v_3 stretch" corresponding to the IR band at 343 cm^{-1} (Table 3.4, Section 3.11.2), depends strongly on the F$^-$ ion content of F,OHAp, and decreases to about 30 % of its intensity in pure OHAp at ~40 % replacement of OH$^-$ by F$^-$ ions [556].

A Cl$^-$ ion next to an OH$^-$ ion in Cl,OHAp causes an additional OH stretch to appear at 3498 cm^{-1} which has been assigned to a weak hydrogen bond OH$\cdots$Cl [600]; this can be seen in Holly Springs OHAp and synthetic samples with small percent of OH$^-$ replaced by Cl$^-$ ions (Figs 3.4c and d) and in dental enamel (Fig. 4.8, Section 4.6.3). The shift is larger than for F,OHAp, whereas a smaller shift might be expected for the less electronegative chlorine. The explanation given for this was that the Cl$^-$ ion was closer to the OH$^-$ ion because it was located near $z = \frac{1}{2}$, rather than $z = \frac{1}{4}$ as for the F$^-$ ion [600]. This difference in behaviour has also been associated with the fact, that in F,Cl,OHAps, the z parameter of both the Cl$^-$ and OH$^-$ ions are primarily dependent on the chlorine content [429] (Section 2.3.1). Changes in the OH stretching and librational region have been studied as a function of chlorine content [833]. The Raman spectra of Cl,OHAps have been investigated [556].

The OH stretching frequency has been measured in a variety of non-calcium F-, Cl- and BrAps containing a few per cent OH$^-$ ions [823]. For example, the OH stretch is at ~3405 cm^{-1} in $Cd_5(PO_4)_3(Br,OH)$.

3.11.6 Other "hydroxyapatites"

Oxyapatite

The IR spectra of apatites that have a divalent ion and a vacancy instead of two monovalent ions on the *c*-axis have conspicuously more complex phosphate absorption bands than FAp [85,86,659]. Examples include: OAp (preparation in Section 3.4.3), A-CO$_3$Ap (IR in Section 4.3.2), perOAp (preparation in Section 3.4.3) and sulphoapatite; these contain the ions O^{2-}, CO$_3^{2-}$, O$_2^{2-}$ and S^{2-} respectively. The phosphate bands are different for the various divalent substitutions; a clear feature of the OAp spectrum, for example, is the presence of two ν_1 PO$_4$ bands about 20 cm^{-1} apart. The explanation for these additional PO$_4$ bands is that, although the structure may statistically still have the space group $P6_3/m$ because of disordering of vacancies and divalent ions, locally, the PO$_4^{3-}$ ions in the unit cell are no longer equivalent as in FAp [85,86,659]. Thus, from an IR point of view, there will be at least two nonequivalent ions, one next to a vacancy, and the other next to a divalent ion (the same will apply if the divalent ions are locally ordered, as is rather likely).

OH,OAp, prepared by heating OHAp in vacuum at 800 to 1000 °C, has two new principal bands in the 500 to 400 cm^{-1} region (~475, ~433 cm^{-1}) whose intensities depend on the composition [654]. For less than ~50% dehydroxylation, the ~433 cm^{-1} band was more intense than the ~475 cm^{-1} band, and *vice versa* for greater than ~50% dehydroxylation. As the dehydroxylation increased, the OH librational band at 631 cm^{-1} and translation band area at ~350 cm^{-1} (*i.e.* the Ca$_3$-(OH) "ν_3 stretch", Table 3.4, Section 3.11.2) decreased. The 434 cm^{-1} band was assigned to a Ca$_3$-O "ν_3 stretch", on the basis of isotopic band shifts in separately enriched ^{18}O and ^{48}Ca OH,OAps, and because a high frequency shift from the corresponding band in OHAp was expected from the increased electrostatic bonding of a doubly charged ion and a slight decrease in mass. Detailed isotopic studies were not carried out on the ~475 cm^{-1} band, but it was suggested that it probably originated from Ca$_3$-O group vibrations with a more strongly bound Ca^{2+} ion than the vibration responsible for the 434 cm^{-1} band. These studies enabled the IR bands in the same region in the previously described [85,86,449,655] (Section 3.4.3) Ca$_{10}$(PO$_4$)$_6$O$_{0.75}$(OH)$_{0.5}\square_{0.75}$ to be assigned as follows: the strong band (doublet) at ~475 cm^{-1} probably to a Ca$_3$-O "ν_3 stretch", and two weak bands at ~440 and ~430 cm^{-1}, probably from ν_2 PO$_4$ and/or a (Ca)$_3$-O "ν_3 stretch" with a more weakly bound Ca^{2+} ion (frequencies from ref. [654]).

Calcium-deficient precipitated apatites

The IR spectra of Ca-def OHAps are similar to that of s-OHAp, the major difference being the addition of HPO_4 bands. The IR spectrum of Ca-def OHAp (Ca/P molar ratios 1.5 to 1.6) precipitated from $(NH_4)_2HPO_4$ and $Ca(NO_3)_2$ in the pH range 7 to 8, has three bands additional to those in s-OHAp [189]. A weak band at 1210 cm^{-1} was assigned to a δ_{OH} mode of hydrogen-bonded HPO_4^{2-} ion, a weak shoulder at 1133 cm^{-1} to one component of the ν_3 HPO_4 vibration, and a band at 870 cm^{-1} to a P-OH ν_5 HPO_4 mode. In terms of C_{3v} symmetry for the HPO_4^{2-} ion [187], these are the doubly degenerate ν_5 P-OH in-plane deformation, a component of the doubly degenerate ν_6 P-O stretch, and the ν_3 P-OH stretch, respectively. At -180 °C, some increased absorption was seen in the region 4000 to 2000 cm^{-1}, with some indication of separate peaks, but no well-defined structure of the type seen in OCP [189] (Section 1.3.8).

Study of the deconvoluted FT IR spectrum of a fairly well-crystallised Ca-def OHAp in the ν_3 PO_4 region showed a number of bands (1144 and 1102 cm^{-1} were the clearest) that were absent from OHAp [190]. Poorly crystallised samples also had bands at 1125 and 1110 cm^{-1} that rapidly vanished if the apatite was subjected to partial dissolution (the high frequency one first); they also disappeared progressively during maturation [190]. These bands were assigned to HPO_4^{2-} ions in two different environments that are first to be solubilised in poorly crystallised BCaPs [190].

A medium intensity band at 550.9 cm^{-1} on the low frequency side of the two main ν_4 PO_4 absorption bands (at 600 and 560 cm^{-1}) in the deconvoluted FT IR spectrum of a Ca-def OHAp has been assigned to HPO_4^{2-} ions; a weak shoulder on the high frequency side at 610 to 615 cm^{-1} was assigned to PO_4^{3-} ions in a labile environment [191]. In part, these assignments were based on the observation that, during maturation and aging of the apatite, the 610 to 615 cm^{-1} band gradually decreased in intensity and the two main ν_4 PO_4 bands moved to slightly higher frequencies and changed their relative intensities. Partial dissolution experiments in EDTA modified the appearance of the ν_4 PO_4 bands, improving their resolution, which was indicative of an increase in the crystallinity of the solid phase. After the EDTA treatment, the integrated intensity of the HPO_4^{2-} bands remained practically unchanged, but the 610 to 615 cm^{-1} shoulder diminished in intensity. The studies of ν_3 and ν_4 PO_4 with FT IR spectroscopy have been extended to include dental enamel and bone [190,191] (Section 4.6.3).

The intensity of the ν_3 P-OH stretch increases as the Ca/P ratio falls [698,707,834]. This change has been used to estimate HPO_4^{2-} ion concentrations in various apatites (Section 3.6.2). Initial attempts to demonstrate a shift to lower frequencies on deuteration were unsuccessful, even though the conditions that were used caused the exchange of structural OH$^-$

ions [707]. However, a new band at 850±2 cm^{-1} was seen after heating a Ca-def OHAp in D$_2$O vapour in a hydrothermal bomb at 2000 bar (200 MPa) and 250 °C [698]. The ratio of P-(OD) to P-(OH) stretching frequencies (0.977) was close to the theoretical value (0.982).

The peak intensity of the librational mode at 630 cm^{-1} in Ca-def OHAps is lower than in s-OHAp (compare for example Fig. 4a and 4c in ref. [189]). It has been reported [698] that the *integrated* intensity of the OH librational band in Ca-def OHAps is independent of the Ca/P molar ratio when this in the range 1.667 to 1.533, although for a ratio of 1.533, the *peak* intensity was reduced by 10% and the band width increased by 9%. The proposal was made that, when the OH$^-$ ion content fell with increased calcium-deficiency, the librational band intensity was made up by water in a similar position in the lattice (Section 3.7.3).

The use of the OH stretching band intensity to measure the OH$^-$ content of Ca-def and other OHAps is discussed in Section 3.6.1.

Calcium-rich apatites

Apatites with Ca/P molar ratios of 1.67 to 1.83 that were prepared by heating OHAp with CaCO$_3$ in moist air at 1000 °C for ten days (Section 3.4.2) showed new bands (the main one at 3544 cm^{-1} and others at 745, 715 and 680 cm^{-1}). These bands increased in intensity as the Ca/P ratio increased, whilst the bands found in OHAp at 3570 and 630 cm^{-1} decreased in intensity [537]. On deuteration by heating in D$_2$O vapour at 900 °C, the 3544 cm^{-1} band moved to 2300 cm^{-1}, and the 745 cm^{-1} band (the most intense of the new low frequency bands) disappeared, probably moving into the v$_4$ PO$_4$ region. The possibility that these observations might be due to the accumulation of F$^-$ ions from the furnace has already been noted (Section 3.4.2).

Non-calcium apatites

Detailed IR assignments for Ba and SrOHAp have been given [549,557,675] and spectra-structure-composition correlations discussed [628]. Band shifts for Ca,SrOHAp [628] and Ba,CaOHAp [664], as a function of composition, have been studied. The frequencies were intermediate between the end-members and the bands of Ca,SrOHAp were slightly broadened, particularly the OH bands, which was attributed to heterogeneity introduced by the different cations [628]. As discussed earlier in relation to the additional high temperature OH stretch bands (Section 3.11.3), a second OH band at 20 to 30 cm^{-1} below the main OH IR band, is seen at room temperatures in a number of hydroxyl apatitic compounds [823]. Intensity ratios of the two OH IR bands in Pb$_5$(AsO$_4$)$_3$OH and Pb$_5$(VO$_4$)$_3$OH have been found to vary significantly with the preparation method [823].

The P-OH stretch in Sr-def OHAp is at 850 cm^{-1}, with an extinction coefficient comparable to that of the calcium salt [698].

3.12 NMR spectroscopy

The use of ^{1}H NMR for the quantitative determination of OH$^-$ ion (Section 3.6.1) and HPO$_4^{2-}$ ion (Section 3.6.2) contents has already been discussed. ^{1}H NMR MAS studies of a precipitated OHAp with a Ca/P molar ratio of 1.63 have been published [183]. The preparation (Section 3.3.3) and OH$^-$ ion content of this sample, determined by NMR, have been reported [637] (Section 3.6.1). At 500.13 MHz and a spinning speed of 7.3 kHz, a sharp resonance with a chemical shift of 0.2 ppm from TMS (trimethyl silane) was assigned to structural OH$^-$ ions, and a resonance with a ~5.6 ppm shift to surface adsorbed water [183]. This latter assignment was based on the fact that the peak intensity was reduced on exchange with D$_2$O in the vapour phase, and was not seen in samples with significantly lower surface areas. A weak peak at 1.5 ppm shift was tentatively assigned to occluded or surface OCP (Section 1.3.8), and a weak peak at 8.7 ppm shift to small amounts of HPO$_4^{2-}$ ions that the sample was known to contain. This latter shift is slightly higher than the 6 ppm shift reported for a large band assigned to HPO$_4^{2-}$ ions [193,700] in a well-crystallised Ca-def OHAp. This apatite had a Ca/P molar ratio of 1.5 and was prepared [644] (Section 3.4.1) by the hydrolysis of DCPA or DCPD in the presence of urea at 80 to 90 °C. Relative areas of the HPO$_4^{2-}$ ion and OH$^-$ ion resonances fitted that expected from Formula 2, Table 3.2 [700] (Section 3.7.3).

Multiple-quantum NMR dynamics have been reported for the quasi-one-dimensional distribution of the uniformly spaced proton spins in OHAp using a phase-incremented even-order selective time-reversal pulse sequence [692]. Two samples were investigated: the first was the precipitated apatite with an OH$^-$ ion content of 3.13 wt % (theoretical, 3.4 wt %), as determined by ^{1}H NMR [637] (Section 3.6.1); and the second was a OHAp prepared by a high temperature method that was known to be highly stoichiometric because ~70 % was in the monoclinic form (estimated by XRD). In both samples, the initial increase of the effective size with preparation time for creation of multiple-quantum coherences was linear; an upward curvature at longer preparation times was observed and qualitatively ascribed to the incomplete isolation of the linear chains of OH$^-$ ions. The slope of the linear section for the precipitated OHAp was measurably less than for the s-OHAp; this was attributed to lack of coherence within the OH OH OH chains introduced by a greater number of OH$^-$ ion vacancies in the precipitated OHAp, compared with the s-OHAp [692].

A single ^{31}P resonance at 2.8±0.2 ppm shift (with respect to 85% H$_3$PO$_4$) with weak sidebands has been seen in the NMR spectrum of OHAp at 68 MHz with proton decoupling and spinning at 1.3 KHz [192]. Monoclinic OHAp gave a similar spectrum. Weak sidebands in both samples showed that there was little chemical shift anisotropy, indicating only slight distortion of the PO_4^{3-} ions from the undistorted tetrahedral symmetry, as expected. The fact that there was only one chemical shift seen in the monoclinic sample indicated that the three crystallographically nonequivalent PO_4^{3-} ions in the unit cell [609] (Section 3.2) had very similar environments, in agreement with the fact that the main difference between the two OHAp samples was in the long range ordering of the OH$^-$ ions. Similar ^{31}P NMR MAS studies [835] showed an identical chemical shift for OHAp, but a smaller chemical shift anisotropy; a recent publication [193] gives a shift of 2.9 ppm for OHAp. The same resonance has been seen from colloidal suspensions of OHAp stabilised with a diphosphonate in which Brownian motion instead of spinning the sample was used to reduce the dipolar coupling and chemical shift anisotropy [836].

The ^{31}P NMR MAS spectra of CaPs that contain HPO_4^{2-} ions, compared with those with only PO_4^{3-} ions, generally show negative (upfield) shifts, and have significant shift anisotropies and obvious sidebands that are markedly changed in proton decoupling experiments [192,835] (see ^{31}P NMR of OCP, DCPD and DCPA in Sections 1.3.8, 1.4.8 and 1.5.7, respectively). Comparison of these spectra with those from Ca-def OHAps enables useful information to be obtained about the HPO_4^{2-} ion environment in synthetic Ca-def OHAps [192,193,700,835] and biological apatites [209,837,838] (Section 4.6.6). ^{31}P NMR MAS spectra of Ca-def OHAps (Ca/P molar ratios of 1.46 and 1.33) [192] gave very similar spectra and the same chemical shifts as OHAp (see above) under the same conditions. For a sample with a Ca/P molar ratio of 1.33, a marked increase in sideband intensities was seen at -187 °C with proton decoupling (cross-polarisation). The longitudinal (or spin-lattice) relaxation time T_1 of the apatitic samples was variable from ~1 s for stoichiometric samples, to about ~30 s for nonstoichiometric and disordered samples [192]. The interpretation given to these results was that the nonstoichiometry involved Ca^{2+} and OH$^-$ ion vacancies in a lattice similar to OHAp, rather than additional acid phosphate phases. Resonances attributed to DCPA were seen in a sample with a Ca/P molar ratio of 1.14 [192]. A OHAp with ~12 % of the phosphorus present as HPO_4^{2-} ions has been studied at 119 MHz with MAS (2 kHz) [835]. Again, the spectrum was very similar to OHAp, but with proton decoupling (cross-polarisation), three small changes from OHAp were observed: (1) a small additional contribution to the central band which itself had a small shift of -0.5 ppm from OHAp, (2) the appearance of a second set of rotational sidebands, and (3) a more asymmetric first set of sidebands. These differences were attributed to HPO_4^{2-} ions in a OHAp-like lattice, as in the earlier study [192] discussed above. The ^{31}P NMR MAS (5 kHz) spectrum at 40 MHz of a well-crystallised Ca-def OHAp (Ca/P molar ratio of 1.5, preparation in ref. [644], Section 3.4.1) was similar to OHAp, but with a much longer T_1 (27.5

instead of 2.0 s) and a slightly broader resonance [193]. These differences were attributed to a disordered phosphate environment due to Ca^{2+} ion vacancies.

The ^{31}P NMR spectra of ACPs are rather similar to OHAp, but with more intense sidebands [363] (Section 1.8.5); the line-width decreases during the solid state hydrolysis to Ca-def OHAp [345] (Section 1.8.4).

There is additional detailed information about the F^- ion local environments from ^1H and ^{19}F NMR studies of single crystals of Holly Springs OHAp [828,839], and single crystals of FAp containing small amounts OH^- ions [840,841]. The ^{19}F NMR spectrum of Holly Springs OHAp with the magnetic field parallel to the c-axis has five well-separated components, with intensity ratios approximately 1:1:2:1:1, which can be interpreted as a superposition of a triplet and doublet from F^- ions in two different environments [828,839]. The triplet was assigned to F^- ions in the sequence OH OH···F···HO HO with an H···F distance of 2.18±0.012 Å, which was consistent with the F^- ion being on the mirror plane, and adjacent OH^- ions in essentially their normal OHAp positions [839]. The doublet was assigned to F^- ions in the sequence OH OH···F OH OH with an H···F distance of 2.12±0.03 Å, in which the F^- ion was displaced towards the proton of the OH^- ion with the OH^- ion again in its normal OHAp position. Intensity ratios of the lines showed that the probability that an F^- ion would occur in OH OH···F···HO HO with OH^- ion direction reversal was 1.7 to 1.8 times the probability for OH OH···F OH OH without reversal. There was no evidence for F F F F sequences.

The ^1H NMR spectrum of FAp containing small amounts of OH^- ions has a doublet due to interaction between ^1H and ^{19}F nuclei on the c-axis [840], whose splitting enabled the H···F distance for F F OH···F F to be calculated (2.035 Å); however, the accuracy of this calculation has been questioned [183]. There was no evidence for the sequence OH OH OH OH [840]. The ^{19}F spectrum consisted of a triplet with two outer doublets [840]. The triplet was assigned to F^- ions in the configuration F F F F because of similarity with the FAp NMR spectrum, and the two outer doublets were assigned to an F^- ion adjacent to a hydrogen atom in the sequence F F OH···F F. Atomic distances in this configuration were studied in detail [840]. The F^- ion resonances are affected by the adjacent proton and F^- ion; the proton alone leads to a doublet, but the F^- ion then causes each component to split to give the two doublets seen experimentally [840]. The magnitude of this splitting enabled the F F distance to be calculated (3.704 Å). The ^{19}F spectrum calculated for this model using the H···F distance (2.035 Å), determined from the ^1H NMR studies above, agreed well with the experimental ^{19}F spectrum. Thus, NMR has established nearest neighbour H···F and F F distances for the OH OH···F···HO HO, OH OH···F OH OH, and F F OH···F F configurations. Correlation between these NMR results and the IR spectra of OH perturbed by F^- ions is discussed in Section 3.11.5 and in ref. [183].

The changes in the ^{1}H and ^{19}F NMR spectra with temperature from 20 to 400 °C have been observed in single crystals of mineral OH,FAp (20% replacement of F$^-$ by OH$^-$ ions), and interpreted in terms of changes in the F F OH$\cdots$F F configuration in a study following on from that described in the previous paragraph [577]. For the ^{19}F spectra, beginning at 100 °C, there was a gradual decrease in the splittings for the two outer doublets, but the triplet did not change. At 160 °C, the entire spectrum narrowed until at 205 °C, there was only a single line. However, if the specimen was kept at this temperature, the triplet reappeared after ~1 h. In a second thermal cycle, a single line was not observed until 290 °C, and in a third cycle, the triplet persisted to 400 °C (the maximum observable temperature). In the ^{1}H spectra, heating produced no observable change in the splitting of the doublet that originated from interaction between the proton and adjacent fluorine atoms. However, the ratio of the intensity of the central singlet ^{1}H line to the intensity of the doublet increased continuously from 120 to 360 °C. The interpretation given to these results was that increased diffusional motion of F$^-$ ions along the *c*-axis was responsible for the breakdown of the fine structure of the ^{19}F spectrum, leading to a narrow singlet line. The activation energy for this process was determined from the onset of the narrowing of the spectrum and was 16.0. 20.0 and >25.0 kcal mol^{-1} (67.0, 83.7 and >104.7 kJ mol^{-1}) for thermal cycles one to three respectively. It was suggested that the change in activation energy with the cycle number was associated with annealing of defects. The ^{1}H spectra were interpreted as indicating static disorder of the OH$^-$ ions between the two symmetrically related positions above and below the mirror plane at room temperature which changed to dynamic disorder at higher temperatures. This took place over the temperature range 120 to 360 °C with an activation energy of 15.0 kcal mol^{-1} (62.8 kJ mol^{-1}).

Perturbation of ^{1}H resonances in OH,FAps by F$^-$ ions have also been seen in MAS spectra of three samples of precipitated OH,FAp powders [183]. A peak with a 1.5 to 1.6 ppm shift from TMS was assigned to F F OH$\cdots$F F because it was the most pronounced resonance in OH,FAp at the highest F concentration (81 % replacement of the OH$^-$ ions), and because it was seen in FAps containing traces of OH$^-$ ions. In OH,FAp containing the least F$^-$ ion (24 % replacement of the OH$^-$ ions), a peak also seen at 1.5 ppm was assigned to OH OH$\cdots$F OH OH, and a peak slightly upfield of this at 1.2 ppm, indicating slightly weaker hydrogen bonding, was assigned to OH OH$\cdots$F$\cdots$HO HO. A peak at 2.5 ppm, observed in all three samples, remained sharp at spinning speeds as low as 0.5 kHz which suggested assignment to a highly mobile "structural H$_2$O" group. A peak at 0.3 to 0.4 ppm shift, due to OH$^-$ ions that were not hydrogen bonded to F$^-$ ions, was seen in all three samples; this indicated a statistical mixing of the OH$^-$ and F$^-$ ions, rather than a scheme in which hydrogen bonding between these ions was maximised.

^{19}F MAS (3 to 4 kHz) NMR has been used to study the F$^-$ ion uptake by OHAp and is able to distinguish between FAp, F,OHAp and CaF$_2$ [770,842] (Section 3.10.2); the isotropic chemical shift moves progressively upfield from 64.0 to 56.6 ppm relative to C$_6$H$_6$ respectively [770]. These NMR studies exploited ^{19}F homopolar dipolar interactions to distinguish the phases. At the spinning speeds used (3 to 4 kHz), the CaF$_2$ spectrum is very broad (30 kHz) because of strong homogeneous F-F dipolar interaction; however, in apatites, the spectrum of ^{19}F is much narrower and has spinning sidebands because the combined anisotropic chemical shift and F-F dipolar interactions behave inhomogeneously at these speeds [769]. However, in NMR MAS spectra at 282 MHz at very much higher speeds (15.5 kHz), the homogeneous broadening of the ^{19}F resonance of CaF$_2$ can be greatly reduced, so the resonances and sidebands of FAp and CaF$_2$ are clearly resolved with isotropic chemical shifts from C$_6$F$_6$ of 64 and 58 ppm respectively [769]. Studies [769] (Section 3.10.2) at these higher spinning speeds of the products of the reaction of F$^-$ ions for different times with OHAp, showed that a sharp resonance at 44 ppm appeared before the ^{19}F peak from F,OHAp at 64 to 60 ppm. The 44 ppm peak was assigned to non-specifically adsorbed F$^-$ ions [769].

^{19}F NMR spin-echo studies have been used to deduce that there are at least two different local fluorine environments in the mineral in rat bone [843,844] (Section 4.6.6).

3.13 Other physical and chemical studies

A thermoluminescent glow peak occurs at 85 °C in synthetic OHAp, ACP and deproteinated rat bone that have been heated to 400 °C and then exposed to UV radiation [845]. The peak intensity from bone increased with the age of the animal and changes in bone chemistry as a result of thyroparathyroidectomy. Thermoluminescent properties of hydrothermally grown OHAp [677], FAp (mineral and synthetic) [542] and ClAp (mineral and hydrothermal) [846] after X-irradiation have been reported. Peaks specific to particular types of lattice were observed (OHAp 132 K, FAp 177 K and ClAp 118 K). Low energy electron emission (exoemission), thermally or optically stimulated, has been reported from β-irradiated (^{90}Sr source) synthetic and Durango FAps, and from dental enamel [847].

ESR studies have been made on hydrothermally grown monoclinic single (but twinned) crystals of OHAp (Section 3.5), after exposure to ionising radiation [612]. A doublet was seen at 92 K, but not at room temperature. On recooling, the original centre reappeared somewhat weaker, accompanied by two other centres. These three centres had the same relative intensities in a number of crystals and were stable at room temperature over a period of months, but could only be seen on cooling. It was thought that the centres were intrinsic in origin, even though they exhibited saturation with increasing radiation dose. Only the original centre was studied in detail. With the external magnetic field at 90° to the *c*-axis, there was a doublet centred at an effective *g* value of 2.0683±0.0002 with a separation of 5.9±0.2 G (590±20 µT). With the external field at 0° to the *c*-axis, a doublet, which did not change on rotation about the *c*-axis, was centred at an effective *g* = 2.0018±0.0002 with a 5.6±0.2 G (560±20 µT) separation. However, a complex pattern of apparently overlapping lines was seen at intermediate angles. The centre was believed to be an O⁻ ion formed by the loss of hydrogen from an OH⁻ ion and it was thought that the hyperfine interaction originated from interaction with the hydrogen atom of an adjacent OH⁻ ion tilted at 6 to 7° to the *c*-axis, with six possible orientations about this axis.

Pseudo-single crystal elastic constants of OHAp have been derived from anisotropic elastic constants of FAp [567] (published values used, not the corrected values given in Section 2.7). This estimation was made through comparison of elastic moduli of polycrystalline OHAp and FAp [848]. Calculated stiffness coefficients were c_{11} = 13.7, c_{12} = 4.25, c_{13} = 5.49, c_{33} = 17.2 and c_{44} = 3.96 (units of 10^{11} dynes cm⁻² or 10^{10} Pa). Rayleigh wave velocity and attenuation measurements have been made on single crystals of Holly Springs OHAp [568]. The wave velocities were from 4 to 7 % higher than those calculated from the above elastic constants.

The heat content (H_T) of OHAp has been measured between 298.16 and 1472.5 K and is given by

$$H_T - H_{298.16} = 228.52T + 19.81 \times 10^{-3}T^2 + 50.00 \times 10^5 T^{-1} - 86670 \quad 3.38$$

where the units of H_T are cal mol⁻¹ K⁻¹ (1 cal = 4.187 J) of $Ca_{10}(PO_4)_6(OH)_2$ and *T* is the temperature in K [849]. Equivalent expressions for the heat capacity and entropy were derived [849]. The heat capacity, thermal diffusivity and thermal conductivity of dense blocks of sintered O,OHAp (sintering temperature 1050 to 1450 °C) have been measured at 130 to 1000 K by the laser flash method [642]. Typical room temperature values were 0.73 J g⁻¹ K⁻¹, 0.0057 cm² s⁻¹ and 0.013 J s⁻¹ cm⁻¹ K⁻¹ respectively. Literature values of ΔH_0, ΔG^0 and *S* at 298.15 K have been compiled [570] (see also Table 1.3).

The thermal expansion of OHAp (prepared at 900 °C) between 20 and 900 °C is linear with $\Delta a/a$ = 13.5 × 10⁻⁶ and $\Delta c/c$ = 12.7 × 10⁻⁶ per °C (relations determined from Fig. 2 of ref. [655]). (Different values of 12.9 × 10⁻⁶ and 8.6 × 10⁻⁶ respectively have been calculated from this figure [850].) A precipitated OHAp gave a similar linear thermal expansion between 20 and 600 °C of $\Delta a/a$ = 12.2 × 10⁻⁶ and $\Delta c/c$ = 11.5 × 10⁻⁶ per °C [850].

The electrical conductivity of compressed pellets of OHAp has been measured between 200 and 800 °C; OH⁻ ions were proposed as the charge carriers [851]. Similar measurements have been made between 250 to 500 °C, except that H^+ ions were suggested as the charge carriers [852]. The conductivity increased with the addition of F^- ions up to a 50 % replacement of the OH^- by F^- ions, but decreased at higher levels [852]. Electrical properties (conduction, dielectric constants and loss) of hot Y^{3+}-substituted OHAp sinters have been measured and conduction mechanisms discussed [658].

Detailed studies of the sorption (Ar, N_2, and CO_2) and flow (He, H_2, Ne, Ar, N_2 and CO_2) of various gases through membranes of compacted OHAp powders in the temperature range -196 to 30 °C have been made [853].

The EXAFS spectrum above the Ca K edge has been recorded and interpreted out to 6 Å in terms of the local environment surrounding a Ca(1) + Ca(2) averaged Ca^{2+} ion [368]. Initial EXAFS parameters were calculated from the structure and refined to obtain the best agreement with experiment. Calculated and experimental spectra agreed well [368].

The diffusion of OH^- and OD^- ions in FAp, diffusion of ^{45}Ca, ^{85}Sr, and ^{32}P in polycrystalline OHAp and the heat of formation of OHAp, are discussed in Section 2.7.

Thermal decomposition

Many aspects of the thermal decomposition of Ca-def OHAps and OHAp have already been discussed and are summarised here for convenience. Lattice water, if present, is lost on heating to ~400 °C, accompanied by a slight contraction of the *a*-axis parameter (Section 3.7.3). Acid phosphate, as determined by the HPO_4 band at 875 cm^{-1}, may be lost in this same temperature range (Section 3.7.3), although the temperature for maximum formation of pyrophosphate from HPO_4^{2-} ions is generally from 400 to 700 °C (Equation 3.15, Section 3.6.2). Quantitation of the pyrophosphate formed on heating is the basis of an important method of determining the HPO_4^{2-} content of Ca-def OHAps (Section 3.6.2) Very faint lines from crystalline α-$Ca_2P_2O_7$ have been detected in Guinier powder XRD patterns of a well-crystallised Ca-def OHAp (preparation, Section 3.4.1) heated for 18 h at 650 °C [643]. On heating above ~700 °C, pyrophosphate is lost by reaction with OHAp to give β-TCP (Equation 3.16, Section 3.6.2). At 900 °C, only OHAp and β-TCP will be present, as given in the overall Reaction 3.22 (Section 3.6.3). The formation of β-TCP at 900 °C is a sensitive indicator of a composition with a Ca/P ratio lower than s-OHAp; measurement of the ratio of the β-TCP to OHAp from XRD powder patterns (Figs A.7 and A.10) forms the basis of a method for the quantitative determination of the Ca/P ratio (Section 3.6.3). The above reactions of Ca-def OHAps do not generally occur for apatites with a Ca/P molar ratio greater than 1.667. On heating these at 900 °C, OHAp and CaO are formed. On cooling, the CaO may react in the atmosphere to form $Ca(OH)_2$ and/or $CaCO_3$, depending on the conditions.

Above ~1000 °C, OHAp decomposes into O,OHAp (Equation 3.4, Section

3.4.3) and at higher temperatures, into α-TCP and TetCP (Equation 3.7, Section 3.4.3). These reactions all depend on the temperature and water vapour partial pressure. Other phases might form, depending on the overall composition, as given in the CaO-P_2O_5 phase diagram (Fig. 1.11, Section 1.6.6), modified in the presence of water vapour. The most thermally stable CaP, $\overline{\alpha}$-TCP, melts between 1756 and 1777 °C (Section 1.6.6), above this temperature, CaO is the only possible solid phase.

Chapter 4

MINERAL, SYNTHETIC AND BIOLOGICAL CARBONATE APATITES

4.1 Introduction

4.1.1 Occurrence, nomenclature and early structural work

The relation between carbonate and apatites is important because carbonate increases their chemical reactivity, particularly by increasing the solubility product and the rate of dissolution in acids, and by reducing the thermal stability. Carbonate apatites (CO_3Aps) occur as francolite (often as rock-phosphates or phosphorites) and dahllite minerals, as various synthetic preparations, and as the inorganic component of bones and teeth. General reviews on these have already been cited (Section 1.1).

Dahllite is much rarer than francolite and is differentiated from it on the basis of a F^- ion content of less than 1 wt % [854]. Possible confusion as a result of this definition has been discussed [43]. Staffelite, quercerite and podolite are obsolete names for carbonate apatite minerals.

According to Henry [855], the name francolite was given by Mr. Brooke and Mr. Nuttall to a mineral from Wheal Franco, Tavistock, Devon, some years prior to 1850. The name was also used in a printed sale catalogue by H. Heuland on 15th May, 1843 [59]. A specimen, probably from this sale, consisted of a "colourless, thin, confusedly crystalline crust on a fibrous quartz mass, heavily impregnated with haematite" [59]. The specimen contained 3.3 wt % CO_2 and 3.71 wt % F, and the refractive indices were O = 1.629 and E = 1.624. The optical properties of francolite are often anomalous (Section 4.2.1).

Dahllite was named after the Swedish mineralogist brothers Dahll by Broeger and Bäckstroem in 1888 [856]. The original mineral was described as a pale yellow crust on apatite, uniaxial negative, with a birefringence somewhat larger than apatite (presumably FAp). It was reported [856] to contain 6.3 wt % CO_2, but without visible calcite to account for this. A more recent analysis [45] (Section 4.2.1) of this sample gave 3.70 wt % CO_2 and 0.035 wt % F, so the original CO_2 analysis was probably in error.

CO_3Aps can be synthesised by high temperature methods (Sections 4.3.1 and 4.4), or by precipitation and other solution reactions (Section 4.5.1). Attempts have been made to grow single crystals, but these have generally been unsuccessful (Section 4.1.2). The designations A-type and B-type have been given [857] to two classes of CO_3Aps based on their CO_3^{2-} ion IR absorption spectra (A-type Section 4.3.2; B-type Section 4.2.2) and their *a*-axis parameters by comparison with those of OHAp and FAp. A-type carbonate apatite (A-CO_3Ap) has an increased a-axis parameter with the CO_3^{2-} ions on the six-fold screw axis (Table 4.5, Section 4.3.1), and the B-type (B-CO_3Ap), a reduced a-axis parameter and CO_3^{2-} ion position corresponding to that in francolite, *i.e.*, in PO_4^{3-} ion sites (Section 4.2.1). A-CO_3Aps (calcium, strontium, barium, *etc.*) form a structurally well-defined series of compounds, but the B-CO_3Aps, even if only the calcium compound is considered, appear to be a class of structures with subtle differences in the way the CO_3^{2-} ion occupies the PO_4^{3-} site and in the charge compensation mechanisms that are used.

Review of early structural work[1]

The structure of the CO_3Aps has been controversial, sometimes extremely so, and remains subject to uncertainty. As early as 1886, Solly [858] wrote about francolite: "but whether we are to consider the CO_2 as chemically combined, and therefore replacing either the P_2O_5 or F_2, or merely as a mechanical mixture of $CaCO_3$, such as is found in varying and microscopic quantities in the apatites from Canada, is a question which cannot be settled satisfactorily, since no simple chemical formula can be given for the proportions found by Mr. Robinson". It took nearly 100 years to unravel this puzzle, and details are still to be clarified.

By 1937, it had become generally accepted that the CO_3^{2-} ion was located either at 0,0,¼ [396] or 0,0,0 [859] with its plane parallel to the basal plane to conform to the three-fold rotational symmetry of the apatite space group, $P6_3/m$. Small differences observed between the powder XRD patterns of francolite (including various rock-phosphates) and FAp disappeared after heating at 1000 °C with the loss of CO_2, an observation consistent with a lattice position for the ion [322]. The differences were confirmed [860] by the observation of a small reduction in the a-axis parameter for francolite containing 3.36 wt % CO_2 and 4.11 wt % F ($a = 9.34\pm0.01$; $c = 6.88\pm0.01$ Å) compared with FAp ($a = 9.36\pm0.01$; $c = 6.88\pm0.01$ Å). However, it was argued [860] that the CO_3^{2-} ion could not be located on the six-fold screw axis because there were already sufficient F⁻ ions to occupy all these positions, and furthermore, if the plane of the ion were parallel to the basal plane, as had been proposed, there would be insufficient space without a very considerable expansion of the lattice, which was not observed. Based on lattice parameters, density and chemical analyses, a double substitution was proposed [860]: the

---

[1]This is based largely on an earlier review of the relationship between CO_3^{2-} ions and the apatite structure in mineral, synthetic and biological apatites [77].

replacement of a PO_4^{3-} ion by a CO_4^{4-} ion, accounting for most of the carbon; and Ca(1) by a CO_3^{2-} ion with its plane parallel to the basal plane and its oxygen atoms derived from the PO_4^{3-} ions. These proposals were criticised [771,772] because of the improbability of a CO_4^{4-} ion. Instead, it was suggested that a CO_3^{2-} ion replaced a PO_4^{3-} ion with the fourth oxygen site occupied by an OH⁻ or F⁻ ion. This idea was endorsed in 1939 [861], in part because it was also believed that a CO_3^{2-} ion could not occupy a six-fold screw axis site for steric reasons. From a consideration of P-O bond distances, the proposal was made that the CO_3^{2-} ion was inclined to the *c*-axis, occupying one of the sloping faces of a tetrahedral PO_4^{3-} ion site. Before this idea was generally accepted to be the location of the majority of the CO_3^{2-} ions in mineral, biological and many, but not all, synthetic apatites, there were extensive digressions.

Other more complex types of substitution were proposed, such as three PO_4^{3-} ions replaced by four CO_3^{2-} ions [862] and a modified form of this substitution [863] in which three-quarters of the CO_3^{2-} ions replaced PO_4^{3-} ions with their planes parallel to the *c*-axis, whilst the other quarter were on the axes passing through the columns of Ca(1) atoms with their planes perpendicular to the *c*-axis. This latter model was criticised on structural grounds [864], and because of inconsistency with birefringence data [718,865] (Section 4.2.1) and polarised IR spectra [77,866,867] (Section 4.2.2).

Between 1940 and 1964 and in parallel with these discussions of possible structures, some authors doubted that the CO_3^{2-} ion formed part of the apatite lattice at all [44,648,868]. Despite these views, clear evidence was accumulating that, at least in mineral apatites, the CO_3^{2-} ion is in the lattice. Thus, as the CO_2 content increased, a systematic reduction in the *a*-axis parameter from 9.36 to 9.30 Å was observed in a series of ten francolites containing 0 to 6.9 wt % CO_2 [869]. In addition, literature values showed [44] there was a systematic fall in the refractive indices and an increase in the negative birefringence as the CO_2 content increased (O = 1.634, E = 1.630 for 0 % CO_2; to O = 1.613, E = 1.596 for 8.5 wt % CO_2).

By analogy with mineral apatites, early workers considered that the CO_3^{2-} ions replaced OH⁻ ions in biological apatites [870], but this view was discarded in 1937 after the work discussed above on francolite [860]. This substitution for biological apatites was not seriously reconsidered until much later [77,866,871], and then only for a small fraction of the CO_3^{2-} ions. The resurrection of this idea was based on polarised IR spectroscopy and was used to explain the small increase in the *a*-axis parameter of dental enamel over OHAp (~0.2 %) that had been clearly observed in 1955 [648,718] (Section 4.6.2). The prevailing view from 1937 until 1967, when LeGeros [79,872] (see later in this section) synthesised precipitated CO_3Aps with a substantial *a*-axis parameter decrease, was that biological apatites had most, or all, of their CO_3^{2-}

ions adsorbed on the crystal surfaces [77,120,648,866,871] or present as a second, possibly amorphous, $CaCO_3$ phase [873-875]. The main reasons for these beliefs were the apparent differential solubility of CO_3^{2-} over PO_4^{3-} ions in biological apatites, particularly bone salts (Section 4.6.5), coupled with the absence of a significant change in lattice parameters from OHAp; both these findings seemed to indicate that major lattice substitutions were not involved. On the other hand, others maintained vigorously that the double lattice substitution of CO_3^{2-} ions envisaged for francolite and dahllite applied equally to biological apatites [2,876,877].

In 1961, the formation of pyrophosphate on heating biological apatites showed the presence of significant amounts of HPO_4^{2-} ion [875], indicating that these apatites had features in common with the Ca-def OHAps (Section 3.7.3).

It might have been thought that investigations of synthetic CO_3Aps would have clarified the situation. Initially, this was certainly not the case. Some authors maintained that apatites precipitated in the presence of CO_3^{2-} ions contained poorly crystallised $CaCO_3$, which explained its absence from the XRD pattern [878]. The apatitic product of the reaction of calcite and boiling alkaline phosphate solution was considered to contain the substitution of one CO_3^{2-} for two OH^- ions because no IR OH stretching band could be seen [879]; on the other hand, chemical analyses of similar apatites produced at room temperature were thought to show the replacement of PO_4^{3-} by CO_3^{2-} ions [880]. The reversible adsorption of CO_3^{2-} (or possibly HCO_3^-) ions onto precipitated apatites was demonstrated in 1956 [881], which supported a surface location for the CO_3^{2-} ion.

In an important paper in 1954, Wallaeys [656] reported that $CaCO_3$ and $Ca_3(PO_4)_2$, when heated at 850 °C for eight days in the presence of CO_2 dried over concentrated H_2SO_4, partially reacted to form an apatite with about a 1.2 % expansion of the *a*-axis parameter. A similar product was produced if OHAp was heated in a stream of dry CO_2 at 900 °C for two days [656]. X-ray studies gave lattice parameters, $a = 9.518$ and $c = 6.851$ kX (9.537 and 6.863 Å respectively), without evidence of other phases. The gain in weight during the reaction, loss in weight on heating with CaF_2 to yield FAp, and the CO_2 content were consistent with the formation of $Ca_{10}(PO_4)_6(OH)_{0.8}(CO_3)_{0.6}$. OHAp was formed on heating this compound in air at 800 °C. These results were interpreted in terms of a reversible reaction between 800 and 900 °C

$$Ca_{10}(PO_4)_6(OH)_2 + xCO_2 \rightleftharpoons Ca_{10}(PO_4)_6(OH)_{2-2x}(CO_3)_x + xH_2O \qquad 4.1$$

with miscibility between the CO_3^{2-} and OH^- ions in the lattice [656]. In 1961, a similar reaction and substantially enlarged *a*-axis parameter was reported for SrOHAp (90 % of the OH^- ions replaced) [882] and BaOHAp (100 % of the OH^- ions replaced) [883] (Table 4.5, Section 4.3.1). Despite the clear evidence that two OH$^-$ ions could be replaced by a CO$_3^{2-}$ ion in the apatite lattice, little note of this work was taken outside the laboratory originally concerned with the investigations until more than a decade after the initial discovery.

During the early 1960's, there was a clear realisation that part of the difficulty in understanding the structures of the CO$_3$Aps arose because there were two modes of substitution in the apatite lattice that could be clearly distinguished by IR spectroscopy [611,857,867] (Sections 4.2.2 and 4.3.2). One corresponded to francolite (B-type, Section 4.2.2), in which PO$_4^{3-}$ ions were replaced by CO$_3^{2-}$ ions, and the other (A-type, Section 4.3.2) to a CO$_3$Ap in which two OH$^-$ ions were replaced by a CO$_3^{2-}$ ion on the hexad axis, as discussed above.

The substitution of two OH$^-$ ions by a CO$_3^{2-}$ ion in the apatite lattice was confirmed in 1964 by measurement of the mass lost and mass of water evolved in Equation 4.1, and by accurate chemical analyses [77,866,884]. Before the reaction, chemical analysis gave: CaO 55.68, P$_2$O$_5$ 42.32, CO$_2$ < 0.1 and H$_2$O (by subtraction) 2.00 wt % (theoretical for OHAp: 55.79, 42.39, 0.0 and 1.80 wt % respectively) and afterwards: CaO 54.63, P$_2$O$_5$ 41.45, CO$_2$ 3.69, and H$_2$O (by subtraction) 0.23 wt % which corresponded to the formula Ca$_{10}$(PO$_4$)$_{5.99}$(CO$_3$)$_{0.86}$(OH)$_{0.26}$O$_{0.02}$. The IR OH stretch at 3570 cm^{-1} in the original OHAp could no longer be seen, but IR CO$_3$ bands appeared at 1528, 1463 and 878 cm^{-1} (Fig. 4.1). The lattice expanded as the CO$_2$ content increased (from a = 9.412, c = 6.878 Å for the original OHAp to a = 9.544, c = 6.859 Å for the compound cited above). Polarised IR showed that the plane of the CO$_3^{2-}$ ion was approximately parallel to the c-axis (Fig. 4.4, Section 4.3.2). The new, nearly parallel, orientation [77] (Section 4.3.3) gave much less steric interference than the more obvious perpendicular orientation conforming directly to the crystallographic symmetry of the $P6_3/m$ space group which had been [860] originally considered. This finding provided an answer to the original objection [860] to this type of substitution that there was insufficient space for CO$_3^{2-}$ ions on the hexad axis.

In 1960, treatment of amorphous calcium carbonate phosphate precipitates in an autoclave at 360 °C and 200 atm (20.2 MPa) steam pressure, or in a 60 % KOH solution at 100 °C, was reported to produce an apatite with an a-axis parameter less than that of OHAp [885]. The apatite produced was thought to be similar to dahllite because mineral apatites also had a reduced a-axis parameter. Subsequently in 1967, LeGeros synthesised well-crystallised CO$_3$Aps at 95 °C by slow precipitation between calcium and phosphate solutions. In these compounds, the a-axis parameter decreased linearly from 9.44 to 9.30 Å as the CO$_3^{2-}$ ion content increased from 0 to 23 wt % [79,872] (Section 4.5.4). A CO$_3$FAp with a reduced a-axis parameter was synthesised by a similar reaction [79,872]. In 1964, a synthetic francolite-like CO$_3$Ap with a = 9.35$_9$ and c = 6.89$_3$ Å was synthesised by heating a calcium

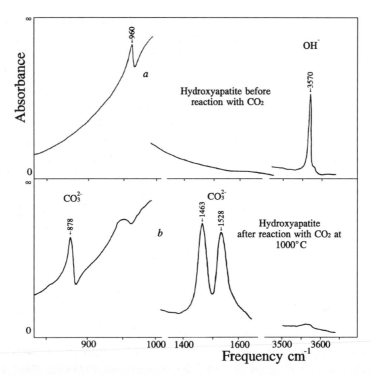

Fig. 4.1 Infrared spectrum of OHAp before and after reaction with dry CO_2 at 1000 °C for 15 h to form A-CO_3Ap. The low absorption on the high frequency side of the v_2 CO_3 band is due to mismatch in refractive index between the apatite and bromoform mull. (Fig. 6.4 from Elliott [77])

carbonate/phosphate coprecipitated in the presence of F^- ions, followed by a second heating with CaF_2, both times in dry CO_2 at 900 °C [857,886] (Section 4.4.1).

   With the clear appreciation of the existence of apatites with two types of CO_3^{2-} ion lattice substitution and the development of methods to synthesise them, further discussion is divided into detailed consideration of the structure of francolite and dahllite (Section 4.2), A-CO_3Ap (Section 4.3), high temperature B-CO_3Aps (Section 4.4), CO_3Aps formed in aqueous systems (Section 4.5) and biological apatites (Section 4.6).

4.1.2 Growth of single crystals

   This introduction will conclude with an account of attempts to grow single crystals of CO_3Aps which, unlike ClAp, FAp and OHAp, have been rather unsuccessful, and thus have so far contributed little to our understanding of CO_3Aps.

The growth of 3 mm needle crystals of CO_3Ap, thought to have the composition $Ca_{10}(PO_4)_6CO_3$, was reported in 1924 [887]. In one preparation, 4 g of $CaCO_3$ and 2 g of $Ca_3(PO_4)_2$ were slowly cooled from 1220 °C under a CO_2 pressure of 92 kg cm^{-2} (9 MPa). Refractive indices for sodium light were: O = 1.635±0.002 and E = 1.626±0.002. Chemical analyses were not reported for the starting materials or for the synthesised crystals, but the apatite slowly evolved CO_2 on dissolution in dilute acids. As XRD methods were not used for characterisation, it is possible the starting material was a BCaP, with the consequence that OH$^-$ ions were available for incorporation into the apatite lattice. These high pressure and temperature experiments were repeated in 1963 [888] with precipitated OHAp as the starting material, but it was impossible to obtain crystals from the solidified melt which were suitable for single crystal X-ray investigations. Chemical analyses and optical measurements were not reported, but powder XRD gave lattice parameters a = 9.38 and c = 6.88 Å.

There have been a number of attempts at hydrothermal bomb syntheses of CO_3Aps. When OHAp was heated at 400 °C in various solutions of sodium phosphate and carbonate, apatites with a CO_2 content of 0.06 wt % or less were produced [674]; the crystals were much smaller than those grown in the absence of carbonate. In similar experiments, 0.2 and 0.1 μm long crystals were prepared that contained 0.7 and 1.5 wt % CO_2 with lattice parameters a = 9.39 and 9.38 Å, and c = 6.88 and 6.89 Å respectively [888]. From powder XRD photographs of the products from phase diagram studies at 1000 bar (100 MPa) between 600 and 950 °C in the system $Ca(OH)_2$-$CaCO_3$-$Ca_3(PO_4)_2$-H_2O, it was concluded that there was no evidence for the entry of CO_3^{2-} ions into the apatite lattice [23]. A mixture of DCPA and $CaCO_3$ with a Ca/P molar ratio of 1.67, heated at 400 °C at 1 kbar (100 MPa) for 20 days, yielded an apatite with average dimensions 1 μm (max 3 μm). The chemical analysis was: CO_2 2.29, CaO 53.87 and P_2O_5 38.74 wt %; and lattice parameters a = 9.43$_6$ and c = 6.88$_3$ Å; and O = 1.637±0.002 (E and wavelength used not reported) [678].

Hydrothermal growth of ~4 mm needles of CO_3Ap from a flux of $Ca(OH)_2$ with temperature oscillation has been reported [684]. The charges used were in the range 10 to 50 wt % $Ca_3(PO_4)_2$ + 90 to 50 wt % $Ca(OH)_2$ + 1 to 25 wt % $CaCO_3$ + 20 wt % H_2O [532]. The temperature and pressure were oscillated from 750 to 880 °C and 14,500 to 17,500 psi (100 to 121 MPa) respectively, followed by slow cooling from 639 °C [684]. No significant differences in lattice parameters from OHAp were found, but the IR spectra showed CO_3^{2-} bands at 1450 and 1418 cm^{-1}, from which it was inferred that CO_3^{2-} ions had replaced PO_4^{3-} ions in the lattice. A sample prepared without temperature oscillation at 800 °C and 15,000 psi (103.4 MPa) had an analysis of CaO 55.5±0.1 and P_2O_5 41.6±0.1 wt % (CO_2 not reported) [684]. From this analysis, the formula, $Ca_{9.9}\square_{0.1}(PO_4)_{5.8}(CO_3)_{0.2}(OH)_2$, was deduced using the assumption that the PO_4^{3-} and CO_3^{2-} ions together occupied all the PO_4^{3-} ion sites and that charge balance was achieved through vacant Ca^{2+} ion sites. Refractive indices were not reported.

Other hydrothermal preparations, at between 2,200 and 15,000 psi (15.1 and 103.4 MPa) and 260 and 850 °C, of CO_3Aps have been studied. In these syntheses, various mixtures of $Ca_3(PO_4)_2$, MCPM, $(NH_4)_2HPO_4$, $Ca(OH)_2$, $CaCO_3$ and water were used (not all simultaneously) under CO_2 pressure [532]. IR spectroscopy showed that CO_3^{2-} ions could replace only OH^- ions, or OH^- and PO_4^{3-} ions, depending on the conditions used. However, the lattice parameters were similar to those of OHAp, with the exception of one sample prepared under CO_2 pressure which had lattice parameters $a = 9.476$ and $c = 6.867$ Å. Crystal sizes and refractive indices were not given. Some hydrothermally grown OHAp crystals discussed earlier [681] (Section 3.5) contained small amounts of CO_3^{2-} ions.

Discs of single crystals of aragonite, ~4 mm in diameter and 0.3 to 0.6 mm thick, can be transformed by hydrothermal treatment in phosphatic solutions into highly oriented microcrystalline (~1 μm diameter) OHAp crystals that contain CO_3^{2-} ions [412] (Section 4.5.6). Although not single crystals, these preparations could, in principle, provide oriented samples for polarised IR, ESR and XRD studies of the environment of CO_3^{2-} ions in CO_3Aps at various stages of the conversion process.

Although the growth of single crystals of calcium CO_3Aps has not been very successful, two analogous compounds have been grown and their structures determined. The first compound, $Ba_{10}(ReO_5)_6CO_3$, is an analogue of A-CO_3Ap. Crystals of this compound of μm dimensions have been grown by heating a mixture of metallic rhenium and $BaCO_3$ in air at 700 °C, followed by cooling (4 °C per h) to room temperature [433,528]. The crystals grew at the edge of the crucible via a process that involved gaseous transport of Re_2O_7 produced by oxidation of rhenium by the air. The lattice parameters were $a = 10.938(1)$, $c = 7.788(3)$ Å and the space group was $P6_3cm$ [528]. The structure is briefly discussed in Section 4.3.3.

The second compound is an analogue of B-CO_3Ap that involves two modes of combination of boron. In the system CaO-P_2O_5-B_2O_3, single crystals with the general formula

$$Ca_{9.5+0.5x}\{(PO_4)_{6-x}(BO_3)_x\}\{(BO_2)_{1-x}O_x\} \qquad 4.2$$

have been grown and used for a structure determination [437,889]. The crystals were grown by the standard flux-growth technique with an excess of B_2O_3 as the flux [437]. A mixture of CaO, P_2O_5 and B_2O_3 (35, 5 and 60 wt % respectively) was heated at 1200 °C for 10 h, then cooled at 8.3 °C per h. Hexagonal prisms (0.3×0.3×11 mm^3) were obtained with a composition corresponding to $Ca_{9.64}(P_{5.73}B_{0.27}O_{24})(BO_2)_{0.73}$ (this corresponds to $x = 0.27$ in

Formula 4.2). The space group was $P\bar{3}$ with lattice parameters $a = 9.456(1)$ and $c = 6.905(1)$ Å [437]. The structure investigation [889] gave no information about the mode of incorporation of BO_3^{3-} ions in the PO_4 sites, but linear BO_2^- ions were located along the c-axis with the boron atom at $0,0,\frac{1}{2}$, as described in Section 4.5.7 in relation to the analogous linear cyanamide and cyanate ions. The IR spectra of the rhenium [529] and boron [889] compounds have been studied.

4.2 Francolite and dahllite
4.2.1 X-ray diffraction, chemical and optical studies

Detailed correlations have been made [42,43,890,891] for various models (Table 4.1) of the relation between lattice parameters and unit cell contents for over 100 phosphate ores, mostly francolites of sedimentary origin. All but 22 of the 110 samples studied [42] contained more F^- ions per unit cell than the theoretical maximum of two for FAp. The unit cell contents were determined from the chemical compositions and on the basis of an idealised formula,

$$Ca_{10-x-y}Na_xMg_y(PO_4)_{6-z}(CO_3)_zF_wF_2 \qquad 4.3$$

(the variables have been changed to make them consistent with later publications of the same authors). z varied from ~0 to 1.5 and, although w and y varied widely, average values were $w = 0.4z$ and $y = 0.4x$; electrical neutrality was preserved when $x = z - w$ [42]. The minimum a-axis parameter was 9.322 Å and the maximum c-axis parameter 6.900 Å (in the same sample with a composition CaO 42.90, P_2O_5 27.74, F 3.92, CO_2 4.70, Na_2O 1.2, MgO 0.33 wt %). Models 3 and 4 (Table 4.1), which gave the highest correlation coefficients, showed that CO_3^{2-} ions caused about twice the change in the a-axis parameter as F^- ions on a molar basis.

The presence of an $F \times CO_3$ interaction term in Model 4 was taken as statistical evidence for the replacement of a PO_4^{3-} ion by a CO_3^{2-} ion with a F^- ion occupying the fourth oxygen position, following an earlier proposal for this coupled substitution [771,772] (Section 4.1.1). Assuming that all the c-axis sites were filled by F^- ions and none by OH^- ions, the chemical composition indicated that about half of the CO_3^{2-} ions were accompanied by F^- ions. This type of substitution (with either an F^- or OH^- ion) has also been proposed for precipitated B-CO_3Aps [82,99,892,893] (Section 4.5.3) and B-CO_3OH,FAps [99,538] (Section 4.5.5). However, the existence of $[CO_3^{2-},F^-]$ groups has been thought improbable because the close proximity of CO_3^{2-} and F^- ions, both with negative charges, violates Pauling's rule [45]. Furthermore, no evidence for $[CO_3^{2-},OH^-]$ or $[CO_3^{2-},H_2O]$ groups has been found in ^1H or ^{13}C NMR studies [894] (Section 4.5.8). Thus, there are still unresolved problems about how the difference in charge between the CO_3^{2-} and PO_4^{3-} ions is balanced.

Table 4.1 Correlations for various models for sedimentary francolites between unit cell lengths (Å) and the number of F⁻ and CO_3^{2-} ions per unit cell (designated F and CO_3 respectively) [42].

Models for a- and c-axes with standard errors[a]	SE[b]	R[c]
a-axis:		
1. $9.4311(130) - 400(61)F$	(152)	0.57
2. $9.4368(40) - 321(14)(F + CO_3)$	(72)	0.92
3. $9.4170(55) - 205(27)F - 389(19)CO_3$	(64)	0.94
4. $9.4436(100) - 332(48)F - 899(164)CO_3 + 238(76)(F \times CO_3)$	(61)	0.94
5. $9.3769(18) - 439(23)CO_3$	(82)	0.90
6. $9.3842(27) - 737(89)CO_3 + 214(62)(CO_3)^2$	(77)	0.91
c-axis:		
7. $6.8755(34) + 74(16)F$	(43)	0.41
8. $6.8801(24) + 39(8)(F + CO_3)$	(43)	0.44

[a]All numbers multiplied by 10^4 except the first constant term for each model.
[b]Standard error, at the 68 % confidence limit, of the observed lattice parameter minus that calculated from the model. The authors considered this the best measure of the agreement between the calculated and observed values.
[c]Correlation coefficient.

The relationships between the lattice parameters and the chemical composition in Table 4.1 were expressed in terms of the total number of F⁻ ions per unit cell [42]. However, it is more appropriate to consider the excess F⁻ ion content over that in FAp (F'), as has been done in Table 4.2. This shows clearly that when $F' = CO_3 = 0$, in those models which take both these ions into account, the a-axis parameter is very close to 9.367 Å, the expected value for pure FAp [46] (Table 1.4).

The fact that there was a relation between the carbonate and magnesium contents of the sedimentary apatites (Table 4.3) suggested [42] that these substitutions, because of their very different ionic size, physically compensated each other. This possibility, and the evidence that the Mg²⁺ ion can probably replace Ca(1) in the lattice to a limited extent, are discussed in Section 2.3.2.

As a result of these studies, the composition of francolites could be represented by the general formula

Table 4.2 Correlations with the substitution $F' = F - 2$, where F' is the number of excess F⁻ ions per unit cell over that in FAp, for the models of sedimentary francolites in Table 4.1. CO_3 is the number of CO_3^{2-} ions per unit cell and the unit cell lengths are in Å.

Models for a- and c-axes[a]	
a-axis:	c-axis:
1. 9.3511 - 400F'	7. 6.8903 + 74F'
2. 9.3726 - 321(F' + CO_3)	8. 6.8879 + 39F' + 39CO_3
3. 9.376 - 205F' - 389CO_3	
4. 9.377 - 332F' - 423CO_3 + 238(F' × CO_3)	

[a]All numbers multiplied by 10⁴ except the first constant term for each model.

$$Ca_{10-x-y}Na_xMg_y(PO_4)_{6-z}(CO_3)_zF_{0.4z}F_2 \qquad 4.4$$

between the end-members, FAp ($x = y = z = 0$) and an apatite with $z \approx 1.38$ [891]. In Formula 4.4, $y \approx 0.4x$, the excess F⁻ ion content over that in FAp has been set at 0.4z and, for electrical neutrality, $x = 0.6z$. The composition of this idealised end-member was given as CaO 55.1, P_2O_5 34.0, CO_2 6.3, F 5.05, Na_2O 1.4 and MgO 0.7 wt % [891]. The a-axis parameter for this compound is 9.314 Å, using the relationship [43,891]

$$\frac{z}{6-z} = \frac{9.369 - a}{0.185} \qquad 4.5$$

where a is in Å. This equation was based on studies [43] (Table 4.3) extended to 260 sedimentary francolites that contained at least two F⁻ ions per unit cell. Correlations between the other constituents (Table 4.3) enabled corresponding relations to be derived for x and y [43,891]. These were:

$$x = 7.173(9.369 - a) \qquad 4.6$$

and

$$y = 2.784(9.369 - a). \qquad 4.7$$

Table 4.3 Correlations between the major constituents of sedimentary francolites and between the *a*-axis parameter and the carbonate content [43].

Relation[a]	Correlation coefficient squared
$CO_3 = 5.996 - 0.998PO_4$	0.938
$Na = 1.253CO_3/PO_4$	0.942
$Mg = 0.463CO_3/PO_4$	0.928
$Na = 4.019SO_4$	0.944
$a = 9.369 - 0.0337CO_3$	0.925

[a]CO_3, PO_4 etc., are the number of ions per unit cell and *a* is in Å.

Thus measurement of the *a*-axis parameter of a francolite from a sedimentary phosphorite enables determination of its major isomorphous substitutions.

More recently, about 30 francolites (mostly single crystals or crystal fragments) from carbonatites, pegmatites and marbles, with a few from volcanic and metamorphic rocks, have been accurately analysed for fluorine, chlorine, structural water (*i.e.* the OH⁻ ion content), phosphorus and CO_2 [895]. Their lattice parameters were also measured. About six samples had more than the theoretical fluorine content of FAp (3.77 wt %), with the highest being 4.10 wt %. The maximum Cl⁻ ion content was 0.63 wt %, but most values were an order of magnitude less. If Cl⁻ and OH⁻ ions were taken into account, it appeared that most of the francolites had more than two ions per unit cell on the hexad axis (the highest was 2.38). The CO_2 contents were all below 1 wt %, except one which contained 1.29 wt %. The *a*-axis parameters ranged from 9.3640(68) to 9.3993(35) Å and the *c*-axis parameters from 6.8767(70) to 6.8971(27) Å. The phosphorus content was not related to the sum of the number of F⁻, Cl⁻ and OH⁻ ions per unit cell when this was below two, but above this, the phosphorus content appeared to decrease with the excess of these ions over two, the number to fill the hexad axis sites. This excess also seemed to be positively correlated with the CO_2 content. There was a strong negative correlation between the phosphorus and CO_2 contents. The *a*-axis parameter showed a slight negative correlation with the sum of the F⁻, Cl⁻ and OH⁻ ions per unit cell, particularly when this exceeded two. These observations led [895] to the suggestion that phosphorus was replaced by carbon and an associated halide or OH⁻ ion, resulting in a general formula for francolite

$$Ca_{10}[(PO_4)_{6-x}(CO_3 \cdot F)_x] \cdot (F,OH,Cl)_2. \qquad 4.8$$

The two general formulae proposed for francolite (Formulae 4.4 and 4.8) have the common feature of placing "excess" F^- ions with the CO_3^{2-} ions when they replace PO_4^{3-} ions, but Na^+ ions seem to play an essential role in charge compensation only in the sedimentary rock-phosphate model (Formula 4.4). It may be that there are differences between the crystal chemistry of the rock-phosphates and well-crystallised francolites, but it is more likely that the more detailed correlations seen in the rock-phosphates also occur in the well-crystallised francolites, but the correlations are not easy to see because of the much lower CO_2 contents and smaller changes in lattice parameters of the latter. Even though the well-crystallised francolites have the disadvantage of lower CO_2 contents, it should be possible to measure their densities accurately because they are well-crystallised, and so establish their unit cell contents on an absolute basis (Section 3.6.3); furthermore, careful selection from them might provide good specimens for single crystal XRD study.

The properties and synthesis at high temperatures of B-CO_3FAps are considered in Section 4.4.1, and solution preparations in Section 4.5.5. The general formula (4.4) for mineral francolites cannot apply to all synthetic B-CO_3FAps because some can be synthesised at high temperatures [82,896] (Section 4.4.1) and in solution [99,538] (Section 4.5.5) without monovalent cations, so that $x = 0$; this requires $z = 0$, and hence the absence of CO_3^{2-} ions which is a logical inconsistency. A detailed comparison between francolite and precipitated B-CO_3FAps, including the variation in a-axis with carbonate content and their general formulae, is made in Section 4.5.5.

A type specimen of the dahllite from Ödegården, Bamle, Norway [856] (Section 4.1.1) showed a slight contraction in the a-axis parameter from pure OHAp (as does francolite from FAp), but unlike francolite, there was a slight *increase* in the c-axis parameter, so the unit cell volume was unchanged within experimental error [45]. The lattice parameters were $a = 9.413 \pm 0.002$ and $c = 6.895 \pm 0.002$ Å for a composition of CO_2 3.70 and F 0.035 wt % [856]. A negligible change in volume with a CO_3 content of 4.2 wt % has also been reported for an AB-CO_3Ap made at 800 °C [897] (Section 4.4.2).

Optical properties

The mean refractive index, i, of sedimentary francolites is given by the relation [891]

$$\frac{z}{6 - z} = \frac{1.633 - i}{0.126} \qquad 4.9$$

where z is the number of CO_3^{2-} ions per unit cell (Equation 4.3). This relation is analogous to the relations from the same study for the unit cell contents as a function of the a-axis parameter (Equations 4.5, 4.6 and 4.7).

The increase in negative birefringence of francolite with an increase in CO_2 content has already been commented on (O = 1.634 and E = 1.630 for 0 % CO_2; to O = 1.613 and E = 1.596 for 8.5 wt % CO_2) [44] (Section 4.1.1). These changes in birefringence have been used to calculate the orientation of the CO_3^{2-} ion by comparing them with the birefringence of calcite in which all the CO_3^{2-} ion planes are perpendicular to the optic axis [865]. The calculation gave an angle between the plane of the ion and the basal plane of 43°, assuming the ion had a single orientation. As this was close to the angle (35.27°) between the inclined faces of a PO_4^{3-} ion tetrahedron and the mirror plane that passes through the ion, it was envisaged that the CO_3^{2-} ions occupied these faces when CO_3^{2-} ions replaced PO_4^{3-} ions. Orientations up and down from the plane were proposed to preserve the mirror symmetry statistically, and hence the lattice symmetry [865].

Apparently anomalous optical properties of francolite have been reported many times [45,59,649,865,898-901]. In a typical report [898], up to 1 mm diameter hexagonal plates consisting of the basal plane and a low pyramid, twinned in six triangular sectors about the hexagonal axis, were described. The sectors were biaxial (α parallel to the c-axis), with α = 1.620, β = 1.627, γ = 1.628 and γ - α = 0.008 (presumably sodium light, but not specified). The angle between γ and the hexagon edge was either 40° or 50°, often in different sectors of the same plate. $2V$ varied from 25° at the centre to 40° at the edge of the plates. Cross-sections indicated further twinning about a composition plane parallel to the basal plane, but with only slight differences in optical orientation about this plane. Many crystals showed an anomalous extinction in polarised light. As a result of these examples of biaxial francolite, the suggestion has been made that francolite is monoclinic with an angle between the a- and b-axes of ~120° [902]. However, in a detailed study of francolite from Fowey Consols Mine, Cornwall, about twice as much CO_2 was found by IR spectroscopy in the uniaxial core of the crystal plates compared with the surrounding biaxial sectors [45]. The birefringence of the core was also higher. It was concluded that the anomalous optical properties could not be specifically ascribed to CO_3^{2-} ions, so there was no evidence that these ions were responsible for reducing the symmetry from hexagonal to monoclinic [45].

The refractive indices of the type sample of dahllite from Ödegården (lattice parameters, CO_2 and F$^-$ ion contents given above) were O = 1.6435±0.0005, E = 1.6360±0.0005 and E - O = -0.0075 (presumably sodium light, but not stated) [45]. The negative birefringence attributable to its CO_2 content was less than half that expected by comparison with francolite [45].

X-ray diffraction studies (single crystal and Rietveld)

Preliminary single crystal Weissenberg camera studies have been undertaken [45] on parts of some very small (0.3 × 0.07 mm^2), but rather perfect, apatite crystals from Magnet Cove, Arkansas (see ref. [899] for a description of this deposit). One end of these crystals had the optical properties of FAp with E = 1.631, O = 1.634 and E - O = -0.0032±0.0002; the other end, sharply demarcated by a basal plane, had E = 1.623 to 1.626, O = 1.633 and E - O from -0.0069 to -0.0102 (all refractive indices ±0.001 and presumably sodium light) [45]. The CO_2 content of the francolite-end was 2.0 to 2.5 wt % estimated by IR spectroscopy and the lattice parameters were a = 9.362 and c = 6.900 Å; by comparison, the FAp-end had a = 9.376 and c = 6.891 Å (all ±0.001 Å). The francolite-end was slightly more dense and contained 1 to 3 wt % yttrium, which was absent from the FAp-end. The francolite-end also had 2 or 3 % less calcium and phosphorus, with fluctuations in these following the changes in birefringence (high negative values corresponded to low calcium and phosphorus contents). The X-ray studies of the francolite-end showed no departure from the space group $P6_3/m$ and, when compared with the FAp-end, no change in fluorine, oxygen or Ca(1) heights or positions in (00.l) projections. However, Ca(2) had moved by 0.02 Å, causing a rotation of the Ca(2) triangle about the c-axis, and the phosphorus peak was lower. Refinements showed a reduction of 7 % in the phosphorus occupancy. As a result of the X-ray investigations, the studies of the optical properties of francolite and dahllite reported above, and a detailed consideration of earlier chemical analyses in the literature, the proposal was made that when CO_3^{2-} replaced PO_4^{3-} ions in the lattice, the fourth oxygen position was occupied by water and the difference in charge was compensated by vacant cation sites [45]. Thus the formula proposed for the Magnet Cove francolite was

$$Ca_{9.3}Y_{0.3}(PO_4)_{5.5}(CO_3)_{0.5}(H_2O)_{0.5}F_{2.0} \qquad 4.10$$

and for the Ödegården dahllite [856] (Section 4.1.1)

$$Ca_{9.31}Fe_{0.11}Na_{0.28}K_{0.02}(PO_4)_{5.34}(CO_3)_{0.66}(H_2O)_{0.66}(OH)_{1.49}(CO_3)_{0.17}. \qquad 4.11$$

It was recognised [45] that this structural model did not explain the excess F⁻ ion content often found in francolites.

In a single crystal XRD study of a carbonate-bearing OHAp from Durango with lattice parameters a = 9.4282(4) and c = 6.8777(6) Å, it was reported [903] that the Ca(2)-Ca(2) distance was 4.098(3) Å compared with 4.083 Å for Holly Springs OHAp [404] and that all the O-O distances, except O(1)-O(2), were significantly shorter. The mean O-O distance for the PO_4^{3-} ion from the Durango and Holly Springs apatites was 2.4907 and 2.5059 Å respectively. These results were taken as evidence for CO_3^{2-} replacing both PO_4^{3-} and OH⁻ ions. The small change in Ca(2)-Ca(2) distance does not appear to be good evidence for CO_3^{2-} ions on the hexad axis because small changes in interionic distances could be due to other causes. However, the reduction in the O-O distances in the covalent PO_4^{3-} ions does seem to indicate a disturbance that might be caused by replacement by a smaller CO_3^{2-} ion.

Recently, francolite from Epirus, Greece, has been studied by XRD using the Rietveld method of structure refinement from powders [904]. Thin sections of the phosphate ore showed a radial development of the negatively birefringent francolite crystals (grain size 5 to 40 μm, refractive index 1.588) around a submicroscopic core. Microprobe analysis (except CO_2 which was determined by thermoanalysis) gave a formula

$$Ca_{9.56}Mg_{0.08}Na_{0.38}(PO_4)_{4.82}(CO_3)_{0.946}(SO_4)_{0.2}F_{2.34}. \qquad 4.12$$

The structure was determined of the native material and after it had been heated for 5 h at 530, 750 and 1200 °C. The results are summarised in Table 4.4. DTA and mass spectrometry showed that at 530 °C, there was a loss of 3.7 wt % water, and at 745 and 855 °C, a loss of 3.6 and 1.2 wt % CO_2 respectively (all endothermic reactions). The Rietveld analysis showed that 2.09 and 4.78 % (not stated if on a weight basis) CaO was formed at 750 and 1200 °C respectively. The results confirmed that, on strong heating, francolite changes to FAp with the loss of CO_2 and the formation of CaO. The Rietveld structure determinations (Table 4.4) were interpreted in terms of $[CO_3^{2-},F^-]$ replacing PO_4^{3-} [904]. The specific evidence presented for this was: (1) the progressive increase in the phosphorus occupancy with loss of CO_2 on heating; (2) the PO_4^{3-} ion distortion as indicated by the short P-O(1) and P-O(2) bond lengths; and (3) the low value of the O(1) occupancy in native francolite (0.902) compared with the theoretical for FAp (1.000) which correlated with the "excess" fluoride over that in FAp (this implies that O(1) is the site of the "excess" fluoride). The Rietveld refinement gave no information about the location of the 3.7 % water in francolite [904].

The Rietveld structure studies of the Epirus francolite (Table 4.4) give very clear evidence of a disturbance of the PO_4^{3-} ion parameters which diminishes as CO_2 is lost on heating. This can be seen in the marked reduction in the distortion index of the PO_4^{3-} ion, based on P-O bond lengths, to a value very close to that found in flux-grown FAp (Table 4.4). The suggestion that the low occupancy of O(1) indicates that this is the site of the "excess" fluoride implies that the CO_3^{2-} ion must occupy O(2) and two mirror related O(3) positions. This conclusion is inconsistent with the polarised IR results (Section 4.2.2) which show that the CO_3^{2-} ion occupies the O(1), O(2) and O(3) positions. There is in fact direct evidence from the Rietveld structure of the native francolite that the CO_3^{2-} ion occupies this position. The evidence is that the observed P-O(1) and P-O(2) bond lengths are each about 5% shorter than the observed P-O(3) length, whereas in FAp, they are only about 0.5 % shorter

(Table 4.4). This difference is as expected if oxygen atoms belonging to the CO_3^{2-} ion were positioned near O(1) and O(2), but shifted towards the phosphorus atom on account of the smaller size of the CO_3^{2-} ion (the effect on the observed P-O(3) length would be less because only half the number of O(3) atoms would be affected).

Table 4.4 Lattice dimensions, phosphate ion geometry and occupancies determined by Rietveld analysis for Epirus francolite heated for 5 h at different temperatures [904].

	Native	530 °C	750 °C	1200 °C	FAp[a]
a-axis Å	9.3207(5)	9.3230(5)	9.3593(5)	9.3708(5)	9.367(1)
c-axis Å	6.8947(5)	6.8987(5)	6.8919(5)	6.8880(5)	6.884(1)
P-O(1)	1.496	1.507	1.544	1.535	1.534(1)
P-O(2)	1.490	1.500	1.539	1.553	1.534(1)
P-O(3)	1.565	1.567	1.553	1.545	1.541(1)
PO_m[b]	1.529	1.535	1.547	1.544	1.536
DI(TO)[c]	0.023	0.021	0.004	0.003	0.002
Occupancy					
O(1)	0.902(28)	0.922(28)	0.922(26)	0.984(20)	1.000(6)
O(2)	0.986(24)	1.026(24)	0.976(22)	0.998(8)	1.000(6)
O(3)	1.0000	1.0000	1.0000	1.0000	0.998(4)
P	0.788(18)	0.826(18)	0.908(18)	0.992(18)	0.992(2)

[a]Single crystal X-ray structure determination of flux-grown synthetic FAp for comparison [46] (bond lengths from ref. [399]).
[b]Mean P-O bond length.
[c]Phosphate ion distortion index, $(\Sigma|PO_i - PO_m|)/4PO_m$, calculated over the i oxygen atoms [905].

Single crystal XRD [437] and IR spectroscopic [889] studies of an apatite with planar BO_3^{3-} ions, probably replacing PO_4^{3-} ions, yielded no information about the details of this substitution (Section 4.1.2).

4.2.2 Infrared and Raman spectroscopy

The undistorted CO_3^{2-} ion is planar with three equal, symmetrically placed C-O bonds, and thus has point-group symmetry D_{3h} [906]. The four atoms give rise to six normal modes, four of which form two doubly degenerate pairs because of the high symmetry. The normal modes in D_{3h} are: v_1 at 1063 cm^{-1} (A_1'), v_2 at 879 cm^{-1} (A_2''), and the doubly degenerate modes v_3 at 1415 cm^{-1}

(E') and v_4 at 680 cm^{-1} (E'). v_1 is only Raman active and v_2 only IR active, but v_3 and v_4 are Raman and IR active. v_2 will absorb IR radiation maximally when the electric vector is perpendicular to the plane of the ion, and v_3 and v_4 maximally when the electric vector is anywhere in the plane of the ion. The intensities of the IR absorption bands fall in the sequence v_3, v_2 and v_4; v_1 is the most intense Raman line.

The symmetry of the ion will be lowered if it is located in a crystal lattice position with point-group symmetry lower than D_{3h}. v_1 will then become IR active, and the degeneracies may be lost. For the purposes of discussion of the CO_3^{2-} ion in the apatite lattice, it will be assumed that the point-group symmetry is reduced to C_{2v} when all degeneracies are lost. The six normal modes in D_{3h} transform to those in C_{2v} as follows:

$$
\begin{aligned}
v_1\ (A_1') &\rightarrow v_1\ (A_1) \\
v_2\ (A_2'') &\rightarrow v_2\ (B_1) \\
v_3\ (E') &\rightarrow v_{3a}\ (A_1) + v_{3b}\ (B_2) \\
\text{and } v_4\ (E') &\rightarrow v_{4a}\ (A_1) + v_{4b}\ (B_2).
\end{aligned}
\qquad 4.13
$$

The transition moments for the symmetric modes, v_{3a} and v_{4a} are in the direction of the C-O bond that forms the two-fold rotation axis, whilst the transition moments for the antisymmetric modes, v_{3b} and v_{4b}, are also in the plane of the ion, but perpendicular to this C-O bond. Thus the transition moments of v_2, v_{3a} and v_{3b} are orthogonal.

The polarised IR spectra of two samples of francolite are shown in Figs 4.2 and 4.3. For each absorption band, the angle, α, between the optic axis (in this case the c-axis) and the transition moment can be calculated [907] from the equation

$$
R = 2\cot^2\alpha \qquad\qquad 4.14
$$

where the dichroic ratio, R, is given by

$$
R = \frac{D_\parallel}{D_\perp}. \qquad\qquad 4.15
$$

$D_\parallel$ is the maximum parallel optical density and $D_\perp$ the maximum perpendicular optical density. The values of α are 62°, 67° and 37° for the Fowey Consols francolite bands (Fig. 4.2) at 1453, 1429 and 865 cm^{-1} respectively; and 59° and 69° for the Wakefield apatite bands (Fig. 4.3) at 1453 and 1427 cm^{-1} respectively (v_2 could not be observed in the Wakefield apatite above the phosphate absorption because of the low CO_2 content) [77]. These results give

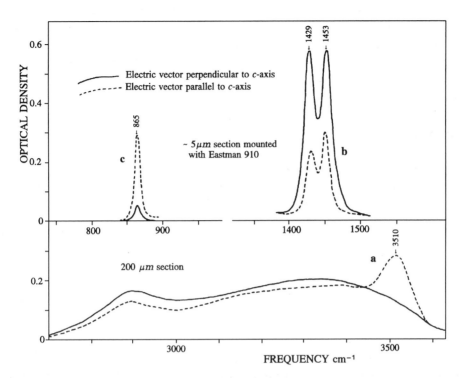

Fig. 4.2 Polarised IR spectrum of francolite from Fowey Consols Mine, St Blazey, Cornwall, British Museum No 55548. (Fig. 9.5 from Elliott [77], sample provided by Mr P G Embrey of the British Museum, Natural History)

average values of α, with estimated errors, of 60.5±3°, 69±4° and 37±4°, respectively [77,866]. The sum of the squares of the cosines of these angles is 1.008 which is nearly equal to 1, a necessary, but not sufficient condition for the three transition moments to be orthogonal. Thus the dichroic ratios are consistent with the assignment of the bands at 1453 and 1429 cm^{-1} to ν_{3b} and ν_{3a} (or *vice versa*) and that at 865 cm^{-1} to ν_2 for a CO_3^{2-} ion with site symmetry C_{2v} or lower. A site symmetry lower than D_{3h}, as indicated by the loss of the degeneracy of ν_3, is also consistent with the observation that the normal to the plane of the ion is not parallel to the c-axis, thus the ion cannot occupy a site with three-fold rotational symmetry. The resolution of the ambiguity in the assignment of ν_{3a} and ν_{3b} requires consideration of the location of the ion in the francolite lattice.

As previously discussed (Sections 4.1.1 and 4.2.1), there is good evidence that CO_3^{2-} ions replace PO_4^{3-} ions in the lattice. The polarised IR spectroscopy described in the previous paragraph shows that the normal to the plane of the

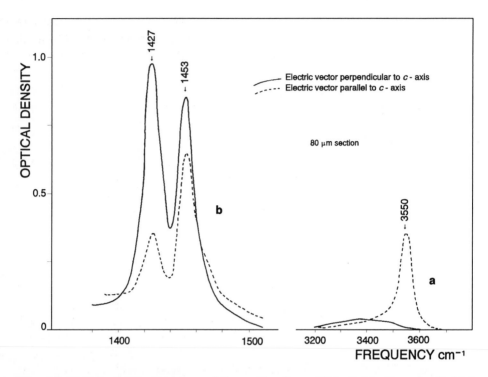

Fig. 4.3 Polarised IR spectrum of francolite from Wakefield, Canada, British Museum No 1917,277. (Fig. 9.4 from Elliott [77], sample provided by Mr P G Embrey of the British Museum, Natural History)

ion is tilted at 37° from the *c*-axis. This value is very close to the angle between the normal to an inclined face formed by O(1), O(2) and O(3) of a PO_4^{3-} ion tetrahedron and the *c*-axis (this angle is 35.27° for a regular tetrahedron), thus it must be this face that is occupied. This is in agreement with earlier deductions from birefringence measurements [865] (Section 4.2.1). The transition moment of v_{3a} will be along a C-O bond; with this location of the ion, this direction is approximately from the centre of the inclined face to O(1), O(2) or O(3). As the direction of the transition moment for v_{3b} is at 90° to that of v_{3a}, it must be in the O(2)-O(3), O(1)-O(3) or O(1)-O(2) bond directions respectively. If the approximation is made that the oxygen atoms form part of a perfect tetrahedron, the angles between the transition moments and the *c*-axis can be calculated for the two possible directions of the transition moment of v_{3a} (there are only two because, in this approximation, O(1) and O(2) are equivalent). For the v_{3a} transition moment along the C-O(3) bond direction, the angles for v_{3a} and v_{3b} are 54.73° and 90° respectively, and when along the C-O(1) bond direction, 73.22° and 60° respectively. Clearly, only the second pair of angles is consistent with the experimental values of 69° and

60.5°. Thus when v_3 splits into two bands in francolite as a result of the loss of its degeneracy, the low frequency component is the symmetric mode, v_{3a}, and the high frequency component the antisymmetric component, v_{3b}. Furthermore, the transition moment of v_{3a} is approximately in the direction from the centre of the sloping face to O(1) (or O(2)) [77]. This interpretation of the francolite IR spectrum is totally inconsistent with the earlier proposal [863] of CO_3^{2-} ions in two different lattice locations with parallel and perpendicular orientations with respect to the *c*-axis.

The CO_3^{2-} ion has been discussed above in terms of a C_{2v} point group, but the position it occupies has only the identity site symmetry, so this must be an approximation. However, detailed birefringence and X-ray studies have shown there was no evidence of a lower lattice symmetry [45] (Section 4.2.1), so the CO_3^{2-} ion probably occupies the one or the other of the sloping faces of the PO_4^{3-} ion tetrahedra randomly.

If some of the O(3) positions that are left vacant by the loss of PO_4^{3-} ions were occupied by F⁻ ions as discussed earlier [42,771,772] (Section 4.2.1), a splitting of at least some of the CO_3 bands would be expected because of the rather different environment of the ion with and without an adjacent F⁻ ion. Even though many francolite specimens have very sharp absorption bands, this effect has not been reported. However in precipitated B-CO_3FAps, bands at 1460 and 1470 cm⁻¹ have been assigned to CO_3^{2-} ions with and without an adjacent F⁻ (or OH⁻) ion respectively [99] (Sections 4.5.2 and 4.6.3).

The IR spectrum [45] of the Ödegården dahllite [856] (Sections 4.1.1 and 4.2.1) was rather similar to that of francolite, but not identical. The frequencies of v_{3b}, v_{3a}, and v_2 were 1455, 1416 and 873.5 cm⁻¹ for dahllite and 1455, 1425 and 866 cm⁻¹ for francolite. The bands in dahllite were also slightly broader than in francolite. These differences, together with others noted earlier in Section 4.2.1 (lattice parameter and birefringence changes smaller than might have been expected from the CO_2 content by comparison with francolite), indicate clearly that the environment of the CO_3^{2-} ion is slightly different in these two minerals [45]. The movement of v_2 from 865 to 873.5 cm⁻¹ in precipitated AB-CO_3F,OHAps [908] (Section 4.5.2) and a reduction in the frequency of v_{3a} from 1429 to 1407 cm⁻¹ and a slight increase in v_{3b} from ~1457 to ~1465 cm⁻¹ in high temperature AB-CO_3F,OHAps [82,896] (Section 4.4.1) have been associated with a decrease in the F⁻ ion content. Thus the differences in the IR spectrum between dahllite and francolite can, for the most part, be ascribed to the lower F⁻ ion content of dahllite. The correspondence between the IR spectrum of dental enamel, whose main CO_3 band frequencies are similar to those quoted above for dahllite, and francolite is discussed in Section 4.6.3.

The smaller than expected reduction in the *a*-axis parameter between dahllite and OHAp by comparison with francolite with the same CO_2 content and FAp [45] (Section 4.2.1) is probably also explained by the much lower F⁻ ion content of dahllite compared with francolite. As will be seen later (Table 4.6, Section 4.5.3), many precipitated fluoride-free CO_3Aps with small CO_2 contents actually have a slightly larger *a*-axis parameter than OHAp synthesised at high temperatures. This comment also applies to dental enamel (Section 4.6.2). Another, probably related, explanation proposed for dahllite [45] (Section 4.2.1) is that there is a slight increase in the *a*-axis parameter caused by some CO_3^{2-} ions in *c*-axis sites (see Formula 4.11 for Ödegården dahllite in Section 4.2.1).

The OH stretching mode in francolite has a parallel dichroism which shows that the OH⁻ ions are oriented parallel to the *c*-axis, in agreement with the structure of OHAp (Fig. 3.2, Section 3.2). There is a water band from ~3,000 to ~3,600 cm⁻¹ (Figs 4.2 and 4.3) which, in the sample from Wakefield (Fig. 4.3), shows evidence of slight perpendicular dichroism. It is most likely that the water is structural rather than adsorbed on internal surfaces as both samples, particularly the Wakefield apatite, are well-crystallised. However, nothing can be deduced about its location from the IR spectra.

The v_1 PO_4 band at 960 cm⁻¹ of Fowey Consols francolite has partial perpendicular dichroism with a dichroic ratio 0.43 [77]. This value is somewhat smaller than the ratio (0.786) found for the same band in dental enamel (Section 4.6.3).

4.2.3 Other physical and chemical studies

Thermal decomposition

Rock-phosphates and francolite lose most, if not all, of their CO_2 with the formation of FAp on heating to 1000 °C. As a result, there is a small expansion of the *a*-axis; indeed, this was one of the original pieces of evidence [322,860] (Section 4.1.1) that the CO_3^{2-} ions in francolite occupied lattice positions. Highly substituted francolites begin to lose CO_2 and fluorine in excess of that in FAp around 700 °C, and there is a systematic increase in the *a*-axis and crystallite size with this loss [43]. Samples with a high organic content may ignite, with a resultant uncontrolled increase in temperature [43]. As these samples usually have a high Ca/P ratio, CaO and CaF_2 may be formed [43]. Rietveld structure analyses of francolite heated at 530, 750 and 1200 °C have confirmed the transformation into FAp and the formation of small amounts of CaO [904] (Section 4.2.1). 3.7 % water (assumed to be adsorbed) was lost at 530 °C, and 3.6 and 1.2 % CO_2 at 745 and 855 °C, respectively [904]. The thermal decomposition of precipitated B-CO_3FAps is discussed in Section 4.5.5. For these, the decomposition temperature falls as the carbonate content rises [99].

Other studies

The equilibrium solubility products of B-CO$_3$FAps have been studied as a model system for the solubility of francolite [751] (Section 4.5.8). This study showed there was a very marked increase in the solubility product of B-CO$_3$F,OHAps compared with FAp.

Fission-track chronothermometry, which allows the dating of apatite minerals and gives information about their thermal history, is discussed in Section 2.7.

4.3 A-type carbonate apatite, Ca$_{10}$(PO$_4$)$_6$CO$_3$

4.3.1 Preparation

Ca$_3$(PO$_4$)$_2$ and CaCO$_3$ were heated together at 900 °C in dry CO$_2$ in the original preparation of A-CO$_3$Ap [656] (Section 4.1.1). The reaction is very incomplete, but goes to 70% completion if the CO$_2$ pressure is increased to 10 to 20 bar (1 to 2 MPa) [58]. A similar high pressure and temperature method can be used to make stoichiometric Ca$_{10}$(AsO$_4$)$_6$CO$_3$ [58] (Table 4.5), but for Ca$_{10}$(PO$_4$)$_6$CO$_3$, the best method is to heat OHAp in dry CO$_2$ at ~900 °C at atmospheric pressure for some days according to the reaction

$$Ca_{10}(PO_4)_6(OH)_2 + xCO_2 \rightleftharpoons Ca_{10}(PO_4)_6(OH)_{2-2x}(CO_3)_x + xH_2O. \quad 4.16$$

The optimum conditions for Reaction 4.16 have been investigated [81]. The most important requirement is that very dry CO$_2$ is used. High purity CO$_2$ was passed through a coil cooled in solid carbon dioxide, over phosphorus pentoxide loaded onto pumice, over magnesium perchlorate, and then over the sample in a platinum boat heated in a combustion tube fitted with a guard bulb containing sulphuric acid at its exit. To optimise drying conditions, the CO$_2$ was first passed through the combustion tube for 3 h at 300 °C to flush out any water vapour, then at the reaction temperature at 0.5 l h^{-1}. The reaction began at 800 °C, but increasing the temperature to 1000 °C made little difference to the rate of reaction. However, increasing the pressure reduced the uptake, so that at 50 bar (5 MPa), only 49±2 % of the theoretical replacement was obtained. The CO$_2$ content did not increase if the reaction time was increased beyond five days. To obtain a near theoretical CO$_2$ content, it was important that the starting OHAp was stoichiometric [81,85]. Often only ~90 % of the theoretical uptake was obtained, which was associated with an OH$^-$ ion deficiency in the starting material [81,85]. The apatite with the theoretical maximum CO$_2$ uptake has been given the formula Ca$_{10-x}$(PO$_4$)$_6$(CO$_3$)$_{1-x}$, where *x* depends on the original stoichiometry [81]. Thus the formula for an apatite with 95% of the theoretical CO$_2$ content would be Ca$_{9.95}$(PO$_4$)$_6$(CO$_3$)$_{0.95}$. A similarly prepared material has been given an alternative formula, Ca$_{10}$(PO$_4$)$_6$(CO$_3$)$_{0.93}$(OH)$_{0.14}$O$_{0.07}$ [85].

Table 4.5 Lattice parameters and transition temperature to hexagonal symmetry (if applicable) for various A-CO_3Aps.

Formula	Lattice Parameters[a] (Å)	Temp. °C
$Ca_{10}(PO_4)_6CO_3$[b]	$a = 9.557(3)$, $b \approx 2a$, $c = 6.872(2)$, $\gamma = 120.36\pm0.04°$ Space group Pb [52]	200 [86,85]
after 40 kbar (4000 MPa) at 950 °C	$a = 9.53_8$, $c = 6.88_1$ [909]	
$Sr_{9.95}(PO_4)_6(CO_3)_{0.95}$[c]	$a = 9.88_2$, $b = 9.72_5$, $c = 7.23_9$, $\gamma = 119.4_3°$ [81,910]	~200 [81,910]
$Ba_{10}(PO_4)_6CO_3$	$a = 10.20\pm0.01$, $c = 7.65\pm0.01$ [883]	
$Ca_{10}(AsO_4)_6CO_3$	slightly distorted from hexagonal [95,58]	~200 [58]
after 40 kbar (4000 MPa) at 950 °C	$a = 9.79_7$, $c = 6.99_5$ [909]	
$Sr_{10}(AsO_4)_6CO_3$	$a = 10.135$, $b = 20.299$, $c = 7.358$, $\gamma = 119.533°$ [95,58]	between 45 and 100 [95,58]
$Ca_{10}(VO_4)_6CO_3$	slightly distorted from hexagonal [85]	not known
$Sr_{10-x}Eu_x(PO_4)_6(CO_3)_{1+x/2}$[d]	$a = 9.877$, $b = 19.713$, $c = 7.223$, $\gamma = 119.45°$ [100]	not known

[a]Structures reported as hexagonal unless otherwise stated. See Table 1.4 for lattice parameters of corresponding OHAps.
[b]Typically has 90 to 95 % of theoretical CO_2. Hexagonal if CO_2 <90 % [81].
[c]Hexagonal if <0.84 CO_3^{2-} ions per hexagonal unit cell [81].
[d]Lattice parameters given for $x = 0.5$; hexagonal if $x > 0.5$ [100].

With increased CO_2 uptake in Reaction 4.16, the a-axis parameter of the CO_3Ap increased and the c-axis parameter decreased (both nonlinearly) [81]. Pseudohexagonal lattice parameters were $a = 9.55_4$ and $c = 6.87_0$ Å for an apatite with nearly the theoretical maximum CO_2 content [81]. Monoclinic lattice parameters for a similar compound, and lattice parameters and references to other A-CO_3Aps, are given in Table 4.5. The synthesis, lattice parameters and space group of single crystals of a rhenium analogue of A-CO_3Ap are given in Section 4.1.2.

A-CO_3Ap, when heated between 700 and 1000 °C forms OHAp, unless heated under high vacuum, very dry nitrogen or helium (both in the absence of oxygen), in which case OAp is formed [86,653,655] (Section 3.4.3). When heated in dry O_2, peroxyapatite is produced [85,87,449] (Section 3.4.3). If A-CO_3Ap is heated in dry nitrogen monoxide, a "nitrated" apatite is formed that contains nitrate, nitrite and NO_2^{2-} ions [439] (Section 2.3.1), and in NH_3 between 600 and 900 °C, an impure cyanamide apatite [453,454] (Section 4.5.7).

A high pressure modification of A-CO$_3$Ap can be synthesised by heating the normal form at 950 °C under CO$_2$ at 40 kbar (4000 MPa) for 15 min, cooling rapidly to room temperature under pressure, followed by a slow release of the gas pressure [95,909]. Lattice parameters changes are given in Table 4.5. The IR spectrum had a weak additional peak at 1410 cm^{-1} and the bands at 1542 and 1455 cm^{-1} moved to 1540 and 1460 cm^{-1}. This more complex spectrum suggested that the form prepared at high pressure had CO$_3^{2-}$ ions in two different environments. The similarity with CO$_3$Aps thought to have CO$_3^{2-}$ ions replacing both OH$^-$ and PO$_4^{3-}$ ions [82] (AB-CO$_3$Aps, Section 4.4.2) was noted, although it was not possible to say if the high pressure form was identical to these [95,909]. Unlike the normal form of A-CO$_3$Ap, the high pressure modification was thermally rather stable; no change in the XRD pattern or IR spectrum could be detected after heating at 970 °C in air, although there was partial decomposition if heated for 5 h at 1070 °C.

A high pressure form of Ca$_{10}$(AsO$_4$)$_6$CO$_3$ was also synthesised by a similar method to that described above [95,909]. The IR spectrum indicated that the CO$_3^{2-}$ ions were in two different environments and the XRD powder pattern showed that the lattice symmetry had increased to hexagonal (Table 4.5). This apatite was also thermally more stable than the low pressure form, but it behaved differently from the phosphate compound on heating in air at 970 °C. After heating, its IR spectrum showed that the CO$_3^{2-}$ ions were in a single environment and XRD showed that the hexagonal symmetry of the lattice had become slightly distorted. However, both its IR spectrum and lattice parameters were different from the original low pressure form, so it had not reverted to the original apatite.

Although the structural changes that take place in these phosphate and arsenate A-type CO$_3$Aps under pressure and subsequent thermal treatment are not known, they clearly are rather complicated, and serve to underline the complexity of the relationship between CO$_3^{2-}$ ions and the apatite lattice.

4.3.2 Infrared and Raman spectroscopy

The CO$_3^{2-}$ bands at 1535 and 1458 cm^{-1} in A-CO$_3$Aps have been assigned to the ν_{3b} and ν_{3a} bands that result from the loss of the degeneracy of ν_3, and the band at 878 cm^{-1} has been assigned to ν_2 [77,867]. The assignment of ν_{3b} and ν_{3a} to the high and low frequency components respectively, rather than *vice versa* is arbitrary [77] (see later discussion in this section). The very weak ν_4 bands at 670 and 760 cm^{-1}, seen in nearly stoichiometric preparations, have been assigned to ν_4 that has lost its degeneracy as a result of distortion of the ion [85]. Frequencies of 675 and 766 have also been reported [911] and there have also been some studies of these bands in precipitated CO$_3$Aps (Section 4.5.2). ν_1 is not seen in the IR spectrum because of the strong ν_3 PO$_4$ band, but the ν_1 Raman line can be seen at 1108 cm^{-1} [85].

The frequencies of v_{3a} and v_2 are independent of the degree of substitution of OH$^-$ by CO_3^{2-} ions within ± 4 cm^{-1}, but v_{3b} decreases in a nonlinear way from about 1565 to 1527 cm^{-1} as the substitution increases [77]. The variability of v_{3b} and near constancy of v_{3a} have been linked [77] to the fact that their transition moments are parallel and perpendicular respectively (Fig. 4.4) to the direction of greatest dimensional change, which is perpendicular to the *c*-axis for this substitution. Note that the frequency decreases when the *a*-axis parameter increases, as would be expected. In studies [82,912] of AB-CO$_3$FAps, AB-CO$_3$OHAps, and A-CO$_3$PO$_4$,AsO$_4$Aps (all calcium salts), it has been shown that the frequency of v_{3b} depends only on the hexagonal unit cell volume; as the volume increases, v_{3b} decreases linearly. The relation deduced from the published graph [82,912] is

$$v_{3b} = 2093.8 - 0.3434V \qquad\qquad 4.17$$

where v_{3b} is the frequency in cm^{-1} and V the unit cell volume in Å^3. The behaviour of v_{3a} and v_{3b} with lattice parameter changes for B-CO$_3$FAps (Section 4.4.1) has some similarities to those for substituted A-CO$_3$Aps, but in the former case, it is v_{3a} that shows the greater change.

The polarised IR spectrum of an oriented sample of A-CO$_3$Ap synthesised from dental enamel is shown in Fig. 4.4 [77,884]. The caption gives details of the preparations which took advantage of the strong fibre-axis texture of enamel resulting from the fact that the *c*-axes of the submicroscopic "OHAp" crystals lie predominantly in the direction of the enamel rods (the optic axis direction). The dichroic ratios can be used to determine the angles between the transition moments and the optic axis (Equations 4.14 and 4.15, Section 4.2.2), but unlike the francolite single crystals (Section 4.2.2), this will not be the true value of the angle between the transition moment and the *c*-axis because of the lack of perfect fibre-texture. The presence of significant disorientation is shown by the fact that the OH band does not have its expected perfect parallel dichroism (compare the OH bands in Figs 4.2 and 4.3, Section 4.2.2, with that in Fig. 4.4). A completely unoriented sample would have dichroic ratios of 1.00 for all its bands, which gives 54° 44′ for the angles between the transition moments and the optic axis. Thus the angles derived from a sample lacking perfect orientation would be biased towards 54° 44′ to an extent reflecting the amount of disorientation. (The results from Fig. 4.4a could not be corrected for lack of perfect orientation using the measured dichroism of the OH band and assuming the OH bonds were parallel to the *c*-axis because the preferred orientation shown by the band at 1548 cm^{-1} is already higher than that shown by the OH band at 3570 cm^{-1} [77].) As a result of these considerations, only limits can be given for the angles between the transition moments and the *c*-axis. These are: >71° for v_{3b} at 1548 cm^{-1}, <32° for v_{3a} at 1457 cm^{-1}, and

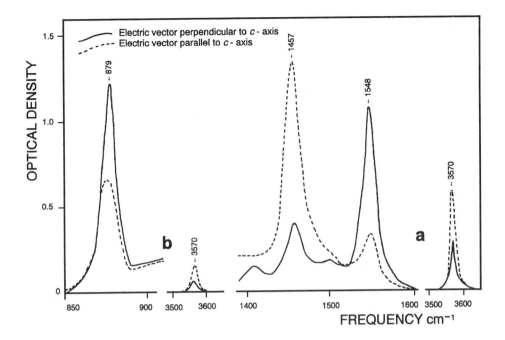

Fig. 4.4 Polarised IR spectrum of A-CO₃Ap derived from heated enamel. (a) 50 μm thick section heated at 900 °C in CO₂ for 30 min; (b) 100 μm thick section heated at 1100 °C in air for 120 min. (Fig. 6.5 from Elliott [77])

>63° for ν_2 at 879 cm⁻¹ [77,884]. The sum of the squares of the cosines of these angles is 1.031 (the theoretical for orthogonal transition moments with perfect orientation is 1). If, in order to take some account of the lack of perfect orientation, the three angles are arbitrarily moved 3° from their limits in a direction away from 54° 44', the angles become 74, 29 and 66°; the sum is then reduced to 1.006. Thus the polarised IR results are consistent with approximately orthogonal transition moments and show that the angle between the c-axis and the plane of the CO_3^{2-} ion is <27°. Like francolite, there are two possible orientations of the C-O bond direction, depending on whether the bands at 1548 and 1457 cm⁻¹ are assigned to ν_{3b} and ν_{3a} respectively, or *vice versa*, alternatives that cannot be distinguished from the IR results alone. These alternatives imply, respectively, an approximately parallel or perpendicular orientation of the bond with respect to the c-axis. The orientation of the plane of the ion away from the basal plane seen from the polarised IR studies means that its site symmetry must be reduced from D_{3h}, which is consistent with the loss of the degeneracies of ν_3 and ν_4.

The CO_3^{2-} ion in the *c*-axis channel causes splitting of the PO_4 absorption bands [85,659]. The most noticeable differences from OHAp are an additional v_3 band at 1130 cm^{-1} and a shoulder on the low frequency side of v_1. Other divalent ions show similar effects (Section 3.11.6) which have been attributed to the two crystallographically different types of PO_4^{3-} ions, one adjacent to the divalent ion (CO_3^{2-}, O^{2-}, S^{2-} *etc.*) and the other adjacent to a vacancy on the *c*-axis [659]. A similar increase in complexity is seen in the ^{31}P NMR spectrum of A-CO$_3$Ap [835] (Section 4.3.4).

The IR spectrum of strontium A-CO$_3$Ap is very similar to that of the calcium compound. v_3 is split into components at 1537 and 1451 cm^{-1}, v_2 is at 874 cm^{-1} and v_4 (seen in stoichiometric preparations) is split into components at 716 and 702 cm^{-1} [81]. A very slight splitting of the nondegenerate v_1 PO_4 can be seen [81,85,86] due to the two types of PO_4^{3-} ions (see above). Many of the references in Table 4.5 present IR spectra of other A-type CO$_3$Aps.

4.3.3 Structure

If A-CO$_3$Ap is nearly stoichiometric (>0.9 [81] or >0.84 [85] moles of CO_3^{2-} ion per "hexagonal" unit cell), the symmetry is lower than hexagonal. In the strontium apatite, the *b*-axis parameter is doubled and the lattice is clearly monoclinic [81,910] (Table 4.5). The monoclinic space group $P2_1/b$ was originally proposed for the calcium [81] and strontium [81,910] compounds. The structure that was suggested [81,910] was derived from monoclinic s-OHAp (Fig. 3.2) by placing every OH⁻ ion exactly at 0,0,¼ or 0,0,¾ and then replacing every alternate OH⁻ ion in the vertical columns by a vacancy, and the remaining OH⁻ by CO_3^{2-} ions. (The vacancies arise because two OH⁻ ions per "hexagonal" cell are replaced by a CO_3^{2-} ion and a vacant site.) There was registration between columns such that all the CO_3^{2-} ions in the planes parallel to the *a*- and *c*-axes, and separated by *b*cos30° (~8.3 Å), alternated between all having *z* = ¼ and all having *z* = ¾, as required by the glide-plane symmetry, thus doubling the *b*-axis parameter.

The space group, $P2_1/b$, as proposed for the structure described above, has a 2_1 screw axis coincident with the vertical columns of CO_3^{2-} ions which is inconsistent with only one ion per *c*-axis repeat. This difficulty was removed by the discovery that there could be no screw axis in the structure because neutron [113] and XRD [52] studies showed the presence of the 001 reflection. The X-ray pattern has been indexed unambiguously in the monoclinic space group *Pb* with *a* = 9.557(3) Å, *b* ≈ 2*a*, *c* = 6.872(2) Å and γ = 120.36±0.04° [52]. This space group has only one glide plane per unit cell which must coincide with the mirror plane at *z* = ¼ or ¾ in $P6_3/m$ from which it is derived. For convenience, parameters relating to *Pb* will be referred to the origin for $P6_3/m$ and the glide plane will be placed at *z* = ¼.

It is known from the polarised IR studies (Section 4.3.2) that the plane of the CO_3^{2-} ion is approximately parallel to the c-axis. Thus changes in its orientation about this axis will have a marked effect on its ease of packing into the structure. The closest approach between the CO_3^{2-} ions is within planes parallel to the a- and c-axes (Fig. 4.5A), where the ions have the same z parameter (the CO_3^{2-} ions in adjacent planes are displaced vertically from these by the operation of the glide plane, so their separation will be greater, Fig. 4.5C). To minimise interaction within these planes, it is likely that the ions will be oriented about the c-axis such that their planes are perpendicular to the a-axis (Fig. 4.5B). However, such an orientation will maximise interaction in the b-axis direction between CO_3^{2-} ions in adjacent planes (these are seen in projection in Fig. 4.5B). This interaction will depend on the distance of the ions from the glide plane; if this is $\sim c/4$, the vertical separation between ions in adjacent planes is maximised at $\sim c/2$ (Fig. 4.5C). This requires the ion to have a z parameter of ~ 0.

In the previous paragraph, interactions between CO_3^{2-} ions were discussed. Other considerations can give further information on the likely position of the ion. Their packing into the c-axis channel in a space filling model showed that the plane of the ion had to be nearly parallel to the c-axis, which agreed with the polarised IR measurements, but there was no obvious position for which interaction with adjacent atoms was minimised [77]. Others have also concluded that space requirements require a similar orientation of the ion [913].

Further information about the CO_3^{2-} ion might come from a consideration of the position of other large anions in the c-axis channels. In ClAp, the Cl⁻ ion is at 0.0016,-0.0014,-0.0561 (parameters referred to origin for $P6_3/m$) [47], and in BrAp, the larger Br⁻ ion is exactly at 0,0,0 [49]. The centre of the CO_3^{2-} ion was found to be at 0,0,-0.0081 in a refinement of single crystal X-ray data for the apatite-like compound, $Ba_{10}(ReO_5)_3CO_3$, in which the CO_3^{2-} ion was represented by a spherical germanium atom in two-fold disorder [528] (Section 4.1.2). A difference Fourier map showed residual electron density on the c-axis spreading as far as $z = \frac{1}{4}$, which indicated the ion was not located exactly at 0,0,0. This compound had a typical A-type IR spectrum with bands at 1445, 1387 and 878 cm⁻¹ which showed that the ion had lost its D_{3h} symmetry (Section 4.3.2), and so could not be positioned to conform strictly to the three-fold rotational symmetry of this site [529]. Thus a consideration of these other structures suggests that large anions have a preference for the 0,0,0 site, so it is likely that the CO_3^{2-} ion is located near this position in A-CO₃Ap.

Initial Rietveld structure refinements in $P6_3/m$ have been undertaken with neutron and X-ray powder diffraction data of A-CO₃Ap [913]. Some intensities of the neutron diffraction data were distinctly different from those of OHAp. For both data sets, the structure was refined with a series of dummy atoms along the c-axis. The refinements using the neutron diffraction data were

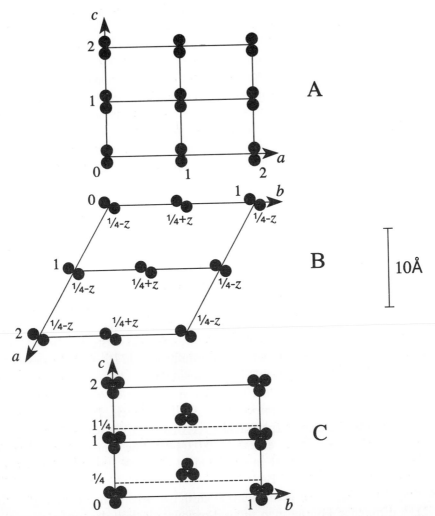

Fig. 4.5 Possible arrangement of the carbonate ions in A-CO$_3$Ap. (A) Closest approach of the carbonate ions which is in the plane parallel to the a- and c-axes. The ions are probably oriented about the c-axis such that they are seen edge-on; (B) view down the c-axis which shows the orientation of the carbonate ions so that their interaction in the **ac** plane (Fig. 4.5A) is minimised; and (C) carbonate ions in the plane parallel to the b- and c-axes showing that their interaction in this plane is minimised if they are displaced ~**c**/4 from the glide plane at $z = $ ¼. Note that the b-axis is drawn coincident with the b-axis of the pseudohexagonal cell.

thought to be more sensitive to the position of the carbon atom because the scattering length of carbon is greater than oxygen. These results suggested that the carbon atom was located near 0.04,0,0.12 (*x* parameter from correction in ref. [914]). On the other hand, the refinements using the X-ray data were thought to be more sensitive to the positions of the oxygen atoms of the CO_3^{2-} ion; these refinements indicated that the ion was oriented so that the oxygen atoms were above and below the carbon atom. Thus one oxygen atom was put directly below the carbon atom on the *c*-axis, and the other two oxygen atoms above the carbon atom, symmetrically distanced from the *c*-axis. The ion was oriented about the *c*-axis so that there was minimal interference between these two oxygen atoms and the Ca(2) atoms at $z = $ ¼. This latter criterion required that the normal to the plane of the ion was approximately parallel to the *a*-axis. The final suggested parameters were: carbon at 0.006,0,0.12; and oxygen at 0,0,-0.046, 0,-0.126,0.203 and 0.126,0.126,0.203. Attempts to refine explicitly the proposed model of the CO_3^{2-} ion in A-CO₃Ap did not provide any more detail about the structure [913]. It was pointed out that the orientation of an ion in this suggested position agreed with the polarised IR results. The proposed orientation about the *c*-axis is also consistent with that derived from the earlier considerations of minimising the interactions between the CO_3^{2-} ions (Fig. 4.5).

The various approaches to the determination of the position of the CO_3^{2-} ion in A-CO₃Ap lead to a rather consistent conclusion. One point of difference is whether the ion is at $z = 0$ (space and packing considerations, and analogy with the rhenium and other compounds) or at $z = 0.12$ (initial Rietveld refinements [913]). In conclusion, the most probable location of the CO_3^{2-} ion is with its carbon atom between 0,0,0 and 0,0,0.12, the normal to the plane of the ion approximately parallel to the *a*-axis of the unit cell in the space group *Pb*, and a C-O bond approximately parallel or perpendicular to the *c*-axis depending on which assignment of the v_3 bands at 1548 and 1457 cm⁻¹ in the polarised IR spectrum (Fig. 4.4, Section 4.3.2) is correct. The Rietveld refinement study [913] favoured a parallel orientation for the C-O bond which would make the antisymmetric v_{3b} mode the high frequency component, as in francolite. Confirmation (or otherwise) of the CO_3^{2-} ion position will have to await detailed Rietveld analyses of X-ray and neutron powder diffraction patterns in the space group *Pb*.

Monoclinic to hexagonal thermal phase transition

Near 200 °C, A-CO₃Ap has a reversible phase transition from the monoclinic to a hexagonal structure, in which the glide plane is lost [85]. Evidence for this was that X-ray diffractometer patterns above 100 °C no longer showed a doubling of the unit cell and that the coefficients of thermal expansion of the *a*- and *c*-axes changed at 200 °C. There do not seem to have been any structural studies above 200 °C, but hexagonal symmetry requires the CO_3^{2-} ions to be in three-fold rotational disorder about the *c*-axis and the CO_3^{2-} ions and vacancies to be disordered along the *c*-axis. (Alternatively, if the ordering of CO_3^{2-} ions and vacancies is preserved within columns, then the sense of the displacements of the ions from the former glide planes must be random between columns). Other A-type CO_3Aps also have thermal phase transitions (Table 4.5).

Relaxation of reorientable electric dipoles in $Sr_{10}(AsO_4)_6CO_3$ has been studied with the thermally stimulated currents method [96,914]. Compensation temperatures were observed at -100, 175, 392 and 573 °C. It was proposed that the reorientable dipoles were primarily CO_3^{2-} ions, and that the cooperative phenomena associated with the 175 and 392 °C temperatures were, respectively, a quasi-statically stabilised monoclinic to hexagonal transition and the onset of dynamical stabilisation of the hexagonal phase [914].

There are possible similarities between the A-CO_3Ap transition and the monoclinic to hexagonal thermal phase transition in ClAp (Section 2.2.3) and OHAp (Section 3.2).

4.3.4 Other physical and chemical studies

The ^{13}C NMR MAS spectrum of A-CO_3Ap observed at 79.9 MHz with spinning rates of 2 to 3.4 kHz has a single isotropic resonance at 166.5 ppm (measured from tetramethylsilane) indicating a single, well-defined, site for the ion [894]. The chemical shift anisotropy covered about 10 kHz, as determined from the rotational sidebands. The presence of a single resonance is consistent with the well-defined simple IR spectrum and with the monoclinic *Pb* structure (Section 4.3.3) which has only one crystallographically distinct CO_3^{2-} ion, neglecting the possibility of subtle forms of CO_3^{2-} ion disorder. (Although it was not stated whether the A-CO_3Ap was monoclinic, the reported mode of preparation suggests it probably was.)

By comparison with the CO_3^{2-} ion, there are six nonequivalent phosphorus atoms in the monoclinic structure, so it is to be expected that the ^{31}P NMR MAS spectrum would be more complex. In fact, four lines at 5.5, 4.6, 3.8 and 2.5 ppm (relative to 85% H_3PO_4, positive shift downfield) have been reported in studies at 119 MHz and 2 kHz spinning frequency [835]. The presence of four, rather than six, lines can be understood if three lines originate from the three PO_4^{3-} ions that surround a CO_3^{2-} ion and the fourth line from the three PO_4^{3-} ions that surround a vacancy. In the former case, the three phosphorus atom environments would be rather different because of the large tilt of the CO_3^{2-} ion so there would be three separate lines; in the latter case, they would be rather similar because they would be related by a pseudo-three-fold rotation axis, so there would be a single line. It is not possible to ascertain if this interpretation fits the relative intensities of the lines (Fig. 3 in ref. [835]) as the splittings are rather small. An analogous increase in the complexity of the IR PO_4 bands in A-CO_3Aps occurs, which has been explained [85,659] (Section 4.3.2) by an argument rather similar to that used here for the ^{31}P NMR spectra.

A density of 3.16 g cm^{-3} has been reported [709] for an A-CO$_3$Ap (preparation in Section 4.4.3) with lattice parameters $a = 9.539$ and $c = 6.865$ Å with a composition Ca 38.98, P 18.14, Na 0.00, CO$_3$ 4.7 wt %. These results gave unit cell contents of Ca$_{10.01}$(PO$_4$)$_{6.03}$(OH)$_{0.32}$(CO$_3$)$_{0.81}$. This density and composition are consistent with the calculated density of 3.160 g cm^{-3} (Table 1.4). OHAp has a very similar calculated value (3.156 g cm^{-3}, Table 1.4) which shows that there is little change in density as a result the substitution of two OH$^-$ ions by a CO$_3^{2-}$ ion in the lattice.

ESR studies of X-irradiated CO$_3$Aps are considered in Section 4.7.

4.4 Synthetic high temperature B-type carbonate apatites
4.4.1 Apatites containing fluoride ions

The first synthetic francolite-like analogue (B-CO$_3$FAp) was prepared by heating a precipitated CO$_3$Ap with CaF$_2$ in dry CO$_2$ at 900 °C [857]. Both the IR spectrum and an a-axis parameter ($a = 9.35_9$, $c = 6.89_3$ Å) that was shorter than the parameter for FAp were characteristics of natural francolites. A similar product could be made by heating either A-CO$_3$Ap and CaF$_2$, or FAp prepared by heating OHAp with CaF$_2$ (but not pure FAp), in dry CO$_2$ at 900 °C [915]. The reaction using FAp made from OHAp was shown to rely on the presence of CaO formed during its synthesis [916]. If CaCO$_3$ was intimately mixed with pure FAp and heated, B-CO$_3$FAp was also formed, but this source of CaO was not as effective as CaO produced in the synthesis of FAp; this was attributed to differences in their dispersions [896].

A change in lattice parameters from $a = 9.37_1$ and $c = 6.89_0$ Å for FAp (made by heating OHAp with CaF$_2$, so CaO would be present) to $a = 9.35_7$ and $c = 6.89_5$ Å after heating in dry CO$_2$ at 900 °C has been reported [82]. The CO$_3$ IR bands were at 1455, 1430 and 862 cm^{-1}, which are close to those in Fowey Consols francolite (1453, 1429 and 865 cm^{-1}, Section 4.2.2). The maximum CO$_2$ uptake corresponded to 0.3 CO$_3^{2-}$ ions per unit cell for which the general formula,

$$Ca_{10-x}(PO_4)_{6-2x}(CO_3,F)_{2x}F_{2(1-x)}, \qquad 4.18$$

was proposed [82]. Alternatively, if the possibility of adjacent CO$_3^{2-}$ and F$^-$ ions is rejected (see Section 4.2.1 for a discussion of this), the reaction between heated FAp and CaCO$_3$ will be

$$yCa_{10}(PO_4)_6F_2 + xCaCO_3 \rightarrow Ca_{10y+x}(PO_4)_{6y}(CO_3)_xF_{2y}. \qquad 4.19$$

If the total number of PO$_4^{3-}$ and CO$_3^{2-}$ ions per unit cell is six, as is probable, then $y = (6 - x)/6$. Assuming this condition, the unit cell contents for the maximum CO_2 uptake ($x = 0.3$) are $Ca_{9.80}(PO_4)_{5.70}(CO_3)_{0.30}F_{1.90}$. If further CaF_2 were available during the reaction, it is likely that it would be taken up to form an apatite with all the c-axis channel sites filled by F^- ions and with fewer calcium site vacancies. The rather low CO_2 content of these synthetic, compared to natural francolites, is attributable to the absence of monovalent cations, particularly Na^+ ions, to replace Ca^{2+} ions. This substitution occurs in mineral francolites and provides a much more effective charge compensation mechanism [43,891] (Section 4.2.1).

AB-CO_3F,OHAps have been prepared [82,896] from F,OHAps made by heating OHAp with a reduced amount of CaF_2 so that only a fraction of the OH^- ions were replaced by F^- ions, and thus with only a fraction of the maximum possible CaO content. The mixture of OH,FAp and CaO was then heated in dry CO_2 at 900 °C. As the F^- ion content in the CO_3Ap was increased, the intensity of the IR bands from CO_3^{2-} replacing OH^- ions decreased, and those from CO_3^{2-} replacing PO_4^{3-} ions increased. The corresponding lattice parameter changes were from $a = 9.53_7$, $c = 6.86_5$ Å (no CaF_2) to $a = 9.35_3$, $c = 6.89_6$ Å (127 % of the stoichiometric CaF_2). The frequency of the band at 1465 cm^{-1} contributed to by both substitutions only moved from 1465 to 1457 cm^{-1}. v_{3a} for CO_3^{2-} replacing PO_4^{3-} ions was first visible with 29 % of the stoichiometric CaF_2 ($a = 9.46_3$, $c = 6.88_6$ Å) at a frequency of 1407 cm^{-1}, which increased with an increasing F^- ion content to 1429 cm^{-1} [896]. This increase was nonlinear with respect to the reduction in unit cell volume [82]. Synthetic "francolites" with Cl^- instead of F^- ions also showed a nonlinear increase in the frequency of v_{3a} from 1405 to ~1429 cm^{-1}, as the unit cell volume decreased with increasing amounts of Cl^- replaced by F^- ions; this relation differed from that for the AB-CO_3F,OHAps [82]. The fact that the frequency of v_{3a} changes much more than the frequency of v_{3b} can at least be partly explained in the same way as the greater shift of v_{3b} compared with v_{3a} in A-CO_3Aps (Section 4.3.2). The angle between the a-axis and the transition moments of v_{3a} and v_{3b} are 21° and 29.5° respectively (Section 4.2.2), so it is v_{3a} that lies closer to the direction of greatest dimensional change: furthermore, the frequency of v_{3a} increases as the a-axis parameter decreases, as expected. Data on v_2 were not reported, but changes in its frequency with F^- ion content from other work are discussed in Section 4.5.2.

There do not seem to have been any extensive studies of high temperature sodium-containing B-CO_3FAps. They would be of interest as model compounds for the study of the crystal chemistry of francolite, particularly as the more efficient charge compensation mechanism from the presence of sodium would lead to higher carbonate contents. On the other hand, extensive studies have been made of B-CO_3FAps synthesised in aqueous systems (Sections 4.5.5 and 4.5.8). Barium and strontium B-CO_3FAps have been prepared at high temperatures with a maximum of 0.3 CO_3^{2-} ions per unit cell [82].

4.4.2 Fluoride-free compounds

Precipitated B-CO$_3$Aps [722] (Section 4.5.3) with a high CO$_3$ content (13.8 wt %), when heated in dry CO$_2$ for several days at 800 °C, yield AB-CO$_3$Aps with only a small fraction of the CO$_3^{2-}$ ions replacing OH$^-$ ions as determined by IR [722,917]. (Low carbonate content starting apatites give AB-CO$_3$Aps with a larger fraction of their CO$_3^{2-}$ ions replacing OH$^-$ ions, [722,917], Section 4.5.7.) The IR spectrum [722,917] of such a preparation had v$_3$ bands at 1430 and ~1405 cm^{-1} (broad and poorly resolved) and v$_2$ at ~873 cm^{-1} (the last two frequencies estimated from published spectra). The v$_3$ bands are clearly rather different from those in dahllite (1453 and 1416 cm^{-1}) [45] (Section 4.2.2), indicating significant differences in the CO$_3^{2-}$ ion environments, perhaps linked to the much higher carbonate content of the synthetic material. Lattice parameters were not given. The absence of the IR OH stretching band has been reported in similar preparations whose IR spectra likewise showed little evidence of A-type carbonate substitution [892].

Single phase AB-CO$_3$Aps have been prepared [897] by sintering DCPA and CaCO$_3$ at 850 °C in a CO$_2$ atmosphere with a partial water vapour pressure of 5 mm Hg (1.67 kPa). Before heating, the reactants were ball milled and pressed into tablets. After heating for from one to five days, the tablets were quenched and checked by X-ray powder diffractometry to see if they were a single phase, if not, the process was repeated. Molar P/Ca ratios were in the range 0.55 to 0.62. Densities (determined with a helium pycnometer with an accuracy of 1%), chemical analyses and lattice parameters were used to determine the unit cell contents. At the extremities of the range, one apatite had a composition of Ca 38.98, P 18.14, CO$_3$ 0.8 wt %, density 3.15 g cm^{-3}, and lattice parameters $a = 9.45$ and $c = 6.88$ Å; and at the other Ca 39.58, P 17.34, CO$_3$ 4.2 wt %, density 3.02 g cm^{-3}, and lattice parameters $a = 9.43$ and $c = 6.90$ Å. These lattice parameters showed that the unit cell volumes were the same within experimental error. The IR spectrum had bands around 1547, 1468, 1415, 880 and 874 cm^{-1}, indicating there were CO$_3^{2-}$ ions in PO$_4^{3-}$ and OH$^-$ ion sites. The CO$_3^{2-}$ ions were allocated between these so that all six PO$_4$ sites per unit cell were occupied by either a PO$_4^{3-}$ or a CO$_3^{2-}$ ion. This led to the formula

$$Ca_{9.59}[(PO_4)_{5.44}(CO_3)_{0.56}][(OH)_{1.55}(CO_3)_{0.12}] \qquad 4.20$$

for the apatite with the maximum carbonate content [897]. Neglecting the small number of CO$_3^{2-}$ ions per unit cell that replaced OH$^-$ ions (all within the range 0.12 to 0.20), the results fitted the general formula

$$Ca_{10-2x/3}(PO_4)_{6-x}(CO_3)_x(OH)_{2-x/3}. \qquad\qquad 4.21$$

Note that these AB-CO$_3$Aps have, in common with dahllite (Section 4.2.1), a virtually unchanged unit cell volume following the CO$_3^{2-}$ ion substitution.

The effect of heating B-CO$_3$Aps under pressure (40 kbar, 4000 MPa) at 950 °C for 15 min has been investigated [95,918]. (The starting material was precipitated by adding a calcium acetate solution to an ammonium phosphate plus carbonate solution, followed by drying at 400 °C [722], Section 4.5.3.) The IR spectra had CO$_3$ bands at 1430 and 1455 cm^{-1}. The sample with the most carbonate had, before treatment, lattice parameters of $a = 9.402$ and $c = 6.895$ Å, and density 2.72 ± 0.02 g cm^{-3} which, with the chemical composition, gave Ca$_{8.60}$(PO$_4$)$_{4.02}$(CO$_3$)$_{1.98}$(OH)$_{1.18}$ for the unit cell contents. After heating under pressure, the lattice parameters were $a = 9.430$, $c = 6.908$ Å and the density 3.04 ± 0.02 g cm^{-3} which gave unit cell contents Ca$_{9.69}$(PO$_4$)$_{4.53}$(CO$_3$)$_{2.23}$(OH)$_{1.33}$. The IR spectrum now had CO$_3$ bands at 1405, 1445, 1500, 1535 and 1560 cm^{-1}, with the first two being the most intense. These results were thought to show that the pressure and heat treatment had reduced the number of Ca^{2+} ion vacancies and moved CO$_3^{2-}$ ions (0.76 per unit cell for the above apatite) from PO$_4$ sites into the c-axis channels which increased the a-axis parameter. Although these high pressure apatites appeared to be similar to precipitated AB-CO$_3$Aps (Section 4.5.3), it was noted [95,918] that the IR CO$_3$ spectra were more complex. This indicated a distribution of CO$_3^{2-}$ ions between different environments along the channels which was responsible for the three bands at 1500, 1535 and 1560 cm^{-1}. Some of the samples decomposed on heating to 625 °C in air, whilst others were stable at 950 °C. These latter showed no change in lattice parameters or densities, but their IR spectra had changed so they now had three distinct bands at 1540, 1450 and 1410 cm^{-1}. The single band at 1540 cm^{-1}, rather than the three seen previously, showed that the CO$_3^{2-}$ ions in the channels had moved to a single environment. The thermal instability of some preparations was thought to be related to the presence of more CO$_3^{2-}$ ions in the channels than Ca^{2+} ion vacancies to match them. For example, the AB-CO$_3$Ap discussed above had 0.076 CO$_3^{2-}$ ions per unit cell on the c-axis, but only 0.31 Ca^{2+} ion vacancies per unit cell to match them; it was therefore thermally unstable. The stable reheated apatites have similarities with the AB-CO$_3$Aps discussed in the previous paragraph [897]: these include densities, an a-axis parameter larger than OHAp, IR spectra (a full comparison is not possible because of incomplete publication of the spectra), and structural formulae. However, the CO$_2$ contents of the high pressure samples are much higher.

4.4.3 Sodium-containing compounds

Sodium-containing AB-CO_3Aps have been prepared [709] at 870 °C by the same method used for AB-CO_3Aps [897] (Section 4.4.2) except that the CO_2 was dried by passing over 95% sulphuric acid and with the addition of Na_2CO_3 to the reaction mixture. The P/Ca and Na/Ca molar ratios ranged from 0.50 to 0.62 and from 0 to 0.2 respectively. Single phases were formed within a triangular field defined by P/Ca and Na/Ca molar ratios of 0.600±0.003 and 0; 0.574±0.003 and 0; and 0.530±0.003 and 0.18±0.01, respectively. The composition, lattice parameters and density of the apatite (A-CO_3Ap) at the first point of the triangle are given in Section 4.3.4. The sample at the third point of the triangle had the most CO_3 (15.4 wt %) with lattice parameters a = 9.367±0.003 and c = 6.934±0.002 Å and density 3.01±0.03 g cm^{-3}. These values, together with the chemical composition, gave unit cell contents

$$Ca_{8.50}Na_{1.57}[(PO_4)_{4.52}(CO_3)_{1.48}][(OH)_{0.10}(CO_3)_{0.97}].$$ 4.22

The IR spectra had bands around 1452, 1415 and 873 cm^{-1}, assigned to CO_3^{2-} ions in PO_4^{3-} ion sites; for samples with low Na/Ca ratios, additional bands around 1549, 1472 and 880 cm^{-1} were assigned to CO_3^{2-} ions in OH$^-$ ion sites. The absence of IR bands typical of CO_3^{2-} ions in the c-axis channels in samples with a high Na/Ca ratio (*e.g.* the apatite with Formula 4.22), even though the analyses indicated a major fraction of the CO_3^{2-} ions were in these sites, was thought to indicate that the CO_3^{2-} ions had a different environment from that in pure A-CO_3Aps.

Excluding the replacement of OH$^-$ by CO_3^{2-} ions (from 0.43 to 0.97 CO_3^{2-} ions per unit cell), the results fitted the general formula

$$Ca_{10-x}Na_x(PO_4)_{6-x}(CO_3)_x(OH)_2.$$ 4.23

Thus for every negative charge lost by replacing a PO_4^{3-} ion by a CO_3^{2-} ion, a positive charge was lost by replacing a Ca^{2+} ion by a Na^+ ion. This charge compensation mechanism, without lattice vacancies, is similar to the mechanism in mineral CO_3Aps [43,45,891] (Formula 4.4, Section 4.2.1); this clearly allows a much higher degree of carbonate substitution than that found in high temperature sodium-free AB-CO_3Aps in which Ca^{2+} and OH$^-$ ion vacancies are required (Formula 4.21, Section 4.4.2). However, the high temperature sodium-containing AB-CO_3Aps appear to be significantly different from precipitated sodium-containing CO_3Aps: the latter have a more complex charge compensation mechanism in which there are vacant OH$^-$ ion sites [88,917] (Formula 4.27, Section 4.5.4).

Although lattice parameters have been published for high temperature sodium-containing CO_3Aps with differing carbonate contents [709,819], these are difficult to interpret because the ratio of the number of CO_3^{2-} ions replacing PO_4^{3-} ions to the number of CO_3^{2-} ions replacing OH^- ions is variable, and is not easily controlled. This complication differs from aqueous preparations in which this ratio can be controlled to a certain extent (an increase in the pH suppresses the replacement of OH^- by CO_3^{2-} ions, Section 4.5.3).

Raman [818] and IR [819] spectra have been reported for sodium-containing AB-CO_3Aps. The samples for the Raman study were prepared by sintering Na_2CO_3, $CaCO_3$ and MCPM at 900 to 950 °C in wet CO_2 for 8 h. Calcite, CaO and $Ca(OH)_2$ impurities were removed by extraction of the ground apatite with neutral triammonium citrate (0.3 mol l^{-1}) at 50 °C. In a sample containing 11.8 wt % CO_3, four bands at 758, 714, 694 and 671 cm^{-1}, seen very clearly in the Raman spectrum, were assigned to v_4 bands from CO_3^{2-} ions in two different environments (assignments and identification of the environments are discussed in ref. [911], Section 4.5.2). v_1 occurred at 1070 cm^{-1} in the Raman spectrum.

High pressure forms of sodium-containing AB-CO_3Aps have been made [95] by heating B-CO_3Aps at 900 °C for 15 min at 50 kbar (5000 MPa). The starting apatites were precipitated at 100 °C by adding a solution of sodium phosphate and sodium carbonate to a solution of $Ca(NO_3)_2$, followed by drying at 400 °C as described in refs [88,917] (Section 4.5.4). The sample with the most carbonate had, before treatment, unit cell contents $Ca_{7.48}Na_{1.69}(PO_4)_{3.55}(CO_3)_{2.45}(OH)_{1.10}$ and lattice parameters $a = 9.306$ and $c = 6.916$ Å. After heating under pressure, the lattice parameters were $a = 9.410$ and $c = 6.895$ Å. A weak IR band appeared at 1540 cm^{-1}, the main v_3 bands at 1450 and 1410 cm^{-1} (the former the more intense) and v_2 became doubled. These results showed that CO_3^{2-} ions had moved from PO_4^{3-} ion sites onto the c-axis, so that the treatment had produced a sodium-containing AB-CO_3Ap. Compared with the sodium-free compounds after high pressure treatment (Section 4.4.2), the increase in the a-axis parameter was much greater and there were fewer v_3 CO_3 IR bands.

Heating precipitated sodium-containing CO_3Aps at 1000 °C leads to decomposition to OHAp, Na_2O and CaO [919]. On further prolonged heating at 1100 °C, the 3560 and 633 cm^{-1} OH bands disappeared and new bands at 3520, 710 and 670 cm^{-1} appeared (the latter two merged to a single strong band at 745 cm^{-1} after 63 h) [919]. The a-axis parameter decreased from 9.42_0 to 9.37_6 Å as a result of this heating. The interpretation given [919] to these results was that sodium had diffused into the apatite and had replaced some of the Ca(2) atoms adjacent to OH^- ions; this resulted in the observed changes in the OH IR bands and decrease in the a-axis parameter. However, it should be borne in mind that somewhat similar changes to these can be observed from the accumulation of F^- ions from a furnace muffle from prolonged heating (BO Fowler, personal communication, 1991, Section 3.4.2).

The conditions required for sintering sodium-containing CO_3Aps under CO_2/H_2O or N_2/H_2O atmospheres between 825 and 1050 °C to produce specific carbonate contents and physical and chemical compositions have been studied [920].

4.5 Carbonate apatites from aqueous systems
4.5.1 Introduction

There have been more extensive investigations of precipitated CO_3Aps, by comparison with high temperature CO_3Aps, presumably because the former have a closer relationship with biological apatites. CO_3Aps prepared in aqueous systems have also been investigated as models for the genesis of phosphorite deposits [880,921] (Section 4.5.6) and [751] (Section 4.5.8).

CO_3Aps are easily prepared by precipitating apatites or hydrolysing ACP, OCP or DCPA in the presence of CO_3^{2-} or HCO_3^- ions. The most obvious effect of these ions is to reduce the crystallinity of the apatite, or even to prevent its formation, so that an amorphous product results. Generally, better crystallised apatites are formed at 80 to 100 °C. In the precipitation of CO_3Aps, different products are obtained if the calcium solution is added to the phosphate plus carbonate solution ("direct" apatites) rather than *vice versa* ("inverse" apatites) [83,722]. Na_2CO_3 solutions and DCPA form well-crystallised CO_3Aps if reacted at 95 °C, but calcite and aragonite are also likely to be precipitated [79] (Section 1.5.5).

Well-crystallised CO_3Aps, precipitated in hot solutions, do not yield pyrophosphate on heating to 400 to 500 °C [83,88,99,722], a result that is usually taken to indicate the absence of HPO_4^{2-} ions in the apatite before heating. However, as has been pointed out [99], any pyrophosphate that formed from HPO_4^{2-} ions might be lost via reaction with carbonate to produce CO_2 and PO_4^{3-} ions (Section 3.6.2), thus invalidating this conclusion. Alternatively, the HPO_4^{2-} ions might be lost directly by reaction with carbonate (Reaction 3.17, Section 3.6.2). However, in one study of sodium-containing B-$^{13}CO_3Aps$ precipitated at pH 9.0 to 9.3 and at 100 °C, no HPO_4 band at 875 cm^{-1} was seen (v_2 $^{13}CO_3$ had moved to 850 cm^{-1}, so this region was clear) [922]. Whether the same result would be obtained for apatite synthesised under more acid conditions is uncertain.

Much of the more recent work (up to 1984) on precipitated CO_3Aps has been reviewed [99]. The main emphasis since about 1973 has been on the study of well-crystallised CO_3Aps whose unit cell contents could be determined from their chemical composition, lattice parameters and densities. It should be noted that many CO_3Aps formed in aqueous systems are poorly crystallised and may have significantly different structures and chemistry. Finely precipitated CO_3Aps [923] (Section 4.5.4) and CO_3Aps prepared by the reaction between alkaline phosphate solutions and $CaCO_3$ (Section 4.5.6), which both have rather labile CO_2 contents, are examples.

Conditions for the precipitation of CO_3Aps, ACa,CO_3P, calcite and aragonite under near physiological conditions have been investigated [924,925]. In other studies [926], it was found that Mg^{2+} ions stabilised the ACa,CO_3P (presumably forming AMg,Ca,CO_3P) and the apatitic precipitates were more poorly crystallised, while F^- ions slightly improved the crystallinity of the apatitic phases, as judged by their XRD patterns. The synthesis of apatites by the hydrolysis of ACP in the presence of CO_3^{2-} ions has been studied [927]. The formation of ACa,CO_3Ps is discussed in Section 1.8.2.

OHAp aqueous suspensions will readily take up CO_3^{2-} (or HCO_3^-) ions. In experiments at pH 7.4 and 25 °C, OHAp with a surface area of 68 m^2 g^{-1} was equilibrated with a solution of $KHCO_3$ and KCl (both 1 mol l^{-1}) [881]. Within 5 min, 0.2 wt % CO_2 was taken up which did not increase after a further 130 h. The suggestion made was that two HCO_3^- ions exchanged with two HPO_4^{2-} ions plus a Ca^{2+} ion. The CO_2 was lost on suspension (2 or 24 h) in potassium phosphate buffer (0.4 or 0.01 mol l^{-1}) at pH 5 or 6 with CO_2-free nitrogen gas bubbled through.

4.5.2 Assignment of IR and Raman carbonate bands

Synthetic [77,866] and biological [928,929] (Section 4.6.3) CO_3Aps often have at least three v_3 and two v_2 CO_3 bands, usually interpreted as showing the presence of CO_3^{2-} ions in two environments. The bands at 1545, 1450 and 880 cm^{-1}, were thought to be due to a small fraction of the total CO_3^{2-} ions replacing OH^- ions in the lattice [77,866,871]. This assignment was based on the similarity of the frequencies and dichroic ratios of two bands in native dental enamel (1545 and 878 cm^{-1}, Fig. 4.7, Section 4.6.3) with two bands in dental enamel heated at high temperatures (1548 and 879 cm^{-1}, Fig. 4.4, Section 4.3.2), known to have this substitution. Further evidence for this assignment came from the observation that, in various CO_3Aps synthesised in aqueous systems, the intensities of the bands at 1545, 1450 and 880 cm^{-1} moved as a whole with respect to the other CO_3 bands, which indicated that these three bands had a common origin [77,884]. The idea that precipitated CO_3Aps contained CO_3^{2-} ions replacing OH^- ions has been criticised [872] because this substitution was originally demonstrated in an apatite made at 900 °C under very dry conditions (Section 4.3.1). However, it is now generally accepted that this A-type substitution can occur to a limited extent in CO_3Aps prepared in aqueous systems.

The CO_3 bands at 1465, 1412 and 873 cm^{-1} from CO_3^{2-} ions in the other, and major, environment were originally assigned to surface CO_3^{2-} ions rather than to CO_3^{2-} ions in the lattice replacing PO_4^{3-} ions [77,866,871]. The reasons for this were: (1) dental enamel had a slightly increased *a*-axis parameter compared with high temperature OHAp (Section 4.6.2), whereas a contraction was anticipated if most of the CO_3^{2-} ions replaced PO_4^{3-} ions; and (2) the CO_3 bands did not match those in francolite (1453, 1429 and 865 cm^{-1}, Section 4.2.2) in which this substitution was known to take place. The assignment to surface ions was subsequently shown to be incompatible with work on precipitated CO_3Aps [79,872] and these absorption bands, at least for the major part, are now believed to originate from CO_3^{2-} ions that replace PO_4^{3-} ions in the lattice (B-type substitution, Section 4.1.1). The differences between the frequencies of the bands in francolite and fluoride-free CO_3Aps (including dental enamel) have been attributed to differences in composition [872]. This problem has already been discussed for francolite and dahllite (Section 4.2.2) in which similar frequency differences were attributed to differences in F$^-$ ion contents. It is not clear if v_2 moves continuously from 873 to 865 cm^{-1} as the F$^-$ ion content increases, or if one band decreases whilst the other increases, but studies of synthetic preparations suggest that the latter is more likely (see Fig. 2 of ref. [908]). Analysis of this region is sometimes complicated by the presence of a band from CO_3^{2-} ions replacing OH$^-$ ions [699]. The shifts in the frequency of v_3 that occur when the F$^-$ ion content changes in high temperature preparations are discussed in Section 4.4.1. The correspondence between the spectra of francolite and dental enamel is further considered in relation to their polarised IR spectra in Section 4.6.3.

The concept that the CO_3^{2-} ions were in two well-defined environments in precipitated CO_3Aps was early recognised to be an over-simplification because of the multiple bands sometimes seen in the v_3 CO_3 region of synthetic CO_3Aps (Fig. 4.6) and because of inconsistencies in the relative intensities of the IR bands in the spectra of synthetic CO_3Aps and the polarised spectrum of enamel [77]. Other examples of this multiple and variable structure can be seen in many published spectra of precipitated AB-CO_3Aps [77,94,126,533,625,699]. For example, in addition to the main bands at 1465 and 1412 cm^{-1}, shoulders at 1568, 1545, 1500, 1470, 1452 and 1416 cm^{-1} have been reported [699]. Bands at 1460 and 1470 cm^{-1} have been attributed to CO_3^{2-} replacing PO_4^{3-} ions with and without an adjacent OH$^-$ (or F$^-$) ion respectively (both environments were thought to contribute to a band at 1420 cm^{-1}), and another weak band at 1500 cm^{-1} to a different A-type environment [99]. The assignments of the v_3 CO_3 bands to different sites will be considered further in Section 4.6.3 when the polarised IR spectrum of enamel is discussed. A similar complexity is seen in the ^{13}C NMR spectra of AB-CO_3Aps in which at least five different CO_3^{2-} ion environments may exist [894] (Section 4.5.8).

More recent quantitative IR studies of the amount of A-type substitution in dental enamel (Section 3.6.3) have confirmed that the relative intensities of the CO_3 bands are not entirely consistent with the concept of well-defined single A- and B-type substitutions [703,704] (Section 4.6.3). Similar difficulties have

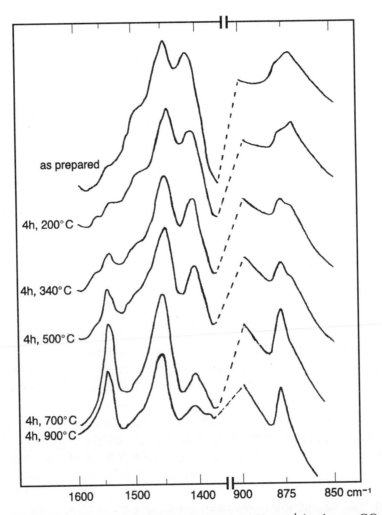

Fig. 4.6 Multiple structure seen from 1580 to 1450 cm^{-1} in the v_3 CO$_3$ bands of a CO$_3$Ap prepared by reaction between CaCO$_3$ and triammonium phosphate solution at ~100 °C for 24 h, followed by drying at 130 °C for 24 h. X-ray powder diffractometer patterns showed only the presence of well-crystallised apatite. The IR spectra demonstrate that, on heating in air, the CO$_3^{2-}$ ions move to sites on the hexad axis. (Fig. 6.9 from Dowker [94], with permission)

been noted for high temperature AB-CO$_3$Aps (absent ~1540 cm^{-1} band, but visible ~879 cm^{-1} band [699,819]; and absence of both bands, but significant A-type substitution according to the unit cell contents [709], Section 4.4.3).

    The assignment of the v_2 and v_3 bands to CO$_3^{2-}$ ions has been confirmed from the IR spectrum of an AB-CO$_3$Ap prepared by refluxing calcite containing carbon with 59 atomic % ^{13}C enrichment in an ammoniacal ammonium phosphate solution [77,126] (Section 4.5.6). Bands from $^{12}CO_3^{2-}$ ions occurred at 1545, 1452, 1412, 879 and 872 cm^{-1}, while the corresponding bands from $^{13}CO_3^{2-}$ ions were at 1502, 1412, 1372, 852 and 845 cm^{-1} (note that the 1412 cm^{-1} band has contributions from both isotopes).

A weak shoulder at 866 cm^{-1}, seen in deconvoluted FT IR spectra, has been assigned to a labile carbonate environment in synthetic precipitated CO_3Aps [699], bone [699,930] and enamel [704] (Section 4.6.3). The assignment to v_2 was confirmed by the preparation of an AB-$^{13}CO_3$Ap using $Na_2^{13}CO_3$ which showed the expected isotopic shift. The intensity of the 866 cm^{-1} shoulder decreased during maturation of the apatite crystals. In addition, partial dissolution of the CO_3Ap crystals (10 min in EDTA at pH 7.4) led to a decrease in the intensity of the 866 cm^{-1} band relative to the other v_2 bands. $^{13}CO_3^{2-}$ - $^{12}CO_3^{2-}$ exchange of the AB-$^{13}CO_3$Ap was studied by immersing it in a solution containing $^{12}CO_3^{2-}$ ions. Deconvoluted IR spectra showed that the shoulder assigned to the labile CO_3^{2-} species was relatively greater in the $^{12}CO_3$ group of bands compared with the $^{13}CO_3$ bands. Quantitative measurement of the amount of exchange was not possible, but it did appear to be less than expected [699]. As a result of the dissolution and exchange experiments, it was suggested that the CO_3^{2-} ions responsible for the weak shoulder at 866 cm^{-1} might be located in poorly organised positions in the crystals which were more reactive than more organised sites in the lattice. The labile CO_3^{2-} ions seemed to be related to an early stage of apatite crystal formation [699].

When the degeneracy of v_3 is lost in many mineral francolites (Figs 4.2 and 4.3, Section 4.2.2) and high temperature A-CO_3Aps (Fig. 4.4, Section 4.3.2), its two components, v_{3a} and v_{3b}, are very clearly separated. This also applies to a slightly lesser extent to dental enamel (Fig. 4.7, Section 4.6.3) and other CO_3Aps prepared in aqueous systems that contain only a few wt % CO_3. However, as the carbonate content of these latter CO_3Aps increases, published spectra [79,99,533,625,872] show that the resolution of the two components decreases and may be lost altogether. Presumably this is a result of increased interaction between the ions as their number per unit cell increases.

The IR v_4 CO_3 bands are very weak, but have been seen in precipitated B-CO_3Aps at 710 and 692 cm^{-1} [88] (Section 4.5.4). Detailed assignments of CO_3 bands in this region have recently been made [911]. Sodium-containing CO_3Aps have been prepared at pH 10 to 11.5 after the method of LeGeros [79] (Section 4.5.4) with and without added F$^-$ ions. In the fluoride-containing compounds dried at 70 °C, bands in the v_4 region were only seen at 716 and 695 cm^{-1}. These were assigned to v_4 components of CO_3^{2-} ions replacing PO_4^{3-} ions, the reasoning being that F$^-$ ions effectively prevented the occupation of *c*-axis sites by CO_3^{2-} ions, so this alternative was not possible. On the other hand, in the absence of F$^-$ ions, bands occurred at 757, 740, 718, 692 and 670

cm^{-1}, the 740 cm^{-1} band only being seen in samples heated to 400 °C. As this apatite was thought to have CO_3^{2-} ions in both OH^- and PO_4^{3-} ion sites, the bands at 718 and 692 were assigned to CO_3^{2-} ions replacing PO_4^{3-} ions, by analogy with the CO_3FAps above, and the bands at 757 and 670 cm^{-1} to CO_3^{2-} ions replacing OH^- ions [911]. This latter assignment was supported by the observation of bands at 766 and 675 cm^{-1} in high temperature A-CO_3Aps [911].

The v_1 CO_3 IR band near 1063 cm^{-1} is very weak and is always obscured by the v_3 PO_4 band. However v_1 CO_3 bands at 1105 cm^{-1} and part of the band at ~1070 cm^{-1} in the Raman spectrum of enamel have been assigned to CO_3^{2-} ions replacing OH^- and PO_4^{3-} ions respectively [560] (Section 4.6.3). This assignment was based on a comparison of enamel with synthetic A- and B-CO_3Aps. Comparison of the Raman spectra of OHAp and dental enamel also led to the assignment of part of the 1070 cm^{-1} band to v_1 CO_3 [818].

4.5.3 Precipitated apatites (fluoride- and alkali-free)

Kühl and Nebergall in 1963 [714] appear to have been the first to propose the formula

$$Ca_{10-x+y}(CO_3)_x(PO_4)_{6-x}(OH)_{2-x+2y} \quad (0 \leq x \leq 2 \text{ and } 2y \leq x) \qquad 4.24$$

which is now generally accepted to represent the unit cell of precipitated CO_3Aps in which the only cations are Ca^{2+} ions. This general formula was derived from a similar formula (No 3, Table 3.2, Section 3.7.3) that these authors had proposed for precipitated Ca-def OHAps in which they replaced HPO_4^{2-} by CO_3^{2-} ions. For every negative charge lost when a CO_3^{2-} ion replaces a PO_4^{3-} ion, another negative charge is lost by the loss of an OH^- ion: the loss of these two negative charges is then compensated by the loss of two positive charges through the formation of a calcium site vacancy. However, some of the OH^- and Ca^{2+} ion vacancies are filled because of the addition of y formula units of $Ca(OH)_2$ per unit cell. They reported chemical analyses for one sample prepared by adding 20 ml of a $Ca(NO_3)_2$ solution (1 mol l^{-1}) to a mixture of solutions comprising 120 ml K_2HPO_4 (0.1 mol l^{-1}), 120 ml KOH (0.1 mol l^{-1}) and 55 ml K_2CO_3 (0.1 mol l^{-1}). This was boiled for 16 h and the precipitate washed with water, alcohol and ether. The chemical analysis was Ca 38.0, PO_4 50.45, CO_3 3.89 wt %. This gave the formula

$$Ca_{9.56}(CO_3)_{0.655}(PO_4)_{5.345}(OH)_{1.775} \qquad 4.25$$

assuming that all the PO_4 sites were filled. The XRD pattern was apatitic and the IR spectrum "characteristic of carbonate apatites", but further details were not given.

A series of sodium-free B-CO$_3$Aps has been prepared [83,722,917] by modifying the method developed earlier for the preparation of sodium-containing CO$_3$Aps (calcium acetate solution added slowly to a sodium phosphate plus sodium carbonate solution at 95 °C [79,872,931], Section 4.5.4). The sodium salts were replaced by ammonium compounds and, in order to reduce the loss of NH$_3$, the temperature was reduced to 75 °C and ammonia solution added during the precipitation. Despite these precautions, the pH rapidly fell to 6.5. This method, in which the calcium acetate was added to the ammonium carbonate and ammonium phosphate solution, was referred to as "direct". If the CO$_3$/PO$_4$ ratio (presumably molar) exceeded 40, calcite was formed in addition to apatite. The maximum CO$_3$ content without calcite formation was 13.8 wt % when the lattice parameters were a = 9.373±0.002 and c = 6.897±0.002 Å; the CO$_3$ absorption bands were at 872, 1412 and 1462 cm^{-1}, without bands attributable to A-type substitution [83] (these have been reported in lightly carbonated samples [99,533], see later). A different "inverse" series was made by adding, during 3 h, 300 ml of a solution at pH 11 and with various ratios of (NH$_4$)$_2$HPO$_4$ to ammonium carbonate (both typically 0.07 mol l^{-1}) to 800 ml of a calcium acetate solution (0.08 mol l^{-1}) at 90 °C and pH 11. If the CO$_3$/PO$_4$ ratio (presumably molar) exceeded 1.5, calcite was formed in addition to apatite. The maximum CO$_3$ content, without calcite formation, was 10.20 wt % when the lattice parameters were a = 9.354 and c = 6.897 Å [83,722]. Thus the maximum possible carbonate content was greater in the "direct", compared with "inverse" apatites, although the a-axis parameter of each "inverse" apatite was always less than the a-axis parameter of the "direct" apatite with the same carbonate content [83,722] (Table 4.6, Formulae 1 and 3). Subsequently, it was found [625] that "inverse" preparations had lower NH$_4^+$ ion contents, as indicated by their lower nitrogen contents (Sections 4.5.4 and 4.5.7). As it was thought that NH$_4^+$ ions replaced Ca^{2+} ions in the lattice, the smaller a-axis parameter for the same carbonate content was attributed to the smaller diameter of the Ca^{2+} ion (0.99 Å) compared with the NH$_4^+$ ion (1.43 Å) [625]. However, this substitution will require another one to compensate for the difference in charge between these ions, which might in itself contribute significantly to the observed a-axis parameter difference.

Both "direct" and "inverse" B-CO$_3$Aps had CO$_3$ absorption bands at 872, 1412 and 1462 cm^{-1}, without bands attributable to A-type substitution [83,722]. As the carbonate content increased, the OH bands decreased in intensity (the librational mode faster than the OH stretch), so that eventually the OH stretch could not be seen for CO$_3$ contents greater than 10 wt %. On heating in air, these B-CO$_3$Aps decomposed in two stages; water was lost below ~500 °C with a small decrease in the a-axis parameter (~0.003 Å) without a significant change in the c-axis [83], and between 700 and 1000 °C with the loss of CO$_2$ and the formation of OHAp and CaO. The first stage resulted in an increased density (details of the measurement technique are in Section 3.6.3) that was more marked as the carbonate content increased [83,722,917]. On heating in dry CO_2 at 900 °C, A-CO_3Ap was formed to an extent determined by the OH⁻ ion content in the c-axis channels of the original B-CO_3Ap (see following paragraph).

The unit cell contents of the "direct" and "inverse" apatites dried at 200 or 500 °C were determined from their densities, lattice parameters and chemical composition [83,722,917]. The contents fitted the general formula

$$Ca_{10-x+u}(PO_4)_{6-x}(CO_3)_x(OH)_{2-x+2u} \qquad\qquad 4.26$$

with $0{\leq}x{\leq}2$ and $0{\leq}2u{\leq}x$ which is similar to that proposed earlier by Kühl and Nebergall [714] (Formula 4.24). This formula with $x = 2$ and $u = 0$ gives a theoretical maximum CO_3 content of 14.62 wt %. Based on the concept that the vacant oxygen site formed when a CO_3^{2-} ion replaced a PO_4^{3-} ion and that the vacancies formed by the loss of the Ca^{2+} and OH⁻ ions should all be as close as possible, it was proposed that neighbouring O(3), Ca(2) and OH sites (Fig. 2.4, Section 2.2.1) became vacant. (This need not necessarily be the case, for example, in $CaCl_2$-def ClAp, both Ca(1) and Ca(2) sites have vacancies [421] (Section 2.2.3), and not just the Ca(2) sites adjacent to the Cl vacancies.) This mode of vacancy clustering required that the CO_3^{2-} ion occupied a sloping face of the PO_4 tetrahedron, as in francolite (Section 4.2.2). The water lost below 400 °C was thought to be located in the vacancy clusters formed by the vacant O(3), Ca(2) and OH sites [83,722,917]. This explained why the water loss increased with an increase in the carbonate content.

Another mode of substitution has been proposed that does not require vacancies [82,892,893]. In this, a CO_3^{2-} ion, accompanied by an OH⁻ ion, replaces a PO_4^{3-} ion. This is analogous to the earlier proposal made for francolite [771,772] (Section 4.2.1) and precipitated B-CO_3FAps [99,538] (Section 4.5.5) which has been used to explain the often found excess of F⁻ ions over that required to fill the c-axis channel sites. In this second mode of substitution, it was suggested that there were u [CO_3^{2-},OH⁻] ion pairs per unit cell, where u is the parameter in Formula 4.26. IR bands at 1460 and 1470 cm⁻¹ have been assigned to the groups [CO_3^{2-},OH⁻] and [CO_3^{2-},□] respectively [99] (Section 4.5.2, also discussed in Section 4.6.3). Evidence for [CO_3^{2-},OH⁻] replacing PO_4^{3-} ions also derived from ESR studies. The ESR signal assigned to an F⁺ centre from electrons thought to be trapped in vacant O(3) sites initially rose, then fell, with an increasing carbonate content [99,893]. On the other hand, the substitution of adjacent negative ions (CO_3^{2-} and F⁻) for a PO_4^{3-} ion in francolite has been criticised [45] (Section 4.2.1).

Further indirect evidence for the structural formula (4.26) came from the observation that the intensity of the IR OH band at 3570 cm⁻¹ fell to zero, as the CO_3 content approached the theoretical maximum of 14.62 wt % [83,722]. Similar measurements have been made on the intensity of the Raman line at 3570 cm^{-1} from CO_3Aps precipitated at 80 °C by the simultaneous addition of a solution of calcium acetate and a solution of ammonium phosphate plus ammonium carbonate to an ammonium acetate buffer at pH of 7.2 [691]. The intensity fell very nonlinearly to zero for a CO_3 content of 12.4 wt %; intermediate intensities were always less than would have been given by a linear relation.

The influence of preparation conditions on the lattice parameters and unit cell contents of AB-CO_3Aps has been investigated [99,533]. This, and earlier results (Table 4.6, top section), show clearly that preparation conditions have a marked effect on the dependence of the *c*- and particularly the *a*-axis on the carbonate content. The NH_4^+ ion concentration has a major influence, which is not surprising, as there is evidence to be discussed later (Sections 4.5.4 and 4.5.7), that NH_4^+ ions can replace Ca^{2+} ions in the lattice to a limited extent. Nevertheless, the results in Table 4.6 (top section) are rather consistent. As the CO_3 concentration increases to about 3 to 4 wt %, the *a*-axis parameter increases and the *c*-axis parameter decreases. The initial increase has been attributed [99,533] to an initial small replacement of OH$^-$ ions by CO_3^{2-} ions, as this substitution would cause such a change (Table 4.5, Section 4.3.1). Evidence for this was that, in lightly carbonated samples, the IR spectrum showed the presence of CO_3^{2-} ions in *c*-axis channel sites (IR bands at 882 and 1545 cm^{-1}) and there was evidence that the total number of PO_4^{3-} plus CO_3^{2-} ions per unit cell slightly exceeded six [99,533]. More prominent A-type bands at low CO_3^{2-} ion concentrations can also be seen in the published spectra of sodium-containing precipitated CO_3Aps [79,819,872] (Section 4.5.4). Similar proposals to explain the slight increase in the *a*-axis parameter of enamel [77] (Section 4.6.2) over that of OHAp, and to explain the smaller decrease in the *a*-axis parameter for a given carbonate content in dahllite compared with francolite [45] (Section 4.2.1), had been made earlier. For precipitated CO_3Aps with a CO_3 concentration exceeding 3 to 4 wt %, the *a*-axis parameter decreases markedly and the *c*-axis parameter increases slightly (Table 4.6, top section), as expected for the replacement of PO_4^{3-} by CO_3^{2-} ions.

The results in Table 4.6 (top section) should be rather similar to those for high temperature B-CO_3Aps [897] (Section 4.4.2). The latter B-CO_3Aps showed a slight decrease in the *a*-axis parameter and increase in the *c*-axis parameter without a change in unit cell volume for the preparation with the maximum CO_3 content (4.2 wt %). This observation is not inconsistent with any of the data in Table 4.6, given the small carbonate content of the high temperature samples. In any case, any differences could be explained by: (1) different values of *u* (Formula 4.26), (2) different amounts of CO_3^{2-} ions in the *c*-axis channels, or (3) the presence of NH_4^+ or HPO_4^{2-} ions in the lattice of the

Table 4.6 Lattice parameter relations for precipitated CO_3Aps with *m* wt % CO_3[a]. Top section from ammonium salts, middle from sodium salts, and the bottom section from sodium salts in the presence of F⁻ ions.

Prep.[b]	$a \times 1000$ Å		$c \times 1000$ Å		
1. DAA	9458-6.18m 7<m<13.8		6891+0.44m 7<m<13.8		[83]
2. DNA	9427+4.8m 0<m<3	9457-6.1m 3<m<13			[99,533]
3. IAA	9448-9.23m 5.2<m<10.2		6887+1.00m 5.2<m<10.2		[83]
4. INA	9427+2.1m 0<m<3	9472-11.6m 4<m<9			[99,533]
5. Sim	9430+1.5m 0<m<3	9470-10m 4<m<7	6891-1.25m 0<m<3	6882+2.5m 4<m<7	[624]
6. Direct Na	9440-6m 0<m<22		6880+2.2m 0<m<22		[79,872]
7. Direct Na	9435-7.2m 7<m<22		6882+1.7m 0<m<22		[88]
8. Direct Na,F	9385-4.5m 0<m<22		6880+1.8x 0<m<22		[79,872]
9. Direct Na,F	9368-6.7m 0<m<22		6886+1.37m 0<m<22		[88]
10. Direct F	0<m<12.8 9368-7.4x		0<m<12.8 6888+2.4m		[99]

[a]The relations, which were mostly determined from published graphs or lattice parameters, are reasonably linear over the ranges of *m* given below them.

[b]Preparation methods: DAA, "direct" with added ammonia solution; DNA, "direct" with no added ammonia; IAA "inverse" with added ammonia; INA, "inverse" with no added ammonia; Sim, calcium acetate and ammonium carbonate plus phosphate solutions added simultaneously at pH 9 to 9.3. All samples, except 5, 6 and 8, were dried at 400 to 470 °C.

precipitated apatite.

Turning now to the influence of preparation conditions on other unit cell constituents, the two proposed modes for CO_3^{2-} ions to replace PO_4^{3-} ions discussed earlier will be considered; these were [CO_3^{2-},OH⁻] ion pairs replacing PO_4^{3-} ions, or CO_3^{2-} ions replacing PO_4^{3-} ions with adjacent OH⁻ and Ca^{2+} ion vacancies. In a study of the relative importance of these, it was found that the number of vacancies decreased with either, or both, excess Ca^{2+} ions or ammonia in the reaction medium, which indicated a preference for the $[CO_3^{2-}, OH^-]$ form of substitution under these conditions [533]. The explanation given for this was that both an increase in OH^- (from the excess NH_4OH) and Ca^{2+} ion concentrations would result in a reduction in the number of OH^- and Ca^{2+} ion vacancies [533]. An increased OH^- ion concentration, from the increased ammonia, also favoured OH^- ions in the competition between OH^- and CO_3^{2-} ions for *c*-axis channel sites [533].

Heating ACa,CO_3Ps (Section 1.8.2), which have been previously dried under vacuum at 70 °C, at 500 to 600 °C yields $AB-CO_3Aps$ [92]. Similar products are formed if undried ACa,CO_3Ps are left at 20 °C in air for from 20 h (1.25 wt % CO_3) to 50 h (5 wt % CO_3) [92] (Section 1.8.4).

4.5.4 Precipitated apatites containing monovalent cations

The first systematic study of well-crystallised precipitated CO_3Aps was made by LeGeros in 1965 [79,872,931]. They were prepared by the dropwise addition (0.05 ml s^{-1}) of 250 ml of a calcium acetate solution (0.02 mol l^{-1}) into 750 ml of a Na_2HPO_4 solution (0.016 mol l^{-1}) containing variable concentrations of $NaHCO_3$ (0 to 0.54 mol l^{-1}) at 95 °C. The precipitates were filtered and air-dried. The *a*-axis parameter decreased linearly from 9.44 to 9.30 Å and the *c*-axis parameter increased linearly from 6.88 to 6.93 Å as the CO_3 content increased from 0 to 22 wt % [79,872] (Table 4.6, Formula 6). There was a decrease in the *a*-axis parameter of 0.01 to 0.02 Å on heating to 400 °C without loss of CO_2. The IR spectra had CO_3 bands at 1470, ~1420 and 871 cm^{-1}. A shoulder at 879 cm^{-1} could be seen if there was less than 15 wt % CO_3. (More prominent A-type bands at low CO_3^{2-} ion concentrations can also be seen in other studies of sodium-containing precipitated $B-CO_3Aps$ [819] and in sodium-free precipitated CO_3Aps, Section 4.5.3). The OH librational mode at 633 cm^{-1} was weak or absent if there was more than ~2.5 wt % CO_3, and the OH stretching mode absent if there was more than ~11 wt % CO_3 [79,872]; a similar behaviour was found in sodium-free precipitated CO_3Aps [83,722] (Section 4.5.3). Progressive changes in several PO_4 absorption bands with an increased carbonate content were seen [79,872]. The major changes were: v_1 at ~960 cm^{-1} became broader, the component of v_3 at ~1050 cm^{-1} diminished in intensity and eventually disappeared, and the relative intensities of the two bands due to the v_4 mode at 604 and 567 cm^{-1} changed, so that the high frequency component became the more intense. The composition of the most highly substituted sample was Ca 34.5, Na 6.3, P 10.7 and CO_3 22.2 wt % [79]. The results were interpreted as showing that there was a coupled equimolar substitution of Na^+ for Ca^{2+} ions and CO_3^{2-} for PO_4^{3-} ions in the lattice [79,872].

The loss of the OH IR absorption bands noted in the previous paragraph, together with the same observation [82,892] on a similarly prepared sample, led to the conclusion [82,892] that, in heavily carbonated samples, the *c*-axis channels contained few OH⁻ ions; instead, it was suggested that OH⁻ ions paired with some of the CO_3^{2-} ions that replaced PO_4^{3-} ions. This proposal is similar to that made for the sodium-free CO_3Aps [82,892,893] (Section 4.5.3).

Detailed investigations have been undertaken [88,917] on the sodium-containing CO_3Aps similar to those on the sodium-free compounds [83,722,917] (Section 4.5.3). The maximum CO_3 content for "direct" preparations was 22.10 wt % for which the lattice parameters after heating at 400 °C were *a* = 9.275 and *c* = 6.920 Å [88]. There was some evidence that the decrease in the *a*-axis parameter with carbonate content was smaller below 7 wt % CO_3, compared with above 7 wt %, but nevertheless, the change in lattice parameters with CO_3 (Table 4.6, Formula 7) was very similar to the results from the earlier work on sodium-containing B-CO_3Aps (Formula 6). Comparison of "direct" CO_3Aps made with sodium salts with those made with ammonium salts shows that, for the same carbonate content, those with sodium have shorter *a*-axes (Table 4.6, Formulae 7 and 2 respectively, and Fig. 43 of ref. [88]); this difference increases with the carbonate content.

Relatively more carbonate was incorporated in "direct" preparations and in the presence of Na⁺ ions [88,917]. In the sodium-containing preparations (presumably both "direct" and "inverse"), IR CO_3 bands typical of B-CO_3Aps were reported at 1462, 1412, 872 cm⁻¹ and weak, but quite clear, bands at 710 and 692 cm⁻¹ were assigned to v_2 (presumably a misprint for v_4) CO_3 bands [88]. As the carbonate content increased, the OH bands at 3560 and 640 cm⁻¹ decreased, but never entirely disappeared [88], unlike the sodium-free CO_3Aps [83,722] (Section 4.5.3). However, others [79,872] (see above) have reported the disappearance of the OH bands in sodium-containing B-CO_3Aps, but in these, the OH band at 3560 cm⁻¹ would have been less visible because the samples were not dried at 400 °C. On heating in air, "direct" CO_3-Aps [88,917] prepared in the presence of Na⁺ ions behaved like the sodium-free CO_3Aps [83,722] (Section 4.5.3). When dried at 400 °C, there were decreases of 0.020 Å in the *a*-axis and 0.0044 Å in the *c*-axis parameters for the most highly carbonated sample; adsorbed water was lost between room temperature and 120 °C and lattice water between 120 and 300 °C; carbonate loss began at 550 °C; at 1000 °C, Na_2O was formed, in addition to CaO and OHAp [88]. The lattice water lost on heating at 400 °C was equal to the carbonate content on a molar basis and was proportional to the contraction of the lattice.

The unit cell contents of the sodium-containing CO_3Aps fitted the formula

$$Ca_{10-x}Na_{2x/3}(PO_4)_{6-x}(CO_3)_x(H_2O)_x(OH)_{2-x/3} \qquad 4.27$$

where $0 \le x \le 3$ [88,917]. In this, for every CO_3^{2-} ion, the difference in charge between the CO_3^{2-} and PO_4^{3-} ions is compensated by replacing Ca^{2+} by a Na⁺

ion; then one-third NaOH is lost from the unit cell to create vacancies. Formula 4.27 is consistent with the IR results that show that OH$^-$ ions are always present. The theoretical maximum CO_3 content is 22.27 wt %, which compares with 14.62 wt % for the sodium-free B-CO_3Aps (Formula 4.26 with $u = 0$), assuming the absence of water in both cases. This difference is consistent with the often made observation that more carbonate is incorporated in CO_3Aps prepared in aqueous systems when Na$^+$ ions are present.

The above formula led to a structural model [88,917] with clustered ion vacancies filled with water similar to that proposed [83,722,917] (Section 4.5.3) for sodium-free B-CO_3Aps. In addition, the suggestion was made that the Na$^+$ ions occupied Ca(2) sites, but were always adjacent to a vacant OH$^-$ ion site. As has been noted earlier, the charge compensation mechanism in these precipitated CO_3Aps is different from the high temperature sodium-containing CO_3Aps [709] (Formula 4.23, Section 4.4.3) in which Ca^{2+} and PO$_4^{3-}$ ions are replaced by Na$^+$ and CO_3^{2-} ions, without the formation of any vacancies. This difference can be illustrated by comparing the lattice parameters of the high temperature CO_3Ap with the most carbonate ($a = 9.367$, $c = 6.934$ Å for 15.4 wt % CO_3 [709], Section 4.4.3) with those for a precipitated CO_3Ap with the same carbonate content calculated from Formula 7, Table 4.6 ($a = 9.324$, $c = 6.908$ Å). The very significant differences in lattice parameters can be attributed to the different unit cell contents; in particular, the larger a-axis parameter of the high temperature apatite is probably mainly due to CO_3^{2-} ions in the c-axis channels.

Strontium [79,99] and barium [79] B-CO_3Aps have been prepared and have an a-axis parameter that decreases with an increase in the carbonate content.

Potassium, ammonium and other monovalent cations

CO_3Aps, precipitated [79,82,88,99,917], or formed by the reaction of alkaline phosphatic solutions with CaCO$_3$ [932] or by the hydrolysis of OCP [177] in the presence of K$^+$ instead of Na$^+$ ions, generally contain less alkali and carbonate. The a-axis parameter of a precipitated potassium-containing CO_3Ap ($a = 9.37_0$, $c = 6.92_2$ Å) containing 16.69 wt % CO_3 was larger than that of the equivalent sodium compound with the same carbonate content ($a = 9.334$ Å as calculated from Table 4.6, Formula 6) [79]. This difference was attributed to the larger radius of the K$^+$ ion compared with the Na$^+$ ion (1.38 and 1.02 Å respectively, Table 2.1) [79]. If both Na$^+$ and K$^+$ ions are present during precipitation, the resultant preferential uptake of sodium has also been attributed to the difference in ionic radii [88,99]. (Apatites formed in the reaction between alkaline phosphate solutions and calcium carbonate also have a preference for Na$^+$ over K$^+$ ions, Section 4.5.6.)

In experiments on the hydrolysis of OCP in alkali carbonates [177], the incorporation of CO_3^{2-} and alkali ions was found to decrease in the sequence lithium > sodium > potassium > rubidium which was linked to an increasing ionic radius. The hydrolysis products had the characteristics of B-type substitution (v_3 CO_3 bands reported only at 1420 and 1470 cm^{-1}). Nevertheless, because a wide range of different sizes of alkali ions were accepted, it was suggested that the alkali and CO_3^{2-} ions occupied disparate defects formed during the irreversible hydrolysis of OCP, and were therefore confined mostly to the "water layer" (Section 1.3.2). (See also Section 1.3.6 on the hydrolysis of OCP.)

CO_3Aps prepared in solutions containing NH_4^+ ions retain small amounts of these ions (typically 0.12 wt % N) which can react with CO_3^{2-} ions on heating at 450 to 650 °C to yield cyanate (NCO$^-$) and cyanamide (NCN^{2-}) ions [94,434,435] (Section 4.5.7). This observation, together with differences in the change in lattice parameters with carbonate content between the sodium and ammonium compounds (Table 4.6, middle and top sections respectively), led to the suggestion that small amounts of NH_4^+ ions could substitute for Ca^{2+} ions in the apatite lattice [624]. The fact that the *a*-axis parameter of "direct" preparations is larger than "inverse" ones with the same carbonate content has also been attributed to an increased replacement of Ca^{2+} by NH_4^+ ions [625] (Section 4.5.3). The affinity of the NH_4^+ ion for the apatite lattice is very small, as is shown by the low nitrogen content of CO_3Aps precipitated in the presence of NH_4^+ ions (see above and ref. [625]), and by the fact that, if Na^+ or K^+ ions are also present, they are preferentially incorporated, so that no nitrogen can be detected [625].

By contrast with the well-crystallised CO_3Aps described above, which were generally prepared at ~95 °C, CO_3Aps with much smaller crystal sizes have been precipitated (presumably at 37 °C, but not stated) by the simultaneous addition, during 3 to 4 h, of 100 ml each of "KHPO$_4$" (presumably KH_2PO_4) (0.1 mol l^{-1}) and $CaCl_2$ (0.16 mol l^{-1}) solutions to 500 ml of a solution (0.15 mol l^{-1} KCl plus 0.025 mol l^{-1} KHCO$_3$) maintained at pH 7.4 by the addition of KOH [923]. Gas (95% O_2 plus 5% CO_2) was flushed through the system. The product had a Ca/P molar ratio of 1.61±0.05 and contained 6.6 wt % CO_2 when wet and 4.8 wt % after freeze drying (lyophilization). A sample that had been aged for 21 days in the system before separation and then dried to constant weight exchanged 18±1 % of its CO_2 with H$^{14}CO_3^-$ ions in a buffer solution at pH 7.4, a percentage that did not increase with further exposure. The interpretation given was that 60 % of the CO_2 was present, presumed to be as CO_3^{2-} ions, within the crystal lattice and 40 %, presumed to be as HCO$_3^-$ ions, within the "hydration shell". About 50 % of the "hydration shell" CO_2 was lost on drying (temperature not reported, but presumably at room temperature).

4.5.5 Precipitated apatites containing fluoride

The original syntheses of sodium-containing CO_3Aps [79] (Section 4.5.4) were also undertaken with a constant F^- concentration in the Na_2HPO_4 plus $NaHCO_3$ solutions used for each preparation, whilst the carbonate/phosphate ratios were varied [79]. The F^- ion content of the samples (2.79 to 2.18 wt %) was insufficient to fill all the *c*-axis channel sites with F^- ions, and did not appear to affect the incorporation of CO_3^{2-} ions. The variation of the *a*-axis parameter with carbonate content (Formula 8, Table 4.6, Section 4.5.3) was rather similar to that for the fluoride-free system (Formula 6, Table 4.6), except for an additional constant contraction of 0.055 Å. This figure is slightly greater than the difference in *a*-axis parameter between OHAp and FAp (0.0505 Å, Table 1.4). The most highly carbonated sample had a composition Ca 36.00, Na 3.33, P 11.15, F 2.18, CO_3 22.13 wt %, and lattice parameters $a = 9.26_8$ and $c = 6.92_4$ Å [79]. In the IR spectrum, v_2 bands occurred at 879 and 869 cm^{-1}, the 869 cm^{-1} band being the more intense (the reverse of that found in the fluoride-free preparations). The presence of an 879 cm^{-1} band shows that some CO_3^{2-} ions occupy *c*-axis sites, which is consistent with there being insufficient F^- ions to fill these positions.

Detailed studies of sodium-containing CO_3FAps [88,917], similar to those on sodium-containing CO_3Aps (Section 4.5.4), have been undertaken. NH_4F was added [88] to the phosphate plus carbonate solution to give F^- ion concentrations from 0.005 to 0.1 mol l^{-1} (above 0.06 mol l^{-1}, CaF_2 was precipitated). "Direct" preparations contained much more CO_3^{2-} and Na^+ ions than the "inverse" ones. The sample with most carbonate had a composition Ca 36.13, PO_4 35.97, CO_3 22.70, Na 5.49 wt %. There were 1.17 F^- ions per unit cell, as deduced from the requirement of charge balance; the lattice parameters (determined from the published graphs) were $a = 9.229$ and $c = 6.922$ Å [88]. The lattice parameters varied linearly with carbonate content (Table 4.6, Formula 9). Preparations with less carbonate had an increased F^- ion content (up to 1.94 F^- ions per unit cell for 1.35 wt % CO_3) [88]. Studies of the unit cell contents showed that the formula derived for the fluoride-free system (Formula 4.27, Section 4.5.4) could be applied [88,917]. Thus it was generalised to

$$Ca_{10-x}Na_{2x/3}(PO_4)_{6-x}(CO_3)_x(H_2O)_x(OH,F)_{2-x/3} \qquad 4.28$$

where $0 \le x \le 3$ [88,917].

Sodium-free $B\text{-}CO_3FAps$ have also been studied [99,538,796,908]. Measurements of lattice parameters, composition and densities of apatites precipitated at 100 °C, then dried at 480 °C, gave a general formula [538]

$$Ca_{10-x+u}(PO_4)_{6-x}(CO_3)_{x-u}(CO_3,F)_uF_{2-x+u}. \qquad\qquad 4.29$$

Formula 4.29 is conceptually the same as that originally proposed for fluoride-free precipitated B-CO_3Aps [714] (Formula 4.24, Section 4.5.3), except that, after compensating for the difference in charge between CO_3^{2-} and PO_4^{3-} ions by the loss of Ca^{2+} and *c*-axis channel ions, the number of vacancies is reduced by the addition of CaF_2, with the F⁻ ions split equally between *c*-axis sites and sites adjacent to CO_3^{2-} ions. IR bands were reported [538] at 862, 1435 and 1450 cm⁻¹; the highest CO_3 content was ~16 wt % for an *a*-axis parameter of 9.25 Å.

Further detailed investigations have been made of sodium-free B-CO_3Aps [99] prepared by a method based on the original sodium-containing preparations [79] (Section 4.5.4). In the new study, ammonium and acetate or nitrate compounds were used. For the "direct" preparations, 250 ml of a solution containing 0.01 mol of Ca^{2+} ion was added slowly to 750 ml of a boiling solution containing 1.68 g NH_4F, 0.012 mol PO_4^{3-} ion and between 0 and 0.72 mol CO_3^{2-} ion. For the "inverse" preparations, 700 ml of a solution containing 1.12 g NH_4F, 0.01 mol PO_4^{3-} ion and between 0 and 0.067 mol of CO_3^{2-} ion was added slowly to 300 ml of a boiling solution containing 0.01 mol Ca^{2+} ion [99]. No OH absorption bands could be seen in the IR spectra of the products. For samples with a low carbonate content, the ν_3 bands at 1435 and 1450 cm⁻¹ were well-separated and sharp, similar to those found in mineral francolites (Figs 4.2 and 4.3), but they lost their resolution as the carbonate content increased. (A similar loss of resolution with an increase in carbonate is seen in many other CO_3Aps, Section 4.5.2.) A weak band at 1470 cm⁻¹ was assigned to CO_3^{2-} ions replacing PO_4^{3-} ions unaccompanied by a F⁻ or OH⁻ ion, and the stronger band at 1460 cm⁻¹ to [CO_3^{2-},X⁻] with the X⁻ ion being an F⁻ or OH⁻ ion; both types of CO_3^{2-} ion were thought to contribute to the band at 1420 cm⁻¹. (These assignments are discussed further in Section 4.6.3.) ESR studies [933] of X-irradiated samples of these apatites were also thought to support the idea of the simultaneous presence of CO_3^{2-} ions with and without adjacent F⁻ ions. These two CO_3^{2-} ion environments have also been used to explain aspects of the decarboxylation and loss of nitrogen from B-CO_3FAps prepared in the presence of NH_4^+ ions (Section 4.5.7). (See also the discussion of this form of substitution in relation to francolite, Section 4.2.1, and precipitated B-CO_3Aps, Section 4.5.3.)

The changes in lattice parameters with carbonate content are given in Table 4.6 (Formula 10) [99]. Samples heated to 480 °C lost 2 to 3 wt % water, but the lattice did not contract, unlike most other CO_3Aps prepared in aqueous systems (Sections 4.5.3 and 4.5.4). On heating above this temperature, CO_2 was lost, and FAp and CaO were formed. Decomposition started at 520 °C for the most heavily carbonated samples; this temperature rose to about 800 °C for those with the lowest carbonate content [99].

The unit cell contents of these B-CO_3FAps have been investigated as a function of their carbonate content [99]. For samples dried at 480 °C with up to one CO_3^{2-} ion per unit cell, only the channel sites were occupied by F$^-$ ions; the charge balance was maintained by a deficiency of Ca^{2+} ions. Above this carbonate content, the deficiency of Ca^{2+} ions decreased slightly, and an excess of F$^-$ ions was taken up to maintain charge balance. In this region ($1 \leq x \leq 2$), the general formulae that were given were:

$$Ca_{9.65+0.1x}(PO_4)_{6-x}(CO_3)_{0.35-0.1x}(CO_3,F)_{1.1x-0.35}F_{1.65+0.1x} \qquad 4.30$$

for "direct" preparations and

$$Ca_{9.3+0.2x}(PO_4)_{6-x}(CO_3)_{0.7-0.2x}(CO_3,F)_{1.2x-0.70}F_{1.30+0.2x} \qquad 4.31$$

for "inverse" B-CO_3FAps [99].

The IR CO_3 bands for AB-CO_3OH,FAps change with their F$^-$ ion content. As this increases, v_2 changes from 873 to 865 cm^{-1} (either the band moves continuously or the 873 cm^{-1} band decreases in intensity and the 865 cm^{-1} band increases) [908] (Section 4.5.2). In high temperature preparations, v_{3a} shifts from 1407 to 1429 cm^{-1} as the F$^-$ content increases, whereas v_{3b} changes rather little [82,896] (Section 4.4.1).

It has been reported that treatment at 50 kbar (5,000 MPa) of B-CO_3Aps causes a reduction of their vacancy content and an increase in the *a*-axis parameter (no details of probable simultaneous thermal treatment) [99]. Of the two samples studied, the larger change was from $a = 9.317$ and $c = 6.893$ Å with unit cell contents of $Ca_{9.70}(PO_4)_{4.80}(CO_3)_{1.20}F_{2.60}$ to $a = 9.346$ and $c = 6.900$ Å with unit cell contents of $Ca_{9.95}(PO_4)_{4.92}(CO_3)_{1.23}F_{2.67}$. The reason for the increase in the *a*-axis parameter is not clear.

The equilibrium solubility properties of B-CO_3FAps, precipitated under constant composition conditions, have been studied [751] (Section 4.5.8).

AB-CO_3OH,FAps can be prepared by reducing the quantity of F$^-$ ions available for incorporation in the lattice during precipitation [99] (sodium-containing AB-CO_3OH,FAps can be similarly prepared [79]).

Detailed studies of strontium B-CO_3FAps have been undertaken [99].

Comparison of synthetic CO_3F,OHAps and francolite

An important question is the extent to which synthetic B-CO_3FAps are similar to mineral francolites. The high temperature B-CO_3FAps (Section 4.4.1) and precipitated preparations (both with and without sodium) have IR absorption bands that are very similar to those of francolite (sharp carbonate bands at 1453, 1429 and 865 cm^{-1} for high fluoride samples). The synthetic precipitated CO$_3$FAps also are like francolite in that their *a*-axis parameter becomes shorter as the carbonate content increases.

The general formula for francolite (Formula 4.4, Section 4.2.1) is, for a magnesium-free mineral,

$$Ca_{10-0.6x}Na_{0.6x}(PO_4)_{6-x}(CO_3)_xF_{0.4x}F_2 \qquad\qquad 4.32$$

which indicates that the mineral appears to have neither calcium nor *c*-axis site vacancies. This is different from the synthetic preparations which generally seem to have vacancies (high temperature sodium-free, Formula 4.18; precipitated, Formulae 4.28, 4.30 and 4.31). However, the "direct" precipitated B-CO$_3$Aps with high carbonate contents (Formula 4.30) are similar to francolite (Formula 4.32), in that they seem to have more than two F$^-$ ions per unit cell.

Sodium is important in the charge compensation mechanism in francolite (Section 4.2.1), so a more detailed comparison has to be directed to the synthetic sodium-containing B-CO$_3$FAps. As explained in Section 4.4.1, there do not seem to be any extensive studies of these made at high temperatures, so only precipitated B-CO$_3$FAps are considered. The variation in the *a*-axis parameter (Å) with *m*, the wt % CO$_3$, obtained from the mean of Relations 8 and 9 in Table 4.6 is

$$a = 9.376 - 0.0056m. \qquad\qquad 4.33$$

The corresponding relation for francolite is

$$a = 9.369 - 0.0054m. \qquad\qquad 4.34$$

This is derived from the last relation in Table 4.3 (Section 4.2.1) by converting the number of CO$_3^{2-}$ ions per unit cell to wt % CO$_3$ by a factor derived from Formula 4.32 with *x* = 1. It is seen that there is satisfactory agreement between Equations 4.33 and 4.34 which shows that the changes in *a*-axis with carbonate content of francolite and precipitated sodium-containing B-CO$_3$FAps are very similar, and are not noticeably affected by their apparent difference in vacancy contents.

4.5.6 Reaction between alkaline phosphate solutions and calcium carbonate

Equilibrium and rate of reaction studies between calcite and KH$_2$PO$_4$ or H$_3$PO$_4$ solutions (2×10^{-4} to 10^{-6} mol l^{-1}) in a 5% CO$_2$ plus 95% N$_2$ atmosphere at room temperature indicated the formation of OHAp [934]. Under different conditions (Na_3PO_4 at 0.3 mol l^{-1} and 28 °C), an apatitic phase was formed in which it was thought that CO_3^{2-} for PO_4^{3-} and Na^+ for Ca^{2+} lattice substitutions took place [880]. Only apatite, without unreacted calcite, was present if the CO_3 content of the product was less than 10 wt %. In similar experiments with sodium and potassium salts [921,932], the alkali content, and to a lesser extent the carbonate content, depended on the pH of the solution, being higher under more alkaline conditions. About five times more sodium than potassium was taken up under comparable conditions. This difference was attributed to the greater difficulty of replacing Ca^{2+} ions in the apatite lattice by the larger K^+ ion, compared with the similarly sized Na^+ ion (radii of Ca^{2+}, K^+ and Na^+ ions, 1.00, 1.38 and 1.02 Å respectively, Table 2.1). (See also the preference for Na^+ compared with K^+ ions during the precipitation of CO_3Aps, Section 4.5.4.) XRD showed that the CO_3Aps were generally poorly crystallised; no lattice parameters were reported [921,932]. DCPD [932] was formed at pH 6.1, and OCP [921] as a transient phase (several days) at pH 7.5. On heating, one sample containing 5.55 wt % CO_2 lost nearly 1 wt % CO_2 at 120 °C; this loss increased continuously as the temperature was raised, so that nearly all had gone by 710 °C [932]. The sample also continuously lost water, so that the total loss of volatiles was about 11.5 wt % at 800 °C. This thermal behaviour contrasts markedly with the much better crystallised sodium-containing CO_3Aps (Section 4.5.4) which lose the majority of their water (including lattice water) below 400 °C and do not lose significant CO_2 until above this temperature (but note the formation of retained molecular CO_2 below this temperature in some CO_3Aps, Section 4.5.7).

CO_3Aps have also been prepared by the reaction between calcium carbonate and alkaline phosphate solutions at 95 to 100 °C. In one of the first such preparations, 1 g of reagent grade calcium carbonate, 10 ml water, 60 ml KH_2PO_4 solution (0.02 mol l^{-1}) and sufficient NaOH solution (2 mol l^{-1}) so that the suspension contained 0.5 mol l^{-1} of NaOH were shaken overnight at 90 °C [879]. The product was separated, washed with water then ethanol, and dried at 110 °C, followed by 2 h at 300 °C. The XRD pattern was very similar to OHAp, but the IR spectrum had CO_3 bands at 1460, 1418 and 875 cm^{-1}, but no water or OH absorption bands: chemical analyses were not reported. These results were thought to indicate that the CO_3Ap had the formula $Ca_{10}(PO_4)_6CO_3$, but later work on the synthesis of A-CO_3Ap (Section 4.3.1) showed this could not be correct. However, the IR spectra indicate that PO_4^{3-} ions are replaced by CO_3^{2-} ions and that there is a reduced OH$^-$ ion content which are both consistent (but not the reported *absence* of OH$^-$ ions) with the general formula for precipitated sodium-containing CO_3Aps (Formula 4.27, Section 4.5.4). CO_3Aps prepared by the reaction between calcite and sodium phosphate solutions at 95 °C are better crystallised than those synthesised at room temperatures, the improvement being greater at pH 5 than at pH 12 [79].

Refluxing calcite with an ammoniacal ammonium phosphate solution produces an apatite with an IR spectrum [77,126] typical of CO_3^{2-} ions in both PO_4^{3-} and OH⁻ ion sites (Section 4.5.2).

The reaction between calcite and alkaline phosphate solutions produces pseudomorphs after the form of the original calcite crystals [880]. There do not seem to be any reports of a specific relation between the crystallographic axes of the calcite and apatite, but this is not so when aragonite is used as the starting material [412]. Single crystals of aragonite, about 4 mm in diameter and 0.3 to 0.6 mm thick, were heated between 260 and 400 °C at 1 kbar (100 MPa) in $(NH_4)_2HPO_4$ or MCPA solutions for one to three weeks, the longer time at the lower temperature [412]. The original shape was preserved, but the specimen became more or less turbid. XRD, polarised light microscopy and SEM of treated discs cut perpendicular to the original aragonite *c*-axis showed that the apatite was fibrous (diameter ~1 μm), with the apatite and aragonite *c*-axes parallel. Most of the apatite crystals had their *a*-axes parallel to each other and oriented with respect to the *a*- and *b*-axes of aragonite as expected from the structural relation between apatite and aragonite (Fig. 2.7, Section 2.2.2). The remaining crystals were randomly oriented about the aragonite *c*-axis. The lack of perfect orientation was ascribed to the slight mismatch between the dimensions of the aragonite and apatite lattices (the pseudohexagonal *a*-axis parameter of aragonite is 9.378 Å. Section 2.2.2, compared with 9.418 Å for the *a*-axis parameter of OHAp). IR spectroscopy showed the presence of CO_3^{2-} ions in both PO_4^{3-} and OH⁻ ion sites [412]. Coral skeletal calcium carbonates (both aragonite and calcite forms) have also been subjected to similar hydrothermal conditions with a view to the production of bone implants [303] (Section 4.6.7). The products were strong, polycrystalline replicas of the original materials comprising OHAp with a minor replacement of OH⁻ and PO_4^{3-} ions by CO_3^{2-} ions; occasionally, but not always, whitlockite was formed when calcite corals were used. The reason for the formation of whitlockite was not completely clear [303], but was thought to be associated with the high magnesium content of the samples, as magnesium is known to stabilise the whitlockite structure (Section 1.6.4).

The reaction between gypsum ($CaSO_4.2H_2O$) and sodium phosphate solutions more concentrated than 0.01 mol l⁻¹ and with a pH above 7.5 leads to the formation of CO_3Aps [935].

4.5.7 Thermal decomposition

Summary

Many aspects of the thermal decomposition of CO_3Aps prepared in aqueous systems have already been discussed for individual compounds. The biological CO_3Aps behave in a rather similar, but not identical, manner (summarised in Section 4.6.4). For the well-crystallised synthetic apatites, these can be summarised as the loss of surface-bound water up to ~120 °C, followed by lattice water up to 400 to 500 °C. This latter loss is accompanied by a decrease of ~0.02 Å in the *a*-axis parameter for sodium-containing CO_3Aps (Section 4.5.4) and a much smaller contraction (~0.003 Å, Section 4.5.3) in the absence of Na^+ ions. Dental enamel (Section 3.7.3), but not CO_3FAps (Section 4.5.5), also shows a small contraction after heating to this temperature range. The small reduction in the *a*-axis parameter of ns-OHAps formed in aqueous systems on heating to 600 °C, attributed to loss of water and HPO_4^{2-} ions (Section 3.7.3), is also pertinent to the behaviour of the CO_3Aps.

For the well-crystallised CO_3Aps, CO_2 loss in air generally starts at 550 °C and is usually complete at 900 to 1000 °C, depending on the time and composition (Sections 4.5.3 and 4.5.4). CaO (and Na_2O if sodium is present) and OHAp are formed. If the Ca/P molar ratio is less than 1.667, β-TCP will also be present (Equation 3.22).

The temperature at which CO_2 is lost in precipitated B-CO_3FAps increases from 520 to ~800 °C as the CO_2 content is reduced [99] (Section 4.5.5). It is a feature of poorly crystallised precipitated synthetic (Section 4.5.3) and some biological (Section 4.6.4) CO_3Aps that they lose CO_2 at much lower temperatures (even room temperature) than well-crystallised preparations. However in the latter, IR spectra often show the formation of retained molecular CO_2 at temperatures below 500 °C (see later in this section). CO_3Aps precipitated in the presence of Mg^{2+} ions lose CO_2 in the temperature range 350 to 905 °C [936].

Pyrophosphate is formed on heating apatites that contain acid phosphate in the temperature range 400 to 700 °C [693] (Equation 3.15, Section 3.6.2), but the yield can be considerably reduced if carbonate is present [355,694] (Section 3.6.2). The well-crystallised CO_3Aps do not contain pyrophosphate after heating [83,88,99,722] (Section 4.5.1), which is probably due to the absence of HPO_4^{2-} ions in the unheated apatite, but could also be due to reaction of pyrophosphate ions with carbonate [99] (Section 3.6.2) or reaction of HPO_4^{2-} ions with carbonate (Reaction 3.17, Section 3.6.2). In contrast, pyrophosphate is found in biological apatites after heating [875] (Section 4.6.1). Other aspects of the thermal decomposition of biological apatites are discussed in Section 4.6.4. Thermal changes at 1000 °C and above are the same as for carbonate-free apatites; these are summarised in Section 3.13.

Detailed studies of the thermal changes involving carbonate are complicated. These can conveniently be divided into five groups: studies of the loss of CO_2 (summarised above) and nitrogen from CO_3Aps; relocation of CO_3^{2-} ions within the structure; formation of retained CO_2; the reactions between carbonaceous and nitrogenous species within the solid that yield cyanate (NCO^-) and cyanamide (NCN^{2-}) ions; and formation of hydrogen and carbon monoxide from retained acetate.

Loss of carbon dioxide and nitrogen

Loss of CO_2 and N_2 on heating CO_3FAps (presumably essentially B-CO_3FAps) precipitated in the presence of ammonium salts has been investigated by gas chromatography and thermogravimetry [102,672,937,938]. Decarboxylation occurred between 450 and 950 °C; three stages were seen for "direct" CO_3FAps [102,672,937], but only the last two stages were present in "inverse" CO_3FAps [102,672,938]. The first stage was attributed to the decomposition of carbamate ($NH_2CO_2^-$) ions derived from the ammonium carbonate used in the preparation and retained by the apatite [102,937]. This stage occurred at 480 °C (580 °C for higher rates of heating) in DTG (differential thermogravimetric analysis) and in CO_2 emission curves. The reason for attributing the CO_2 loss to carbamate decomposition was that this stage was present in "direct" CO_3FAps prepared with ammonium carbonate (all other salts were sodium compounds), but absent in "direct" CO_3FAps if prepared with $NaHCO_3$ (all other salts were ammonium compounds): in the first preparation, carbamate ions would be present in the apatite, but ammonium ions would be displaced by the Na^+ ions; in the second preparation, no carbamate ions would be present because of the absence of ammonium carbonate. The second and third stages at ~700 and ~800 °C (temperatures depended on rate of heating) were assigned [102,938] to the decarboxylation of $[CO_3^{2-},\square]$ and $[CO_3^{2-},F^-]$ ion pairs respectively. Fluoride-free "direct" and "inverse" CO_3Aps behaved in a similar way to CO_3FAps, but in this case, the last stage of decomposition was attributed to the decarboxylation of $[CO_3^{2-},OH^-]$ ion pairs [102].

Changes in the CO_3^{2-} ion location

On heating fluoride-free precipitated AB-CO_3Aps or B-CO_3Aps at temperatures from 500 to 800 °C, the CO_3 bands at 1545, 1450 and 880 cm^{-1} from CO_3^{2-} replacing OH^- ions become more prominent with respect to the other CO_3 bands (Fig. 4.6, Section 4.5.2). This effect was first reported [929] for enamel and assumed to result from increased crystal perfection, but was subsequently ascribed [871] to the replacement of OH^- by CO_3^{2-} ions induced by heating. An account of a detailed study [721] of the thermally induced changes in the location of CO_3^{2-} ions in enamel is given in Section 4.6.4.

Detailed studies of lattice parameter and IR spectral changes on heating CO_3Aps have been made [99]. For sodium-free CO_3Aps prepared in ammoniacal solutions at 100 °C, the IR bands from CO_3^{2-} ions in the c-axis channels became, on heating to ~500 °C, more prominent than the bands from CO_3^{2-} ions replacing PO_4^{3-} ions. This was most noticeable in "inverse" preparations and "direct" preparations with a low carbonate content; high carbonate content "direct" preparations showed no change [99]. All samples showed a small (~0.1 Å) increase in the a-axis parameter which was, in part, attributed to the increased number of CO_3^{2-} ions in the *c*-axis channels. However, it was thought that the formation of cyanate (NCO^-) and cyanamide (NCN^{2-}) ions, which probably also occupy *c*-axis sites, contributed to the expansion (see later discussion on the formation and loss of these species); this seemed to apply particularly to the high carbonate "direct" preparations that showed no evidence for relocation of CO_3^{2-} ions to *c*-axis sites. AB-CO_3FAps precipitated at 100 °C and then dried at 400 °C behaved in a rather similar manner [99]. Weak shoulders in the IR spectrum at 882 and 1545 cm^{-1}, which were assigned to CO_3^{2-} ions on the *c*-axis, increased in intensity after further heating at 480 °C: there was also a small increase in the *a*-axis parameter. These changes were ascribed to CO_3^{2-} ions moving from PO_4^{3-} to OH^- ion sites [99]. They decreased as the F^- ion content increased, as expected with an increased number of F^- ions on the *c*-axis sites which would prevent CO_3^{2-} ions moving to these positions. Again it was thought that part of the expansion was caused by the formation of cyanate ions (cyanamide was not observed) [99].

It has been suggested that the number of CO_3^{2-} ions on the *c*-axis sites is limited by the number of OH^- ions in these positions before heating [722,917]. When precipitated CO_3Aps were heated at 800 °C in dry CO_2 for some days, they had IR CO_3 bands characteristic of AB-CO_3Aps [722,917] (Section 4.4.2). The 1542 and 883 cm^{-1} bands of CO_3^{2-} ions in *c*-axis sites were much less prominent in apatites for which the starting material had a high carbonate content (see Section 4.4.2 for a discussion of the IR of such a preparation). This was linked to a fall in the OH^- content of the precipitate as its carbonate content rose (see proposed general Formula 4.26 in Section 4.5.3).

The ESR spectra of precipitated CO_3Aps, dried at different temperatures and then X-irradiated, differ with the drying temperature used when this is in the range 25 to 400 °C; this is predominantly a reflection of differing contributions from a variety of carbon-containing radicals [939] (Section 4.7).

Formation and loss of retained carbon dioxide

An absorption band at 2340 cm^{-1} in the IR spectrum of CO_3Aps (prepared by reaction of ammoniacal ammonium phosphate solution and $CaCO_3$ at 100 °C), which is seen after heating at 300 °C, has been assigned to v_3 of molecular CO_2 [77,866]. This was confirmed by a shift to 2277 cm^{-1} in the ^{13}C compound [77,866]. The same band was seen in "direct" but not "inverse" sodium-free CO_3Aps after heating to 600 °C in dry CO_2 for 3 h [83].

Detailed investigations of the effect of time and temperature on the intensity of the CO_2 band in CO_3Aps prepared by refluxing $CaCO_3$ and an ammonium phosphate solution, or by "direct" precipitation at 100 °C using calcium and ammonium salts (both preparations with added ammonia) have been made [94,534]. After heating in air for 4 h at fixed temperatures, CO_2 bands typically appeared at 150 °C, reached a maximum at 400 °C and were not seen above 800 °C.

Sodium-free CO_3Aps, after heating in air at 400 °C, have a very much stronger 2340 cm^{-1} band in "direct", compared with "inverse" preparations, and in high, compared with low, carbonate content preparations [99]. The same band was seen in similarly treated "direct" CO_3FAps, but not "inverse" preparations, and in $SrCO_3$Aps, but never in sodium-containing preparations [99].

A band at 2360 cm^{-1}, seen in several apatite minerals after heating, has been assigned to CO_2 within the lattice [940]. This band appeared simultaneously with a band at 740 cm^{-1} that was attributed to pyrophosphate formed by the condensation of two HPO_4^{2-} ions [940]. A CO_2 band at 2340 cm^{-1} was seen in only three of six phosphorites that were studied [94,534]; the appearance of this band could not be associated with any particular characteristic of the minerals [94,534]. This CO_2 band has also been seen in some mineralised tissues after heating [94,534] (Section 4.6.4). A weak band at 2421 cm^{-1} with partial perpendicular dichroism (dichroic ratio = 0.383) has been seen in unheated Fowey Consols francolite, but this is probably not due to molecular CO_2 (JC Elliott, unpublished results, 1962). The spectrum of the carbonate bands from the same section is shown in Fig. 4.2; for the purposes of comparison, the perpendicular component of the 2421 cm^{-1} band had an optical density of 0.113.

There do not seem to have been any attempts to estimate the absolute quantity of CO_2 present in CO_3Aps from the optical density of the 2340 cm^{-1} band. The v_2 CO_2 bending mode at ~667 cm^{-1} has been looked for, but could not be detected [94,534].

The CO_2 band at 2340 cm^{-1} showed little dichroism in the polarised IR spectrum of an oriented sample (made by heating a 250 μm thick longitudinal section of enamel in CO_2 at 950 °C for 30 min) [77,534]. From this, it was concluded [534] that it was very unlikely that the linear CO_2 molecules were located in the c-axis channels as others [83,427] had proposed. Two suggestions were made [534] to explain the absence of dichroism: either the CO_2 molecules had an orientation of ~56° to the c-axis, possibly because they originated from a CO_3^{2-} ion which occupied the sloping face of a PO_4^{3-} ion site, or the molecules were randomly oriented (α = 54.7° when R = 1 in Equation 4.14), possibly in pores within the structure. The CO_2 was not lost under high vacuum or when the sample was kept in boiling water for a week [534]. A vacant PO_4^{3-} site has also been proposed as the location [721] (Section 4.6.4).

Various mechanisms for CO_2 production have been discussed [94,534]. These included formation from HCO_3^- ions, as proposed earlier [923,941] to explain the loss of CO_2 at relatively low temperatures (300 °C) from some poorly crystallised CO_3Aps made in aqueous systems. Another suggestion

[94,534] was the reaction between CO_3^{2-} and HPO_4^{2-} previously put forward [355] (Formula 3.17, Section 3.6.2) as one possible cause of the low pyrophosphate yield from Ca-def OHAps in the presence of CO_3^{2-} ions. CO_2 has been proposed as an intermediary in the thermally induced relocation of CO_3^{2-} ions from PO_4^{3-} to OH⁻ ion sites [721] (Section 4.6.4).

Formation and loss of cyanate (NCO⁻) and cyanamide (NCN²⁻) ions

IR bands at 2200 and 2012 cm⁻¹, as seen in some CO_3Aps after heating at temperatures up to 900 °C, were first, but incorrectly, assigned to CO_2 [77] and later, again incorrectly, to PO_4 overtone bands [83]. Subsequently, it was shown that the synthetic CO_3Aps, prepared from ammonium salts, contained small amounts of nitrogen (typically 0.1 wt %) [94,434]. This led to the assignment of the 2200 and 2012 cm⁻¹ bands to v_3 of the cyanate (NCO⁻) and cyanamide (NCN²⁻) linear ions respectively, formed by reaction between retained NH_4^+ ions and carbonate [94,434]. This assignment was confirmed by observing the expected IR shifts for ¹³C and ¹⁵N compounds [94,434]. The NCO⁻ and NCN²⁻ bands were observed in dentine, bone and shark vertebra, but not anorganic shark vertebra; only the NCO⁻ band was seen in enamel [94,435]. In these cases, the nitrogen presumably originated from protein. A very weak v_2 NCN²⁻ band was seen at 697 cm⁻¹ in dentine heated in vacuum at 1000 °C and at 699 cm⁻¹ in a synthetic CO_3Ap heated in NH_3 at 700 °C [94,435]. Spectra of samples, after heating for 4 h in air at fixed temperatures, showed that NCO⁻ and NCN²⁻ ions generally were first observed at ~350 °C and no longer at ~800 °C [94,435]. For one synthetic sample, the frequency of v_3 for NCO⁻ increased slightly as the temperature of treatment increased, and shoulders were seen at 2172 and 2179 cm⁻¹. Small shifts in the frequency of v_3 for NCN²⁻ were also observed.

The v_3 NCO⁻, but not the v_3 NCN²⁻, band has been seen in "direct" B-CO_3FAps precipitated at 100 °C from solutions rich in F⁻ ions and then heated to 500 °C [99]. Neither of these bands were seen in sodium-containing B-CO_3Aps (fluoride-free). This was consistent with the displacement of NH_4^+ ions by Na^+ ions, as indicated by the low nitrogen content [99] (Section 4.5.4). On heating "direct" B-CO_3FAps precipitated in the presence of NH_4^+ ions and absence of Na^+ ions, nitrogen is evolved between 500 and 700 °C, which was presumed to come from the thermal decomposition of the NCO⁻ ions [102,937].

The polarised IR spectrum of enamel heated in CO_2 at 950 °C showed that the NCO⁻ ion was highly oriented in the c-axis direction, which led to the conclusion that the ion was located in the c-axis channels [94,435]. This location is analogous to that of the linear BO_2^- ion in boron-containing apatites [436,437] (Sections 2.3.1 and 4.1.2). The BO_2^- ion is coincident with the c-axis with the boron atom at 0,0,½ and oxygen atoms at 0,0,0.315 and 0,0,0.685 [437] so that the negatively charged oxygen atoms point towards the centres of the positively charged Ca(2) triangles at z = ¼ and ¾. Because of the absence of a centre of symmetry in the NCO⁻ ion, the carbon atom must be slightly displaced from 0,0,½ along the c-axis. No corresponding polarised IR data were available for the NCN²⁻ ion, but a similar position was proposed [94,435]. This suggestion was consistent with the demonstration [453,454] of the formation of impure cyanamide apatite, $Ca_{10}(PO_4)_6CN_2$, on heating A-CO₃Ap in NH₃ at 600 to 900 °C (the pure compound with lattice parameters of $a = 9.47_0$ and $c = 6.87_7$ Å could be prepared by heating $CaCN_2$ and OHAp at 900 °C in vacuum). Because of the centre of symmetry of the NCN²⁻ ion, it is very likely that the carbon atom is exactly at 0,0,½.

The formation of compounds containing NCO⁻ and NCN²⁻ ions is to be expected on heating calcium compounds in the NH_3-CO_2-H_2O system [942]. Possible reactions in this system include the oxidation of NCN²⁻ to NCO⁻ ions and the direct formation of NCO⁻ ions from NH₃ and CO_3^{2-} ions [435]. The suggestion has been made that carbamate ions ($NH_2CO_2^-$) are not involved in the formation of NCO⁻ ions [102,937]. The evidence for this came from study of a "direct" B-CO₃FAp prepared in the presence of Na⁺ ions and ammonium carbonate that appeared to contain carbamate but not ammonium ions (see earlier subsection on the loss of carbon dioxide and nitrogen). This B-CO₃FAp did not form NCO⁻ or NCN²⁻ ions on heating, despite the apparent presence of nitrogen in the carbamate.

A small decrease of ~0.01 Å in the a-axis parameter and a weight loss not accounted for by the loss of carbonate in certain CO₃Aps heated in air between 500 and 700 °C has been attributed to loss of nitrogenous species by oxidation [625]. The CO₃Aps that showed these effects were well-carbonated "direct" apatites precipitated in the presence of added ammonia and with the highest nitrogen contents (typically 3.00 wt % C and 0.30 wt % N after drying at 400 °C).

Evolution of hydrogen and carbon monoxide

These two gases have been detected between 650 and 800 °C by gas chromatography when B-CO₃FAps or CO₃Aps precipitated from solutions containing acetate ions were heated [672] (Section 3.4.4). It was suggested these gases came from acetate ions incorporated into the apatite structure.

4.5.8 Other physical and chemical studies
EXAFS studies

The EXAFS spectra of the Ca K adsorption edge in CO₃Aps have been measured over the range 4.0 to 4.4 keV at a resolution of 3 to 5 eV [943]. The samples were prepared at 100 °C and pH 10 from ammonium phosphate and calcium nitrate solutions. The a-axis parameter of the CO₃Aps became shorter with an increase in the carbonate content (details not published) and the IR absorption bands were at 1460, 1410 and 869 cm^{-1} which indicated that PO$_4^{3-}$ ions had been replaced by CO$_3^{2-}$ ions. This substitution of carbonate into the structure caused a marked change in the EXAFS spectrum compared with OHAp. As the carbonate content increased, a small peak at 90 eV above the absorption edge weakened, and peaks at 170 and 190 eV broadened, and finally merged for a CO$_3$ content of 6 wt %. Fourier transforms of the spectra showed that the coordination of the Ca^{2+} ions by the nearest neighbour oxygen atoms was not detectably affected, but marked changes in the transform occurred beyond 0.3 nm. The interpretation given to this was that the oxygen atoms of the planar CO$_3^{2-}$ ion occupied three of the four vacant oxygen sites left by a PO$_4^{3-}$ ion, and that the fourth oxygen site was directed away from the Ca^{2+} ion [943].

The interpretation of these EXAFS results can be extended. Per unit cell, the four Ca(1) atoms are nine-fold coordinated (three O(1), three O(2) and three O(3)) and the six Ca(2) atoms seven-fold coordinated (O(1), O(2), four O(3) and OH) by oxygen atoms (Figs 2.2 and 2.4, Section 2.2.1). As both the Ca(1) and Ca(2) atoms are coordinated by every type of oxygen atom, if the oxygen coordination to calcium is not to change when a CO$_3^{2-}$ replaces a PO$_4^{3-}$ ion, ideally, every calcium site adjacent to a vacant oxygen site must itself be vacant. This condition can never be precisely met because, whichever oxygen site is left vacant, both Ca(1) and Ca(2) sites would have to become vacant, which is incompatible with the maintenance of charge balance. However, the model proposed [83,722] (Formula 4.26, Section 4.5.3) for precipitated sodium-free CO$_3$Aps gives a reasonably close approximation. In this model,

$$Ca_{10-x+u}(PO_4)_{6-x}(CO_3)_x(OH)_{2-x+2u} \qquad 4.35$$

was proposed for the unit cell contents. Taking $u = 0$ for convenience, each CO$_3^{2-}$ ion occupies the sloping face of a PO$_4$ tetrahedral site leaving O(3) vacant, and with adjacent OH and Ca(2) sites also vacant (there are no vacant Ca(1) sites). Assuming this model, the substitution of a CO$_3^{2-}$ ion will not greatly change the nine-fold coordination of the Ca(1) atoms because most change will be in the three O(3) atoms: these are about 0.4 Å further away from Ca(1) than the other six oxygen atoms (Section 2.2.1), and in any case, only one quarter of the Ca(1) atoms will be affected per CO$_3^{2-}$ ion in the unit cell. The Ca(2) atoms will also not be greatly affected by the substitution of a CO$_3^{2-}$ ion. In the unsubstituted structure, every O(3) atom is coordinated to two Ca(2) atoms (these are in adjacent mirror planes). Per CO$_3^{2-}$ ion in a unit cell, only one of the Ca(2) sites will be occupied, so that only one-sixth of the Ca(2) atoms will be affected. Each missing OH$^-$ ion will affect only two of the Ca^{2+} ions in a Ca(2) triangle as the third Ca(2) site is vacant, so that two-sixths of the Ca(2) atoms per CO$_3^{2-}$ ion in a unit cell will be affected. As each of these changes is in a fraction of the coordinating oxygen atoms and only a fraction of the calcium atoms are affected, the total of the changes in the coordination of the calcium atoms by the nearest neighbour oxygen atoms will not be very noticeable, in agreement with the EXAFS results.

NMR Investigations

Proton decoupled ^{31}P NMR MAS spectra (referenced to 85% H_3PO_4) of B-CO_3Aps containing 3.2 or 14.5 wt % CO_3 have been studied [835]. From the reference given for their preparation (ref. [722]), it is presumed they were sodium-free and precipitated at 75 °C. The apatites with 3.2 and 14.5 wt % CO_3 had resonances with chemical shifts of 2.8 and 3.0 ppm respectively. The line-widths (~5 ppm) were much larger than those seen in carbonate-free apatitic precipitates (0.5 to 1 ppm). There were clearly discernible differences between the Bloch decay and cross-polarisation spectra (this distinguishes between protonated and non-protonated species) for both B-CO_3Aps, showing there were at least two different types of phosphate ions in the structure [835].

^{13}C and ^{1}H NMR MAS spectra of various CO_3Aps have been studied at 79.9 and 317.7 MHz and with rotor speeds of 2 to 3.4 and 11 kHz respectively [894]. All resonances were reported with respect to tetramethylsilane. B-CO_3Ap (an "inverse" preparation precipitated at 80 °C in ammoniacal solution and effectively sodium-free, and dried at 70 °C) had an asymmetric ^{13}C line at 170.2 ppm, indicative of several environments, and a very small peak at 166.5 ppm. An AB-CO_3Ap prepared under less basic conditions had a strong ^{13}C line at 170.2 ppm (assigned to CO_3^{2-} replacing PO_4^{3-} ions) and weaker lines at 168.2 and 166.5 ppm. The 166.5 ppm line was assigned to CO_3^{2-} replacing OH⁻ ions as it matched the line at the same position in pure A-CO_3Ap (Section 4.3.4). The 168.2 ppm line corresponded to the position found for CO_3^{2-} ions in ACa,CO_3P. On heating the CO_3Aps at 400 °C, the lines became better resolved, a characteristic of higher crystallinity; the 170.2 ppm line shifted to 169.7 ppm, and the 168.2 ppm line became weaker. This latter change was taken as evidence that the 168.2 ppm line originated from CO_3^{2-} ions in an unstable environment. Dipolar suppression experiments were undertaken to identify those resonances from ^{13}C nuclei that were close to protons, as these would weaken with respect to the other ^{13}C resonances. Only the line at 168.2 ppm showed evidence of nearby protons. This line also showed evidence of associated protons in AB-CO_3Aps prepared by heating ACa,CO_3P at 520 °C. The only source of protons seemed to be OH⁻ ions: other protonated species would have been lost in the thermal treatment. However, this idea was inconsistent with the observation that the 168.2 ppm line was strongest in CO_3Aps prepared in such a manner that their OH⁻ ion content would be minimal. It was therefore proposed [894] that the protons associated with CO_3^{2-} ions came from water adsorbed after heating. This idea was supported by the observation that the ^{1}H NMR signal from water was strongest in those samples that also gave a strong 168.2 ppm signal. This led to the suggestion that these thermally unstable CO_3^{2-} ions were located on the surfaces, or in perturbed regions, of the crystals. The ^{13}C spectrum of the AB-CO_3Ap recrystallised at 520 °C had shoulders on the 169.7 and 168.4 ppm peaks which showed that there must be a total of at least five different carbonate environments. A similar complexity has been noted earlier (Section 4.5.2) in the IR spectra of precipitated AB-CO_3Aps. No evidence was found from the ^{13}C or ^{1}H NMR studies for water or OH$^-$ ions occupying the fourth oxygen site when a CO_3^{2-} ion replaces a PO_4^{3-} ion in the lattice: this proposal had been made earlier in the case of the OH$^-$ ion [82,99,892,893] (Section 4.5.3).

Electron Microscopy

High resolution electron microscopy [944] of sodium-containing precipitated CO_3Aps show that they have planar defects parallel to (100) that correspond to the "central dark line" seen in biological apatites (Section 4.6.6).

Solubility Product

There have been surprisingly few equilibrium solubility studies of model systems of CO_3Aps, considering the importance of the solubility product in precipitation and dissolution processes of mineral and biological CO_3Aps. The solubility product of enamel, dentine and OHAp under constant partial pressures of carbon dioxide is discussed in Section 4.6.5. In one study of precipitated CO_3Aps [751], B-CO_3F,OHAps were prepared under constant composition conditions at fixed pH values in the range 5 to 8 by the simultaneous addition of Ca(NO$_3$)$_2$ (~0.2 mol l^{-1}) and phosphate solutions (K$_2$HPO$_4$ ~0.12 mol l^{-1} plus KF 0.048 mol l^{-1}) at ~8 ml per hour into ~1.8 l of a KNO$_3$ solution (0.1 mol l^{-1}) at 70 °C with 10 % CO$_2$ plus 90 % N$_2$ or 100 % N$_2$ bubbled through. The pH was kept constant by controlled addition of alkali (KOH 1 mol l^{-1}). After precipitation of ~3 g of CO_3Ap, the addition of the reactants was stopped and the pH was maintained constant for a further 24 h. The unit cell contents were estimated from an analysis of the solid assuming the sum of the negative charges per unit cell was 20. This gave up to 1.45 CO_3^{2-} ions and from 2.02 to 2.59 F$^-$ ions per unit cell. Powder XRD showed the apatites to be well-crystallised, with peak positions that changed in the same way as mineral francolites with an increase in the carbonate content. The values of log$_{10}$(ion activity product), pI, were calculated for each sample from the unit cell contents and the composition of the solution in equilibrium with it (either the final solution in the preparation or a solution of similar composition equilibrated with the separated crystals). The pI values from the final solution were ~2.3 units lower than from the separately equilibrated solution. There was a linear fall of pI with an increase in the carbonate content that followed the equation,

$$pI = 123.7 - 11.0x \qquad\qquad 4.36$$

where x is the number of CO_3^{2-} ions per unit cell. (This equation is derived from the published graph [751] of the straight line that was fitted by linear regression to all the results.) This showed that B-CO_3FAps were very much more soluble than FAp so that, for example, it was concluded [751] that the seawater phosphate concentration in equilibrium with francolite with 1.4 carbonate ions per unit cell is 90 to 100 times greater than would be in equilibrium with pure FAp. From this, it was suggested [751] that francolite was the thermodynamically stable apatite phase in high carbonate solutions, such as seawater.

Rate of Dissolution and Reaction in Solution

Many of the theoretical and experimental studies of the dissolution of OHAp (Section 3.10.3) are directly applicable to CO_3Aps. The dissolution and solution reactions of biological CO_3Aps are discussed in Section 4.6.5.

The "solubility", as indicated by the calcium concentration in solution, of CO_3Aps precipitated at 40, 60 and 80 °C with a variety of carbonate contents has been studied by equilibrating 50 mg of the apatite in 2 ml of acetate buffer at pH 4.0, 6.0 and 8.0 for one month [945]. The calcium concentration increased markedly as the carbonate content increased from 0 to 9 wt % CO_3 for the 40 and 60 °C preparations, and from 0 to 3 wt % for the 80 °C preparation. Similar dissolution experiments [796] at pH 4.0 and 6.0 in acetate buffers at 37 °C with CO_3F,OHAps precipitated at 80 °C showed that the "solubility" decreased with an increasing F⁻ ion content and approached that of F,OHAps at high F⁻ ion contents (1.9 mmol g⁻¹).

Studies [793] of the initial rates (within the first 10 min) of the release of Ca^{2+} ions, in rotating disc experiments at 50 °C, from precipitated and high temperature OHAps and CO_3Aps in acetate buffer (0.01 mol l⁻¹) at pH 5.0 showed that the CO_3Aps were more reactive than the respective OHAps. Structurally incorporated F⁻ ions at a level of 1,000 μg g⁻¹ in the precipitated CO_3Ap (note that this is much lower than those discussed in the previous paragraph) did not affect the initial rate, but F⁻ ions at a concentration of 1 μg ml⁻¹ in the buffer reduced the rate by 20 to 30 %. The explanation given for this was that the F⁻ ion concentration at the crystal surface would be much higher when it originated from the solution than the solid. After 10 min, diffusion processes within the pellet were found to influence the reaction rate [793]. In later similar studies of dissolution at pH 4.5 of CO_3Aps precipitated at 37, 60 and 85 °C and pH values from 6 to 9.5 [792], the initial rate of release of calcium decreased markedly with an increase in the temperature of preparation after correcting the rates for the different carbonate contents, particularly the low carbonate content of the 37 °C preparation. This decrease in dissolution rate was attributed to the increase in particle size (from XRD line broadening and electron microscopy) that was seen with an increase in the temperature of preparation. Thus it seemed that, at least for these precipitated CO_3Aps, the reactivity was mainly dependent on the particle size of the precipitates, rather than directly on the presence of carbonate in the structure [792].

Just as Ca-def OHAps change towards s-OHAp in composition and improve their crystallinity in hot or boiling water (Section 3.3.3), precipitated CO_3Aps lose CO_2 and become better crystallised. For example, a CO_3Ap precipitated at 60 °C and containing 9 wt % CO_3, when incubated at 37 °C in water for one month, lost ~2 wt % CO_3 and showed a significant sharpening of the 002 reflection [946]. Changes in carbonate content on refluxing a precipitated CO_3Ap for 120 h, when extrapolated to zero carbonate content, showed that it would essentially all be lost in a year [947]. The preferential loss of carbonate over phosphate on the dissolution of biological CO_3Aps in acids has been reported on several occasions (Section 4.6.5).

A mathematical model has been developed for the early diagenesis of phosphorus and fluorine in marine sediments that includes modelling of the precipitation of carbonate fluorapatite [948].

4.6 Biological apatites
4.6.1 Introduction

Reviews relevant to biological apatites have already been cited (Section 1.1). Typical compositions of the inorganic component of enamel, dentine and bone are given in Table 4.7; as discussed in the following paragraphs, significant departures from these can occur. Features to be noted are: (1) the CO_2 contents of dentine and bone are similar and significantly higher than that of enamel; (2) the magnesium contents of bone and dentine are higher than that of enamel, but there is almost twice as much magnesium in dentine as in bone; and (3) the Ca/P molar ratio of enamel is significantly lower than the ratio for OHAp (1.667) whilst the Ca/P ratio for dentine and bone is approximately equal to the ratio for OHAp (and can even be higher).

As would be expected for a biological material, the composition of the inorganic component of teeth and bones is rather variable, depending on the part of the tissue sampled. There are often also changes during formation and maturation, and with age and disease. The composition can also be species dependent. Changes in the mineral during the formation of calcified tissues that can be observed by IR spectroscopy are discussed in Section 4.6.3.

Table 4.7 Composition of the inorganic components of enamel, dentine and bone (wt percent).

Component	Enamel[a]	Dentine[b]	Bone[c]
Ca	37.6	40.3	36.6
P	18.3	18.6	17.1
CO_2	3.0	4.8	4.8
Na	0.7	0.1	1.0
K	0.05	0.07	0.07
Mg	0.2	1.1	0.6
Sr	0.03	0.04	0.05
Cl	0.4	0.27	0.1
F	0.01	0.07	0.1
Ca/P molar	1.59	1.67	1.65

[a]Based on Tables 1 and 2 in ref. [949], but expressed as a percentage of the ash content of whole enamel (96 wt %).
[b]As for footnote (a), with an ash content of whole dentine of 73 wt %.
[c]Bovine cortical bone, based on Table 3 in ref. [950], but expressed as a percentage of the mineral fraction (taken as 73 wt % of dry fat-free bone).

In enamel, the F⁻ ion content is sharply peaked at the surface and depends principally on the F⁻ ion content in the drinking water during its formation [949]. The finding that shark enamel is rather similar to FAp is a conspicuous example of species dependence [383], although recent studies [385] have shown that its significant carbonate content gives it an even greater similarity to francolite. The carbonate content of human enamel increases from the surface going inwards [949].

There appears to have been rather little study of the variability in composition of dentine, but changes for bone have been investigated. Density fractionation of rat bone particles (<5 μm) gave a range of densities from 1.65 to 2.25 g cm⁻³ whose Ca/P molar ratios increased from a lower limit between 1.33 and 1.41 to an upper limit of 1.67 respectively [950]. The proportion of high density particles increased with the age of the rat. This increase correlated with a reduction in the acid phosphate content as demonstrated by a fall from 11 to 2.7 % in the fraction of the phosphorus present as pyrophosphate on heating unfractionated bones from newborn and adult rats respectively at 300 °C for 15 h [950]. (See Section 3.6.2 for a discussion of this method of acid phosphate determination and of the doubts about its validity when carbonate is present, as in biological apatites.) An inverse relation between the acid phosphate (pyrophosphate method) and carbonate contents has been seen in chick bone from early embryonic mineral to fully matured chickens which was consistent with the authors' view that CO_3^{2-} ions substituted for HPO_4^{2-} ions during the synthesis of bone apatite [951]. However, during the maturation of pig enamel, the carbonate content has been found to remain constant, whilst the acid phosphate content rose [704]. Other studies of the formation of pyrophosphate on heating mineralised tissues are given below.

Another rather interesting example of the heterogeneity of bone mineral is the finding of calcite as a second crystalline phase in the medullary bone of laying hens [952] (Section 4.6.2). This observation shows that the concept of $CaCO_3$ as a separate phase in CO_3Aps (Section 4.1.1) may, in some circumstances, have some validity. On the other hand, a separate ACP phase is not now thought to exist in bone (see discussion below on biomineralisation).

The F⁻ ion content of bone increases with age. In one study [953], the rate of accumulation of F⁻ ion in human iliac crest bone was found to be 26 μg g⁻¹ of bone ash per year for individuals in London drinking nonfluoridated water so that, for example, the value for a 99 year old individual was 0.34 wt %. It was estimated that this rate of accumulation required a total daily F⁻ ion intake of 0.87 mg. An F⁻ ion content of 2.0 wt % has been reported in the bone of cows grazing on grass contaminated with F⁻ ions [386] (Section 4.6.2). Strontium can also accumulate in bone (and presumably dentine and enamel) mineral to a substantial extent if an animal is fed a strontium-rich diet [954] (Section 4.6.2). The fall in the CO_2 and calcium content of bone that occurs in long-standing uraemia and acidosis is an example of a change in composition associated with disease [955].

Unlike synthetic precipitated CO_3Aps [83,88,99,722] (Section 4.5.1), enamel, dentine and bone mineral yield pyrophosphate on heating [875] through the condensation of HPO_4^{2-} ions with the loss of water (Equation 3.15, Section 3.6.2). The temperature for the maximum yield of pyrophosphate on heating in air for 1 h was 325 °C for dentine and bone (4.4 % of the phosphorus as pyrophosphate) and 600 °C for enamel (2.2 % of the phosphorus as pyrophosphate). Two other studies of pyrophosphate formation on heating bone were given above. A value of ~7 % for the phosphorus as pyrophosphate has been reported for human enamel after heating, but none could be detected in carious enamel, although there was evidence of HPO_4^{2-} ions from IR spectroscopy [697]. The HPO_4^{2-} content of the basal opaque region of nascent rat incisor enamel has been found to fall with time [695]. $NaH_2^{32}PO_4$ was injected for varying times before sacrifice and the enamel removed and pyrolysed at 500 °C for 48 h. The ortho- and pyrophosphate were then separated by anion-exchange chromatography and the fractions assayed by scintillation counting (Section 3.6.2). In these determinations of pyrophosphate after heating, it should be again emphasised that the estimate is likely to be low because of interference from CO_3^{2-} ions [355] (Section 3.6.2).

Biomineralisation is a complex process involving subtle interactions between inorganic ions, crystals and organic molecules [956]. Apatitic systems are particularly complicated by the number of calcium phosphates that occur in the phase diagram (Fig. 1.1). It has been proposed that biological apatites are formed via an OCP intermediate [107,957] which is well-established as one of the modes for the formation for OHAp in model aqueous systems (Section 3.8). The most compelling evidence for the participation of OCP in biomineralisation comes from the morphology of the apatite crystals (Section 4.6.6). In enamel, the first formed crystals are lath-like, and do not have a hexagonal cross-section as expected for a hexagonal apatite phase. Shark enamel crystals, which have hexagonal cross-sections [383] (Section 4.6.6), are an exception to this but, as mentioned earlier, they are essentially FAp and would be expected to grow as F,OHAp crystals rather than via OCP [738] (Section 3.8) (see also Section 1.3.5 for the effect of F⁻ ions on OCP hydrolysis). An alternative explanation [400], that the crystals have the monoclinic space group, is very improbable as they would have to have a high degree of stoichiometry. The presence of a thin central core of OCP as a result of the crystal growth process also provides one possible explanation for the central dark line seen in enamel crystals [944] (Section 4.6.6). Although the morphology of the apatite crystals in bone seems rather uncertain (Section 4.6.6), it is likely that at least some are plate-like crystals with the *c*-axis in the plane of the plate. Again, this morphology can be understood if the crystals grow via an OCP intermediate. OCP has been reported in an X-ray diffraction study [145] (Table 4.8, Section 4.6.2) of bone from a two-day old rabbit, which if found to be a normal occurrence, would confirm the role of OCP as a precursor in biomineralisation.

DCPD has also been suggested as an intermediate in bone mineralisation [120]. It was reported in the same X-ray diffraction study of two-day old rabbit bone mentioned above in which OCP was seen [145] (Table 4.8, Section 4.6.2). However, it was not possible to detect crystalline DCPD in embryonic bovine and chick bone [208,209] (Sections 1.4.1 and 4.6.2). Despite this, ^{31}P NMR studies have shown that the HPO_4^{2-} ions in embryonic bone have a brushite-like configuration, possibly located on the surface of the apatite crystals [209] (Section 4.6.6).

ACP can also be an intermediate in the formation of OHAp (Section 1.8.4) and might therefore have this role in biological apatites. At one time it was thought to form a significant component of bone mineral [325], particularly in young animals, but this now does not seem to be so [209,326] (Sections 1.8.1 and 4.6.2). Conceptually, ACP would seem to be an unlikely intermediate because its lack of three-dimensional order would make the required control of the nucleation process difficult during the formation of calcified tissues.

In summary, there is strong circumstantial evidence for the participation of OCP in the nucleation of biological apatites, but it seems improbable, on the present evidence, that ACP or crystalline DCPD have a role.

There is no separate section on the structure of the biological apatites as most aspects of these have been dealt with elsewhere. Much of the following discussion is devoted to the inorganic component of dental enamel. The reasons for this are that enamel mineral is better crystallised than the other biological apatites, so more detailed XRD detail is available. Furthermore, the good orientation of the crystals in this tissue allows the use of polarised IR spectroscopy which yields much more information than spectra of powders. However, the presumption must be that the mineral in dentine, cementum and bone has similarities to the mineral of enamel, although effects attributable to the surface adsorption of ions will be very much more marked because of the much smaller crystal size in these tissues.

4.6.2 X-ray diffraction studies

The accurate measurement of the lattice parameters of enamel presents no real problems. They have been measured with a 190 mm diameter Debye-Scherrer camera using chromium radiation [648] (see below), but diffractometer methods are to be preferred if there is sufficient sample. Although lattice parameters can be calculated from the powder XRD pattern by standard least squares fitting of calculated to observed peak positions, the Rietveld whole pattern fitting method (Section 1.1 and below) that involves simultaneous structure and lattice parameter refinements probably gives more accurate results (compare quoted errors in the following paragraph), particularly for comparison purposes. Unless the mineral in bone and dentine has a high F^- ion content, it is poorly crystallised which makes accurate measurement difficult. One solution to this problem has been to fit a Fourier series to step-scan diffractometer data for each peak and to determine the peak position from this functional representation [958]. Although the Rietveld method does not appear to have been used for bone and dentine, it might provide an attractive alternative.

Dental enamel

Lattice parameters of $a = 9.440 \pm 0.005$ and $c = 6.881 \pm 0.005$ Å for human enamel *versus* $a = 9.421 \pm 0.001$ and $c = 6.882 \pm 0.001$ Å for OHAp ("prepared from very dilute solutions . . . dimensions up to 100 μm") have been reported from measurements of powder photographs [648]). Diffractometer measurements gave $a = 9.441 \pm 0.006$ and $c = 6.884 \pm 0.006$ Å for human enamel *versus* $a = 9.421 \pm 0.003$ and $c = 6.881 \pm 0.003$ Å for OHAp (prepared by repeated autoclaving β-TCP and $Ca(OH)_2$ at 360 °C and 200 atm (20 MPa), followed by ignition at 900 °C) [718]. Lattice parameters from a Rietveld structure refinement [959] (see below) of the dense fraction (density >2.95 g cm^{-3}) of human enamel gave $a = 9.441(2)$ and $c = 6.878(1)$ Å.

The results in the previous paragraph clearly show that the *a*-axis parameter of dental enamel is ~0.02 Å greater than in OHAp. The various lattice substitutions proposed to explain this include: (1) an expansion due to 5 to 10% of the CO_3^{2-} ions replacing OH$^-$ ions in the lattice [77,866] (Section 4.5.2); (2)

a contraction of 0.02 Å due to all the CO_3^{2-} replacing PO_4^{3-} ions with the now greater expansion to be explained caused by HPO_4^{2-} replacing PO_4^{3-} ions [960]; (3) a 0.0096 Å expansion from Cl^- replacing OH^- ions with the remaining required expansion, for example, from HPO_4^{2-} replacing PO_4^{3-} ions and water replacing OH^- ions [668]; and (4) "disorder in some of its constituents and because of the presence of structural water" [721].

The expected increase, from OHAp, of the *a*-axis parameter of enamel attributable to the CO_3^{2-} and Cl^- ions can be calculated making the following assumptions: (1) the composition of enamel is as given in Table 4.7 with 11 % of the CO_3^{2-} replacing OH^- ions [703] (Section 4.6.3) and the balance replacing PO_4^{3-} ions; (2) the changes can be calculated from model compounds and are additive, without synergistic effects; (3) the changes for Cl^- and CO_3^{2-} ions replacing OH^- ions can be calculated by linear interpolation between the *a*-axis parameters for the appropriate end-members as given in Table 1.4 (Section 1.1); and (4) the change in the *a*-axis parameter per wt % CO_3 replacing phosphate is -0.006 Å [872] (Formula 6, Table 4.6, Section 4.5.3). This calculation gives expansions of +0.0123, +0.0108 and -0.0218 Å for the Cl^- for OH^- ion, CO_3^{2-} for OH^- ion, and CO_3^{2-} for PO_4^{3-} ion substitutions respectively. Together, these give an expected expansion of +0.0013 Å, which compares with the observed expansion of +0.023 Å, taking the *a*-axis parameter for OHAp as 9.418 Å (Table 1.4). The difference between these leaves an expansion of 0.022 Å as yet unaccounted for. The corresponding figure for the CO_3Aps prepared in aqueous systems is given by subtracting 9.418 from the constant term in the formulae for the *a*-axis parameter applicable to small carbonate contents (Table 4.6, Section 4.5.3); this difference ranges from 0.009 to 0.054 Å. Many precipitated ns-OHAps also have an *a*-axis parameter which is from 0.015 to 0.03 Å greater than OHAp (Section 3.4.1). This expansion has been attributed to HPO_4^{2-} ions and/or water in the lattice (Section 3.7.3). It seems very probable that the unaccounted increase of 0.022 Å in the *a*-axis parameter of enamel and the expansion of from 0.009 to 0.054 Å in synthetic CO_3Aps have the same origin. Thus the difference in the *a*-axis parameter between enamel and OHAp seems mainly due to HPO_4^{2-} and/or water, whilst the effects of the lattice substitutions of Cl^- and CO_3^{2-} ions almost cancel each other out. This interpretation is strengthened by the observation of a sudden decrease in the *a*-axis parameter of ~0.014 Å on heating enamel from 250 to 300 °C which has been attributed to the loss of water and HPO_4^{2-} ions [721] (Section 3.7.3).

The structure of the dense fraction of human dental enamel (density >2.95 g cm^{-3}) has been studied by Rietveld whole-pattern fitting structure refinement (Section 1.1) using neutron and X-ray diffraction data [959]. (The refined lattice parameters were given above.) Not surprisingly, the refined structure was very similar to that of OHAp, but the observed differences from OHAp in the distribution of the scattering density along the *c*-axis, including differences between the X-ray and neutron estimations, would accommodate water in some form of orientational disorder centred between $z = 0.11$ to 0.14, as well as the ~0.3 wt % chlorine that enamel contains. No unambiguous evidence was found for the positions of the CO_3^{2-} or HPO_4^{2-} ions.

The high angle XRD patterns of human dental enamel showed no measurable change in lattice parameters as the F^- ion content increased from 70 to 670 ppm, but the 002, 211, 300 and 202 peaks became slightly sharper [961]. If inhomogeneous strain effects were ignored, the crystal sizes for the low and high fluoride enamels, calculated from the 002, 300 and 211 peaks, were 1420 ± 100 and 2740 ± 240 Å, 780 ± 25 and 1030 ± 25 Å, and 780 ± 20 and 1000 ± 20 Å respectively [961]. Lattice parameters $a = 9.385\pm0.005$ and $c = 6.885\pm0.005$ Å for shark enamel containing 3.3 wt % F and 0.55 wt % CO_2 were attributed [383] to the nearly complete replacement of OH by F^- ions (for FAp, $a = 9.367(1)$ and $c = 6.884(1)$ Å, Table 1.4). Shark dentine had a reduced *c*-axis parameter ($a = 9.42\pm0.03$, $c = 6.84\pm0.01$ Å) which was not clearly understood, but thought possibly due to a very low Ca/P molar ratio (~1.5) and high magnesium content (~3 wt % on an ash basis) [383]. (See Section 2.3.2 for a discussion of Mg^{2+} substituting for Ca^{2+} ions in apatite.) *Lingula* shell apatite also has a short *c*-axis parameter, low Ca/P ratio and high magnesium content [962] (Section 4.6.4).

Diffractometer measurements (Cu Kα radiation) have been made of line profiles from blocks of hippopotamus enamel which has a particularly good preferred orientation of the apatite crystals (*c*-axis in the direction of the enamel rods). Measurements were made with the apatite *c*-axes either predominantly parallel or perpendicular to the diffraction vector [963]. The crystal size in the *a*-axis direction, calculated from the Scherrer formula (this neglects line broadening from inhomogeneous strain), was independent of the order of diffraction and was 410 Å (approximate precision limits 370 and 450 Å). On the other hand, the sizes in the *c*-axis direction calculated from the 002, 004, 006 and 008 lines were 741 ± 100, 466 ± 40, 361 ± 35 and 295 ± 25 Å respectively. A function, $\beta_{1/2} = k_1/\cos\theta + k_2\tan\theta$, where θ is the Bragg angle and $\beta_{1/2}$ is the half-breadth at half-height in degrees 2θ, was fitted to the results and gave $k_1 = 0.05$ and $k_2 = 0.26$. The first term derives from the Scherrer formula for crystal size and the second term was to account for inhomogeneous strain. The function gave a crystal size, corrected for inhomogeneous strain, in the *c*-axis direction of 1600 Å (approximate precision limits 1040 and 2700 Å).

Bone

The *a*-axis parameter of the apatite mineral from bones with different F^- ion contents, from cows grazing on grass contaminated with F^- ions, has been studied [386]. The parameter changed from 9.416 ± 0.003 to 9.350 ± 0.003 Å for a change in composition from 0.064 wt % F and 3.635 wt % CO_2 to 2.050 wt % F and 3.07 wt % CO_2 (F expressed in terms of bone mineral and CO_2 in terms of dry bone). The reduction in the *a*-axis parameter was greater than could be attributed to the F⁻ ion content alone (note that the minimum *a*-axis parameter was less than for pure FAp (9.367 Å), even though the F⁻ ion content was only 53 % of that for pure FAp). By analogy with francolite minerals (Section 4.2.1), this result indicated that CO_3^{2-} replaced PO_4^{3-} ions in the apatite lattice [386]. In a study of human iliac crests, ribs and vertebrae from post-mortem samples from individuals where the water supplies contained from 0.1 to 4.0 ppm F⁻, a marked increase in crystallinity with greater F⁻ content was shown by the improved resolution of the 211, 112, 300 and 202 lines in XRD patterns [964,965]. A detailed study of the line profiles indicated that the improvement took place in a direction perpendicular to the *c*-axis [964,965]. On the other hand, no difference was seen in the halfwidths of the 002 and 130 profiles (Cu K radiation, step-scanning with steps of 0.02° and 100 s counting time at each step) between the cortical bone of rats fed a fluoride-supplemented diet and controls [966]. The experimental rats had 6 mmol l⁻¹ fluoride in their drinking water and 16 μg fluoride per mg of calcium in their food for up to six months; at this time, the experimental and control bones contained 0.55 and 0.16 wt % F respectively. Density fractionation studies showed that the fluoride was not homogeneously distributed, but that it accumulated in the denser fractions [966].

An *a*-axis parameter of 9.482±0.007 Å for bone with a Sr/(Ca + Sr) molar ratio of 0.1008 from mice fed a strontium-rich diet has been reported [954]; this compared with 9.4334±0.0016 for controls with a Sr/(Ca + Sr) molar ratio less than 10^{-3}. The expansion of 0.049 Å was attributed to the replacement of Ca^{2+} ions by Sr^{2+} ions in the lattice. This figure is in fair agreement with 0.034 Å, the expansion calculated from the lattice parameters of OHAp and SrOHAp (Table 1.4), assuming Vegard's Law is obeyed.

X-ray powder diffractometer studies with Cu Kα radiation of bone from the distal metaphysis of a two-day old rabbit have shown a number of diffraction peaks that could not be attributed to apatite, but were attributed to OCP and DCPD [145] (Table 4.8). The spacings of the three strongest lines assigned to OCP and DCPD correspond to published values (Table 4.8), so there seems little doubt that the sample contained these phases, but the significance of this observation is not clear until the experiment is repeated and further studies are undertaken; however, both OCP and DCPD have been proposed as precursors of bone mineral, Section 4.6.1.

X-ray radial distribution function studies [326] of 16 day old embryonic chicks and one year old chickens gave no evidence for the presence of significant quantities of ACP, a phase at one time thought to occur in bone [325] (Sections 1.8.1 and 4.6.1). Recent ³¹P NMR (Section 4.6.6) and similar

Table 4.8 Lattice spacings of the strongest lines assigned [145] to OCP and DCPD in the X-ray diffraction pattern of bone from the distal metaphysis of a two-day old rabbit compared with published [16] values.

θ°ᵃ	*d* Å	*d* Å [16]	Intensityᵇ [16]
OCP			
2.4	18.38	18.6	100
4.6	9.60	9.46	40
4.8	9.20	9.07	40
DCPD			
5.9	7.49	7.57	100
10.6	4.18	4.24	100
14.8	3.01	3.05	75

ᵃAssuming that the scale on the published diffractometer trace [145] is in degrees θ, and not degrees 2θ.
ᵇ100 corresponds to the most intense line.

XRD studies of 11 to 17 day old embryonic chick bones showed the absence of any substantial amounts of ACP (upper limit 8 to 10 wt %) and crystalline DCPD (upper limit 1 wt %) [209].

Calcite has been detected in X-ray diffraction patterns (Guinier camera) from pulverised medullary bone samples from one-year old laying hens [952]. Identification was based on the presence of 104, 113, 116 and less distinct 110 reflections. The calcite lines were broadened which made quantitation difficult, but it was estimated that calcite constituted about 10% of the bone.

4.6.3 Infrared and Raman spectroscopy

The IR and Raman assignments of the CO_3 bands for precipitated CO_3Aps, including some consideration of dental enamel, have already been discussed in Section 4.5.2.

Dental enamel

The polarised IR spectrum of a section of human enamel, in which the *c*-axes of the crystals are predominantly oriented in the direction of the enamel rods, is shown in Fig. 4.7. The spectrum of the OH region before and after deuteration is shown in Fig. 4.8. As the OH⁻ ion absorbs maximally when the electric vector is parallel to the O-H bond direction, these spectra show that the bond is parallel to the *c*-axis. This orientation agrees with the polarised IR spectroscopy of francolite (Figs 4.2 and 4.3) and the neutron diffraction [599] study of OHAp (Fig. 3.2).

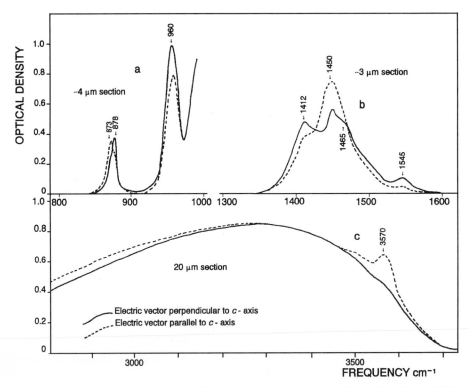

Fig. 4.7 Polarised IR spectrum of human dental enamel. (Fig. 12.4 from Elliott [77])

Partial deuteration of enamel (Fig. 4.8b) reveals an additional band at 3500 cm^{-1} that has been assigned [600] (Section 3.11.5), by comparison with model compounds, to an OH^- ion perturbed by an adjacent Cl^- ion. Single crystal X-ray structure studies of F,Cl,OHAps [429] (Section 2.3.1) suggest that in enamel (which has typically 0.2 to 0.3 wt % Cl) the Cl^- ion should have a z parameter between 0.36 and 0.37 (the z parameter in ClAp is 0.44).

F^- ions adjacent to OH^- ions in F,OHAp also cause additional OH stretch and librational bands (Section 3.11.5). Such weak bands have been seen in the OH stretching and librational regions of enamel with a high F^- ion content [549,827]. Bands with frequencies of ~3545, ~735 and ~715 cm^{-1} were reported in enamel containing 0.3 wt % F [549].

The OH librational mode at 630 cm^{-1}, seen strongly in OHAp (Section 3.11.2), is much weaker in the spectrum of enamel. This is consistent with the reduced librational intensity in Ca-def OHAps (Section 3.11.6), precipitated sodium-free CO_3Aps [83,722] (Section 4.5.3) and sodium-containing CO_3Aps [79,872] (Section 4.5.4). The broad OH stretching band of water from 2800 to

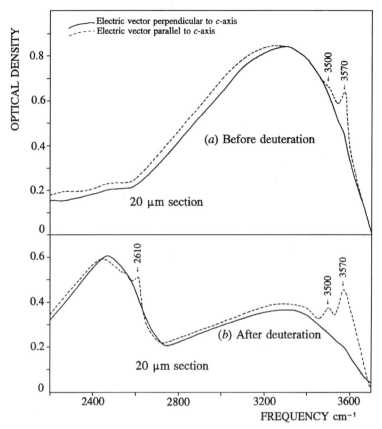

Fig. 4.8 Polarised IR spectrum of 20 μm section of human enamel (a) before and (b) after deuteration in D_2O vapour for 2 h at 80 °C. (Fig. 12.5 from Elliott [77])

3600 cm⁻¹ (Figs 4.7 and 4.8) and water bending mode at 1640 cm⁻¹ (not shown) have no significant dichroism [77]. This suggests that the water in enamel is randomly oriented (the alternative interpretation of orientation of both transition moments with $\alpha = 54°\ 44'$ for $R = 1$ in Equation 4.14 would seem to be very improbable).

ν_1 PO_4 occurs at 960 cm⁻¹ with partial perpendicular dichroism (Fig. 4.7) with a dichroic ratio of 0.786 (JC Elliott, unpublished results, 1963). This ratio is significantly different from 0.43, the value found for Fowey Consols francolite [77] (Section 4.2.2). The triply degenerate ν_3 PO_4 mode at ~1071 cm⁻¹ is much too absorbing to be seen in normally prepared sections. However, treatment of a piece of human dental enamel in buffer, originally at pH 3.5, for three months produced sufficient demineralisation that 10 to 20 μm thick sections could be cut with a sledge microtome (JC Elliott, unpublished results, 1962). Measurements on such a section showed that the v_3 band at 1025 cm^{-1} had partial parallel dichroism with a dichroic ratio of 1.45 and a shoulder at 1077 cm^{-1} had partial perpendicular dichroism. Some recrystallisation probably took place as a result of the acid treatment, hence the preferred orientation might have been reduced. This would have lowered the observed dichroic ratio of the 1025 cm^{-1} band.

HPO$_4$ and PO$_4$ bands in the v_3 PO$_4$ [190] and v_4 PO$_4$ [191] regions in enamel have been investigated by deconvoluting FT IR spectra, the same method these authors used to study Ca-def OHAps (Section 3.11.6). Samples at various stages of enamel maturation from six-month old pigs and mature enamel from two-year old pigs were studied. The position and intensities of the bands in the v_3 region showed considerable variation from sample to sample [190]. Bands at 1020, 1100, 1110, 1125 and 1145 cm^{-1} were seen, which were absent in well-crystallised OHAp. The bands at 1020, 1100 and the weak band at 1145 cm^{-1} were assigned to HPO$_4^{2-}$ ions. The bands at 1110 and 1125 cm^{-1} disappeared progressively during maturation of the enamel. They were assigned to HPO$_4^{2-}$ ions in poorly crystallised regions that were easily solubilised by analogy with synthetic Ca-def OHAps [190] (Section 3.11.6). In the studies of v_4 PO$_4$ [191], two bands at 520 to 530 and 540 to 550 cm^{-1} on the low frequency side of v_4 were assigned to HPO$_4^{2-}$. A third band at 610 to 615 cm^{-1} on the high frequency side was assigned to labile PO$_4^{3-}$ ions unrelated to HPO$_4^{2-}$ ions. During aging of the enamel, the integrated intensity of the HPO$_4$ bands and the intensity of the PO$_4$ band at 610 to 615 cm^{-1} fell; in fully mature enamel, the HPO$_4$ bands were hardly visible [191].

As explained in Section 4.5.2, the assignment [77,866,884] of the bands at 1545, 1450 and 880 cm^{-1} in precipitated CO$_3$Aps and dental enamel to a small fraction of the CO$_3^{2-}$ ions replacing some of the OH$^-$ ions was based on comparison with synthetic compounds and consideration of the dichroism of the v_{3b} band at 1545 cm^{-1} and v_2 band at 880 cm^{-1} (summarised in Table 4.9): both these bands have almost complete perpendicular dichroism in A-CO$_3$Ap (Fig. 4.4, Section 4.3.2) and enamel (Fig. 4.7). Thus the deduction was made that a small fraction of the CO$_3^{2-}$ ions in enamel replace OH$^-$ ions with the plane of the ion nearly parallel to the c-axis, as in A-CO$_3$Ap (Section 4.3.3). Examination of the polarised IR spectrum of A-CO$_3$Ap (Fig. 4.4) shows that the v_{3a} band in enamel, from CO$_3^{2-}$ ions replacing OH$^-$ ions, must be near 1457 cm^{-1} and have parallel dichroism; the only likely candidate for this assignment is the band at 1450 cm^{-1} (Fig. 4.7 and Table 4.9).

Comparison of enamel and francolite. The major fraction of the CO$_3^{2-}$ ions in enamel are thought to replace PO$_4^{3-}$ ions and to be responsible for the bands at 1465, 1412 and 873 cm^{-1} (Section 4.5.2) which have perpendicular, perpendicular and parallel dichroism respectively (Fig. 4.7 and Table 4.9).

These bands should therefore be similar to those in francolite in which this substitution takes place (Section 4.2.2). In francolite, the CO_3 bands are at 1453, 1429 and 865 cm^{-1} with perpendicular, perpendicular and parallel dichroism respectively (Figs 4.2 and 4.3, and Table 4.9). Thus the signs of the dichroic ratios of the enamel and francolite CO_3 bands match, which is consistent with the major fraction of the CO_3^{2-} ions in enamel occupying the sloping faces of vacant PO_4 sites, as in francolite. There are however significant differences in band frequencies between enamel and francolite (Table 4.9). These have been attributed to differences in composition [872] (Section 4.5.2), their different F$^-$ ion contents probably being the most important factor. Similar differences exist between the IR spectra of dahllite and francolite [45] which have also been attributed to the higher F$^-$ ion content of francolite compared with dahllite (Section 4.2.2).

Table 4.9 Assignments of carbonate bands in enamel and comparison with francolite and A-CO_3Ap (see text for references).

Frequency cm^{-1}	Dichroism	Assignment
Enamel:		
1465	$\perp$	v_{3b}
1412	$\perp$	v_{3a} } CO_3^{2-} for PO_4^{3-}
873	$\parallel$	v_2
1545	$\perp$	v_{3b}
1450	$\parallel$	v_{3a} } CO_3^{2-} for OH$^-$
880	$\perp$	v_2
Francolite:		
1453	$\perp$	v_{3b}
1429	$\perp$	v_{3a} } CO_3^{2-} for PO_4^{3-}
865	$\parallel$	v_2
A-CO_3Ap:		
1548	$\perp$	v_{3b}
1457	$\parallel$	v_{3a} } CO_3^{2-} for OH$^-$
879	$\perp$	v_2

There are a number of problems with this simple interpretation of the IR spectrum of enamel and its extension to other precipitated CO_3Aps, as has already been discussed in some detail in Section 4.5.2. The first is that additional bands are often seen in the v_3 CO_3 region of synthetic CO_3Aps prepared in aqueous systems [77] (see Fig. 4.6, Section 4.5.2 for an example). These bands are presumably due to other minor environments, although there can be alternative explanations of additional bands in IR spectra [125]. An example of a possible minor environment can be seen in the polarised spectrum of heated enamel (Fig. 4.4). The weak IR bands of similar intensity and perpendicular dichroism at 1500 and 1410 cm^{-1} might be assigned to v_{3b} and v_{3a} respectively of CO_3^{2-} ions in this minor environment. The corresponding v_2 band must have parallel dichroism in order that the transition moments of the three bands are orthogonal (Section 4.2.2). Although there is only one v_2 band at 879 cm^{-1} (Fig. 4.4), it clearly has a low frequency component with parallel dichroism which is responsible for the lower frequency of the peak position when measured with the electric vector parallel to the *c*-axis. Thus this minor environment has the plane of the CO_3^{2-} ion approximately perpendicular to the *c*-axis, with IR absorption bands at 1500, 1410 and ~870 cm^{-1}. Whether this environment occurs in precipitated CO_3Aps is unknown. Complex CO_3 spectra are also seen in B-CO_3Aps after heating 15 min at 950 °C and 40 kbar (4000 MPa) [95,918] (Section 4.4.2).

The second difficulty with the assignments given in Table 4.9 concerns the relative intensities of the IR bands in synthetic precipitated CO_3Aps [77]. In a series of such CO_3Aps with an increasing fraction of the CO_3^{2-} ions replacing OH⁻ ions, the increase in the intensity of the ~1450 cm^{-1} band was greater than the increase in the intensity of the band at ~1542 cm^{-1}. It was pointed out [77] that the additional absorption of the band at ~1450 cm^{-1}, due to the replacement of OH⁻ ions by CO_3^{2-} ions, should have been comparable with the total absorption of the band at ~1545 cm^{-1}. Thus a third CO_3^{2-} ion environment seemed to be contributing to the ~1450 cm^{-1} band. This difficulty could be seen more clearly from an examination [77] of the polarised IR spectrum of enamel (Fig. 4.7). The strongest CO_3 peak at 1450 cm^{-1} with strong, but not complete parallel dichroism, was assigned to CO_3^{2-} ions replacing OH⁻ ions; the weak shoulder at 1465 cm^{-1}, with perpendicular dichroism, to CO_3^{2-} ions replacing PO_4^{3-} ions (Table 4.9). However, only a small fraction of the CO_3^{2-} ions are thought to replace OH⁻ ions (about 11 percent, see following paragraph), which is inconsistent with these intensities. Furthermore, the band at 1450 cm^{-1} is much more intense than the band at 1545 cm^{-1}, whereas a comparable intensity would have been expected for the two components of the same v_3 mode.

Further insight into the problem has come from two measurements of the fraction of the CO_3^{2-} ions that replaces OH⁻ ions in enamel [703,704] (experimental details in Section 3.6.3). In the first study, the v_{3b} band at 1545 cm^{-1} in enamel was compensated by a lattice parameter matched A-CO_3Ap in the reference beam of the spectrometer [703]. Knowledge of the carbonate contents of the enamel and compensating A-CO_3Ap indicated that about 11±1 % of the carbonate in the enamel replaced OH⁻ ions. However, the 880 cm^{-1} band was not quite compensated [703] and the discrepancy was ascribed to absorption from HPO_4^{2-} ions in the enamel. The compensated spectrum should have been similar to that obtained from precipitated B-CO_3Ap, but the ~1465 cm^{-1} band was more intense than the ~1410 cm^{-1} band, which is the opposite way round to that usually seen in precipitated B-CO_3Aps. This result was explained [703] by postulating, either that there were some CO_3^{2-} ions in a third site in enamel, or that the molar extinction coefficients for the CO_3^{2-} ion depended not only on whether PO_4^{3-} or OH⁻ ions were replaced, but also on the type of compound in which the substitution took place. The second study [704] was based on intensity ratios of the ν_3 CO_3 band at 1540 cm⁻¹ with the ν_4 PO_4 band at 560 cm⁻¹ in deconvoluted FT IR spectra. It was found that 15 % of the CO_3^{2-} ions replaced OH⁻ ions in fully mineralised enamel from two-year old pigs. This amount of CO_3^{2-} ions in the hexad axis could only account for 38±5 % of the intensity of the ν_2 band at 878 cm⁻¹, a result in qualitative agreement with the earlier study [703] (see above). However, instead of ascribing this difference to HPO_4^- ions [703], the additional absorption was assigned [704] to CO_3^{2-} ions on the *c*-axis that were distorted by the presence of CO_3^{2-} replacing PO_4^{3-} ions.

Having discussed the assignments of the ν_3 CO_3 bands in the polarised spectrum of enamel, it is appropriate to consider some slightly different assignments that have been made [99] (Table 4.10) for synthetic precipitated CO_3Aps. These CO_3Aps often have more ν_3 CO_3 bands (Fig. 4.6, Section 4.5.2) than are seen in enamel. The assignment of the 1500 cm⁻¹ band to CO_3^{2-} ions that replaced OH⁻ ions (Table 4.10) was based on the observation of the concurrent loss of bands at 1500 and 1545 cm⁻¹ in a CO_3Ap heated with calcium fluoride at 700 °C. (The implication of this observation was that the F⁻ ions displaced CO_3^{2-} ions from the *c*-axis channels.) Although not stated explicitly, the CO_3Ap used for the experiment was probably a B-CO_3Ap heated at high temperature and pressure in which bands at 1500, 1535 and 1560 cm⁻¹ were assigned to CO_3^{2-} ions in channel sites [95] (Section 4.4.2). The assignment of the 1470 cm⁻¹ band (Table 4.10) was less certain: assignment to CO_3^{2-} ions replacing OH⁻ ions was rejected because a shoulder at this frequency was seen in a sample of FAp containing a little carbonate. On the other hand, the intensity of the 1420 cm⁻¹ band seemed to be correlated simultaneously with the intensities of the bands at 1460 and 1470 cm⁻¹. Thus, the 1420 and 1460 cm⁻¹ bands were assigned to ν_3 from one environment, and the 1420 and 1470 cm⁻¹ bands to ν_3 from another environment (Table 4.10). The assignment of the latter pair of bands to $[CO_3^{2-},F^-]$ was based on the observation that the 1470 cm⁻¹ band fell in intensity in a series of synthetic precipitated B-CO_3OH,FAps with a reducing F⁻ ion content.

The assignments discussed in the previous paragraph (Table 4.10) can now be compared with those for dental enamel (Table 4.9), particularly to see how consistent they are with the added information from the measurements with polarised IR radiation. The assignments of the 1545 and 1420 cm⁻¹ bands to CO_3^{2-} ions replacing OH⁻ and PO_4^{3-} ions respectively agree with the assignments in dental enamel. However, there is no significant band seen at 1500 cm⁻¹, so it seems unlikely that this can be a correct assignment for the

Table 4.10 Proposed assignments of v_3 CO_3 bands in precipitated CO_3Aps [99].

Frequency cm^{-1}	Assignment
1545	CO_3^{2-} for OH$^-$
1500	CO_3^{2-} for OH$^-$
1470	$[CO_3^{2-},\square]$ for PO_4^{3-}
1460	$[CO_3^{2-},X^-]$ X$^-$ = F$^-$ or OH$^-$ for PO_4^{3-}
1420	$[CO_3^{2-},\square]$ and $[CO_3^{2-},X^-]$ for PO_4^{3-}

majority of the CO_3^{2-} ions that replace OH$^-$ ions in enamel. In any case, it seems at too high a frequency (1458±4 cm^{-1} is found in A-CO$_3$Ap, Section 4.3.2). (A band at 1500 cm^{-1} has been assigned to a minor distorted A-type CO_3 environment [95], see above.) The 1470 and 1460 cm^{-1} bands (Table 4.10) might correspond to the bands at 1465 and 1450 cm^{-1} seen in dental enamel, in which case, they would be both assigned to CO_3^{2-} ions in PO_4^{3-} ion sites. Indeed, the 1465 cm^{-1} band was given this assignment in enamel (Table 4.9). However, the nearly complete parallel dichroism of the 1450 cm^{-1} band requires that the associated v_2 mode should have a perpendicular dichroism; this is contrary to that observed for the 873 cm^{-1} band which is generally accepted as originating from CO_3^{2-} ions in PO_4^{3-} ion sites. The conclusion from this discussion is that, although there are difficulties with the assignments in Table 4.9, the difficulties seem to be even greater for those in Table 4.10. It follows that it is probably too early to make fine distinctions between $[CO_3^{2-},\square]$ and $[CO_3^{2-},OH^-]$ environments in the assignment of components of the v_3 spectrum, and likewise for $[CO_3^{2-},\square]$ and $[CO_3^{2-},F^-]$.

Changes during maturation. Changes in intensity of the v_2 CO_3 band at 866 cm^{-1} due to CO_3^{2-} ions assigned to a labile environment [699] (Section 4.5.2) have been studied [704] in pig enamel of increasing age and maturity. Whilst there was a progressive increase in the amount of mineral, the concentration of the labile CO_3^{2-} ions decreased. A similar fall was seen in model systems during the maturation of precipitated CO_3Aps [699] (Section 4.5.2). The distribution of the CO_3^{2-} ions between the different apatitic sites, as deduced from a detailed analysis of the deconvoluted spectra, appeared to vary randomly during the formation and maturation of the mineral phase in a way that related to changes in the composition [704]. However, parameters assessing the degree of crystallinity based on v_2 CO_3 and v_4 PO_4 IR data revealed a significant discrepancy which was thought to be related to an inhomogeneous partition of the CO_3^{2-} ions in the mineral phase [704].

There do not seem to be any reported observations of the weak v_4 CO_3 IR bands in apatitic biological minerals, although they have been seen in synthetic precipitated CO_3Aps [911] (Section 4.5.2). The observation of Raman spectra of dental enamel is hindered by strong fluorescence [556,818]; nevertheless, assignments have been made for v_1 CO_3 bands [560] (Section 4.5.2).

Bone

Observation of v_3 CO_3 in bone is difficult because of absorption from the organic phase. The v_2 CO_3 region has been studied by deconvoluting FT IR spectra of bone from various young adult species (rat, rabbit, chicken, cow and human) [699]. The 866 cm^{-1} band, assigned to CO_3^{2-} ions in a labile environment in synthetic precipitated CO_3Aps [699] (Section 4.5.2) and in pig enamel (see above), varied in its intensity relative to the other v_2 CO_3 bands in the different species [699]. However, the relative intensity of the bands at 871 and 878 cm^{-1}, due to CO_3^{2-} ions in PO_4^{3-} and OH^- ion sites, respectively, was remarkably constant in all the bone samples. These studies were extended with an investigation of v_2 CO_3 in density fractions from the midshaft of the diaphysis of the long bones of embryonic, ten week, and three year old chickens [930]. The intensity ratio of the 871 and 878 cm^{-1} bands was also constant between the different density fractions, but the intensity ratio of the 866 cm^{-1} to the 871 cm^{-1} band decreased with the age of the animal and, for bone of the same age, was highest in the most abundant density fraction [930].

4.6.4 Thermal decomposition

The thermal decomposition of synthetic CO_3Aps has been discussed in detail in Section 4.5.7; this included cross-references to specific types of synthetic CO_3Ap and some consideration of the thermal decomposition of biological CO_3Aps. Well-crystallised precipitated synthetic CO_3Aps do not form pyrophosphate on heating [83,88,99,722] (Section 4.5.1) and, in this respect, differ from calcified tissues. Examples of pyrophosphate measurement, as an indicator of the HPO_4^{2-} ion content in calcified tissues, were given in Section 4.6.1; a major difficulty with this assay is that a quantitative reaction (Equation 3.15) is difficult to achieve because of incomplete or side reactions, particularly in the presence of CO_3^{2-} ions (Section 3.6.2). Another significant difference between biological apatites and the well-crystallised precipitated CO_3Aps is that biological apatites start to lose CO_2 at 180 °C or lower (see below), but the temperature has to be 550 °C or above for the synthetic CO_3Aps (Sections 4.5.3, 4.5.4 and 4.5.5). Other sections relevant to the thermal decomposition of biological apatites are: Section 3.6.2 (thermal formation of pyrophosphate from HPO_4^{2-} ions in Ca-def OHAps); Section 3.4.3 (loss above ~1000 °C of constitutional water from OHAp and the formation of O,OHAp or OAp, and at higher temperatures, the formation of TetCP and α-TCP); Section 1.6.6 (high temperature phase diagram for the system $CaO-P_2O_5$); and Section 3.13 (summary of the thermal decomposition of Ca-def OHAps and OHAp). Both the temperature and the water vapour partial pressure determine the reactions and occurrence of phases above 1000 °C.

Enamel

Thermal changes on heating dental enamel up to ~1500 °C have been reviewed [967]. The thermal decomposition of human dental enamel (fraction with density >2.95 g cm^{-3}), heated for 24 h at temperatures from 25 to 1000 °C in nitrogen or vacuum, has been thoroughly investigated by TGA, IR spectroscopy (of both the enamel and evolved gases) and by lattice parameter measurements [721]; this included detailed discussion of the results in relation to earlier work. The following paragraphs are mainly a synopsis of this work.

Thermogravimetric analysis. All the TGA curves showed an initial weight loss with a reduced slope around 100 to 110 °C and an increased slope between 300 and 350 °C. The slope then decreased very gradually up to ~900 °C, at which point the total weight loss was about 4.8 wt % [721].

Evolved gases. IR spectra were made of the gases evolved when a sample was heated for 24 h successively at 180, 250, 360, 460 and 570 °C [721]. Evolution of carbon dioxide started at 180 °C and was strong at 360 and 460 °C. Substantial amounts of water were also evolved (see next paragraph). In another study [672], the gases evolved from enamel, previously heated at 300 °C in air, were investigated with gas chromatography as the enamel was heated to 1000°C (heating rate 150 or 300 °C h^{-1}). Maxima in the CO_2 evolution occurred at 450, 750 and 900 °C and carbon monoxide and nitrogen were detected between 700 and 1000 °C.

Water content. The amount of water in enamel after heating, determined by the height of the peak at 3300 cm^{-1}, depended on the atmosphere in which it was heated and cooled and its subsequent exposure to air [721]. In nitrogen dried over P_2O_5, nearly one-third of the amount initially present was lost between 270 and 300 °C, but a small fraction was retained, or resupplied by decomposition of other components, up to 800 °C.

Hydroxyl and chloride ions. When enamel was heated to 400 °C and cooled to room temperature in nitrogen dried over P_2O_5, there was a ~70 % increase in the area of the OH peak at 3569 cm^{-1} [721]. The peak also became about 30 % narrower. The OH content fell to essentially zero if the enamel was heated at 900 °C and cooled. This fall was not seen if cylinder nitrogen was used directly, without drying over P_2O_5, either because OH$^-$ ions were not lost, or the sample became rehydroxylated on cooling. With direct cylinder nitrogen, the OH content did not begin to fall until the enamel had been heated at 1000 °C. The behaviour of the Cl-perturbed OH peak at 3495 cm^{-1} was very similar to that of the main OH peak except that, when heated above 400 °C in direct cylinder nitrogen, the peak fell with treatment temperature much more slowly, so that it did not fall essentially to zero until the enamel was heated at

~1200 °C. It was suggested that a likely explanation for the rehydroxylation effects being greater for the Cl-perturbed OH peak, compared with the main OH peak, was that Cl⁻ ions had been driven out of the structure [721].

Carbonate and carbon dioxide. These have already been discussed in relation to synthetic CO_3Aps and, to a limited extent, enamel, in Section 4.5.7. The relative amounts of CO_3^{2-} ions in phosphate and hydroxyl sites and molecular CO_2 were estimated from the areas of the bands at 1415, 1546 and 2340 cm⁻¹ [721] in samples heated at various temperatures and then cooled. There was an apparent loss from both carbonate sites at temperatures ≤200 °C at which temperature the CO_2 band became visible. The CO_2 band reached a maximum at ~400 °C and fell to zero at ~800 °C, whilst the band due to CO_3^{2-} ions in phosphate sites fell linearly, becoming zero at ~1000 °C. On the other hand, the band from CO_3^{2-} ions in hydroxyl sites started to fall at ~100 °C to a minimum at ~300 °C, then rose to a maximum at ~700 °C, and to zero at ~1100 °C. The mechanisms proposed for the thermal evolution of these species was that CO_3^{2-} ions in both sites begin to decompose to CO_2 at 200 °C or below; the CO_2 then diffused out of the crystals via the hexad axis channels, but near 400 °C, the production rate built up so that CO_2 accumulated, probably in vacant PO_4 sites, and began to form more CO_3^{2-} ions in the hexad channels; finally at 1100 °C, all the carbon-containing species were lost, at which point a small amount of β-TCP was detected by XRD, although the principal crystalline phase was thought to be an oxyapatite provided rehydroxylation on cooling was prevented. CO_3^{2-} ions could form in the hexad channels via the reaction $CO_2 + 2OH^- \rightarrow CO_3^{2-} + H_2O$ which, in part, accounted for the observed loss of OH⁻ ions above 400 °C. The thermal behaviour of the CO_2 band in enamel is similar to that seen in synthetic CO_3Aps for which several reactions for the formation of molecular CO_2 have been suggested [94,534] (Section 4.5.7). A shoulder at ~2356 cm⁻¹ has been reported on the main CO_2 band in enamel after heating for 4 h at 670 to 765 °C [94,534]. The increased prominence of IR bands due to CO_3^{2-} ions replacing OH⁻ ions compared with those replacing PO_4^{3-} ions is also seen in synthetic apatites on heating (Fig. 4.6, Section 4.5.2), though their exact behaviour depends on the type of sample, particularly the F⁻ content (Section 4.5.7).

Acid phosphate and pyrophosphate. The IR estimation of HPO_4^{2-} content in CO_3Aps from the absorption at ~875 cm⁻¹ is complicated by the presence of ν_2 CO_3 bands at 873 and 880 cm⁻¹ (Section 3.6.2). Thus an attempt to follow quantitatively the thermal loss of HPO_4^{2-} ions in enamel also involved study of the associated CO_3 bands [721]. In this study, it was found that the IR bands for β-$Ca_2P_2O_7$ at 725 cm⁻¹ [182] and γ-$Ca_2P_2O_7$ at 715 cm⁻¹ [189,824] were too weak to provide useful information about the formation of pyrophosphate. However, X-ray powder diffractometer patterns of enamel heated in dry nitrogen at 400, 600 and 800 °C indicated the absence of crystalline pyrophosphate phases at 400 °C, the presence of both γ-$Ca_2P_2O_7$ and β-$Ca_2P_2O_7$ at 600 °C and essentially only β-$Ca_2P_2O_7$ at 800 °C [721]. The intensities of both bands fell on heating to a minimum at ~350 °C for the 879 cm^{-1} band and ~450 °C for the 872 cm^{-1} band; both then increased to maxima at 800 °C before falling to zero at ~1000 °C [721]. Up to ~400 °C, the intensity of the 872 cm^{-1} band correlated well with that of the 1415 cm^{-1} band on the assumption that they both originated from CO_3^{2-} ions in phosphate sites; but at higher temperatures, the 872 cm^{-1} band increased, whereas the 1415 cm^{-1} band decreased. Thus, it was suggested that something else was contributing to the 872 cm^{-1} peak, possibly the tail of the 879 cm^{-1} peak or PO_3F^{2-} ions. The 879 cm^{-1} peak qualitatively followed the intensity of the 1546 cm^{-1} peak, assuming that they both originated from CO_3^{2-} ions in hydroxyl sites. However, above 600 °C, the intensity of the 879 cm^{-1} band was less than expected, possibly due to loss of absorption from HPO_4^{2-} ions [721]. These difficulties in correlating the intensities of different bands from CO_3^{2-} ions thought to be in the same locations is very reminiscent of similar, but not identical, difficulties discussed in Sections 4.5.2 and 4.6.3 for unheated CO_3Aps. This emphasises that the assignments given in Table 4.9 for enamel can only be a first approximation.

Species of organic origin. The thermal formation of cyanate (NCO^-) and cyanamide (NCN^{2-}) ions in CO_3Aps containing nitrogenous species has already been discussed in detail [94,435] (Section 4.5.7). In samples of enamel heated for 24 h in nitrogen, the cyanate band at 2200 cm^{-1} began developing at ~400 °C, was a maximum at ~600 °C, and disappeared above 750 to 800 °C [721]. A band at 756 cm^{-1}, possibly of organic origin but otherwise unidentified, had much the same temperature dependence as the 2200 cm^{-1} band. There was no report of a band attributable to cyanamide ions.

Lattice parameter changes. The decrease in the *a*-axis parameter, which occurs on heating enamel to 400 °C, has been attributed to loss of lattice water [719,721]. The *a*-axis parameter of enamel, measured at temperature on samples heated in vacuum, but corrected to room temperature values, showed a "sudden" reduction of ~0.014 Å between 250 to 300 °C [721]. This contraction is mainly due to loss of lattice water, although there might be a small contribution from loss of HPO_4^{2-} ions [720] (Section 3.7.3). Lattice parameters of enamel heated in dried nitrogen for 12 to 16 h at various temperatures, but measured at room temperature, are given in Table 4.11.

The expansion of the *a*-axis parameter in the 600 to 800 °C range corresponded to the increase of CO_3^{2-} ions in hydroxyl sites seen in IR spectra [721] (see above). The IR results suggested that about 20 % of the original CO_3^{2-} ions in phosphate sites was still present at 800 °C, which gave an estimated contraction of ~0.003 Å in the *a*-axis parameter. Taking as a base line the *a*-axis at 1200 °C when all the carbonate was lost, there was an expansion of ~0.030 Å at 800 °C due mainly to CO_3^{2-} ions in hydroxyl sites. If this substitution was

Table 4.11 Lattice parameters of human tooth enamel after heating for 12 to 16 h in dry nitrogen at various temperatures [721].

Temperature °C	a Å	c Å
Room	9.440(6)	6.872(4)
400	9.428(3)	6.880(2)
800	9.441(3)	6.875(2)
1200	9.414(2)	6.885(1)

the sole cause of the expansion, it would imply that about 20 % (error limits 12 to 28 %) of the OH⁻ ion sites were occupied by CO_3^{2-} ions [721]. This estimate was in satisfactory agreement with estimates from the IR band intensities [721].

The thermal effects of laser irradiation on the composition and phases present in dental enamel have been investigated [654]. The teeth were irradiated for 1 s at an energy density of ~10,000 J cm⁻² with a 10.6 μm CO_2 laser with an output of 20 W focused to a 0.5 mm diameter spot. This treatment melted the enamel to produce a modified apatite and minor phases of α-TCP and TetCP. The OH⁻ ion content of the apatite was little changed. The apatite modifications and other changes were: (1) reduction in water, protein, carbonate, and chloride (or a rearrangement of the chloride); (2) possible incorporation of some O^{2-} ions in place of OH⁻ ions; and (3) an uptake of traces of carbon dioxide and cyanate. The presence of O^{2-} ions in the apatite was indicated by the presence of a band at 434 cm⁻¹ assigned to oxide translation (*i.e.* a Ca_3-O "v_3 stretch" [654], Section 3.11.6). The probable effects of these thermally induced changes on the solubility of enamel have been discussed [967].

The thermal decomposition products, between room temperature and 1000 °C, of forming, maturing and mature enamel (human and bovine) have been compared by IR and powder XRD [968]. Carbonate (measured from the IR spectra [701], Section 3.6.3) was lost more quickly in forming and maturing enamel, compared with mature enamel; the loss was complete by 800 °C. XRD showed that forming and maturing enamel had transformed almost completely into β-TCP at 1000 °C, whereas mature enamel only formed a small amount of β-TCP. This finding was explained by assuming that developing enamel had a formula close to $Ca_9(HPO_4)(PO_4)_5OH$ [968].

The thermal decomposition of *Lingula unguis* and *Lingula shantoungensis* shell apatites and dental enamel has been compared [962]. As the behaviours of the two types of shell apatite were very similar, only details for the first will be quoted. Studies were undertaken on a fraction with a density between 2.5 and 2.7 g cm⁻³. This fraction contained 2.5±0.2, 2.0±0.04 and 0.98±0.01 wt %

CO_3, F and Mg respectively, and "Cl was detected by titration with $AgNO_3$". The Ca/P molar ratio was 1.58 and the lattice parameters were $a = 9.383(1)$ and $c = 6.859(1)$ Å. Lattice parameters of samples heated in air for 24 h at 200, 400, 500, 700 and 1000 °C were measured. At 200 °C, the a- and c-axis parameters had contracted by ~0.027 and ~0.007 Å respectively. The a-axis parameter showed little further change at 400 °C, but then increased nearly linearly to ~9.392 Å at 1000 °C; the c-axis parameter increased approximately linearly from 200 °C to ~6.890 Å at 1000 °C. The contraction of the lattice at 200 °C was accompanied by a loss of ~35 % of the water (estimated from the IR absorption at 3300 cm^{-1}) and was not affected by the quantity of water in the atmosphere in which the sample was heated. β-TCa,MgP formed above 500 °C, and comprised about one-third of the sample at 1000 °C. Carbonate bands were reported at 1430 and 1465 cm^{-1} which became broader at 700 °C, and were not seen in samples heated at 1000 °C. Bands attributable to CO_3^{2-} ions in hydroxyl sites were not seen in either the unheated, or heated samples. The OH stretching band was absent from unheated samples, but was seen at 3545 cm^{-1} in samples heated in air (but not dry nitrogen) at 700 °C and above. The results were discussed in terms of the proposal that the shell mineral was a "CO_3-containing, OH-deficient, and hydrated F + Cl-apatite". The shorter c-axis of the shell apatite compared with human enamel was attributed to the replacement of OH⁻ ions by Cl⁻ ions. This interpretation was based on the hypothesis that only F⁻ and Cl⁻ ions occupied c-axis sites, and the fact that this substitution causes a reduction of the c-axis. The CO_3^{2-} ions were thought to replace PO_4^{3-} ions and to decrease slightly the a-axis parameter and increase the c-axis parameter. There was no discussion of quantitative aspects of these effects. The contraction of the lattice at 200 °C was ascribed to the loss of structural water that was thought to be more loosely bound than in dental enamel because the loss occurred at a slightly lower temperature than in enamel.

In the absence of a quantitative chlorine analysis, it is difficult to assess whether the low c-axis parameter of the unheated shell apatite, compared with FAp ($a = 9.367$ and $c = 6.884$ Å), can be attributed to Cl⁻ ions, particularly because of the uncertain contributions from the CO_3^{2-} ions and water. The c-axis parameter is even shorter after heating at 200 °C when the contribution of the water will be much less and that from other ions should be unchanged; thus, it is this sample that will be discussed. Taking the hypothesis that only F⁻ and Cl⁻ ions occupy c-axis sites, ignoring the CO_3^{2-} ions, and assuming a linear interpolation from the lattice parameters of FAp and ClAp (Table 1.4), the lattice parameters of the shell apatite with 2 wt % F (47 % occupancy of the hexad axis sites) should be $a = 9.505$ and $c = 6.820$ Å. These values can be compared with the observed parameters $a = 9.356$ and $c = 6.852$ Å for the 200 °C sample. Although the c-axis parameter is in reasonable agreement, the *a*-axis parameter is much too large, even if the reduction from the neglected 2.5 wt % CO_3 in phosphate sites is taken into account. It seems that other factors must be mainly responsible for the short *c*-axis parameter. There is some similarity between the shell apatite and shark dentine which also has a short *c*-axis parameter ($a = 9.42\pm0.03$, $c = 6.84\pm0.01$ Å), high magnesium content (~3 wt %) and low Ca/P molar ratio (~1.5) [383] (Section 4.6.2). For shark dentine, it was suggested that the short *c*-axis parameter might be due to the low Ca/P ratio and the high magnesium content; this may also be the explanation for the shell apatite.

Bone

The carbon dioxide (as carbonate or hydrogen carbonate) in bone is much more labile than in enamel and well-crystallised precipitated CO_3Aps. For example, the femurs of young rats injected with ^{14}C and sacrificed an hour later lost nearly half of the labelled CO_2 on drying for 24 h at 110 °C [923]. Crushed rat femurs, which were oven dried at 110 °C for 18 h, lost ~15 % of their carbonate; similar unheated samples increased their carbonate content by ~16 % when exposed to pure CO_2 gas at 760 torr (98.8 kPa) for 60 minutes or longer, but with prior drying, there was no uptake [969]. These results, and other experiments on the uptake and loss of CO_2 in different atmospheres and for various times, led to the conclusion that the CO_2 in bone consisted of 30 % hydrogen carbonate, probably located on the crystal surfaces, and 70 % carbonate deeper within the lattice [969]. A different view has been expressed as a result of studies of the thermal decomposition of rabbit periosteal bone that had been dried at room temperature under vacuum for five days [970]. The CO_2 began to be lost at 150 °C, but was mainly lost in two stages; about 30 % was lost between 200 and 300 °C, and the rest between 500 and 800 °C. The proposal was made that all the CO_3^{2-} ions were located in phosphate sites and that the low temperature loss arose from carbon dioxide formed by the reaction between CO_3^{2-} and HPO_4^{2-} ions (or $P_2O_7^{4-}$ formed from HPO_4^{2-} ions) [970].

The IR band at 2340 cm^{-1} from retained CO_2 that was seen in heated dental enamel (above), was not detected in heated dentine or bone, a difference attributed to the smaller crystal size in the latter two tissues which would allow more ready loss of CO_2 [94,534].

4.6.5 Reactions in solution

Many aspects of the chemical reactions of biological apatites in solution have already been discussed in relation to synthetic analogues. For OHAp, this includes solubility and interfacial phenomena (Section 3.9), adsorption and surface reactions (Section 3.10.1), and reactions with F$^-$ ions (Section 3.10.2). Rates of dissolution of OHAp and enamel are considered in Section 3.10.3 and some aspects of the dissolution of enamel were included in Section 4.5.8 on the solubility and rates of dissolution of CO_3Aps. Preferential loss of the central cores of enamel crystals is a characteristic feature seen in EM studies of crystals collected from early carious lesions in enamel (Section 4.6.6). Subsurface demineralisation in dental caries, which involves the study of the spatial distribution of mineral loss with time as a result of acid on the enamel surface [817,971,972], is not considered here. However, similar subsurface demineralisation in permeable OHAp aggregates is illustrated in Fig. 3.3 (Section 3.10.3). The surface structure of bone apatite crystals, including a discussion of adsorption and exchange processes, has been reviewed [973].

Solubility product

The equilibrium solubility properties of enamel have been studied on the assumption that the inorganic phase is OHAp; thus the solubility product was expressed in terms of the activity product $(Ca^{2+})^5(PO_4^{3-})^3(OH^-)$ [974,975]. In one study [975], an activity product constant of 5.5×10^{-55} mol^9 l^{-9} was reported (presumably at 25 °C), but in the other [974], the solubility product was found to vary between 7.2×10^{-53} to 6.4×10^{-58} mol^9 l^{-9}, depending on the cumulative amount of the dissolution of the solid in a sequence of experiments using phosphoric acid in the pH range 4.5 to 7.6. All of these values are significantly higher than 3.04×10^{-59} mol^9 l^{-9}, the activity product of OHAp at 25 °C [38] (Section 3.9). This increase in the solubility product parallels that seen for precipitated $B-CO_3F$,OHAps compared with FAp [751] (Section 4.5.8), and presumably is also due to CO_3^{2-} ions in the lattice.

More recently, it has been pointed out that it is inappropriate to ignore the presence of the CO_3^{2-} ions, generally accepted as occupying lattice positions in biological apatites, in the calculation of the activity product [749,750,976]. This means that the equilibration experiments must be undertaken under a constant partial pressure of carbon dioxide. Taking carbonate into account, the equilibrium solubility behaviour of OHAp, and enamel and dentine from human permanent teeth, have been compared [750]. The samples were equilibrated in dilute phosphoric acid solutions (0.066 to 1.54 mmol l^{-1}) under various CO_2/N_2 mixtures (CO_2 partial pressures from 5.06 to 5060 Pa, 0.005 % to 5 %) at 25 °C. The OHAp results for partial pressures of carbon dioxide from 0.005 % to 0.13 % were as expected for the equilibrium of a solid with the composition of OHAp. However, the results for partial pressures from 1.8 % to 3.33 % (the 5 % results were excluded because of the possibility of calcite precipitation) displayed quite different behaviour [750]. They could be understood if the system was regarded, for simplicity, as a quaternary one defined in terms of the components H_3PO_4, $Ca(OH)_2$, CO_2 and H_2O in which there were two equilibrating solids, the original OHAp and a new precipitated CO_3Ap with a composition $Ca_{5-p}(OH)_{1-p}(PO_4)_{3-p}(CO_3)_p$. An analysis of the two equilibria in this system, which has one degree of freedom, allowed two variables $(\log[(Ca^{2+})(OH^-)^2]$ and $\log P_{CO2} - \log[(H^+)^3(PO_4^{3-})]$, where the round brackets imply activities and P_{CO2} is the CO_2 partial pressure) to be defined that should be linearly related with a slope of minus unity; note that the slope does not depend on p. The experimental points lay reasonably close to a straight line, regardless of the partial pressure of CO_2, with a slope (0.95) that was not significantly different from unity. Thus the results were consistent with the proposed model [750]. For the equilibration studies with enamel and dentine, it was necessary to derive a more general model in which OHAp was replaced by the biomineral with a formula

$$Ca_{5-x-y}(HPO_4)_v(CO_3)_w(PO_4)_{3-x}(OH)_{1-x-2y} \qquad 4.37$$

with the restrictions that $0 \leq x \leq 2$, $y \leq (1 - x/2)$, and, for electrical neutrality, $v + w = x$. It was assumed that the precipitating CO_3Ap had the same composition as in the equilibrium with OHAp. Again, a straightforward algebraic manipulation showed that there was a linear relation between the same two variables that had been defined for the OHAp equilibrium. However, the slope was determined by the stoichiometry of the two phases in equilibrium with the solution. Thus, the more the stoichiometry of the biomineral departed from that of OHAp, the more the slope departed from unity. The intercept of the straight line depended on the solubility product constants of the biomineral and precipitating bioapatite, as well as their stoichiometries [750]. The equilibration studies for enamel and dentine minerals at a partial CO_2 pressure of 1.86 %, gave reasonable fits to straight lines with slopes of 0.95 and 0.75 respectively, with the line for enamel displaced from the line expected for OHAp. These results showed that the solubility behaviour of enamel and dentine minerals could not be adequately described in terms of a solid with the stoichiometry of OHAp. The results fitted the more general model for enamel and dentine, but further assessment of the model was not possible because the parameters describing the stoichiometries of the bioapatites and precipitating CO_3Ap were not known. These experiments showed that, in principle, solubility product constants for enamel and dentine mineral could be defined [750].

The studies described above have been extended to porcine enamel at various stages of development [749]. However, in the new study, chemical analysis of the enamel samples enabled calculation of the stoichiometric coefficients in the general formulae (Formula 4.38) used to model the solubility behaviour. (This was slightly different from Formula 4.37 as used in the earlier study.). A solubility product, K_{EN}, was defined in terms of the stoichiometric coefficients in Formula 4.38. This is given in Equation 4.39, in which the brackets refer to activities.

$$Ca_{5-x}(HPO_4)_v(CO_3)_w(PO_4)_{3-x}(OH)_{1-x} \qquad\qquad 4.38$$

$$K_{EN} = (Ca^{2+})^{(5-x)}(HPO_4^{2-})^v(CO_3^{2-})^w(PO_4^{3-})^{(3-x)}(OH^-)^{(1-x)} \qquad 4.39$$

Neglecting the formation of a precipitating CO_3Ap, two variables could be defined that should have a linear relation with a slope of negative unity. The variables were the same as given above, except that the second variable included K_{EN} and the stoichiometric coefficients. The intercept of the linear relation enabled K_{EN} to be determined, because the stoichiometric coefficients were known. Values of K_{EN} could also be calculated directly from Equation 4.39 for the individual samples from the composition of the solution in equilibrium with the enamel; the mean of these gave another estimate of K_{EN}. Equations were also developed for the situation when the precipitation of a CO_3Ap phase was present. In order to take into account the different stoichiometries, comparison between the solubility products of the enamel samples was based on the mean activity, $a^s\pm$, defined as $(K_{EN})^{1/t}$ where $t = 9 - 3x + v + w$ and s indicates the saturation condition. The formulae derived from the chemical analyses and the mean activities at saturation are given in Table 4.12. The formulae show that, during enamel development, the Ca/P molar ratio increases, and the acid phosphate and carbonate contents decrease. The OH⁻ content appears to increase, but this was not determined directly by experiment, but by the condition for electrical neutrality; its value will therefore depend on the precise details of the model chosen. The results also show that the outer (younger) secretory mineral is the most soluble and that the solubility of the enamel decreases with advancing developmental stages. Plots of the two variables for the model without precipitating CO_3Ap fitted straight lines with slopes within the range -0.84 to -1.18. The departure from the theoretical value of -1 was small, but not negligible [749]. This difference was attributed to the formation of a precipitating CO_3Ap phase during equilibration. EM studies gave direct evidence for this second phase. In addition, the linear fits predicted by the model that took the CO_3Ap into account had far smaller residual sums of squares than the model without this second phase. Values of $a^s\pm$, determined from the intercepts of the linear relation for the model without the precipitating CO_3Ap, were not in very good agreement with the values in Table 4.12; when the precipitating CO_3Ap was included, very good agreement was attained [749]. Thus it is possible to make self-consistent measurements of the solubility products of CO_3Aps, provided the CO_2 atmosphere is controlled and a model is used in which the equilibrium includes two processes: dissolution of the original apatite and precipitation of a new CO_3Ap.

Table 4.12 Stoichiometry of developing pig enamel [749].

Sample[a]	Formulae for half a unit cell	Mean activity at saturation[b], $a^s\pm.$ $\times 10^6$
S1	$Ca_{4.16}(HPO_4)_{0.41}(CO_3)_{0.42}(PO_4)_{2.16}(OH)_{0.16}$	3.28(±0.20)
S2	$Ca_{4.29}(HPO_4)_{0.31}(CO_3)_{0.40}(PO_4)_{2.29}(OH)_{0.29}$	2.20(±0.16)
M1	$Ca_{4.45}(HPO_4)_{0.23}(CO_3)_{0.32}(PO_4)_{2.45}(OH)_{0.45}$	1.56(±0.11)
M2	$Ca_{4.63}(HPO_4)_{0.11}(CO_3)_{0.25}(PO_4)_{2.63}(OH)_{0.63}$	1.23(±0.08)

[a]From top to bottom, outer and inner secretory enamel, and early and late mature enamel.
[b]Average (standard deviation) of the ion activity products in the saturated solutions on the basis of the stoichiometries in column two.

Preferential loss of carbonate

The preferential loss of carbonate over phosphate on the dissolution of biological apatites (particularly bone) in dilute acids has been observed many times [121,805,977,978]. Preferential loss of carbonate is also seen in early carious lesions [979]. These findings have often been taken as evidence for a surface location of the CO_3^{2-} ions (Section 4.1.1). However, it is now generally accepted that most of the CO_3^{2-} replaces PO_4^{3-} ions in the lattice; there must therefore be another explanation. It seems most likely that, during dissolution, there is reprecipitation of an apatitic phase containing little or no carbonate. This is likely, given the higher solubility product of CO_3Aps compared with OHAp, as discussed above. There is also evidence for this type of process from the detailed investigations of the equilibrium solubility of enamel [749,750] (see above). Other, more direct, evidence comes from the reported slight changes in the apatite XRD patterns (sharpening or slight changes in lattice parameters) of samples removed from the surface of enamel or OHAp blocks exposed to acid in model systems for dental caries [980-983]. Preferential loss of magnesium is also seen during the acid dissolution of enamel powder [204] and in early carious enamel lesions [984]. Although reprecipitation processes seem the most likely explanation of the preferential dissolution of CO_3^{2-} and Mg^{2+} ions, it has been suggested that many of the observations can be understood if biological mineral contained a number of different phases with different solubility properties [117,122,123,204]. Clearly further investigations are required to resolve these alternatives and to understand the complexities of the dissolution process.

Reactions with F⁻ and FPO₃²⁻ ions

Reactions with F^- and FPO_3^{2-} ions

The reactions of OHAp with F^-, FPO_3^{2-} ions and with SnF_2 have been discussed in Section 3.10.2 and, in general, similar reactions occur with enamel and other biological apatites.

X-ray diffraction studies have shown that CaF_2 is the predominant phase formed, after 1 h at room temperature, in the reaction between powdered enamel and either commercial acidulated phosphate fluoride solutions or 2 % neutral sodium fluoride solutions [985]. When intact enamel surfaces were exposed to these solutions for 24 h at 37 °C, no changes in the diffraction patterns were observed; this was attributed to the limited penetration of the F^- ions into the tissue, so that the diffraction pattern was dominated by unreacted material [985]. However, more recent XRD and electron diffraction studies have confirmed the formation of CaF_2 on human enamel surfaces when they were treated with an acidic silane fluoride lacquer, a neutral NaF lacquer or an acidulated phosphate fluoride solution during clinically used application times [986]. The CaF_2 crystals were 4 to 15 nm particles without evidence for large amounts of precipitated FAp.

CaF_2 or CaF_2-like material has been detected by micro-Raman spectroscopy of bovine enamel treated with a commercial topical fluoride gel agent (1.2 F^- as NaF and HF in H_3PO_4 at pH 5.3) for 72 h [774,987]. The method depended on measurement of the ratio of the intensity of the Raman lines at 322 cm⁻¹, from a CaF_2 lattice mode, to the intensity of the v_2 PO_4 lines at 432 and 447 cm⁻¹ in the apatite; quantitative analysis was possible with an accuracy of 3 wt % within a spot of 5 μm diameter [774]. The line-width of the 322 cm⁻¹ line in the CaF_2 in the treated enamel was about twice as wide as in CaF_2 standards, which implied that the lattice dynamics of the two were different [987].

Reaction of enamel powder at 20 °C in an acetate buffer at pH 4 for a year led to the initial precipitation of CaF_2, followed by the formation of F,OHAp; the reduction in the a-axis parameter depended on the initial amount of F^- ion added (the smallest a-axis parameter was 9.383 Å) [988].

Pretreatment of enamel (powders and whole tissue) with a saturated solution of DCPD at pH 2.1 for a few minutes enhances the F^- ion uptake from an acidulated phosphate fluoride solution [206,989]. The DCPD formed from the enamel mineral in the pretreatment subsequently reacts with F^- ions to yield FAp [206,989]. The suggested advantages [206,989] of this pretreatment were that only insoluble FAp was formed, whereas the more soluble, and hence more readily lost, CaF_2 was the dominant phase with neutral sodium fluoride solutions and conventional acidulated phosphate fluoride treatments (see above). However, this may not be very significant as there may be an advantage from a more labile CaF_2 phase in enamel in the reduction of dental caries, as is discussed in the more recent literature [990].

Synergistic effects of sodium fluoride and Sr^{2+} ions in acidic buffered solutions in reducing the rate of dissolution of enamel pellets have been observed [991]. Bones containing up to 0.57 wt % F^- from rats fed a fluoride-rich diet dissolved more slowly than bones from control animals [992].

4.6.6 NMR spectroscopy and other physical and chemical studies

Previous discussions of particular interest to the interpretation and understanding of the NMR spectra of biological apatites are to be found in Section 3.12 (^1H and ^{31}P NMR of OHAp and Ca-def OHAps, and ^1H and ^{19}F NMR of OH,FAps) and Section 4.5.8 (^1H, ^{31}P and ^{13}C NMR of precipitated CO_3Aps). Because of the possibility of the presence of other phases (or environments related to these), the following sections are also of relevance: Sections 1.3.8 (OCP), 1.4.8 (DCPD), 1.8.4 (transformation of ACP into OHAp), 1.8.5 (ACP), 3.10.2 (reaction of OHAp with F^- ions) and 4.5.8 (A-CO_3Ap).

^1H NMR spectroscopy

The ^1H NMR spectrum of bone and dentine is dominated by contributions from the organic phase, but study of the inorganic component of enamel is possible. Spin-spin, T_2, spin-lattice, T_1 and magnetisation fractions of natural and deuterated enamel have been studied [993]. It was possible to identify solid-like interstitial water (5 wt %), enamel apatite and semi-liquid-like water (~1 to 2 wt %) components. Neither the solid-like nor semi-liquid water exchanged with deuterium after 8 h exposure to D_2O. T_2 for the apatite was ~61 µs.

^1H NMR studies at 317 MHz and ~2 kHz spinning rate have been made of porcine enamel at various stages of maturation [837]. The ^1H-NMR Hahn echo spectra were complex and not fully analysed, but they showed what appeared to be a characteristic pattern of rotational sidebands, similar to those seen in s-OHAp. The intensity of these bands was relatively weak in the outer "cheesy" enamel, but increased in intensity in the more mature embryonic enamel; there was little difference between spectra from intermediate maturity enamel from half-erupted molars and fully mature enamel. It was suggested that these changes might reflect the increase in OH^- ion content inferred from parallel FTIR studies of similar samples.

^{13}C NMR spectroscopy

As with ^1H NMR studies, ^{13}C NMR spectra of bone are dominated by contributions from the organic component, but it is possible to observe a weak signal from enamel that can be unambiguously assigned to CO_3^{2-} ions [894] (experimental details in Section 4.5.8). Although the signal-to-noise ratio was poor, the main resonances seen in precipitated AB-CO$_3$Aps [894] (Section 4.5.8) could be identified. These resonances had chemical shifts of 166.9 ppm and 170.5 ppm which were assigned to CO$_3^{2-}$ ions in OH$^-$ and PO$_4^{3-}$ ion sites respectively by analogy with spectra of synthetic compounds (A-CO$_3$Ap, Section 4.3.4; B-CO$_3$Ap, Section 4.5.8).

^{19}F NMR spectroscopy

^{19}F NMR spin echoes have been used to study the local environment of F$^-$ ions in bone mineral from rats at low F$^-$ ion concentrations [843,844]. The studies at 26.9 MHz used 0.5 g samples of rat bone; one (0.865 wt % F) was made by the reaction of bone powder with F$^-$ ions and the other (0.184 wt % F) came from rats with F$^-$ ions added to their drinking water [844]. Following a 90°-τ-180° pulse sequence, the echo envelopes from both samples showed two distinct components, indicating at least two different environments. The shorter component (Gauss decay time ~90 μs) was assigned, for the most part, to fluorine nuclei with only fluorine as nearest neighbours, and the longer component (decay times ~300 and ~400 μs) to fluorine nuclei with only hydrogen as nearest neighbours [844].

In other ^{19}F NMR studies of various bones [994], the relation between the F$^-$ ion relaxation rate and the F$^-$ ion concentration was found to be linear in cortical and trabecular bone, but with different slopes. This suggested that the microscopic F$^-$ ion concentration in trabecular bone, in the samples studied, was nearly twice the concentration in the cortical bone, even though the directly measured macroscopic concentration was about three times greater [994].

^{31}P NMR spectroscopy

The main interest in ^{31}P NMR spectroscopy of mineralised tissues [209,837,838,995] has been in attempts to identify precursors (if any) of biological apatite and in studies of changes in the mineral phase with maturation. The ^{31}P NMR MAS spectrum of mature bone is very similar to that of OHAp. A chemical shift for bone of 3.1 ppm, compared with 2.9 ppm for OHAp has been reported, but T_1 was significantly longer (~30 s for bone compared with 2.0 s for OHAp) [193]. However, the low density fraction of bone has much more prominent rotational sidebands than OHAp [995]. The intensity of these falls with the maturation of the bone [209,838]. As these sidebands were rather similar to those seen in DCPD, they were initially assigned to DCPD as a separate phase [995]. In later experiments at 119 MHz and MAS of 2 kHz, cross-polarisation techniques were used to enhance the contributions from phosphate groups with directly bound or nearby protons [838]. Bloch decay and dipolar suppression pulse sequences were also used. These experiments showed that two separate ^{31}P resonances could be identified:

a major fraction with weak sidebands, isotropic and anisotropic chemical shifts and dipolar-suppression behaviour similar to apatite; and a minor component that gave a wide sideband pattern that was strongly attenuated in dipolar suppression experiments and was typical of CaPs containing HPO_4^{2-} ions [838]. Comparison of the spectral features of the major component with those obtained earlier from a number of apatite standards [835] suggested that this component was closest to a B-CO_3Ap containing ~3.2 wt % CO_3^{2-} ions. The minor component was compared with ^{31}P NMR MAS spectra [835] of various CaPs containing HPO_4^{2-} ions (OCP, DCPD and DCPA), a Ca-def OHAp with 12 % of its phosphorus in the form of HPO_4^{2-} ions and ACP; computer additions of these spectra with that from the B-CO_3Ap were also examined. There were one or more spectral features (Bloch decay sideband intensities, dipolar-suppression behaviour or chemical shift anisotropy) that effectively ruled out all these except DCPD or HPO_4^{2-} ions in a DCPD-like configuration. As crystalline DCPD could not be detected in very young bone [208,209] (Section 4.6.2), the minor component seen in ^{31}P NMR MAS spectra was assigned to HPO_4^{2-} ions in a DCPD-like environment [209,838].

^{31}P NMR MAS spectroscopy has been used to follow the resorption of $2\times4\times10$ mm^3 β-TCP bioceramic plates that had been attached in the femoral notch of rabbits for times up to 7 weeks [193]. The quantity of remaining β-TCP was measured from cross-polarisation experiments in which β-TCP was distinguished from the mineral of bone by its much shorter T_1 relaxation time (0.9 s compared with ~30 s).

As a model system for bone, one dimensional projections of OHAp packed into a glass tube with an internal diameter of 5.7 mm have been obtained in ~40 min in a ^{31}P NMR imaging experiment [996]. The spectrometer operated at 128 MHz with a field gradient of 22 G cm^{-1} (2.2 mT cm^{-1}); the resolution obtained was ~400 μm. The solid state nature of the specimen introduces severe difficulties, in part from the long T_1 and short T_2 times. In order to reduce the effects of these, pulsed gradient phase-encoding of the spatial dimension, using a compensating gradient pulse to cancel the distorting effects of gradient waveform transients, was used.

EM studies of lattice structure and defects

There have been extensive EM studies of biological apatites, particularly of enamel apatite. These have been mostly concerned with the process of dissolution of the crystals in acids, of importance because of dental caries, and the structure of the "central dark line", dislocations and crystal morphology, because of information these might yield about crystal growth processes. These studies show that crystals from embryonic enamel are (100) ribbons or lath-like crystals that are very long in the *c*-axis direction. Dimensions of $263\pm22 \times 683\pm34$ Å^2 with lengths up to 100 μm and beyond have been reported [997] for early human enamel crystals. The apatite crystals in bone are much smaller; the smallest dimension is about 50 Å and the dimension in the c-axis direction about 350 Å, although there is uncertainty as to whether the crystals have a plate- or needle-like habit [998].

A very characteristic feature of crystals obtained from early carious lesions in enamel is the preferential acid dissolution of a central core parallel with the c-axis [999-1001] (further references are given in ref. [807]). This same feature has been seen in hydrothermally grown CO_3Ap crystals of mm dimensions with a CO_3 content more than 1000 ppm when dissolved in lactic, citric, perchloric or nitric acids [813]. At pH 4.5, the acid formed a local hole in the basal plane with a rate of penetration of ~2 nm s^{-1}; at pH 2, the penetration rate was 50 nm s^{-1}. This dissolution behaviour was thought to be almost certainly due to the presence of a dislocation, or group of dislocations, parallel to the c-axis [813]. The strained cylindrical region at the centre of the dislocation which dissolved more readily in acid was thought to have a radius of ~5 to ~10 nm. Crystals with lower carbonate contents hardly dissolved in the above acids during the same time periods [813].

Lattice periodicities of precipitated synthetic and biological apatites have frequently been observed in high resolution EM studies [166,792,944,1002-1006] (see also references cited in these papers). Images have been recorded along various zone axes and compared with computer simulations [944,1002,1005,1006]. Good correlation is obtained between the atomic detail in the simulations and the experimental results in images of enamel crystals [1002] aligned along [0001] and synthetic CO_3Aps [944] aligned along [011] and [010] (Fig. 4.9).

Various lattice defects have been observed, including screw dislocations, and high- and low-angle boundaries [1005]. Particular attention has been directed to the study of the "central dark line" of width 0.8 to 1.5 nm often seen in the centre of cross-sections cut perpendicular to the c-axis running in a direction parallel to (100), or sometimes seen running down the length of the crystals. This feature was first seen in enamel crystals and, more recently, in some synthetic precipitated apatites. Various explanations of its origin have been proposed, including a remnant of the original organic matrix, a screw dislocation, twin or grain boundary, or a localised defect containing CO_3^{2-} ions [944].

Detailed high resolution EM investigations of the "central dark line", including diffraction and image calculations, have been made [792,944]. OHAp and CO_3Aps precipitated in the pH range 6.0 to 9.5 and at 37, 60 and 85 °C were studied. At under-focus, the line appeared white and electron-lucent and bounded by Fresnel-like fringes (Fig. 4.9) and at over-focus, it was dark relative to the surrounding apatite crystal; the line was more difficult to see near to Gaussian focus [944]. Because of this dependence of the appearance on

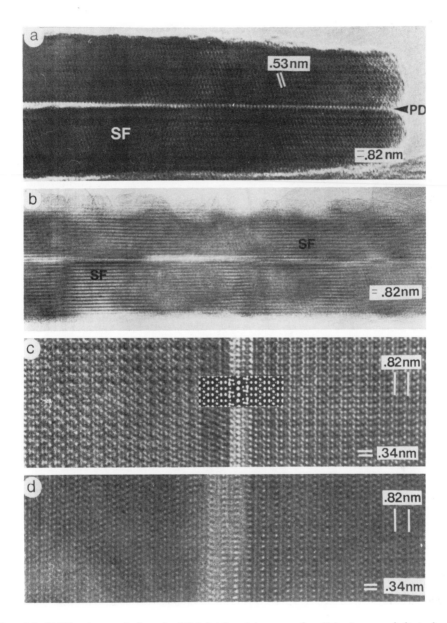

Fig. 4.9 (100) planar defect in EM lattice images of a CO_3Ap precipitated at pH 6.0 and 85 °C (Ca/P mole ratio = 1.61, 2.5 wt % CO_3). (a) 200 kV image along [011]. SF regions are strain fields associated with the defect. (b), (c) and (d) were taken at 400 kV along [0$\overline{2}$1] for (b), and along [010] for (c) and (d). The insert in (c) is a computer simulation of the apatite lattice with a core of a unit cell of OCP embedded in it. (Fig. 1 from Nelson *et al.* [944], with permission)

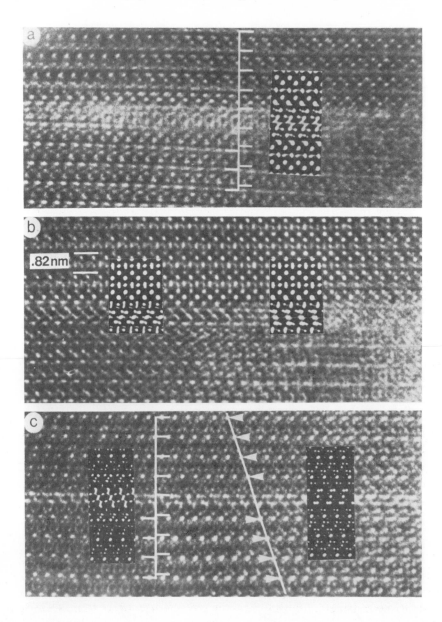

Fig. 4.10 400 kV images at a resolution of 0.18 nm of a CO_3Ap, oriented in the [011] zone axis, taken over two min. The inserts are computer simulations of the OCP defect model except the bottom right-hand one which is "collapsed OCP". (a) White marks are 0.82 nm apart and show an expansion of ~[¼00] across the planar defect. (b) left-hand and right-hand inserts are [0$\overline{1}$1] and [011] zone axes respectively. (c) The marks now show an expansion of less than [¼00] across the defect. (Fig.3 from Nelson *et al.* [944], with permission)

focus, the name "(100) planar defect" was preferred to "central dark line". Two-dimensional images of the planar defects were obtained in images aligned down the [011] and, occasionally, the [010] zone axes. Computer simulations of four models for the defect were considered [944]. These involved (1) Ca^{2+} ion vacancies, (2) a stacking fault, (3) a unit cell thickness of OCP, and (4) a unit cell of "collapsed OCP". In the OCP models, the OCP and OHAp lattices were oriented with their *b*- and *c*-axes aligned, as during the partial hydrolysis of OCP to OHAp [157] (Section 1.3.6). This alignment is in agreement with their close structural relation (Fig. 1.4). The computer simulations were consistent only with the OCP core models, as is shown in Fig. 4.9c. Some crystals with a planar defect appeared to have no displacement vector, whereas others (Fig. 4.10a) appeared to have an expansion vector of approximately [¼00]. During EM examination, the expansion vector reduces, indicating some collapse of the defect core, as is seen in the sequence Fig. 4.10a to Fig. 4.10c taken over 2 min. The planar defect often disappeared completely when irradiated with a focused electron beam [944]. In electron diffraction studies, the 300 spacing was measured from sequential diffraction patterns made by scanning a 1 nm beam across a ~36 nm wide CO_3Ap crystal with a (100) planar defect [944]. This showed that there was ~20% expansion over a ~2 nm wide region in the centre of the crystal. The electron diffraction patterns from this region were streaked in the [100] direction. The planar defects appeared to occur only in CO_3Aps precipitated in the pH range of 7±1 containing physiological concentrations of carbonate (1 to 4 wt %). They were not seen in any of the OHAps. It was concluded that the results of the study strongly suggested that the central planar defect was caused by a single unit cell thickness of OCP embedded in an apatite matrix, and that this layer could be the initial crystallite nucleus during biological apatite precipitation [944].

Central (100) planar defects have also been seen in crystals from model systems used to study enamel crystal growth [166]. In these, Ca^{2+} ions diffused into a phosphate solution through a cation-selective membrane on which the crystals grew [164] (Section 1.3.5). The experiments were carried out at 37 °C and pH 6.5 in the presence or absence of 1 ppm of F⁻ ions [166]. In the presence of F⁻ ions, lath-like apatite crystals grew with a central planar defect, thought to be a lamella of OCP, but in the absence of F⁻ ions, OCP grew without a central planar defect. As the authors pointed out, the function of the F⁻ ions was to catalyse the hydrolysis of OCP to OHAp, a known property of these ions [154] (Section 1.3.6).

A deviation from hexagonal symmetry of enamel crystals has been reported, based on high resolution EM images and their computer simulation [1002,1006]. The reported values of the *a*-axis periodicities measured from images aligned along [0001] in the three different directions that are equivalent under hexagonal symmetry were 9.294±0.007, 9.385±0.009 and

9.430±0.013 nm (presumably Å intended) [1002]. Additional evidence for the deviation from hexagonal symmetry was: "marked lines parallel to the $(10\overline{1}0)$ planes of intensity different from the one of the other $\{10\overline{1}0\}$ and from the $(30\overline{3}0)$ planes". These intensity differences were thought not to be due to poor adjustment of the astigmatism of the microscope objective, nor to crystal tilt. It was suggested that the enamel crystals were monoclinic and that one possible explanation for this was the replacement of two OH^- ions by a CO_3^{2-} ion, a substitution known to have the potential to lower the symmetry to $P2_1/b$ (Section 4.3.3). In a later publication [1006] on the loss of the six-fold symmetry axis in a human foetal enamel crystal, streaking in only one direction in optical diffraction patterns of images of the crystal aligned along [0001] was reported. It was concluded that the foetal enamel crystal did not have six-fold symmetry and that "the symmetry loss arises from a distortion of the crystal matrix due to differences in the atomic sizes of solute and solvent atoms since the streaking of the reflections on the optical diffractogram seems to be more extended for higher order spots, but we cannot rule out the possibility of multiple effects". The authors did not think the structure corresponded to monoclinic OHAp or monoclinic A-CO_3Ap. The suggestion was made that "the twofold symmetry observed...corresponds to a structure not previously described. However, we would like to stress this interpretation is at the moment only a working hypothesis". Whatever the explanation for these observations, ordering of the OH^- or CO_3^{2-} ions in c-axis channel sites to give a monoclinic structure seems very unlikely because nearly complete substitution of either of these ions would be required, whereas enamel apatite is known to be impure and nonstoichiometric. The a-axis values reported above also rule out a substantial number of CO_3^{2-} ions in c-axis sites. It might be possible for there to be an ordering of CO_3^{2-} ions in PO_4^{3-} ion sites that would give rise to a monoclinic structure, but this possibility has been thought unlikely in the case of francolite [45] (Section 4.2.1). A possible cause of the streaking of the diffraction spots is a limited number of lattice periodicities in the direction of the streaking (Fig. 1 of ref. [1006] suggests that, in the crystal illustrated, this might have been only eight). Recently, adult human enamel crystals have been examined by convergent beam electron diffraction [1007]. The symmetries observed led to the conclusion that the crystals had the $P6_3/m$ space group, with no evidence for a monoclinic space group. It was suggested that disorder, most likely in the OH^- ion columns, was the cause of weak intensity in the otherwise forbidden 000l (l odd) reflections and the low visibility of first-order Laue zone reflections in the convergent beam diffraction patterns from crystals orientated along the [0001] zone axis [1007].

An intrinsic birefringence (that is without any contribution from the form birefringence of the structure) of the apatite from the body of a carious lesion of -0.0034±0.0002 (wavelength not stated) has been reported [1008]. This result was uncorrected for the lack of perfect fibre orientation of the crystals and will therefore be a minimum value. In sound enamel, the planar oriented CO_3^{2-} ions would be expected to make a significant contribution to the intrinsic birefringence. However, there is a preferential loss of carbonate in carious enamel [979] (Section 4.6.5), so a value close to the birefringence of OHAp (-0.006, Table 1.4) is expected, as is seen to be the case.

4.6.7 Calcium phosphates in biomaterials

Biomaterials that make use of CaPs are generally, but not exclusively, used for the replacement of bone; the CaP may be used in the form of a bioceramic, as a coating on a metal prosthesis or in some form of composite. Methods for the production, properties and applications of these biomaterials have been reviewed [105,1009-1016]. Recent work is also to be found in the series of proceedings of the International Symposia on Ceramics in Medicine [1017-1022]. The processes that are used in the manufacture of these biomaterials often require temperatures well in excess of 1000 °C, so knowledge of the high temperature chemistry of the CaPs is important (CaO-P_2O_5 phase diagram, Figs 1.10 and 1.11, Section 1.6.6; formation of O,OHAp from OHAp, Section 3.4.3; summary of thermal decomposition of Ca-def OHAps and OHAp, Section 3.13). References to methods for the larger scale production of OHAp for biomaterials manufacture were given in Section 3.3.4.

Dense ceramics

The main use of these dense ceramics is to replace or augment small pieces of bone. OHAp (or O,OHAp, depending on the extent of the loss of constitutional water during the high temperature fabrication), β-TCP and mixtures of OHAp and β-TCP are the usual CaPs that are employed. The starting material is normally preheated at 600 to 1000 °C, ground to a particle size of ~250 μm, pressed uniaxially at ~50 MPa and heated in a moist atmosphere at 1000 to 1300 °C for 1 to 6 h; it is then cooled slowly (100 °C h^{-1}) to room temperature [1010]. Alternatively, the CaP may be hot pressed, but as this method results in no significant improvement in properties and is much more expensive, little work on this technique has been published [1023].

Dense polycrystalline OHAp has also been manufactured by a process somewhat similar to a slip casting technique that does not require the application of external pressure [640,1024]. OHAp was precipitated by the dropwise addition of a solution of ammonium phosphate into a solution of calcium nitrate (both at pH 11 to 12) to produce a milky, somewhat gelatinous precipitate. The solution and precipitate were then stirred for various times up to 48 h and, in some preparations, the solution was boiled. Electron microscopy indicated that the individual crystals had dimensions of ~200 Å. The reaction mixture was then centrifuged, washed and filtered on a Büchner funnel. The sticky filter cake was dried at 90 °C for 15 h. Despite the shrinkage and cracking that took place, ~3 mm thick plates with surface dimensions up to 100 cm^2 could be obtained. These were then sintered at 1000 to 1200 °C for 1 h. The product had a density close to the theoretical for OHAp and was highly translucent. The mechanical properties (average compressive strength 917 MN m^{-2} and average tensile strength 196 MN m^{-2}) were superior to those previously reported for ceramics made by sintering compressed powders. Recently, a slip casting method has been used to prepare dense CO$_3$Ap ceramics [1025]. In this study, the morphology of the initially precipitated CO$_3$Ap was investigated as a function of the HCO$_3^-$ concentration (0 to 640 mmol l^{-1}) and temperature (3, 25, 45, 70 and 90 °C); the dependence of the properties of the final sinter on the temperature (1000 to 1300 °C) and atmosphere (air, dry carbon dioxide, wet carbon dioxide, and wet air) of sintering was also studied. It was found that the size of the precipitated crystals decreased from ~800 to ~10 nm with a decrease in temperature of precipitation. Slip cast pellets (0.78 wt % CO$_3$) made from CO$_3$Ap precipitated at 3 °C formed translucent, dense ceramics when fired at 1150 to 1300 °C in a moist carbon dioxide atmosphere. Firing in other atmospheres resulted in opaque ceramics.

Porous ceramics

Porous ceramics are potentially very attractive because their large interconnecting pore structure may facilitate the ingrowth of surrounding boney or other tissues. Porous ceramics of OHAp, OH,OAp, CO$_3$Ap and β-TCP have all been manufactured. One study of OHAp ceramics used as a bone substitute indicated that the optimum size of the pores for achieving osteogenesis was between 300 and 600 μm [1026]. Porous ceramics of β-TCP generally seem to biodegrade much more rapidly than do ceramics made of OHAp [1027].

One of the first methods for the manufacture of CO$_3$Ap porous structures used the massive scleractinian coral *Porites* or the spines of the asteroid *Acanthaster planci* as starting materials [303]. These corals are formed from aragonite and calcite respectively and were transformed to CaPs with an alkaline phosphate solution. The corals were heated in (NH$_4$)$_2$HPO$_4$ solution, sometimes with added Ca(OH)$_2$, at temperatures up to 350 °C and pressures up to 15,000 lbs in^2 (100 MPa), which resulted in the formation of CO$_3$Ap (occasionally whitlockite from *Acanthaster planci*) (Section 4.5.6). The final product had reasonable mechanical strength and retained the original structure and interconnecting porosity of the coral.

Release of oxygen from hydrogen peroxide on heating has also been used to impart porosity to O,OHAp ceramics [1028]. The hydrogen peroxide was added to an aqueous slurry of OHAp in a plaster mould which was then allowed to rise and dry by gradually increasing the temperature to 80 °C during 18 h. The OHAp was then removed from the mould, heated at 60 °C h^{-1}, and fired at 1100 to 1400 °C for 6 h in an oxygen atmosphere. Ceramics could be prepared with porosities in the range 25 to 60 %, the macroporosity being determined by the quantity of hydrogen peroxide in the slurry and the microporosity by the sintering temperature and time. Hydrogen peroxide can also be used to generate porosity in β-TCP ceramics [1027].

Another method for the production of porous CaP ceramics follows the method for production of dense ceramics, but organic materials, such as naphthalene, polyethylene or camphor are mixed with the CaP prior to pressing. When the pellet is heated, the organic phase is volatilised or oxidised leaving spaces behind [1010].

Ceramic coatings on metal implants

The principal applications of coated metal implants are in dentistry and orthopaedics. The aim of this type of implant is to combine the benefit of a bioactive surface with the strength of the underlying metal. Several different coating methods have been developed (for example, plasma spraying, slip casting/sintering, electrophoretic deposition/sintering, electrochemical deposition, and sputter coating), but plasma spraying is most commonly used [1029,1030]. The usual coating material is OHAp [1031]. In plasma spraying, a powder is mixed with a gas stream which then passes through a plasma generated by a DC arc struck between two electrodes. The high temperature of the plasma melts the surface of the particles and may cause chemical and/or crystallographic modifications of the whole. The item to be coated is placed in the gas stream at a short distance from the plasma. Typical coating thicknesses of OHAp are 50 to 200 μm. An X-ray diffraction study of a sprayed coating of OHAp showed an appreciable line-broadening of the peaks [1032] whilst another [1033], indicated transformation into a mixture of OHAp, α-TCP and TetCP. When β-TCP is plasma sprayed, it changes to α-TCP [1032,1033], but TetCP is essentially unmodified [1032].

Bioactive glasses and glass-ceramics

45S5 Bioglass®, developed in 1970 from the SiO_2-P_2O_5-CaO-Na_2O system, was the first bioactive glass [1034]. The characteristic of bioactivity was found to be confined to a rather limited region of the phase diagram. The "45" refers to the ~45 wt% SiO_2 it contains as a network former and the "5" to its Ca/P molar ratio of ~5 [1034]. The full composition is SiO_2 46.1, Na_2O 24.4, CaO 26.9 and P_2O_5 2.6 wt % [1035]. Implants of 45S5 Bioglass® bond to cortical bone because the glass apparently nucleates OHAp crystals via a 40 to 50 μm thick calcium- and phosphorus-rich layer that develops on the surface of the implant; bone formation then follows in intimate relation to the implant [1035].

A-W is an example of a bioactive glass ceramic in which the crystalline phases of oxyfluorapatite and wollastonite ($CaO.SiO_2$) are precipitated in a glassy matrix [1036]. It has an overall composition MgO 4.6, CaO 44.9, SiO_2 34.2, P_2O_5 16.3 and CaF_2 0.5 wt %, of which 35 wt % is O,FAp and 40 wt % is $CaO.SiO_2$. The fluorine, added in the form of CaF_2, is to prevent the transformation of apatite to β-TCP [1036]. The cortical bone bonds directly to the glass ceramic through a layer of the ceramic that has a decreased silicon and magnesium and increased phosphorus content, without change in the calcium concentration [1037].

Composites

Composites often have superior mechanical properties to those of the unfilled matrix, particularly with regard to toughness, because the filler particles or fibres prevent the propagation of cracks. The rather limited range of composite biomaterials has been reviewed [1038]. Bioactive composites, in which the filler confers bioactive properties on the material, include OHAp or Bioglass® as fillers in a matrix of polyethylene, polyhydroxybutyrate, polysulphonate, polyactinide/glycolide, polymethylmethacrylate or collagen [1038]. The main use of bioactive composites is as bone cements or for bone augmentation.

4.7 Electron spin resonance of X-irradiated carbonate apatites

The phenomenon of electron spin resonance (ESR) originates from the absorption of electromagnetic radiation with an appropriate quantum energy by paramagnetic centres in a strong external magnetic field. Such a centre has one or more unpaired electrons and includes free radicals, various crystal imperfections and most transition metals and rare-earths. ESR spectroscopy is very sensitive and can detect paramagnetic centres at concentrations of the order of 10^{-9} mol l^{-1} [1039].

Paramagnetic centres that have been reported in X-irradiated apatites under various conditions include: electron and hole trapped species in apatite single crystals [389,456,612,1040,1041] (refs [456] and [612] are discussed further in Sections 2.3.1 and 3.13 respectively); F^+ centres (electrons trapped by an oxygen vacancies) [933] (Section 4.5.5); NO_2^{2-} [439] (Section 2.3.1); O^{3-} [427] (Section 2.3.1); O^- [612,1042] and $(H_3C)_2\dot{C}$-R [1043]. Other reported paramagnetic centres that contain carbon are discussed below.

This section is concerned with paramagnetic centres in CO_3Aps generated by ionising radiation. A substantial number of papers have been published on the interpretation of ESR spectra of (mostly) biological and synthetic samples. The main reason is that most, but not all, investigations have, of necessity, been carried out on powders. This makes the interpretation of the spectra rather difficult because the position of the resonance from a paramagnetic centre in a crystal is generally dependent on the orientation of the crystallographic axes with respect to the spectrometer's magnetic field. Thus with a powder specimen, a paramagnetic centre may have nearly superimposed resonances from these different orientations in its spectrum. This is in addition to other causes of multiple lines, such as interaction between the spin magnetic moment of the unpaired electron in the paramagnetic centre with those of neighbouring atomic nuclei (hyperfine splitting) and other effects (*e.g.* spin-lattice interactions). A further considerable complication with carbon-containing paramagnetic centres in X-irradiated CO_3Aps is that there are several species which seem to be located in a number of environments, as well as the possibilities of a number of different types of CO_3Aps and behaviour dependent on the sample drying temperature. Other factors that might influence the observed spectra are the irradiation dose, the time elapsed after irradiation, the temperature and conditions (*e.g.* vacuum or O_2 atmosphere) under which the sample is stored and examined, and the microwave power (saturation effects) of the spectrometer.

A characteristic feature of the ESR spectrum of X-irradiated CO_3Aps is a long-lived species (many months or years [1042]) that gives an asymmetric doublet with a *g* factor of about 2, close to the free-spin value. Two features have led to its assignment to a carbon-containing species: (1) the absence of hyperfine structure with any isotope of naturally high abundance at low amplifier gain, and (2) low intensity satellites observed at higher gain that can be attributed to ^{13}C (natural abundance, 1.1 wt %) [1044]. The assignment to carbon-containing species has been confirmed by studies on ^{13}C enriched CO_3Aps [1044-1047].

The radical responsible for the asymmetric signal was originally thought to be the pyramidal species, CO_3^{3-}, with its three-fold rotation axis parallel with the *c*-axis [1048]. This idea was based on single crystal studies of francolite that contained ~0.3 % CO_2. This same interpretation of the spectrum was later given to X-irradiated enamel with $g_\perp = 2.0036$ and $g_\parallel = 1.9983$ deduced for the centre [1049] and to spectra from oriented samples of sound and carious enamel [1050]. However, in studies of A-CO_3Ap and enamel, it was shown that the asymmetric signal had an unresolved orthorhombic character that was inconsistent with the C_{3v} symmetry of the CO_3^{3-} species [1051]. The *g*-tensor components deduced from the spectra were $g_1 = 2.0034 \pm 0.0002$, $g_2 = 2.0018 \pm 0.0002$ and $g_3 = 1.9971 \pm 0.0002$ for A-CO_3Ap and $g_1 = 2.0032 \pm 0.0003$, $g_2 = 2.0018 \pm 0.0003$ and $g_3 = 1.9972 \pm 0.0002$ for dental enamel. Proton and phosphorus ENDOR (electron nuclear double resonance) studies [1052] of X-irradiated oriented samples of human enamel showed that there were no ^{31}P nuclei within a radius of 6 Å around the paramagnetic centre, the minimum separation between centres and protons was ≥ 9 Å and the local environment of the centre had rhombic character. Based on these ENDOR results, which excluded the paramagnetic centre from being in a phosphorus or *c*-axis channel site, and comparison of the *g*-tensor components with literature values, the species responsible for the asymmetric signal was assigned to CO_2^- [1051]. A similar assignment was given on the basis of ^{13}C-hyperfine coupling studies of X-irradiated precipitated $^{13}CO_3Aps$ and A-$^{13}CO_3Aps$ [1045]. Subsequently, a surface location for the CO_2^- species was proposed [1044] and agreed by others [1042] (see below). Studies of the ESR spectrum of powdered enamel as a function of microwave power, irradiation and storage time were also consistent with the assignment of the main component near $g = 2$ to CO_2^-, but there was evidence for contributions from at least four other species, two of which were assigned to CO_3^{3-}, which contributed a maximum of 10 % to the overall signal [1053]. The results of these and more recent research by the same authors are summarised in the following paragraph.

Detailed studies of the paramagnetic centres near $g = 2$ in a variety of X-irradiated CO_3Aps (precipitated, high temperature and dental enamel) have been made [939,1042,1053-1056]. These studies included ^{13}C-containing preparations, investigations of the effect of the spectrometer microwave power, time of X-irradiation, time between irradiation and examination, sample drying temperature, and carbonate content. The principal components of the g-tensors of the various radicals were determined by numerical methods and complex spectra were analysed by subtraction of simpler spectra from samples with identified radicals or linear combinations of these. The complexity of the situation is illustrated by the fact that the presence of at least nine different radicals was demonstrated in X-irradiated CO_3Aps prepared by hydrolysing DCPA in a sodium carbonate solution at 95 °C [939]. The characteristics of some of these centres are summarised in Table 4.13. A3, as explained above, was assigned to a surface CO_2^- species and is responsible for the main, long-lived, anisotropic signal from X-irradiated precipitated CO_3Aps. Its signal intensity, rather surprisingly for a carbonate-derived radical, decreased with increasing carbonate content. However, this behaviour was consistent with the proposed surface location, as saturation of the surface could be rapidly reached, making the concentration quasi-independent of the carbonate content. A2 was also ascribed to a surface species, in this case CO_3^-. I2 and I3, assigned respectively to a CO_3^- and CO_2^- species, were thought to be situated in a phosphate ion site and to derive from a CO_3^{2-} ion precursor in this position. It was likely that these isotropic species were freely tumbling. ^{13}C studies showed that the A1 centre did not contain carbon and this centre was identified as an O^- radical.

ESR dating of fossil dental enamel

This recently developed technique [1057,1058] for dating fossil remains relies on the measurement of the ESR signals at $g = 2.0018$ and $g = 1.9976$ that arise from paramagnetic centres in the apatite of dental enamel. These centres are caused by radioactive elements in the tooth and its surrounding, the

Table 4.13 Characteristics of paramagnetic species in X-irradiated CO_3Aps prepared by the hydrolysis of DCPA in sodium carbonate solution at 95 °C and dried at 25 and 400 °C. After Tables 1 and 5 in ref. [1056].

Name[a]	g_x	g_y	g_z	g_{iso}	Defect and location	Behaviour as a function of increasing CO_3 content
A1	2.0413	2.0329	2.0023		O^- Surface	Decrease
Z_1	2.0041	2.0041	2.0008		CO_3^{3-} in PO_4 site	Increase
L	2.0035	2.0035	2.0015		CO_3^{3-} in OH site	Decrease
I2				2.0115	CO_3^- in PO_4 site	Increase
I3				2.0006	CO_2^- in PO_4 site	Increase
A2	2.0170	2.0084	2.0060	2.0105	CO_3^- Surface	Decrease
A3	2.0030	2.0015	1.9970	2.0005	CO_2^- Surface	Decrease

[a]A and I refer to anisotropic and isotropic g-tensors respectively

most important being uranium, thorium and ^{40}K, with a contribution from cosmic rays. The dose rate for a particular sample to be dated has to be estimated from a number of factors, including the concentrations of these elements. The rate of production of paramagnetic centres for a given dose rate (*e.g.* from a known ^{60}Co source) is also determined, so that the age of the sample can be estimated from the natural dose rate and paramagnetic centre concentration. The upper limit of the dating range is determined mainly by the thermal stability of the paramagnetic centres, as is also the case for fission-track dating (Section 2.7). The mean life of the paramagnetic centres is in the range of 10 to 100 million years, so the upper limit of the method is about a tenth of this time [1057,1058]. The method is not very applicable to bone for a number of reasons, including the smaller size of the apatite crystals and the higher organic content which result in significant changes in the bone mineral structure with time [1057,1058].

4.8 Summary

There are two accepted locations for the CO_3^{2-} ion in the apatite lattice: one, the site of the phosphate ion and the other on the hexad axis. However, in neither case has it been possible to determine the full structures of apatites in which these substitutions take place.

The replacement of 2OH⁻ ions by a CO_3^{2-} ion in OHAp is well established by chemical and X-ray powder diffraction studies. The problem of the lack of space in the hexad axis channels has been answered by the IR determination of the orientation of the ion which shows that its plane is tilted at a large angle to the basal plane. However, details of the exact position of the ion remain frustratingly uncertain. There seems little prospect of growing single crystals for X-ray study: the best chance of a solution to this problem would seem to lie with Rietveld analysis of highly substituted monoclinic A-CO₃Ap.

Evidence for CO_3^{2-} replacing PO_4^{3-} ions in francolite derives from chemical analyses, changes in lattice parameters and birefringence with CO_3 content, and recently, from Rietveld powder diffraction analyses of francolite before and after heating. Polarised IR gives the orientation of the ion which shows that it lies in the position of the sloping face of the replaced tetrahedron, the same orientation that was determined earlier from the change in the birefringence as a function of the carbonate content. The principal mechanism of compensation for the difference in charge between the CO_3^{2-} and PO_4^{3-} ions is the replacement of Ca^{2+} by Na^+ ions, but there are repeated suggestions in the literature that the difference is also compensated by F⁻ (and also OH⁻) ions accompanying the CO_3^{2-} ion in the phosphate site. This accounts for the presence of more F⁻ ions than can be accommodated on the hexad axis, as is often found in mineral and some synthetic francolites, although from an energetic point of view, it is not satisfactory. There is no direct structural evidence for the proposed [CO_3^{2-},F⁻] group in francolite, so it cannot be regarded as established beyond doubt: perhaps detailed ¹⁹F NMR studies of francolite or synthetic analogues with excess fluorine would clarify the situation. There is also the possibility of repeating the single crystal X-ray diffraction studies which clearly showed differences between francolite and unsubstituted FAp; a synchrotron source would be particularly suitable, as this would allow very small crystals to be used which are more likely to be homogeneous, and still obtain good counting statistics that would enable subtle aspects of the structure to be resolved.

Extensive studies of synthetic CO₃Aps formed in solution have established a self-consistent view of their structures. This has been primarily based on chemical analyses correlated with lattice parameters, densities, and IR spectroscopy. These studies have shown that CO_3^{2-} ions replace PO_4^{3-} ions and can also, to a limited extent, occupy positions in the hexad axis similar to those in A-CO₃Ap. Charge compensation involves the formation of vacancies in calcium sites and, if sodium is present, the replacement of Ca^{2+} by Na^+ ions, in which case there is often a greater carbonate uptake. There also appears to be a loss of OH⁻ ions from the hexad axis, in part evidenced by a fall in the OH IR intensity with an increase in carbonate content. It has been proposed that the CO_3^{2-} ion occupies the sloping face of a vacant PO_4^{3-} ion site and is adjacent to vacant Ca(2) and OH⁻ sites. This orientation of the CO_3^{2-} ion is consistent with the polarised IR studies of enamel, but because the CO_3 bands are broader and more complex in enamel, by comparison with those in francolite, the measurements of dichroic ratios are less precise. Although it is plausible to have clustered vacancies (these would provide a site for interstitial water) there is no direct proof that this happens. Clearly Rietveld structure analysis of powder diffraction patterns might provide this.

A major complication in the understanding of the structure of precipitated CO_3Aps, including biological apatites, is that their IR spectra cannot be fully understood in terms of a model involving only the two types of substitution discussed above, namely for the sloping face of a PO_4^{3-} ion and on the hexad axis with a position similar to that in A-CO_3Ap. The complexity and intensities of the spectra indicate that there are more than two environments, although the additional ones might derive from modifications of the environments already discussed, such as another orientation in a phosphate site. Further ^{13}C NMR studies and repetition of the polarised IR study of dental enamel at the higher resolution that is presently obtainable would probably be most helpful in elucidating these problems. These studies could also include a ^{13}C NMR and polarised IR (and X-ray diffraction) investigation of changes in the carbonate environment during the transformation of aragonite into an oriented CO_3Ap; intermediate structures might correspond to many found in precipitated CO_3Aps and the starting positions of the CO_3^{2-} ions in the aragonite lattice are well-known.

Studies of the thermal decomposition of precipitated CO_3Aps, although less specific than IR spectroscopy, clearly show a wide range of thermal stabilities as regards the carbon dioxide content. These studies imply that, for the very poorly crystallised CO_3Aps, at least some of the carbonate is present in a very labile form, a finding that is inconsistent with a site buried within the crystal lattice.

It is curious that synthetic precipitated CO_3Aps do not yield pyrophosphate on heating, whilst the biological CO_3Aps do. Whether this really indicates that the synthetic CO_3Aps do not contain HPO_4^{2-} ions is uncertain because of the possibility of side reactions during pyrolysis, but assuming it is, how do we synthesise CO_3Aps that contain HPO_4^{2-} ions so they are similar to the biological CO_3Aps?

The B-CO_3Aps and AB-CO_3Aps prepared at high temperatures have not been very extensively investigated. However, it is clear that they, like the mineral francolites and dahllites, have conspicuously lower maximum CO_3 contents than CO_3Aps formed under aqueous conditions. More CO_3 is incorporated if sodium is present, as in the precipitated CO_3Aps. Charge compensation when PO_4^{3-} ions are replaced by CO_3^{2-} seems to involve calcium site vacancies or the replacement of calcium by sodium if the latter is present.

Although there might be some hexad axis vacancies, this does not appear to be nearly as extensive as in the precipitated CO_3Aps, even taking into account the much lower CO_3 contents. The high temperature CO_3Aps can also have CO_3^{2-} ions on the hexad axis.

From the above discussion, it is clear that the CO_3Aps formed in aqueous systems are much more complicated than the synthetic high temperature preparations (no comment can be made on the mineral CO_3Aps because little work has been published on the variability of their CO_3 environment as indicated by IR spectra). Part of this additional complication could originate from the usual method of preparation of CO_3Aps in which one solution is added directly to another. In this mode of preparation, the growth conditions for the inside and outside of a crystal will be different. Thus it is highly likely that the mode of incorporation of the CO_3^{2-} ion will also differ, which clearly complicates their study. The way out of this difficulty is to prepare the CO_3Aps in solution under constant composition conditions (including the CO_2 partial pressure) as has been done for many of the other CaPs in studies of their rates of crystal growth and dissolution.

The physical chemistry of the CO_3Aps has not been very extensively investigated by comparison with the other CaPs. However, it has been clearly demonstrated that the presence of CO_3^{2-} ions in the apatite lattice substantially increases the solubility product and rate of dissolution in acids, and lowers the thermal stability. One the other hand, the presence of F^- ions in the lattice generally has the reverse effect. The rates of these reactions will depend on kinetic as well as thermodynamic considerations. Although kinetic factors may be all important in determining the course of these reactions (as for crystal growth in the carbonate- and fluoride-free $Ca(OH)_2$-H_3PO_4-H_2O system), there have been few studies of the way in which F^- and CO_3^{2-} ions affect the thermodynamic constants of apatite, clearly a prerequisite to a proper understanding of the effects of these ions.

The presence of carbonate has a profound influence on the growth of apatite crystals in solution, resulting in smaller crystals or even amorphous products. Magnesium ions have similar but less marked effects. The presence of CO_3^{2-} and Mg^{2+} ions might explain why biological apatites are rather poorly crystallised, but it is difficult to doubt that there are not other factors at work as well. For example, despite the 3 or 4 wt % CO_3 content of dental enamel, the apatite crystals grow many microns long in the *c*-axis direction whilst bone crystals are much smaller, plate-like crystals. It is much more probable that most of the control of the size, habit and particularly orientation of biological apatite crystals resides with organic molecules synthesised by cells for that specific purpose: the way in which these organic molecules control the growth of CO_3Ap crystals leads directly to one of the central unsolved problems of biomineralisation.

Appendix

CALCULATED X-RAY DIFFRACTION PATTERNS OF THE CALCIUM ORTHOPHOSPHATES

The X-ray diffraction patterns, Figs A.1 to A.10, were calculated with the Rietveld analysis program DBWS-9006PC [1059]. The calculations were for a powder diffractometer with a crystal monochromator and CuKα radiation (α_1 = 1.5405 Å, α_2 = 1.5443 Å, α_2:α_1 intensity ratio = 0.5). The calculated patterns have been scaled so that the intensity of the maximum peak (sometimes off scale) is 1000 and the horizontal scale is in degrees twotheta. Where more than one reflection contributes to the peak, the indices are given in descending order of intensity. The space groups, lattice parameters and references to the structural parameters used are given below.

Fig. A.1 Monocalcium phosphate monohydrate, MCPM. $P\bar{1}$, a = 5.6261(5), b = 11.889(2), c = 6.4731(8) Å, α = 98.633(6), β = 118.262(6), γ = 83.344(6)° at 25 °C [133].

Fig. A.2 Monocalcium phosphate anhydrous, MCPA. $P\bar{1}$, a = 7.5577(5), b = 8.2531(6), c = 5.5504(3) Å, α = 109.87(1), β = 93.68(1), γ = 109.15(1)° at 25 °C [137].

Fig. A.3 Octacalcium phosphate, OCP. $P\bar{1}$, a = 19.692(4), b = 9.523(2), c = 6.835(2) Å, α = 90.15(2), β = 92.54(2), γ = 108.65(1)° [148].

Fig. A.4 Dicalcium phosphate dihydrate, DCPD. Ia [211], a = 5.812±0.002, b = 15.180±0.003, c = 6.239±0.002 Å, β = 116° 25'±2' [210]. Atomic parameters from refinement using neutron data [213].

Fig. A.5 Dicalcium phosphate anhydrous, DCPA. $P\bar{1}$, a = 6.910(1), b = 6.627(2), c = 6.998(2) Å, α = 96.34(2), β = 103.82(2), γ = 88.33(2)° at 25°C [261]. Note that the temperature of this structure determination is in the region of the noncentrosymmetric → centrosymmetric phase transition [259,260] (Section 1.5.2).

Fig. A.6 α-tricalcium phosphate, α-TCP. $P2_1/a$, a = 12.887(2), b = 27.280(4), c = 15.219(2) Å, β = 126.20 (1)° [283].

Fig. A.7 β-tricalcium phosphate, β-TCP. $R3c$, a = 10.439(1), c = 37.375(6) Å (hexagonal setting) [286].

Fig. A.8 Tetracalcium phosphate monoxide, TetCP. $P2_1$, a = 7.023(1), b = 11.986(4), c = 9.473(2) Å, β = 90.90(1)° at 25 °C [316].

Fig. A.9 Fluorapatite, FAp. $P6_3/m$, a = 9.367(1), c = 6.884(1) Å [46]. Atomic parameters from refinement using data from flux-grown FAp [46].

Fig. A.10 Hydroxyapatite, OHAp. $P6_3/m$, a = 9.4176, c = 6.8814 Å (Section 3.3.2). Atomic parameters from refinement using data from Holly Springs OHAp (crystal X-23-4) with OH⁻ ions in two-fold disorder about the mirror plane at z = ¼ [404]. The calculated pattern for monoclinic OHAp is given in Fig. 3.1.

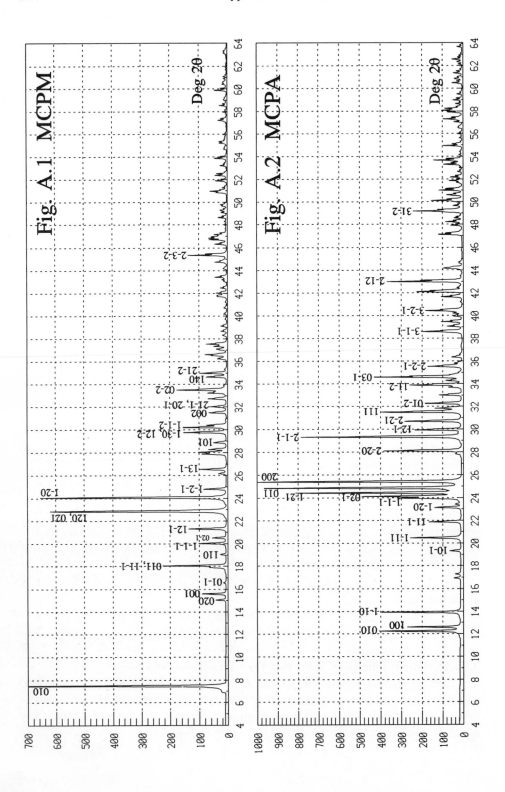

Fig. A.1 MCPM

Fig. A.2 MCPA

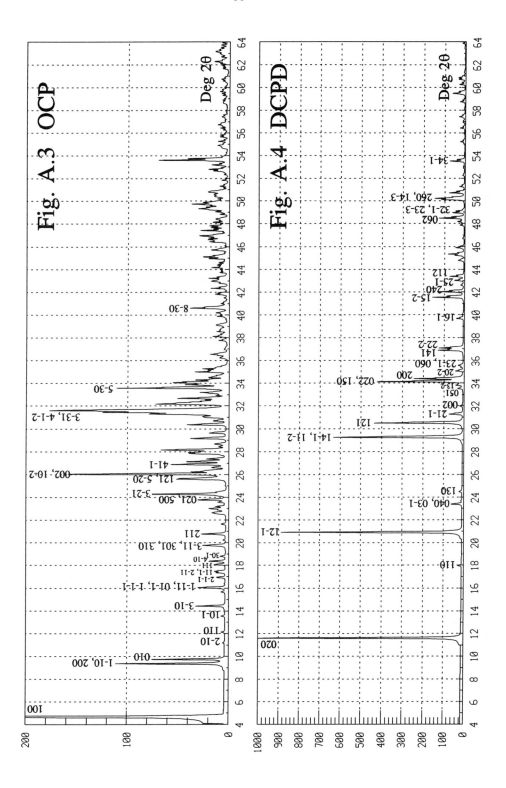

Fig. A.3 OCP

Fig. A.4 DCPD

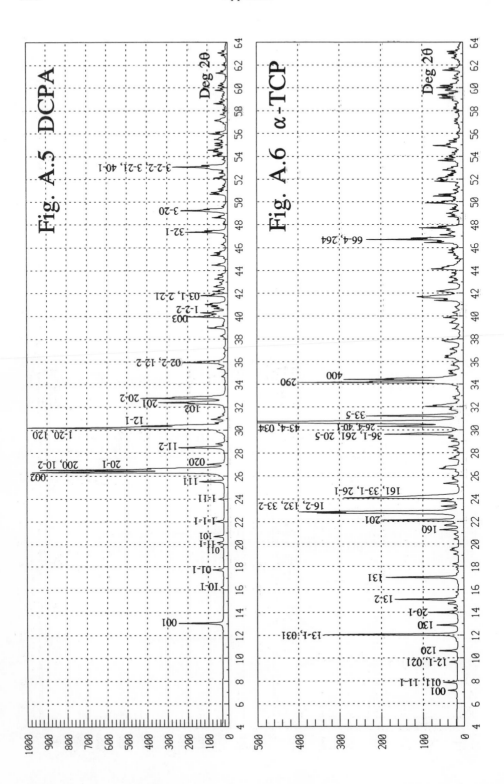

Fig. A.5 DCPA

Fig. A.6 α-TCP

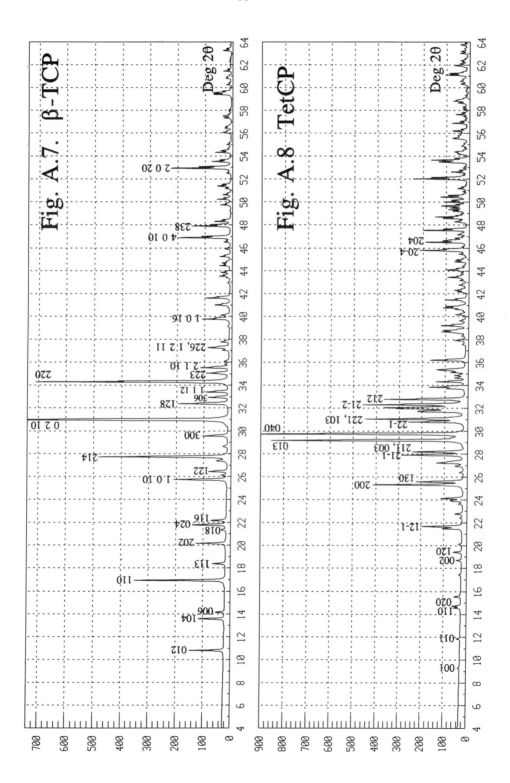

Fig. A.7. β-TCP

Fig. A.8 TetCP

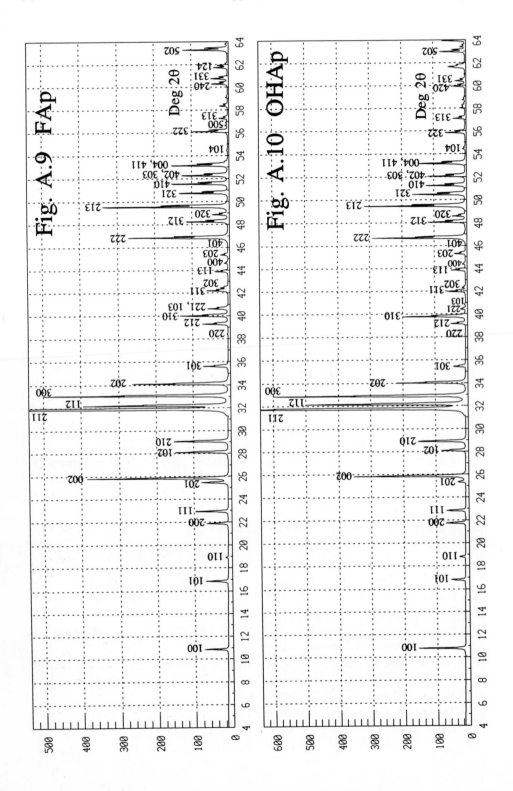

Fig. A.9 FAp

Fig. A.10 OHAp

REFERENCES

1 Frondel C: Mineralogy of the calcium phosphates in insular phosphate rock. Am Miner 1943;28:215-232.

2 McConnell D: Apatite, Its Crystal Chemistry, Mineralogy, Utilization, and Geologic and Biologic Occurrences. New York, Springer-Verlag, 1973.

3 Nriagu JO, Moore PB (eds): Phosphate Minerals. Berlin, Springer-Verlag, 1984.

4 Notholt AJG, Jarvis I (eds): Phosphorite Research and Development. London, Geological Society Special Publication No 52, 1990.

5 Dana JD, Dana ES: System of Mineralogy, 7th Ed Vol II, Rewritten by C Palache, H Berman, C Frondel. New York, John Wiley and Sons, 1944.

6 Eisenberger S, Lehrman A, Turner WD: The basic calcium phosphates and related systems, some theoretical and practical aspects. Chem Rev 1940;26:257-296.

7 Bjerrum N: Calciumorthophosphate. I. Die Festen Calciumorthophosphate. II. Komplexbildung in Lösung von Calcium-und Phosphate-Ionen. Math.-fys Medd Kong Danske Viden Selskab 1958;31:1-79. (Translated by M H Rand, Atomic Energy Res Est, Harwell, UK, Trans 841, 1959).

8 Newesely H: Kristallchemische und mikromorphologische Untersuchungen schwerlöslicher Calciumphosphate. Fortschr Chem Forsch 1966;5:688-746.

9 Gmelins Handbuch der Anorganischen Chemie, Calcium, Part 3, Lieferung 3, System No 28. Weinheim, Verlag Chemie, 1961, pp 1115-1223.

10 Kanazawa T (ed): Inorganic Phosphate Materials, Materials Science Monograph 52. New York, Elsevier, 1989.

11 Benard J: Combinaisons avec le phosphore; in Pascal P (ed): Nouveau Traité de Chimie Minérale, Tome IV, Le Calcium. Paris, Masson et Cie, 1958, pp 455-488.

12 Van Wazer JR: Phosphorus and its Compounds, Vol 1, New York, Interscience Publishers, 1958.

13 Mooney RW, Aia M: Alkaline earth phosphates. Chem Rev 1961;61:433-462.

14 Young RA, Brown WE: Structures of biological minerals; in Nancollas GH (ed): Biological Mineralization and Demineralization, Dahlem Konferenzen, 1981. Berlin, Springer-Verlag, 1982, pp 101-141.

15 Brown EH, Lehr JR, Smith JP, Frazier AW: Preparation and characterization of some calcium pyrophosphates. J Agr Food Chem 1963;11:214-222.

16 Lehr JR, Brown EH, Frazier AW, Smith JP, Thrasher RD: Crystallographic Properties of Fertilizer Compounds, Chemical Engineering Bulletin No 6. Muscle Shoals, Tennessee Valley Authority, 1967.

17 Biggar GM: Experimental studies of apatite crystallization in parts of the system CaO-P$_2$O$_5$-H$_2$O at 1000 bars. Miner Mag 1966;35:1110-1122.

18   Skinner HCW: Phase relations in the $CaO-P_2O_5-H_2O$ system from 300° to 600°C at 2 kb H_2O pressure. Am J Sci 1973;273:545-560.

19   Riboud PV: Composition et stabilité des phase à structure d'apatite dans le système $CaO-P_2O_5$-oxyde de fer-H_2O à haute température. Ann Chim (Paris) 1973;8:381-390.

20   Roth RS, Negas T, Cook LP, Smith G, Clevinger MA (eds): Phase Diagrams for Ceramists, Vol V. Columbus, The American Ceramic Society.

21   Riboud PV: Composition et stabilité des phase à structure d'apatite dans le système $CaO-P_2O_5$-oxyde de fer-H_2O à haute température; in: Physico-Chimie et Cristallographie des Apatites d'Intérêt Biologique, Colloques Internationaux du Centre National de la Recherche Scientifique No 230 Paris, 1973. Paris, CNRS, 1975, pp 473-480.

22   Biggar GM: Apatite composition and liquidus phase relationships on the join $Ca(OH)_2-CaF_2-Ca_3(PO_4)_2-H_2O$ from 250 to 4000 bars. Miner Mag 1967;36:539-564.

23   Biggar GM: Phase relationships in the join $Ca(OH)_2-CaCO_3-Ca_3(PO_4)_2-H_2O$ at 1000 bars. Miner Mag 1969;37:75-82.

24   Zawacki SJ, Koutsoukos PB, Salimi MH, Nancollas GH: The growth of calcium phosphates; in Davis JA, Hayes KF (eds): Geochemical Processes at Mineral Surfaces, Am Chem Soc Symp Series No 323. pp 650-662, 1986.

25   Nancollas GH: *In vitro* studies of calcium phosphate crystallization; in Mann S, Webb J, Williams RJP (eds): Biomineralization, Chemical and Biochemical Perspectives. Weinheim, VCH Verlagsgesellschaft, 1989, Ch 6, pp 157-187.

26   Johnsson MS-A, Nancollas GH: The role of brushite and octacalcium phosphate in apatite formation. Crit Rev Oral Biol Med 1992;3:61-82.

27   Nancollas GH: The nucleation and growth of phosphate minerals; in Nriagu JO, Moore PB (eds): Phosphate Minerals. Berlin, Springer-Verlag, 1984, Ch 2, pp 137-154.

28   Moreno EC, Gregory TM, Brown WE: Solubility of $CaHPO_4.2H_2O$ and formation of ion pairs in the system $Ca(OH)_2-H_3PO_4-H_2O$ at 37.5 °C. J Res Natn Bur Stand 1966;70A:545-552.

29   Leung VW-H, Darvell BW, Chan AP-C: A rapid algorithm for solution of the equations of multiple equilibrium systems—Rameses. Talanta 1988;35:713-718.

30   Darvell BW, Leung VW-H: The Rameses algorithm for multiple equilibria—II Some further developments. Talanta 1990;37:413-423.

31   Leung VW-H, Darvell BW: The Rameses algorithm for multiple equilibria—III Acceleration and standardized formation constants (Rameses II). Talanta 1990;37:425-429.

32   Darvell BW, Leung VW-H. The Rameses algorithm for multiple equilibria—IV. Strategies for improvement (Rameses III). Talanta 1991;38:875-888.

33   Darvell BW, Leung VW-H. The Rameses algorithm for multiple equilibria—V. Error statements. Talanta 1991;38:1027-1032.

34 Gregory TM, Moreno EC, Brown WE: Solubility of $CaHPO_4.2H_2O$ in the system $Ca(OH)_2-H_3PO_4-H_2O$ at 5, 15, 25, and 37.5 °C. J Res Natn Bur Stand 1970;74A:461-475.

35 McDowell H, Brown WE, Sutter JR: Solubility study of calcium hydrogen phosphate. Ion-pair formation. Inorg Chem 1971;10:1638-1643.

36 Tung MS, Eidelman N, Sieck B, Brown WE: Octacalcium phosphate solubility product from 4 to 37°C. J Res Natn Bur Stands 1988;93:613-624.

37 Gregory TM, Moreno EC, Patel JM, Brown WE: Solubility of β-$Ca_3(PO_4)_2$ in the system $Ca(OH)_2-H_3PO_4-H_2O$ at 5, 15, 25, and 37 °C. J Res Natn Bur Stands 1974;78A:667-674.

38 McDowell H, Gregory TM, Brown WE: Solubility of $Ca_5(PO_4)_3OH$ in the system $Ca(OH)_2-H_3PO_4-H_2O$ at 5, 15, 25, and 37 °C. J Res Natn Bur Stand A Phys Sci 1977;81A:273-281.

39 Wagman DD, Evans WH, Parker VB, Schumm RH, Halow I, Bailey SM, Churney KL, Nuttall RL: NBS tables of chemical thermodynamic properties - selections for inorganic and C_1 and C_2 organic substances in SI units. J Phys Chem Ref Data 1982, vol 11, Suppl 2.

40 Viellard P, Tardy Y: Thermochemical properties of phosphates; in Nriagu JO, Moore PB (eds): Phosphate Minerals. Berlin, Springer-Verlag, 1984, Ch 4, pp 171-198.

41 Gerhard CA: Grundriss des Mineral-Systems, 1786, pp 281. Cited by: AH Chester, A Dictionary of the Names of Minerals, New York, John Wiley (also London, Chapman & Hall), 1896, p 15.

42 McClellan GH, Lehr JR: Crystal chemical investigation of natural apatites. Am Miner 1969;54:1374-1391.

43 McClellan GH, Van Kauwenbergh SJ: Mineralogy of sedimentary apatites; in Notholt AJG, Jarvis I (eds): Phosphorite Research and Development. London, Geological Society Special Publication No 52, 1990, pp 23-31.

44 Geiger Th: Beiträge zum Problem der Karbonatapatite. Schweiz Miner Petrog Mitt 1950;30:161-181.

45 Carlström D: Mineralogical carbonate-containing apatites; in Brown WE, Young RA (eds): Proceedings of International Symposium on Structural Properties of Hydroxyapatite and Related Compounds. Gaithersburg, 1968, Ch 10. (unpublished)

46 Sudarsanan K, Mackie PE, Young RA: Comparison of synthetic and mineral fluorapatite, $Ca_5(PO_4)_3F$, in crystallographic detail. Mat Res Bull 1972;7:1331-1338.

47 Mackie PE, Elliott JC, Young RA: Monoclinic structure of synthetic $Ca_5(PO_4)_3Cl$, chlorapatite. Acta Cryst 1972;B28:1840-1848.

48 Elliott JC: Optical properties of synthetic and mineral chlorapatites. J Appl Cryst 1985; 18:384-387.

49 Elliott JC, Dykes E, Mackie PE: Structure of bromapatite, $Ca_5(PO_4)_3Br$, and the radius of the bromide ion. Acta Cryst 1981;B37:435-438.

50   Dykes E: Preparation and characterisation of calcium bromapatite. Mater Res Bull 1974;9:1227-1236.

51   Mengeot M: Hydrothermal Growth and Electron-Spin-Resonance Investigations of Calcium Hydroxyapatite Single Crystals. PhD Thesis, University of Connecticut, 1975.

52   Elliott JC, Bonel G, Trombe JC: Space group and lattice constants of $Ca_{10}(PO_4)_6CO_3$. J Appl Cryst 1980; 13:618-621.

53   Collin RL: Strontium-calcium hydroxyapatite solid solutions: Preparation and lattice constant measurements. J Am Chem Soc 1959;81:5275-5278.

54   Negas T, Roth RS: High temperature dehydroxylation of apatitic phosphates. J Res Natn Bur Stand A 1968;72A:783-787.

55   Hata M, Okada K, Iwai S, Akao M, Aoki H: Cadmium hydroxyapatite. Acta Cryst 1978;B34:3062-3064.

56   Engel G: Infrarotspektroskopische und röntgenographische Untersuchungen von Bleihydroxylapatit, Bleioxyapatit und Bleialkaliapatiten. J Solid State Chem 1973;6:286-292.

57   Kutoglu A von: Structure refinement of the apatite $Ca_5(VO_4)_3(OH)$. Neues Jb Miner Mh 1974;210-218.

58   Roux P, Bonel G: Sur la préparation de l'apatite carbonatée de type A, à haute température par évolution, sous pression de gaz carbonique, des arséniantes tricalcique et tristrontique. Ann Chim (Paris) 1977;2:159-165.

59   Sandell EB, Hey MH, McConnell D: The composition of francolite. Miner Mag 1939;25:395-401.

60   Bauer M: Röntgenographische und Dielektrische Untersuchungen an Apatiten. Dissertation, Fakultät für Physik, Universität Karlsruhe, 1991.

61   Jaffe EB: Abstracts of the literature on synthesis of apatites and some related phosphates. US Geol Survey Circ No 135, Washington, 1951.

62   Wondratschek H: Untersuchungen zur Kristallchemie der Blei-Apatite (Pyromorphite). Neues Jb Miner Abh 1963;99:113-160.

63   Cockbain AG: The crystal chemistry of the apatites. Miner Mag 1968;36:654-660.

64   Elliott JC: Recent progress in the chemistry, crystal chemistry and structure of the apatites. Calcif Tissue Res 1969;3:293-307.

65   Montel G: Sur l'étude physico-chimique des solides à structure d'apatite. Ann Chim (Paris) 14th Series 1969;4:255-266.

66   Sobolev VS (ed): Physics of Apatite (Spectroscopic Investigation of Apatite). Academy of Sciences of the USSR, Siberian Branch, Trans Inst Geol Geophys Issue 50, Novosibirsk, Nauka, 1975. (in Russian)

67   Montel G, Bonel G, Trombe J-C, Heughebaert J-C, Rey C: Progrès dans le domaine de la chimie des composes phosphores solides à structure d'apatite. Application à la biologie et au traitement des minerais. Pure Appl Chem 1980;52:973-989.

68   Elliott JC: The structure and function of hydroxyapatite: Proceedings of 1st International Conference on Applications of Magnetic Resonances to Dental Research, Dubrovnik, 1985. Zobozdrav Vestn 1986;41(Suppl 1):25-42.

69   Brown WE, Young RA (eds): Proceedings of International Symposium on Structural Properties of Hydroxyapatite and Related Compounds, Gaithersburg, 1968. (unpublished)

70   Colloque International sur les Phosphates Minéraux Solides, Toulouse, 1967. Bull Soc Chim Fr (special no) 1968:1693-1847.

71   Physico-Chimie et Cristallographie des Apatites d'Intérêt Biologique, Colloques Internationaux du Centre National de la Recherche Scientifique No 230 Paris, 1973. Paris, CNRS, 1975.

72   Nancollas GH (ed): Biological Mineralization and Demineralization, Dahlem Konferenzen, 1981. Berlin, Springer-Verlag, 1982.

73   Aoki H (ed): Apatite. Proceedings of the Meeting of the Japanese Association of Apatite Science. Tokyo, Japanese Association of Apatite Science. (in press)

74   Ion C (ed): 2nd International Congress on Phosphorus Compounds Proceedings, April 21-25th. Boston, 1980. IMPHOS, Paris, 1980.

75   Wallaeys R: Contribution à l'étude des apatites phosphocalciques. Ann Chim (Paris) 12th series 1952;7:808-848.

76   Montel G: Mécanismes de la synthèse de l'apatite de fluor. Ann Chim (Paris) 13th Series 1958;3:311-369.

77   Elliott JC: The Crystallographic Structure of Dental Enamel and Related Apatites. PhD Thesis, University of London, 1964.

78   Hoekstra AH: The Chemistry and Luminescence of Antimony-containing Calcium Chlorapatite. PhD Thesis, Technische Hogeschool, Eindhoven, 1967.

79   LeGeros RZ: Crystallographic Studies of the Carbonate Substitution in the Apatite Structure. PhD Thesis, New York, 1967.

80   Dykes E: Crystal Chemistry of Mono- and Divalent Anionic Substitutions in Synthetic and Biological Apatites. PhD Thesis, University of London, 1971.

81   Bonel G: Contribution à l'étude de la carbonatation des apatites I. Synthèse et étude des propriétés physico-chimiques des apatites carbonatées du type A. Ann Chim (Paris) 14th Series 1972;7:65-88.

82   Bonel G: Contribution à l'étude de la carbonatation des apatites II. Synthèse et étude des propriétés physico-chimiques des apatites carbonatées de type B. III. Synthèse et étude des propriétés physico-chimiques d'apatites carbonatées dans deux types de sites. Évolution des spectra infrarouge en fonction de la composition des apatites. Ann Chim (Paris) 14th Series 1972;7:127-144.

83   Labarthe J-C: Contribution à l'Étude de la Structure et des Propriétés des Apatites Carbonatées de Type B Phospho-calcique. Doctorat de Spécialité, Université Paul Sabatier, Toulouse, 1972.

84   Tse C: Point Defect Formation and Migration in Apatite. PhD Thesis, Princeton University. 1972.

85   Trombe J-C: Contribution à l'Étude de l'Décomposition et de la Réactivité de Certaines Apatites Hydroxylées, Carbonatées ou Fluorées Alcalino-terreues. PhD Thesis, University of Toulouse, 1972.

86   Trombe J-C: Contribution à l'étude de l'décomposition et de la réactivité de certaines apatites hydroxylées et carbonatées. Ann Chim (Paris) 14th Series 1973;8:251-269.

87   Trombe J-C: Mise en évidence d'oxygène à différentes degrés d'oxydation dans le réseau des apatites phosphocalciques et phosphostrontiques. Ann Chim (Paris) 14th Series 1973;8:335-347.

88   Vignoles C: Contribution à l'Étude de l'Influence des Ions Alcalins sur la Carbonatation dans les Sites de Type B des Apatites Phospho-calciques. Thèsis, Université Paul Sabatier, Toulouse, 1973.

89   Roufosse A: The Defect Structure of Calcium Chlorapatite. PhD Thesis. Institute of Materials Science, University of Connecticut, 1974.

90   Bartholomäi G: Spectroskopische und Modellmäßige Untersuchung der Normalschwingungen von Apatitstrukturen. PhD Thesis, University of Karlsruhe, 1975.

91   Knottnerus DIM: Point Defects in Apatites. PhD Thesis, Rijksuniversiteit, Groningen, 1976.

92   Heughebaert J-C: Contribution à l'Étude de l'Évolution des Orthophosphates de Calcium Précipités Amorphes en Orthophosphate Apatitiques. PhD Thesis. L'Institut National Polytechnique de Toulouse, 1977.

93   Lamson SH: Hydroxyapatite: MS-X-Alpha Calculations of Electronic Structure, Defects and Ion Migration. PhD Thesis, Rensselaer Polytechnic Institute, 1978.

94   Dowker SEP: Infrared Spectroscopic Studies of Thermally-Treated Carbonate-Containing Apatites. PhD Thesis, University of London, 1980.

95   Roux P: Contribution à l'Étude du Comportement et de la Synthèse des Apatites Carbonatées sous Haute Pression. Thèse d'état, Université de Toulouse, 1982.

96   Hitmi N: Étude des Transitions dans les Composantes Minérale et Organique des Tissus Calcifiés par Spectroscopie Diélectrique Basse Fréquence. PhD Thesis, Toulouse, 1983.

97   Lacout J-L: Contribution à l'Étude de l'Extraction par Vapometallurgie du Vanadium et Manganèse des Apatites. PhD Thesis, Toulouse, 1983.

98   Rey C: Étude des relations entre apatites et composés moléculaire. Thèse d'État, Institut National Polytechnique de Toulouse, 1984.

99   Vignoles-Montréjaud M: Contibution à l'Étude des Apatites Carbonatées de Type B. Thèse d'État, Institut National Polytechnique de Toulouse, 1984.

100   Taïtaï A: Étude des Phosphoapatites Calco et Strontioeuropiques qui Contiennent des Anions Bivalent Dans les Tunnels; Synthèse, Caracterisation, Fluorescence. PhD Thesis, Polytechnique de Toulouse, 1985.

101   Terpstra RA: Thermodynamic Stability and Crystal Morphology of Some Calcium Phosphates. PhD Thesis, Katholieke Universiteit, Nijmegen, 1985.

102 Khattech, I: Sur la Décomposition Thermique d'Apatites Carbonatées de Type B Synthétiques, Application aux Apatites Naturelles. Thèse de Spécialité, Tunis, 1986.

103 Chérifa, Ali Ben: Détermination de Certaines Grandeurs Thermochimiques de Phosphates de Calcium Synthétisés par Différentes Voies. Thèse de Spécialité, Tunis, 1988.

104 Bennis A: Contribution à l'Étude des Apatites Non Stœchiometriques par la Technique des Courants Thermo-Stimules. Thèse, Université Mohammed V, Rabat, 1990.

105 Best S: Characterisation, Sintering and Mechanical Behaviour of Hydroxyapatite Ceramics. PhD Thesis, University of London, 1990.

106 Dallemagne MJ, Winand L, Richelle L, Herman H, François P, Fabry C, Cloosen J: Les sels osseux. État actuel de la question. Bull Acad Roy Med Belg 7th Ser 1961;1:749-808.

107 Brown WE: Crystal growth of bone mineral. Clin Orthop Rel Res 1966;44:205-220.

108 Hayek E: Die Mineralsubstanz der Knochen. Klin Wochen 1967;45:857-863.

109 Posner AS: Crystal chemistry of bone mineral. Physiol Rev 1969;49:760-792.

110 Simpson DR: Problems of the composition and structure of the bone minerals. Clin Orthop Rel Res 1972;86:260-286.

111 Elliott JC: The problems of the composition and structure of the mineral component of the hard tissues. Clin Orthop Rel Res 1973;93:313-345.

112 Young RA: Biological apatite vs hydroxyapatite at the atomic level. Clin Orthop Rel Res 1975;113:249-262.

113 Young RA: Some aspects of crystal structure modeling of biological apatites; in: Physico-Chimie et Cristallographie des Apatites d'Intérêt Biologique, Colloques Internationaux du Centre National de la Recherche Scientifique No 230 Paris, 1973. Paris, CNRS, 1975, pp 21-40.

114 Brown WE, Chow LC: Chemical properties of bone mineral. Ann Rev Mater Sci 1976;6:213-236.

115 LeGeros RZ: Apatites in biological systems. Prog Crystal Growth Charact 1981;4:1-45.

116 Montel G, Bonel G, Heughebaert JC, Trombe JC, Rey C: New concepts in the composition, crystallization and growth of the mineral component of calcified tissues. J Cryst Growth 1981;53:74-99.

117 Driessens FCM: Mineral Aspects of Dentistry, Monographs in Oral Science 10. Basel, Karger, 1982.

118 LeGeros R: Calcium Phosphates in Oral Biology and Medicine. Monographs in Oral Science 15, Basel, Karger, 1991.

119 Driessens FCM, Verbeeck RMH: Biominerals. Boca Raton, CRC Press, 1990.

120 Neuman WF, Neuman MW: The Chemical Dynamics of Bone Mineral. Chicago, University of Chicago Press, 1958.

121 Neuman WF, Neuman MW: The nature of the mineral phase of bone. Chem Rev 1953;53:1.

122   Driessens FCM, Verbeeck RMH: The probable phase composition of the mineral in sound enamel and dentine. Bull Soc Chim Belg 1982;91:573-596.

123   Driessens FCM: The mineral in bone, dentin and tooth enamel. Bull Soc Chim Belg 1980;89:663-689.

124   Hawthorne FC (ed): Spectroscopic Methods in Mineralogy and Geology, Reviews in Mineralogy. Washington, Miner Soc Am, 1988, Vol 18.

125   Farmer VC (ed): The Infrared Spectra of Minerals. London, Mineralogical Society, 1974.

126   Elliott JC: Infrared and Raman spectroscopy of calcified tissues; in Dickson GR (ed): Methods of Calcified Tissue Preparation. Amsterdam, Elsevier, 1984, pp 413-434.

127   Young RA: Pressing the limits of Rietveld refinement. Aust J Phys 1988;41:297-310.

128   Young RA (ed): The Rietveld Method. Oxford, The Oxford University Press, 1993.

129   Elmore KJ, Farr TD: Equilibrium in the system calcium oxide-phosphorus pentoxide-water. Indust Eng Chem 1940;32:580-586.

130   Farr TD: Phosphorus, Properties of the Element and Some of Its Compounds. Chem Engin Rept No 8, Wilson Dam, Tennessee Valley Authority, 1950.

131   Bassett H: The phosphates of calcium. Part V. Revision of the earlier space diagram. J Chem Soc 1958;2949-2955.

132   Bassett H: Beiträge zum Studium der Calciumphosphate. III Das System CaO-P_2O_5-H_2O. Z Anorg Chem 1908;59:1-55.

133   Dickens B, Bowen JS: Refinement of the crystal structure of $Ca(H_2PO_4)_2.H_2O$. Acta Cryst 1971;B27:2247-2255.

134   MacLennan G, Beevers CA: The crystal structure of monocalcium phosphate monohydrate, $Ca(H_2PO_4)_2.H_2O$. Acta Cryst 1956;9:187-190.

135   Jones DW, Cruickshank DWJ: The crystal structures of two calcium orthophosphates: $CaHPO_4$ and $Ca(H_2PO_4)_2.H_2O$. Z Krist 1961;116:101-125.

136   Beevers CA, Raistrick B: Properties of the calcium phosphates. Nature 1954;173:542-543.

137   Dickens B, Prince E, Schroeder LW, Brown WE: $Ca(H_2PO_4)_2$, a crystal structure containing unusual hydrogen bonding. Acta Cryst 1973;B29;2057-2070.

138   Smith JP, Lehr JR, Brown WE: Crystallography of monocalcium and dicalcium phosphates. Am Miner 1955;40:893-899.

139   Brown WE, Smith JP, Lehr JR, Frazier AW: Crystallography of hydrated monocalcium phosphates containing potassium or ammonium. J Phys Chem 1958;62,625-627.

140   Mathew M, Takagi S, Brown WE: Planar Ca-PO_4 sheet-type structures: calcium bromide dihydrogenphosphate tetrahydrate, $CaBr(H_2PO_4).4H_2O$, and calcium iodide dihydrogenphosphate tetrahydrate, $CaI(H_2PO_4).4H_2O$. Acta Cryst 1984;C40:1662-1665.

141   Berzelius JJ: Lehrbuch der Chemie Vol 4, Arnoldischen Buchhandlung, Dresden 1836, p 274.

142 Brown WE, Mathew M, Tung MS: Crystal chemistry of octacalcium phosphate. Prog Crystal Growth Charact 1981;4:59-87.

143 Tovborg Jensen A, Gebhard Hansen K: Tetracalcium hydrogen triphosphate trihydrate, a constituent of dental calculus. Experientia 1957;13:311.

144 Schroeder HE: Formation and Inhibition of Dental Calculus. Berne, Hans Huber, 1969.

145 Muenzenberg KJ, Gebhardt M: Brushite, octacalcium phosphate, and carbonate-containing apatite in bone. Clin Orthop Rel Res 1973;90:271-273.

146 Zahidi E, Lebugle A, Bonel G: Sur une nouvelle classe de matériaux pour prothèses osseuses ou dentaires. Bull Soc Chim Fr 1985;523-527.

147 Brown WE: Octacalcium phosphate and hydroxyapatite: Crystal structure of octacalcium phosphate. Nature 1962;196:1048-1050.

148 Mathew M, Brown WE, Schroeder LW, Dickens B: Crystal structure of octacalcium bis(hydrogenphosphate) tetrakis(phosphate)pentahydrate, $Ca_8(HPO_4)_2(PO_4)_4.5H_2O$. J Cryst Spectrosc Res 1988;18:235-250.

149 Brown WE, Lehr JR, Smith JP, Frazier AW: Crystallography of octacalcium phosphate. J Am Chem Soc 1957;79:5318-5319.

150 Chickerur NS, Tung MS, Brown WE: A mechanism for incorporation of carbonate into apatite. Calcif Tissue Int 1980;32:55-62.

151 Tung MS, Brown WE: The role of octacalcium phosphate in subcutaneous heterotopic calcification. Calcif Tissue Int 1985;37:329-331.

152 Monma H: Preparation of octacalcium phosphate by the hydrolysis of α-tricalcium phosphate. J Mater Sci 1980;15:2428-2434.

153 Monma H, Goto M: Succinate-complexed octacalcium phosphate. Chem Soc Jpn 1983;56:3843-3844.

154 Newesely H: Darstellung von "Oktacalciumphosphat" (Tetracalcium-hydrogentrisphosphat) durch homogene Kristallisation. Monatsh Chem 1960;91:1020-1023.

155 LeGeros RZ, Kijkowska R, LeGeros JP: Formation and transformation of octacalcium phosphate, OCP: a preliminary report. Scanning Electron Microscopy 1984;4:1771-1777.

156 LeGeros RZ: Preparation of octacalcium phosphate (OCP): a direct fast method. Calcif Tissue Int 1985;37:194-197.

157 Brown WE, Smith JP, Lehr JR, Frazier AW: Octacalcium phosphate and hydroxyapatite: Crystallographic and chemical relations between octacalcium phosphate and hydroxyapatite. Nature 1962;196:1050-1055.

158 LeGeros RZ, Daculsi G, Orly I, Abergas T, Torres W: Solution-mediated transformation of octacalcium phosphate (OCP) to apatite. Scanning Microscopy 1989;3:129-138.

159 Terpstra RA, Bennema P: Crystal morphology of octacalcium phosphate: theory and observation. J Cryst Growth 1987;82:416-426.

160 Heughebaert JC, Nancollas GH: Kinetics of crystallization of octacalcium phosphate. J Phys Chem 1984;88:2478-2481.

161 Heughebaert J-C, Zawacki SJ, Nancollas GH: The growth of octacalcium phosphate on beta tricalcium phosphate. J Cryst Growth 1983;63:83-90.

162 Frèche M, Heughebaert JC: Calcium phosphate precipitation in the 60-80° C range. J Cryst Growth 1989;94:947-954.

163 Salimi MH, Heughebaert JC, Nancollas GH: Crystal growth of calcium phosphates in the presence of magnesium ions. Langmuir 1985;1:119-122.

164 Moriwaki Y, Doi Y, Kani T, Aoba T, Takahashi J, Okazaki M: Synthesis of enamel-like apatite at physiological temperature and pH using ion-selective membranes; in Suga S (ed): Mechanisms of Tooth Enamel Formation. Tokyo, Quintessence Publishing Co, 1983, pp 239-256.

165 Iijima M, Moriwaki Y: Effects of inorganic ions on morphology of octacalcium phosphate grown on cation selective membrane at physiological temperature and pH in relation to enamel formation. J Cryst Growth 1989;96:59-64.

166 Iijima M, Tohda H, Moriwaki Y: Growth and lamellar mixed crystals of octacalcium phosphate and apatite in a model system of enamel formation. J Cryst Growth 1992;116:319-326.

167 Sharma VK, Johnsson M, Sallis JD, Nancollas GH: Influence of citrate and phosphocitrate on the crystallization of octacalcium phosphate. Langmuir 1992;8:676-679.

168 Shyu LJ, Perez L, Zawacki SJ, Heughebaert JC, Nancollas GH: The solubility of octacalcium phosphate at 37°C in the system $Ca(OH)_2$-H_3PO_4-KNO_3-H_2O. J Dent Res 1983;62:398-400.

169 Heughebaert JC, Nancollas GH: Solubility of octacalcium phosphate at 25°C and 45°C in the system $Ca(OH)_2$-H_3PO_4-KNO_3-H_2O. J Chem Eng Data 1985;30:279-281.

170 Zhang J, Nancollas GH: Kinetics and mechanisms of octacalcium phosphate dissolution at 37 °C. J Phys Chem 1992;96:5478-5483.

171 Verbeeck RMH, Devenyns JAH: The effect of the solution Ca/P ratio on the kinetics of dissolution of octacalcium phosphate at constant pH. J Cryst Growth 1990;102:647-657.

172 Brown WE, Schroeder LW, Ferris JS: Interlayering of crystalline octacalcium phosphate and hydroxylapatite. J Phys Chem 1979;83:1385-1388.

173 Eanes ED, Meyer JL: The maturation of crystalline calcium phosphates in aqueous suspensions at physiological pH. Calcif Tiss Res 1977;23:259-269.

174 Nelson DGA, McLean JD: High-resolution electron microscopy of octacalcium phosphate and its hydrolysis products. Calcif Tissue Int 1984;36:219-232.

175 Brown WE, Mathew M, Chow LC: Roles of octacalcium phosphate in surface chemistry of apatites; in Misra DN (ed): Adsorption on and Surface Chemistry of Hydroxyapatite. New York, Plenum Press, 1984, pp 13-28.

176 Cheng P-T: Formation of octacalcium phosphate and subsequent transformation to hydroxyapatite at low supersaturation: A model for cartilage calcification. Calcif Tissue Int 1987;40:339-343.

177 Tomazic BB, Mayer I, Brown WE: Ion incorporation into octacalcium phosphate hydrolyzates. J Cryst Growth 1991;108:670-682.

178 Brown WE, Tung MS, Chow LC: Role of octacalcium phosphate in the incorporation of impurities into apatite. Proc. 2nd Int Congr Phosphorus Compounds, Boston, 1980, Ed C. Eon, IMPHOS Pub., Paris, 1980, pp 59-71.

179 Monma H, Ueno S: Uptake of fluoride ion by octacalcium phosphate and related calcium salts. Gypsum & Lime 1981;172:11-17. (in Japanese)

180 Newesely H: Thermische Umwandlungsreaktionen der Calciumhydrogen-phosphate und des Oktacalciumphosphats (Papierchromatographische Untersuchungen). Monatsh Chem 1967;98:379-389.

181 Simpson DR: Substitutions in apatite: I. Potassium-bearing apatite. Am Miner 1968;53:432-444.

182 Fowler BO, Moreno EC, Brown WE: Infra-red spectra of hydroxyapatite, octacalcium phosphate and pyrolysed octacalcium phosphate. Arch Oral Biol 1966;11:477-492.

183 Yesinowski JP, Eckert H: Hydrogen environments in calcium phosphates: ^{1}H MAS NMR at high spinning speeds. J Am Chem Soc 1987;109:6274-6282.

184 Anderson CW, Beebe RA, Kittelberger JS: Programmed temperature dehydration studies of octacalcium phosphate. J Phys Chem 1974;78:1631-1635.

185 Berry EE, Baddiel CB: Some assignments in the infra-red spectrum of octacalcium phosphate. Spectrochim Acta 1967;23A:1781-1792.

186 Šoptrajanov B, Petrov I: On the structure and symmetry of the phosphate ions in some calcium phosphates. Croatica Chem Acta 1967;39:37-45.

187 Casciani FS, Condrate RA: The infrared and Raman spectra of several calcium hydrogen phosphates; in Eon C (ed): Proceedings 2nd International Conference on Phosphorus Compounds, Boston, 1980. Paris, IMPHOS, 1980, pp 175-190.

188 Fowler BO, Marković M, Brown WE: Octacalcium phosphate. 3. Infrared and Raman vibrational spectra. Chem Mater 1993;5:1417-1423.

189 Berry EE: The structure and composition of some calcium-deficient apatites. J Inorg Nucl Chem 1967;29:317-327.

190 Rey C, Shimizu M, Collins B, Glimcher MJ: Resolution-enhanced Fourier transform infrared spectroscopy study of the environment of phosphate ion in the early deposits of a solid phase of calcium phosphate in bone and enamel and their evolution with age: 2. Investigations in the v_3 PO_4 domain. Calcif Tissue Int 1991;49:383-388.

191 Rey C, Shimizu M, Collins B, Glimcher MJ: Resolution-enhanced Fourier transform infrared spectroscopy study of the environment of phosphate ions in the early deposits of a solid phase of calcium-phosphate in bone and enamel, and their evolution with age. I: Investigations in the v_4 PO_4 domain. Calcif Tissue Int 1990;46:384-394.

192 Rothwell WP, Waugh JS, Yesinowski JP: High-resolution variable-temperature ^{31}P NMR of solid calcium phosphates. J Am Chem Soc 1980;102:2637-2643.

193 Miquel JL, Facchini L, Legrand AP, Marchandise X, Lecouffe P, Chanavaz M, Donazzan M, Rey C, Lemaitre J: Characterisation and conversion study into natural living bone of calcium phosphate bioceramics by solid state NMR spectroscopy. Clinical Mater 1990;5:115-125.

194 Monma H: The incorporation of dicarboxylates into octacalcium bis(hydrogenphosphate) tetrakis(phosphate) pentahydrate. Chem Soc Jpn 1984;57:599-600.

195 Monma H, Goto M: Complexes of apatitic layered compound $Ca_8(HPO_4)_2(PO_4)_4.5H_2O$. J Inclusion Phenomena 1984;2:127-134.

196 Marković M, Fowler BO, Brown WE: Incorporation of polycarboxylate ions into the octacalcium phosphate structure. Abstract No 1259 J Dent Res (Special Issue) 1988;67:270.

197 Marković M, Fowler BO, Brown WE: Octacalcium phosphate carboxylates. 1. Preparation and identification. Chem Mater 1993;5:1401-1405.

198 Marković M, Fowler BO, Brown WE: Octacalcium phosphate carboxylates. 2. Characterization and structural considerations. Chem Mater 1993;5:1406-1416.

199 Marcović M, Fowler BO, Brown WE: Octacalcium phosphate carboxylates. 4. Kinetics of formation and solubility of octacalcium phosphate succinate, J Cryst Growth (in press)

200 Monma H, Goto M: Thermal alteration of succinate-complexed octacalcium phosphate. J Mater Sci Lett 1985;4:147-150.

201 Mathew M, Brown WE: A structural model for octacalcium phosphate-succinate double salt. Bull Chem Soc Jpn 1987;60:1141-1143.

202 Moore GE: On brushite, a new mineral in phosphatic guano. Am J Sci 1865;39:43-44.

203 Lonsdale K: Human stones: Science 1968;159:1199-1207.

204 Borggreven JMPM, Driessens FCM, Dijk JWE van: Dissolution and precipitation reactions in human tooth enamel under weak acid conditions. Arch Oral Biol 1986;31:139-144.

205 Chow LC, Brown WE: Reaction of dicalcium phosphate dihydrate with fluoride. J Dent Res 1973;52:1220-1227.

206 Chow LC, Guo MK, Hsieh CC, Hong YC: Apatitic fluoride increase in enamel from a topical treatment involving intermediate $CaHPO_4.2H_2O$ formation. Caries Res 1981;15:369-376.

207 Lénárt G, Bidló G, Pintér J: Some basic problems in the examination of the calcium hydrogen phosphates of bone. Clin Orthop Rel Res 1972;83;263-272.

208 Bonar LC, Grynpas MD, Glimcher MJ: Failure to detect crystalline brushite in embryonic chick and bovine bone by X-ray diffraction. J Ultrastruct Res 1984;86:93-99.

209 Roberts JE, Bonar LC, Griffin RG, Glimcher MJ: Characterization of very young mineral phases of bone by solid state [31]phosphorus magic angle sample spinning nuclear magnetic resonance and X-ray diffraction. Calcif Tissue Int 1992;50:42-48.

210 Beevers CA: The crystal structure of dicalcium phosphate dihydrate, $CaHPO_4.2H_2O$. Acta Cryst 1958;11:273-277.

211 Jones DW, Smith JAS: The structure of brushite, $CaHPO_4.2H_2O$. J Chem Soc 1962;1414-1420.

212 Jones DW, Smith JAS: Proton magnetic resonance in brushite. Trans Faraday Soc 1960;56:638-647.

213 Curry NA, Jones DW: Crystal structure of brushite, calcium hydrogen orthophosphate dihydrate: a neutron-diffraction investigation. J Chem Soc A 1971;3725-3729.

214 Atoji M, Rundle RE: Neutron diffraction study of gypsum, $CaSO_4 2H_2O$. J Chem Phys 1958;29:1306-1311.

215 Cole WF, Lancucki CJ: A refinement of the crystal structure of gypsum $CaSO_4.2H_2O$. Acta Cryst Sect B 1974;B30:921-929.

216 Sakae T, Nagata H, Sudo T: The crystal structure of synthetic calcium phosphate-sulphate hydrate, $Ca_2HPO_4SO_4.4H_2O$, and its relation to brushite and gypsum. Am Miner 1978;63:520-527.

217 Tovborg Jensen A, Rathlev J: Calcium hydrogen orthophosphate 2-hydrate and calcium hydrogen orthophosphate. Inorganic Syntheses 1953;4:19-22.

218 Marshall RW, Nancollas GH: The kinetics of crystal growth of dicalcium phosphate dihydrate. J Phys Chem 1969;73:3838-3844.

219 Christoffersen MR, Christoffersen J: The kinetics of crystal growth and dissolution of calcium monohydrogen phosphate dihydrate. J Cryst Growth 1988;87:51-61.

220 St. Pierre PDS: The preparation of dicalcium phosphate dihydrate and calcium pyrophosphate. J Am Chem Soc 1955;77:2197-2198.

221 Moreno EC, Brown WE, Osborn G: Solubility of dicalcium phosphate dihydrate in aqueous systems. Soil Sci Soc Am Proc 1960;24:94-98.

222 Egan EP, Wakefield ZT: Low temperature heat capacity and entropy of dicalcium phosphate dihydrate, 10° to 310 °K. J Chem Engin Data 1964;9:544-545.

223 Aia MA, Goldsmith RL, Mooney RW: Precipitating stoichiometric $CaHPO_4.2H_2O$. Ind Eng Chem 1961;53:55-57.

224 Huffman EO, Cate WE, Deming ME, Elmore KL: Rates of solution of calcium phosphates in phosphoric acid solutions. J Agr Food Chem 1957;5:266-275.

225 LeGeros RZ, LeGeros JP: Brushite crystal growth by diffusion in silica gel and in solution. J Cryst Growth 1972;13/14:476-480.

226 Nancollas GH, Wefel JS: The effect of stannous fluoride, sodium fluoride and stannous chloride on the crystallization of dicalcium phosphate dihydrate at constant pH. J Cryst Growth 1974;23:169-176.

227 Hohl H, Koutsoukos PG, Nancollas GH: The crystallization of hydroxyapatite and dicalcium phosphate dihydrate; representation of growth curves. J Cryst Growth 1982;57:325-335.

228 Hassan KAR: The Microelectrophoretic Behavior of Sparingly Soluble Salts. MS Thesis, SUNY, Buffalo, 1984. (cited in ref. [24])

229 Christoffersen J, Christoffersen MR: Spiral growth and dissolution models with rate constants related to the frequency of partial dehydration of cations and to the surface tension. J Cryst Growth 1988;87:41-50.

230 Barone JP, Nancollas GH, Tomson M: The seeded growth of calcium phosphates. The kinetics of growth of dicalcium phosphate dihydrate on hydroxyapatite. Calcif Tissue Res 1976;21:171-182.

231 Barone JP, Nancollas GH: The growth of calcium phosphates on hydroxyapatite crystals. The effect of fluoride and phosphonate. J Dent Res 1978;57:735-742.

232 Heughebaert JC, de Rooij JF, Nancollas GH: The growth of dicalcium phosphate dihydrate on octacalcium phosphate at 25°C. J Cryst Growth 1986;77:192-198.

233 García-Ramos JV, Carmona P: The effect of some homopolymers on the crystallization of calcium phosphates. J Cryst Growth 1982;57:336-342.

234 Shyu LJ: The Solid/Solution Interface—A Kinetic Study of the Crystallization of Calcium Fluoride and Phosphate Salts. PhD Thesis, SUNY, Buffalo, 1984. (cited in ref. [24])

235 Kemenade van MJJM, Bruyn de PL: The influence of casein on the precipitation of brushite and octacalcium phosphate. Colloids and Surfaces 1989;36:359-368.

236 Nancollas GH, Marshall RW: Kinetics of dissolution of dicalcium phosphate dihydrate crystals. J Dent Res 1971;50:1268-1272.

237 Zhang J, Nancollas GH: Interpretation of dissolution kinetics of dicalcium phosphate dihydrate. J Cryst Growth 1992;125:251-269.

238 Patel PR, Gregory TM, Brown WE: Solubility of $CaHPO_4.2H_2O$ in the quaternary system $Ca(OH)_2$-H_3PO_4-NaCl-H_2O at 25°C. J Res Natn Bur Stand 1974;78A:675-681.

239 Tung MS, Chow LC, Brown WE: Hydrolysis of dicalcium phosphate dihydrate in the presence or absence of calcium fluoride. J Dent Res 1985;64:2-5.

240 Booth DH, Coates RV: The stability of calcium hydrogen phosphate precipitated from solutions of calcium nitrate and phosphoric acid. J Chem Soc 1961;4914-4921.

241 Lemp R: Étude du mécanisme cinétique de l'hydrolyse de l'orthophosphate dicalcique dihydraté en hydroxylapatite. Chimia 1971;25:317-325.

242 Ozawa T, Ujiie T, Tamura K: Dehydration of calcium hydrogenphosphate dihydrate in water. J Chem Soc Jpn 1980;1352-1357. (in Japanese)

243 Perez L, Shyu LJ, Nancollas GH: The phase transformation of calcium phosphate dihydrate into octacalcium phosphate in aqueous suspensions. Colloids and Surfaces 1989;38:295-304.

244 Zhang J, Ebrahimpour A, Nancollas GH: Dual constant composition studies of phase transformation of dicalcium phosphate dihydrate into octacalcium phosphate. J Colloid Interface Sci 1992;152:132-140.

245 Eidelman N, Chow LC, Brown WE: Calcium phosphate phase transformations in serum. Calcif Tissue Int 1987;41:18-26.

246 Duff EJ: Orthophosphates. Part II. The transformations brushite → fluoroapatite and monetite → fluoroapatite in aqueous potassium fluoride solution. J Chem Soc (A) Inorg Phys Theor 1971;33-38.

247 Wei SHY, Tang TE, Wefel JS: Reactions of dicalcium phosphate dihydrate with fluoride solutions. J Dent Res 1974;53:1145-1154.

248 Rowles SL: The precipitation of whitlockite from aqueous solutions. Bull Soc Chim Fr 1968;1797-1802.

249 Arai Y, Yasue T: Mechanical dehydration of brushite. J Chem Soc Jpn, Ind Chem Sec 1971;74:1343-1348. (In Japanese)

250 McIntosh AO, Jablonski WL: X-ray powder patterns of the calcium phosphates. Analytical Chem 1956;28:1424-1427.

251 Rabatin JG, Gale RH, Newkirk AE: The mechanism and kinetics of the dehydration of calcium hydrogen phosphate dihydrate. J Phys Chem 1960;64:491-493.

252 Yasue T, Suzuki T, Arai Y: Dehydration and condensation of calcium hydrogenphosphate dihydrate. J Chem Soc Jpn, Chem and Ind Chem 1983:494-500. (in Japanese)

253 Kanazawa T, Chikazawa M, Sano Y: Change in surface properties during thermal treatment of powderly $CaHPO_4.2H_2O$. J Soc Mater Sci Jpn 1982;31:855-859. (in Japanese)

254 Wikholm NW, Beebe RA, Kittelberger JS: Kinetics of the conversion of monetite to calcium pyrophosphate. J Phys Chem 1975;79:853-856.

255 Berry EE, Baddiel CB: The infra-red spectrum of dicalcium phosphate dihydrate (brushite). Spectrochim Acta 1967;23A:2089-2097.

256 Petrov I, Šoptrajanov B, Fuson N, Lawson JR: Infra-red investigation of dicalcium phosphates. Spectrochim Acta Part A 1967;23A:2637-2646.

257 Casciani F, Condrate RA: The vibrational spectra of brushite, $CaHPO_4.2H_2O$. Spectrosc Lett 1979;12:699-713.

258 Shepard CU: On two new minerals, monetite and monite, with a notice of pyroclasite. 1882;23:400-405.

259 Catti M, Ferraris G, Filhol A: Hydrogen bonding in the crystalline state. $CaHPO_4$ (monetite), $P\overline{1}$ or P1? A novel neutron diffraction study. Acta Cryst 1977;B33:1223-1229.

260 Catti M, Ferraris G, Mason SA: Low-temperature ordering of hydrogen atoms in $CaHPO_4$ (monetite): X-ray and neutron diffraction study at 145 K. Acta Cryst 1980;B36:254-259.

261 Dickens B, Bowen JS, Brown WE: A refinement of the crystal structure of $CaHPO_4$ (synthetic monetite). Acta Cryst 1972;B28:797-806.

262 Egan EP, Wakefield ZT: Low temperature heat capacity and entropy of anhydrous dicalcium phosphate, 10° to 310 °K. J Chem Engin Data 1964;9:541-544.

263 MacLennan G, Beevers CA: The crystal structure of dicalcium phosphate, $CaHPO_4$. Acta Cryst 1955;8:579-583.

264 Vasserman IM, Silant'eva NI: Preparation of calcium hydrogen phosphate of stoichiometric composition. Russ J Inorg Chem 1965;10:717-721.

265 Chughtai A, Marshall R, Nancollas GH: Complexes in calcium phosphate solutions. J Phys Chem 1968;72:208-211.

266 Ishikawa K, Eanes ED: The hydrolysis of anhydrous dicalcium phosphate into hydroxyapatite. J Dent Res 1993;72:474-480.

267 LeGeros RZ, LeGeros JP, Trautz OR, Shirra WP: Conversion of monetite, $CaHPO_4$, to apatites: effects of carbonate on the crystallinity and the morphology of the apatite crystallites; in Barrett CS, Newkirk JB, Ruud CO (eds): Advances in X-ray Analysis. New York, Plenum, 1971, vol 14, pp 57-66.

268 LeGeros RZ, Shirra WP, Miravite MA,, LeGeros JP: Amorphous calcium phosphates: synthetic and biological; in: Physico-Chimie et Cristallographie des Apatites d'Intérêt Biologique, Colloques Internationaux du Centre National de la Recherche Scientifique No 230 Paris, 1973. Paris, CNRS, 1975, pp 105-115.

269 LeGeros RZ, Daculsi G, Kijkowska R, Kerebel B: The effect of magnesium on the formation of apatites and whitlockites; Itokawa Y, Durlach J (eds): Magnesium in Health and Disease. John Libbey, 1989, pp 11-19.

270 Casciani F, Condrate RA: The Raman spectrum of monetite, $CaHPO_4$. J Solid State Chem 1980;34:385-388.

271 Tovborg Jensen A, Rowles SL: Magnesian whitlockite, a major constituent of human dental calculus. Acta Odont Scand 1957;15:121-139.

272 Trömel G: Beiträge zur Kenntnis des Systems Kalziumoxyd-Phosphorpentoxyd. Mitt Kaiser-Wilhelm-Inst Eisenforschg, Düsseldorf 1932;14:25-34.

273 Welch JH, Gutt W: High-temperature studies of the system calcium oxide-phosphorus pentoxide. J Chem Soc 1961;4442-4444.

274 Nurse RW, Welch JH, Gutt W: High-temperature equilibria in the system dicalcium silicate-tricalcium phosphate. J Chem Soc 1959;1077-1083.

275 Roux P, Louër D, Bonel G: Sur une nouvelle forme cristalline du phosphate tricalcique. Compt Rend Acad Sci (Paris) Ser C 1978;286;549-551.

276 Romdhane SS, Bacquet G, Bonel G: Étude des phase β et haute pression du phosphate tricalcique par la RPE de l'ion Cu^{2+}. J Solid State Chem 1981;40:34-41.

277 Frondel C: Whitlockite: a new calcium phosphate, $Ca_3(PO_4)_2$. Am Miner 1941;26:145-152.

278 Calvo C, Gopal R: The crystal structure of whitlockite from the Palermo Quarry. Am Miner 1975;60:120-133.

279 Vahl J von, Höhling HJ, Frank RM: Elektronenstrahlbeugung an rhomboedrisch aussehenden Mineralbildungen in kariösem Dentin. Arch Oral Biol 1964;9:315-320.

280 Kodaka T, Debari K, Abe M: Hexahedrally based crystals in human tooth enamel. Caries Res 1992;26:69-76.

281 Lowenstam, HA: Minerals formed by organisms. Science 1981;211:1126-1131.

282 Ishiyama M, Sasagawa I, Akai J: The inorganic content of pleromin in tooth plates of the living holocephalan, *Chimaera phantasma*, consists of a crystalline calcium phosphate known as β-$Ca_3(PO_4)_2$ (whitlockite). Arch Histol Japon 1984;47:89-94.

283 Mathew M, Schroeder LW, Dickens B, Brown WE: The crystal structure of α-$Ca_3(PO_4)_2$. Acta Cryst 1977;B33:1325-1333.

284 Monma H, Nagai M: Calcium orthophosphates; in Kanazawa T (ed): Inorganic Phosphate Materials, Materials Science Monograph 52. Amsterdam, Elsevier, 1989, Ch 4, pp 79-103.

285 Mackay AL: A preliminary examination of the structure of α-$Ca_3(PO_4)_2$. Acta Cryst 1953;6:743-744.

286 Dickens B, Schroeder LW, Brown WE: Crystallographic studies of the role of Mg as a stabilizing impurity in β-$Ca_3(PO_4)_2$. I. The crystal structure of pure β-$Ca_3(PO_4)_2$. J Solid State Chem 1974;10:232-248.

287 Gopal R, Calvo C, Ito J, Sabine WK: Crystal structure of synthetic Mg-whitlockite, $Ca_{18}Mg_2H_2(PO_4)_{14}$. Can J Chem 1974;52:1155-1164.

288 Schroeder LW, Dickens B, Brown WE: Crystallographic studies of the role of Mg as a stabilizing impurity in β-$Ca_3(PO_4)_2$. II. Refinement of Mg-containing β-$Ca_3(PO_4)_2$. J Solid State Chem 1977;22:253-262.

289 Süsse P, Buerger MJ: The structure of $Ba_3(PO_4)_2$. Z Krist 1970;131:161-174.

290 Koelmans H, Engelsman JJ, Admiral PS: Low-temperature phase transitions in β-$Ca_3(PO_4)_2$ and related compounds. J Phys Chem Solids 1959;11:172-173.

291 Terpstra RA, Driessens FCM, Schaeken HG, Verbeeck RMH: The whitlockite phase in the system CaO-P_2O_5-MgO at 1000 °C. Z Anorg Allg Chem 1983;507:206-212.

292 Clement D, Tristan JM, Hamad M, Roux P, Heughebaert JC: Étude de la substitution Mg^{2+}/Ca^{2+} dans l'orthophosphate tricalcique β. J Solid State Chem 1989;78:271-280.

293 Dickens B, Brown WE: The crystal structure of $Ca_7Mg_9(Ca,Mg)_2(PO_4)_{12}$. Tschermaks Miner Petrogr Mitt 1971;16:79-104.

294 Kostiner E, Rea JR: The crystal structure of manganese-whitlockite, $Ca_{18}Mn_2H_2(PO_4)_{14}$. Acta Cryst 1976;B32:250-253.

295 Schneiderhöhn H: Mikroskopisch-optische Untersuchungen der Schmelzen. Mitt Kaiser-Wilhelm-Inst Eisenforschg, Düsseldorf 1932;14:34-36.

296 Terpstra RA, Driessens FCM, Verbeeck RMH: The CaO-MgO-P_2O_5 system at 1000°C for $P_2O_5 \leq 33.3$ mole %. Z Anorg Allg Chem 1984;515:213-224.

297 Verbeeck RMH, Bruyne PAM De, Driessens FCM, Terpstra RA, Verbeek F: Solubility behaviour of Mg-containing β-$Ca_3(PO_4)_2$. Bull Soc Chim Belg 1986;95:455-476.

298 Heughebaert J-C, Montel G: Sur l'existence d'une série de solides de composition variable, correspondant au phosphate tricalcique précipité. Compt Rend Acad Sci (Paris) 1970;270:1585-1588.

299 Hayek E, Newesely H: Über die Existenz von Tricalciumphosphat in wässriger Lösung. Monatsh Chem 1958;89:88-95.

300 Chickerur NS, Lenka RC, Sabat BB, Nayak GH: Solubility behaviour of synthetic whitlockite containing magnesium in aqueous medium. Ind J Chem 1986:25A;181-182.

301 Chickerur NS, Nayak GH, Lenka RC, Mahapatra PP: Hydrolysis of dicalcium phosphate dihydrate in presence of magnesium & fluoride ions in aqueous media. Indian J Biochem Biophys 1983;20:315-317.

302   Hamad M, Heughebaert JC: The growth of whitlockite. J Cryst Growth 1986;79:192-197.

303   Roy DM, Linnehan SK: Hydroxyapatite formed from coral skeletal carbonate by hydrothermal exchange. Nature 1974;247:220-222.

304   Monma H, Ueno S, Kanazawa T. Properties of hydroxyapatite prepared by the hydrolysis of tricalcium phosphate. J Chem Tech Biotechnol 1981;31:15-24.

305   Monma H, Kanazawa T: The hydration of α-tricalcium phosphate. J Ceram Soc Jpn 1976;84:209-213.

306   Monma H, Ueno S, Tsutsumi M: Effects of water-soluble additives on the properties of hardened apatite. Gypsum & Lime 1978;156:6-11. (in Japanese)

307   Monma H, Goto M, Kohmura T: Effect of additives on hydration and hardening of tricalcium phosphate. Gypsum & Lime 1984;188:11-16. (in Japanese)

308   Fix W, Heymann H, Heinke R: Subsolidus relations in the system $2CaO \cdot SiO_2$-$3CaO \cdot P_2O_5$. J Am Ceram Soc 1969;52:346-347.

309   Kreidler ER, Hummel FA: Phase relationships in the system $SrO-P_2O_5$ and the influence of water vapor on the formation of $Sr_4P_2O_9$. Inorg Chem 1967;6:884-891.

310   Ando J: Tricalcium phosphate and its variation. Bull Chem Soc Jpn 1958;31:196-201.

311   Hilgenstock G: Stähl u Eisen 1883;3:498. (cited in ref. [132])

312   Brown WE, Epstein EF: Crystallography of tetracalcium phosphate. J Res Natn Bur Stand 1965;69A:547-551.

313   Brown WE, Chow LC: A new calcium phosphate, water-setting cement; in Brown PW (ed): Cements Research Progress 1986. Westerfield, American Ceramic Society, 1986, pp 351-379.

314   Brown WE, Chow LC: Combinations of sparingly soluble calcium phosphates in slurries and pastes as remineralizers and cements. US Patent 1986, No 4,612,053.

315   Xie L, Monroe EA: The hydrolysis of tetracalcium phosphate and other calcium orthophosphates; in Yamamuro T, Hench LL, Wilson J (eds): CRC Handbook of Bioactive Ceramics, Vol II, Calcium Phosphate and Hydroxylapatite Ceramics. Boca Raton, CRC Press, 1990, pp 29-37.

316   Dickens B, Brown WE, Kruger GJ, Stewart JM: $Ca_4(PO_4)_2O$, tetracalcium diphosphate monoxide. Crystal structure and relationships to $Ca_5(PO_4)_3OH$ and $K_3Na(SO_4)_2$. Acta Cryst 1973;B29:2046-2056.

317   Bauer H, Balz W: Über Erdalkaliphosphate, -arsenate und -vanadate vom Typus $4MeO.X_2O_5$. Z Anorg Allg Chem 1965;340:225-231.

318   Monma H, Goto M, Nakajima H, Hashimoto H: Preparation of tetracalcium phosphate. Gypsum & Lime 1986;202:17-21. (in Japanese)

319   Cieśla K, Rudnicki R: Synthesis and transformation of tetracalcium phosphate in solid state. Part I. Synthesis of roentgenographically pure tetracalcium phosphate from calcium dibasic phosphate and calcite. Pol J Chem 1987;61:719-727.

320 Cieśla K, Rudnicki R: Synthesis and transformations of tetracalcium phosphate in solid state. Part II. Studies on synthesis of tetracalcium phosphate by the thermal analysis method. Pol J Chem 1988;62:31-39.

321 Bredig MA, Frank HH, Füldner H: Beiträge zur Kenntnis der Kalk-Phosphorsäure-Verbindungen. Z Elektrochem 1932;38:158-164.

322 Bredig MA, Frank HH, Füldner H: Beiträge zur Kenntnis der Kalk-Phosphorsäure-Verbindungen II. Z Elektrochem 1933;39:959-969.

323 Trömel G: Untersuchungen über die Bildung eines halogenfreien Apatits aus basischen Calciumphosphaten. Z Physik Chem 1932;158:422-432.

324 Schleede A, Schmidt W, Kindt H: Zur Kenntnis der Calciumphosphate und Apatite. Z Elektrochem 1932;38:633-641.

325 Termine JD, Posner AS: Infrared analysis of rat bone: age dependency of amorphous and crystalline fractions. Science 1966;153:1523-1525.

326 Grynpas MD, Bonar LC, Glimcher MJ: X-ray diffraction radial distribution function studies on bone mineral and synthetic calcium phosphates. J Mater Sci 1984;19:723-736.

327 Posner AS, Betts F: Synthetic amorphous calcium phosphate and its relation to bone mineral. Accounts Chem Res 1975;8:273-281.

328 Heughebaert JC, Montel G: Sur la transformation des phosphates amorphes en phosphates apatitiques par réaction intracrystalline; in: Physico-Chimie et Cristallographie des Apatites d'Intérêt Biologique, Colloques Internationaux du Centre National de la Recherche Scientifique No 230 Paris, 1973. Paris, CNRS, 1975, pp 283-293.

329 Heughebaert J-C, Montel G: Conversion of amorphous tricalcium phosphate into apatitic tricalcium phosphate. Calcif Tissue Int 1982;34:S103-S108.

330 Larsen MJ, Jensen SJ: Solubility study of the initial formation of calcium orthophosphates from aqueous solutions at pH 5-10. Arch Oral Biol 1986;31:565-572.

331 Christoffersen J, Christoffersen MR, Kibalczyc W and Andersen FA: A contribution to the understanding of the formation of calcium phosphates. J Cryst Growth 1989;94:767-777.

332 Eanes ED, Gillessen IH, Posner AS: Intermediate states in the precipitation of hydroxyapatite. Nature 1965;208:365-367.

333 Termine JD, Posner AS: Calcium phosphate formation *in vitro*. I. factors affecting initial phase separation. Arch Biochem Biophys 1970;140:307-317.

334 Holmes JM, Beebe RA: Surface areas by gas adsorption on amorphous calcium phosphate and crystalline hydroxyapatite. Calcif Tissue Res 1971;7:163-174.

335 Termine JD, Eanes ED: Comparative chemistry of amorphous and apatitic calcium phosphate preparations. Calcif Tissue Res 1972;10:171-197.

336 Boskey AL, Posner AS: Conversion of amorphous calcium phosphate to microcrystalline hydroxyapatite. A pH-dependent solution-mediated, solid-solid conversion. J Phys Chem 1973;77:2313-2317.

337 Aoba T, Moriwaki Y, Doi Y, Okazaki M, Takahashi J, Yagi T: Diffuse X-ray scattering from apatite crystals and its relation to amorphous bone mineral. J Osaka Dent Univ Sch 1980;20:81-90.

338 Harries JE, Hukins DWL, Holt C, Hasnain SS: Conversion of amorphous calcium phosphate into hydroxyapatite investigated by EXAFS spectroscopy. J Cryst Growth 1987;84:563-570.

339 Greenfield DJ, Eanes ED: Formation chemistry of amorphous calcium phosphates prepared from carbonate containing solutions. Calcif Tissue Res 1972;9:152-162.

340 Holt C, Kemenade MJJM van, Harries JE, Nelson LS, Bailey RT, Hukins DWL, Hasnain SS, Bruyn PL de: Preparation of amorphous calcium-magnesium phosphates at pH 7 and characterization by X-ray absorption and Fourier transform infrared spectroscopy. J Cryst Growth 1988;92:239-252.

341 Holt C, Kemenade MJJM van, Nelson LS, Hukins DWL, Bailey RT, Harries JE, Hasain SS, Bruyn PL de: Amorphous calcium phosphates prepared at pH 6.5 and 6.0. Mater Res Bull 1989;23:55-62.

342 Walton AG, Friedman BA, Schwartz A: Nucleation and mineralization of organic matrices. J Biomed Mater Res 1967;1:337-354.

343 Blumenthal NC, Posner AS, Holmes JM: Effect of preparation conditions on the properties and transformation of amorphous calcium phosphate. Mater Res Bull 1972;7:1181-1190.

344 Puech J, Heughebaert J-C, Montel G: A new mode of growing apatite crystallites. J Cryst Growth 1982;56:20-24.

345 Roberts JE, Heughebaert M, Heughebaert J-C, Bonar LC, Glimcher MJ, Griffin RG: Solid state ^{31}P NMR studies of the conversion of amorphous tricalcium phosphate to apatitic tricalcium phosphate. Calcif Tissue Int 1991;49:378-382.

346 Kaufman HW, Kleinberg I: An X-ray crystallographic examination of calcium phosphate formation in $Ca(OH)_2/H_3PO_4$ mixtures. Calcif Tissue Res 1977;22:253-264.

347 Wuthier RE, Rice GS, Wallace JEB, Weaver RL, LeGeros RZ, Eanes ED: *In vitro* precipitation of calcium phosphate under intracellular conditions: formation of brushite from an amorphous precursor in the absence of ATP. Calcif Tissue Int 1985;37:401-410.

348 Füredi-Milhofer H, Purgarić B, Brečević Lj, Pavković N: Precipitation of calcium phosphates from electrolyte solutions. I. A study of the precipitates in the physiological pH region. Calcif Tissue Res 1971;8:142-153.

349 Meyer JL, Eanes ED: A thermodynamic analysis of the amorphous to crystalline calcium phosphate transformation. Calcif Tissue Res 1978;25:59-68.

350 Tung MS, Brown WE: An intermediate state in hydrolysis of amorphous calcium phosphate. Calcif Tissue Int 1983;35:783-790.

351 Christoffersen MR, Christoffersen J, Kibalczyc W: Apparent solubilities of two amorphous calcium phosphates and of octacalcium phosphate in the temperature range 30-42°C. J Cryst Growth 1990;106:349-354.

352 Kibalczyc W, Christoffersen J, Christoffersen MR, Zielenkiewicz A and Zielenkiewicz W: The effect of magnesium ions on the precipitation of calcium phosphates. J Cryst Growth 1990;106:355-366.

353 Termine JD, Peckauskas RA, Posner AS: Calcium phosphate formation *in vitro*. II. Effects of environment on amorphous-crystalline transformation. Arch Biochem Biophys 1970;140:318-325.

354 Fleisch H, Russell RGG, Bisaz S, Termine JD, Posner AS: Influence of pyrophosphate on the transformation of amorphous to crystalline calcium phosphate. Calcif Tissue Res 1968;2:49-59.

355 Greenfield DJ, Termine JD, Eanes ED: A chemical study of apatites prepared by hydrolysis of amorphous calcium phosphates in carbonate-containing aqueous solutions. Calcif Tissue Res 1974;14:131-138.

356 Nawrot CF, Campbell DJ, Schroeder JK, Valkenburg M van: Dental phosphoprotein-induced formation of hydroxylapatite during in vitro synthesis of amorphous calcium phosphate. Biochemistry 1976;15:3445-3449.

357 Root MJ: Inhibition of the amorphous calcium phosphate phase transformation reaction by polyphosphates and metal ions. Calcif Tissue Int 1990;47:112-116.

358 Umegaki T, Yamashita S, Kanazawa T: Heat of solution of amorphous calcium phosphate in phosphoric acid. Gypsum & Lime 1983;183:11-15.

359 Eanes ED, Posner AS: Kinetics and mechanism of conversion of noncrystalline calcium phosphate to crystalline hydroxyapatite. Trans NY Acad Sci 1965;28:233-241.

360 Boskey AL, Posner AS: Magnesium stabilization of amorphous calcium phosphate: a kinetic study. Mater Res Bull 1974;9:907-916.

361 Chapman AC, Thirlwell LE: Spectra of phosphorus compounds-I. The infra-red spectra of orthophosphates. Spectrochim Acta 1964;20:937-947.

362 Termine JD, Posner AS: Infra-red determination of the percentage of crystallinity in apatitic calcium phosphates. Nature 1966;211:268-270.

363 Tropp J, Blumenthal NC, Waugh JS: Phosphorus NMR study of solid amorphous calcium phosphate. J Am Chem Soc 1983;105:22-26.

364 Belton PS, Harris RK, Wilkes PJ: Solid-state phosphorus-31 NMR studies of synthetic inorganic calcium phosphates. J Phys Chem Solids 1988;49:21-27.

365 Nylen MU, Eanes ED, Omnell KÅ: Crystal growth in rat enamel. J Cell Biol 1963;18:109-123.

366 Bienenstock A, Posner AS: Calculation of the X-ray intensities from arrays of small crystallites of hydroxyapatite. Arch Biochem Biophys 1968;124:604-607.

367 Betts F, Posner AS: An X-ray radial distribution study of amorphous calcium phosphate. Mater Res Bull 1974;9:353-360.

368 Harries JE, Hukins DWL, Hasnain SS: Analysis of the EXAFS spectrum of hydroxyapatite. J Phys C: Solid State Phy 1986;19:6859-6872.

369 Nelson LS, Holt C, Harries JE, Hukins DWL: Amorphous calcium phosphates of different composition give very similar EXAFS spectra. Physica B 1989;158:105-106.

370 Eanes ED: Thermochemical studies on amorphous calcium phosphate. Calcif Tissue Res 1970;5:133-145.

371 Umegaki T, Shiba S, Kanazawa T: Thermal change of amorphous calcium phosphate; effect of preparative conditions. Yogyo-Kyokai-Shi (J Ceram Soc Jpn) 1984;92:612-616.

372 Egan EP, Wakefield ZT: Thermodynamic properties of calcium pyrophosphate, 10 to 1700 °K. J Am Chem Soc 1957;79:558-561.

373 Kanazawa T, Umegaki T, Uchiyama N: Thermal crystallisation of amorphous calcium phosphate to α-tricalcium phosphate. J Chem Tech Biotechnol 1982;32:399-406.

374 Houben JL: Free radicals produced by ionizing radiation in bone and its constituents. Int J Radiat Biol 1971;20:373-389.

375 Doi Y, Aoba T, Takahashi J, Okazaki M, Moriwaki Y: Analysis of electron-excess and electron-deficient centers in X-ray-irradiated tricalcium phosphates by electron spin resonance spectroscopy. Calcif Tissue Int 1979;29:239-244.

376 McKeag AH, Ranby PW: Improvements in luminescent materials. British Patent 1942, No 578,192.

377 Davis TS, Kreidler ER, Parodi JA, Soules TF: The luminescent properties of antimony in calcium halophosphates. J Luminescence 1971;4:48-62.

378 Mazelsky R, Ohlmann RC, Steinbruegge K: Crystal growth of a new laser material, fluorapatite. J Electrochem Soc Solid State Sci 1968;115:68-70.

379 Ohlmann RC, Steinbruegge KB, Mazelsky R: Spectroscopic and laser characteristics of neodymium-doped calcium fluorophosphate. Appl Optics 1968;7:905-914.

380 Steinbruegge KB, Henningsen T, Hopkins RH, Mazelsky R, Melamed NT, Riedel EP, Roland GW: Laser properties of Nd^{+3} and Ho^{+3} doped crystals with the apatite structure. Appl Optics 1972;11:999-1012.

381 Young EJ, Myers AT, Munson EL, Conklin NM: Mineralogy and geochemistry of fluorapatite from Cerro de Mercado, Durango, Mexico. US Geological Survey Professional Paper 650-D, 1969, D84-D93.

382 Hounslow AW, Chao GY: Monoclinic chlorapatite from Ontario. Can Miner 1970;10:252-259.

383 Glas J-E: Studies on the ultrastructure of dental enamel. VI. Crystal chemistry of shark's teeth. Odont Revy 1962;13:315-326.

384 Daculsi G, Kerebel LM: Ultrastructural study and comparative analysis of fluoride content of enameloid in sea-water and fresh water sharks. Arch Oral Biol 1980;25:145-151.

385 Clement JG: The Development, Structure and Chemistry of Elasmobranch Skeletal Tissues, PhD Thesis, University of London, 1986.

386 Slatkine S: Étude biocristallographique de l'incorporation du fluor dans la substance minérale osseuse in vivo. Rev Mens Suisse Odonto-stom 1962;72:1068-1087.

387 Mazelsky R, Hopkins RH, Kramer WE: Czochralski growth of calcium fluorophosphate. J Cryst Growth 1968;3,4:360-364.

388 Schleede A, Meppen B, Jörgensen OB: Zur Frage der Citronensäurelöslichkeit von Naturphosphaten (Apatiten). Angew Chem 1939;52:316-319.

389 Prener JS, Piper WW, Chrenko RM: Hydroxide and oxide impurities in calcium halophosphates. J Phys Chem Solids 1969;30:1465-1481.

390 Dzyuba ED, Sokolov MT, Valyukevich LP: Thermal stability of calcium phosphates. Inorganic Materials 1982;18:89-92. Translated from Izvestiya Akademii Nauk SSSR, Neorganicheskie Materialy 1982;18:107-110.

391 Prener JS: Nonstoichiometry in calcium chlorapatite. J Solid State Chem 1971;3:49-55.

392 Monma H, Kanazawa T: The mechanism of thermal reactions in apatite-silica-water vapor systems. Bull Chem Soc Jpn 1975;48:1816-1819.

393 Monma H, Kanazawa T: Effect of hydroxylation on the thermal reactivities of fluorapatite and chlorapatite. Bull Chem Soc Jpn 1976;49:1421-1422.

394 Mehmel M: Über die Struktur des Apatits. Z Krist 1930;75:323-331.

395 Náray-Szabó St: The structure of apatite $(CaF)Ca_4(PO_4)_3$. Z Krist 1930;75:387-398.

396 Mehmel Z: Beziehungen zwischen Kristallstruktur und chemischer Formel des Apatits. Z Phys Chem B 1931;15:223-241.

397 Beevers CA, McIntyre DB: The atomic structure of fluor-apatite and its relation to that of tooth and bone mineral. Miner Mag 1946-46;27:254-257.

398 Henry NFM, Lonsdale K (eds): International Tables for X-ray Crystallography, Vol 1, 3rd ed. Birmingham, The Knoch Press for The International Union of Crystallography, 1969.

399 Engel G, Pretzsch J, Gramlich V, Baur WH: The crystal structure of hydrothermally grown manganese chlorapatite, $Mn_5(PO_4)_3Cl_{0.9}(OH)_{0.1}$. Acta Cryst 1975:B31:1854-1860.

400 Elliott JC, Mackie PE: Monoclinic hydroxyapatite; in: Physico-Chimie et Cristallographie des Apatites d'Intérêt Biologique, Colloques Internationaux du Centre National de la Recherche Scientifique No 230 Paris, 1973. Paris, CNRS, 1975, pp 69-76.

401 Young RA: Dependence of apatite properties on crystal structural details. Trans NY Acad Sci Ser II 1967;29:949-959.

402 Greenblatt M, Pifer JH, Banks E: Electron spin resonance of CrO_4^{3-} in fluorapatite $Ca_5(PO_4)_3F$. J Chem Phys 1977;66:559-562.

403 Pifer JH, Ziemski S: Effect of anion substitution on the electron spin resonance of Cr^{5+} in calcium phosphate apatite. J Chem Phys 1983;78:7038-7043.

404 Sudarsanan K, Young RA: Significant precision in crystal structural details: Holly Springs hydroxyapatite. Acta Cryst 1969;B25:1534-1543.

405 Mathew M, Mayer I, Dickens B, Schroeder LW: Substitution in barium-fluoride apatite: the crystal structures of $Ba_{10}(PO_4)_6F_2$, $Ba_6La_2Na_2(PO_4)_6F_2$ and $Ba_4Nd_3Na_3(PO_4)_6F_2$. J Solid State Chem 1979;28:79-95.

406 Prener JS: The growth and crystallographic properties of calcium fluor- and chlorapatite crystals. J Electrochem Soc 1967;114:77-83.

407 Dickens B, Schroeder LW: Investigation of epitaxy relationships between $Ca_5(PO_4)_3(OH)$ and other calcium ortho-phosphates. J Res Natn Bur Stand 1980;85:347-362.

408 Moore PB, Araki T: Samuelsonite: its crystal structure and relation to apatite and octacalcium phosphate. Am Miner 1977;62:229-245.

409 Dickens B, Brown WE: The crystal structure of $Ca_5(PO_4)_2SiO_4$ (silico-carnotite). Tschermaks Miner Petrogr Mitt 1971;16:1-27.

410 Fayos J, Watkin DJ, Pérez-Méndez M: Crystal structure of the apatite-like compound $K_3Ca_2(SO_4)_3F$. Am Miner 1987;72:209-212.

411 Giuseppetti G, Rossi G, Tadini C: The crystal structure of nasonite. Am Miner 1971;56:1174-1179.

412 Eysel W, Roy DM: Topotactic reaction of aragonite to hydroxyapatite. Z Krist 1975;141:11-24.

413 Wondratschek H, Merker L, Schubert K: Beziehungen zwischen der Apatit-Struktur und der Struktur der Verbindungen vom Mn_5Si_3-$(D8_8)$-Typ. Z Krist 1964;120:393-395.

414 Parthé E, Rieger W: Nowotny phases and apatites: a comparative study. J Dent Res 1968;47:829-835.

415 Boyer LL: A higher symmetry crystal structure for apatite compounds. Phys Stat Sol (a) 1973;20:555-562.

416 Young RA, Elliott JC: Atomic-scale bases for several properties of apatites. Arch Oral Biol 1966;11:699-707.

417 Ruszala F, Kostiner E: Preparation and characterization of single crystals in the apatite system $Ca_{10}(PO_4)_6(Cl,OH)_2$. J Cryst Growth 1975;30:93-95.

418 Rausch EO: Dielectric Properties of Chlorapatite. PhD Thesis, Georgia Institute of Technology, 1976.

419 Elliott JC, Young RA: Dielectric measurements on single crystals of synthetic chlorapatite. Bull Soc Chim Fr (No Spec) 1968;1763-1765.

420 Hitmi N, Lacabanne C, Young RA: TSC study of electric dipole relaxations in chlorapatite. J Phys Chem Solids 1984;45:701-708.

421 Elliott JC, Young RA, Mackie PE, Dykes E: Effect of chlorine ion vacancies on the local atomic parameters in non-stoichiometric chlorapatite. Acta Cryst 1975;A31:S268. (abstract)

422 Schwarz H: Apatite des Typs $M^{II}_{10}(X^{VI}O_4)_3(X^{IV}O_4)_3F_2$ (M^{II} = Sr,Pb; X^{VI} = S,Cr; X^{IV} = Si, Ge). Z Anorg Allg Chem 1967;356:36-45.

423 Schwarz H: Strontiumapatite des Typs $Sr_{10}(PO_4)_4(X^{IV}O_4)_2$ (X^{IV} = Si,Ge). Z Anorg Allg Chem 1968;357:43-53.

424 Warren RW: Defect centres in calcium fluorophosphate. Phys Rev B 1972;B6:4679-4689.

425 Simpson DR: Oxygen-rich apatite. Am Miner 1969;54:560-562.

426 Rey C, Trombe JC, Montel G: Some features of the incorporation of oxygen in different oxidation states in the apatite lattice. III. Synthesis and properties of some oxygenated apatites. J Inorg Nucl Chem 1978;40:27-30.

427 Tochon-Danguy HJ, Very JM, Geoffroy M, Baud CA: Paramagnetic and crystallographic effects of low temperature ashing on human bone and tooth enamel. Calcif Tissue Res 1978;25:99-104.

428 Mackie PE, Young RA: Fluorine-chlorine interactions in fluor-chlorapatite. J Solid State Chem 1974;11:319-329.

429 Sudarsanan K, Young RA: Structural interactions of F, Cl and OH in apatites. Acta Cryst 1978;B34:1401-1407.

430 Engel G: Einige Cadmiumapatite sowie die Verbindungen Cd_2XO_4F mit X = P, As und V. Z Anorg Allg Chem 1970;378:49-61.

431 Sudarsanan K, Young RA, Wilson AJC: The structures of some cadmium 'apatites' $Cd_5(MO_4)_3X$. I. Determination of the structures of $Cd_5(VO_4)_3I$, $Cd_5(PO_4)_3Br$, $Cd_5(AsO_4)_3Br$ and $Cd_5(VO_4)_3Br$. Acta Cryst 1977;B33:3136-3142.

432 Wilson AJC, Sudarsanan K, Young RA: The structures of some cadmium 'apatites' $Cd_5(MO_4)_3X$. II. The distributions of the halogen atoms in $Cd_5(VO_4)_3I$, $Cd_5(PO_4)_3Br$, $Cd_5(AsO_4)_3Br$, $Cd_5(VO_4)_3Br$ and $Cd_5(PO_4)_3Cl$. Acta Cryst 1977;B33:3142-3154.

433 Baud G, Besse J-P, Sueur G, Chevalier R: Structure de nouvelles apatites au rhenium contenant des anions volumineux: $Ba_{10}(ReO_5)_6X_2$ (X = Br,I). Mater Res Bull 1979;14:675-682.

434 Dowker SEP, Elliott JC: Infrared absorption bands from NCO^- and NCN^{2-} in heated carbonate-containing apatites prepared in the presence of NH_4^+ ions. Calcif Tissue Int 1979;29:177-178.

435 Dowker SEP, Elliott JC: Infrared study of the formation, loss and location of cyanate and cyanamide in thermally treated apatites. J Solid State Chem 1983;49:334-340.

436 Calvo C, Faggiani R, Krishnamachari N: The crystal structure of $Sr_{9.402}Na_{0.209}(PO_4)_6B_{0.996}O_2$—a deviant apatite. Acta Cryst 1975;B31:188-192.

437 Ito A, Aoki H, Akao M, Miura N, Otsuka R, Tsutsumi S: Flux growth and crystal structure of borate apatites. J Ceram Soc Jpn Inter Ed 1988;96:302-306.

438 Besse J-P, Baud G, Chevalier R: Mise en évidence de l'ion O_2^- dans l'apatite au rhénium $Ba_5(ReO_5)_3O_2$. Mater Res Bull 1980;15:1255-1261.

439 Dugas J, Bejjaji B, Sayah D, Trombe JC: Etude par RPE de l'ion NO_2^{2-} dans une apatite nitrée. J Solid State Chem 1978;24:143-151.

440 Ito J: Silicate apatites and oxyapatites. Am Miner 1968;53:890-907.

441 Felsche F: Rare earth silicates with apatite structure. J Solid State Chem 1972;5:266-275.

442 Azimov SY, Ismatov AA, Fedorov N F: Synthetic silicophosphates, silicovanadates, and silicoarsenates with the apatite structure. Inorganic Materials 1981;17:1384-1387. Translated from Izvestiya Akad Nauk SSSR, Neorganicheskie Materialy 1981;17:1863-1867.

443 Piriou B, Fahmi D, Dexpert-Ghys J, Taitai A, Lacout JL: Unusual fluorescent properties of Eu^{3+} in oxyapatites. J Luminescence 1987;39:97-103.

444   Taïtaï A, Lacout JL, Bonel G: Sur la dioxyapatite calco-europique et ses solutions solides avec l'oxyapatite phosphocalcique. Ann Chim (Paris) 1985;10:29-35.

445   Lacout JL, Mikou M: Sur les dioxyapatites phosphostrontiques contenant deux ions de terres rares. Ann Chim (Paris) 1989;14:9-14.

446   Maunaye M, Hamon C, L'Haridon P, Laurent Y: Composés à structure apatite. IV. Étude structurale de l'oxynitrure $Sm_8Cr_2Si_6N_2O_{24}$ Bull Soc Fr Minér Crist 1976;99:203-205.

447   Guyader J, Grekov FF, Marchand R, Lang J: Nouvelles séries de silicoapatites enrichies en azote. Rev Chim Minér 1978;15:431-438.

448   Schroeder LW, Mathew M: Cation ordering in $Ca_2La_8(SiO_4)_6O_2$. J Solid State Chem 1978;26:383-387.

449   Trombe JC, Montel G: Some features of the incorporation of oxygen in different oxidation states in the apatite lattice. II. On the synthesis and properties of calcium and strontium peroxiapatites. J Inorg Nucl Chem 1978;40:23-26.

450   Trombe J-C, Montel G: Sur les conditions de préparation d'une nouvelle apatite contenant des ions sulfure. Compt Rend Acad Sci (Paris) Ser C 1975;280:567-570.

451   Suitch PR, Taïtaï A, LaCout JL, Young RA: Structural consequences of the coupled substitution of Eu, S, in calcium sulfoapatite. J Solid State Chem 1986;63:267-277.

452   Taïtaï A, Lacout JL: On the coupled introduction of Eu^{3+} and S^{2-} ions into strontium apatites. J Phys Chem Solids 1989;50:851-855.

453   Trombe JC, Montel G: On the existence of bivalent ions in the apatite channels, a new example—phosphocalcium cyanamido-apatite. J Solid State Chem 1981;40:152-160.

454   Trombe J-C, Montel G: Préparation et mise en évidence de la cyanamido-apatite phosphocalcique. Compt Rend Acad Sci (Paris) 1980;290:141-143.

455   Trombe JC, Montel G: Sur l'introduction d'ions sulfate dans les tunnels du réseau des apatites phosphocalciques. Ann Chim (Paris) 1980;5:443-454.

456   Knottnerus DIM, Hartog HW den, Lugt W van der: Optical and EPR investigations of colour centres in calciumchlorapatite. Phys Stat Sol (a) 1972;13:505-515.

457   Knottnerus DIM, Hartog HW den: Axially non-symmetric hole centers in calcium chlorapatite. Phys Stat Sol 1975;29:183-193.

458   Mathew M, Brown WE, Austin M, Negas T: Lead alkali apatites without hexad anion: the crystal structure of $Pb_8K_2(PO_4)_6$. J Solid State Chem 1980;35:69-76.

459   Pauling L: The Nature of the Chemical Bond, (3rd ed). Ithica, Cornell University Press, 1960.

460   Blasse G: Influence of local charge compensation on site occupation and luminescence of apatites. J Solid State Chem 1975;14:181-184.

461   Kreidler ER, Hummel FA: The crystal chemistry of apatite: structure fields of fluor- and chlorapatite. Am Miner 1970;55:170-184.

462 Shannon RD: Revised effective ionic radii and systematic studies of interatomic distances in halides and chalcogenides. Acta Cryst 1976;A32:751-767.

463 Schneider W: Caracolit, das $Na_3Pb_2(SO_4)_3Cl$ mit Apatitstruktur. Neues Jb Miner Mh 1967;284-289.

464 Engel G: Fluoroberyllate mit Apatitstruktur und ihre Beziehungen zu Sulfaten und Silicaten. Mater Res Bull 1978;13:43-48.

465 Engel G, Krieg F, Reif G: Mischkristallbildung und Kationeordnung im System Bleihydroxylapatit-Calciumhydroxylapatit. J Solid State Chem 1975;15:117-126.

466 Verbeeck RMH, Lassuyt CJ, Heijligers HJM, Driessens FCM, Vrolijk JWGA: Lattice parameters and cation distribution of solid solutions of calcium and lead hydroxyapatites. Calcif Tissue Int 1981;33:243-247.

467 Hata M, Marumo F, Iwai S, Aoki H: Structure of a lead apatite $Pb_9(PO_4)_6$. Acta Cryst 1980;B36:2128-2130.

468 Miyake M, Ishigaki K, Suzuki T: Structure refinements of Pb^{2+} ion-exchanged apatites by X-ray powder pattern-fitting. J Solid State Chem 1986;61:230-235.

469 Khudolozhkin VO, Urusov VS, Tobelko KI: Ordering of Ca and Sr in cation positions in the hydroxylapatite-belovite isomorphous series. Geochem Int 1972;9:827-833.

470 Khudolozhkin VO, Urusov VS, Tobelko KI: Distribution of cations between sites in the structure of Ca, Sr, Ba - apatites. Geochem Int 1973;10:266-269.

471 Heijligers HJM, Driessens FCM, Verbeeck RMH: Lattice parameters and cation distribution of solid solutions of calcium and strontium hydroxyapatite. Calcif Tissue Int 1979;29:127-131.

472 Drifford M, Dalbiez J-P, Feltin C: Étude des chlorapatites de calcium et de barium et de leurs solutions solides par diffusion Raman. J Chim Physique 1976;73:738-744.

473 Sudarsanan K, Young RA: Structure of partially substituted chlorapatite $(Ca,Sr)_5(PO_4)_3Cl$. Acta Cryst 1980;B36:1525-1530.

474 Sudarsanan K, Young RA: Structure refinement and random error analysis for strontium 'chlorapatite', $Sr_5(PO_4)_3Cl$. Acta Cryst 1974;B30:1381-1386.

475 Suitch PR, LaCout JL, Hewat A, Young RA: The structural location and role of Mn^{2+} partially substituted for Ca^{2+} in fluorapatite. Acta Cryst 1985;B41:173-179.

476 Terpstra RA, Driessens FCM: Magnesium in tooth enamel and synthetic apatites. Calcif Tissue Int 1986;39:348-354.

477 Okazaki M: Magnesium action on the stability of fluorapatite. Magnesium 1988;7:148-153.

478 Khudolozhkin BO, Urusov VS, Kurash VV: Mössbauer study of the ordering of Fe^{2+} in the fluor-apatite structure. Geochem Int 1974;11:748-750.

479 LeGeros RZ: Incorporation of magnesium in synthetic and in biological apatites; in Fearnhead RW, Suga S (eds): Tooth Enamel IV. Amsterdam, Elsevier, 1984, pp 32-36.

480 Patel PN: Magnesium calcium hydroxylapatite solid solutions. J Inorg Nucl Chem 1980;42;1129-1132.

481 Chiranjeevirao SV, Hemmerle J, Voegel JC, Frank RM: A method of preparation and characterization of magnesium-apatites. Inorganica Chim Acta 1982;67:183-187.

482 Neuman WF, Mulryan BJ: Synthetic hydroxyapatite crystals IV. Magnesium incorporation. Calcif Tissue Res 1971;7:133-138.

483 Grisafe DA, Hummel FA: Crystal chemistry and color in apatites containing cobalt, nickel, and rare-earth ions. Am Miner 1970;55:1131-1145.

484 Anderson JB, Kostiner E: The crystal structure of cobalt-substituted calcium chlorapatite. J Solid State Chem 1987;66:343-349.

485 Kingsley JD, Prener JS, Segall B: Spectroscopy of MnO_4^{3-} in calcium halophosphates. Phys Rev A 1965;137:A189-A202.

486 Borromei R, Fisicaro E: Crystal growth of $Sr_5(PO_4)_3Cl$ doped with Mn(V) and study of the absorption spectrum. Gazz Chim Ital 1979;109:191-194.

487 Ohkubo Y: Optical and EPR studies of divalent manganese ions in calcium halophosphates. J Phys Soc Jpn 1969;27:1516-1526.

488 Warren RW: EPR of Mn^{+2} in calcium fluorophosphate. I. The Ca(II) site. Phys Rev B 1970;2:4383-4388.

489 Gilinskaya LG, Shcherbakova MJa: Peculiarities in Mn^{2+} entering in apatite structure as revealed by EPR studies; in Proceedings of the 16th Congress Ampère in Bucharest in 1970, Bucharest, 1971, pp 755-758.

490 Ryan FM, Ohlmann RC, Murphy J, Mazelsky R, Wagner GR, Warren RW: Optical properties of divalent manganese in calcium fluorophosphate. Phys Rev B 1970;2:2341-2352.

491 Ryan FM, Hopkins RH, Warren RW: The optical properties of divalent manganese in strontium fluorophosphate: a comparison with calcium fluorophosphate. J Luminescence 1972;5:313-333.

492 Warren RW, Mazelsky R: EPR of Mn^{+2} in calcium fluorophosphate. II. Modified Ca(II) site. Phys Rev B 1974;10:19-25.

493 Mayer I, Fischbein E, Cohen S: Apatites of divalent europium. J Solid State Chem 1975;14:307-312.

494 Baran EJ, Apella MC: The infrared spectra of some cadmium apatites. Rev Chim Minér 1979;16:527-534.

495 Engel G: Einige Apatite des Cadmiums. Z Anorg Allg Chem 1968;362:273-280.

496 Donnay JDH, Sudarsanan K, Young RA: Twinning in relation to structural detail in $Cd_5(PO_4)_3Cl$, 'cadmium chlorapatite'. Acta Cryst 1973;B29:814-817.

497 Roeder PL, MacArthur D, Ma X-P, Palmer GR, Mariano AN: Cathodoluminescence and microprobe study of rare-earth elements in apatite. Am Miner 1987;72:801-811.

498 Mayer I, Roth RS, Brown WE: Rare earth substituted fluoride-phosphate apatites. J Solid State Chem 1974;11:33-37.

499 Budin J-P, Michel J-C, Auzel F: Oscillator strengths and laser effect in $Na_2Nd_2Pb_6(PO_4)_6Cl_2$ (chloroapatite), a new high-Nd-concentration laser material. J Appl Phys 1979;50:641-646.

500 Eiberger B, Greenblatt M: Rare-earth-substituted chloro-phosphate apatites. J Solid State Chem 1982;41:44-49.

501 Mayer I, Cohen S: The crystal structure of $Ca_6Eu_2Na_2(PO_4)_6F_2$. J Solid State Chem 1983;48:17-20.

502 Mayer I, Swissa S: Lead and strontium phosphate apatites substituted by rare earth and silver ions. J Less-Common Metals 1985;110:411-414.

503 Wanmaker WL, ter Vrugt JW, Verlijsdonk JG: Luminescence of alkaline earth yttrium and lanthanum phosphate-silicates with apatite structure. J Solid State Chem 1971;3:452-457.

504 Lang J, Marchand R, Hamon C, L'Haridon P, Guyader J: Composés à structure apatite II. Cristallochimie des silico-apatites azotées. Bull Soc Fr Minér Crist 1975;98:284-288.

505 Khudolozhkin VO, Urusov VS, Tobelko KI, Vernadskiy VI: Dependence of structural ordering of rare earth atoms in the isomorphous series apatite-britholite (abukumalite) on composition and temperature. Geochem Int 1973;10:1171-1177.

506 Mackie PE, Young RA: Location of Nd dopant in fluorapatite, $Ca_5(PO_4)_3F$:Nd. J Appl Cryst 1973;6:26-31.

507 Eaglet RD: Site Selection and Laser Properties of Neodymium and Europium Ions in Fluorapatite Laser Crystals. PhD Thesis, University of Southern California, 1970.

508 Ryan FM, Warren RW, Hopkins RH, Murphy J: Selective site laser excitation and ESR studies of Nd^{3+} ions in $Ca_5(PO_4)_3F$. J Electrochem Soc 1978;125:1493-1498.

509 Mishra KC, Patton RJ, Dale EA, Das TP: Location of antimony in a halophosphate phosphor. Phys Rev B 1987;B35:1512-1520.

510 Moran LB, Berkowitz JK, Yesinowski JP: ^{19}F and ^{31}P magic-angle spinning nuclear magnetic resonance of antimony(III)-doped fluorapatite phosphors: Dopant sites and spin diffusion. Phys Rev B 1992;45:5347-5360.

511 DeBoer BG, Sakthivel A, Cagle JR, Young RA: Determination of the antimony substitution site in calcium fluorapatite from powder X-ray diffraction data. Acta Cryst 1991;B47:683-692.

512 Moran LB, Berkowitz JK, Yesinowski JP: A method for detection of spectral spin diffusion from minor peaks and its application to ^{19}F MAS-NMR of antimony(III)-doped fluorapatite. Solid State Nucl Magnetic Resonance 1992;1:307-311.

513 Keppler U: Monokliner Mimetesit, $Pb_5(AsO_4)_3Cl$. Neues Jb Miner Mh 1968;359-362.

514 Massuyes M, Trombe J-C, Bonel G, Montel G: Étude comparée des structures et des propriétés physico-chimiques de quelques apatites calciques phospho-arseniées. Bull Soc Chim Fr 1969;7:2308-2315.

515 Wilhelmi K, Jonsson O: X-ray studies on some alkali and alkaline-earth chromates(V). Acta Chem Scand 1965;19:177-184.

516 Banks E, Jaunarajs KL: Chromium analogs of apatite and spodiosite. Inorg Chem 1965;4:78-83.

517 Banks E, Greenblatt M, McGarvey BR: Electron spin resonance of CrO_4^{3-} in chlorapatite $Ca_5(PO_4)_3Cl$. J Solid State Chem 1971;3:308-313.

518 Greenblatt M, Kuo J-M, Pifer JH: Electron spin resonance of CrO_4^{3-} in strontium chlorapatite, $Sr_5(PO_4)_3Cl$. J Solid State Chem 1979;29:1-7.

519 Forster K, Greenblatt M, Pifer JH: Electron spin resonance of CrO_4^{3-} in barium chloroapatite, $Ba_5(PO_4)_3Cl$. J Solid State Chem 1979;30:121-124.

520 Schwarz H: Verbindungen mit Apatitstruktur. I. Ungewöhnliche silicatapatite. Inorg Nucl Chem Lett 1967;3:231-236.

521 Harris RK, Leach MJ, Thompson DP: Silicon-29 magic angle spinning nuclear magnetic resonance study of some lanthanum and yttrium silicon oxynitride phases. Chem Mater 1989;1;336-338.

522 Dupree R, Lewis MH, Smith ME: High resolution silicon-29 nuclear magnetic resonance in the Y-Si-O-N system. J Am Chem Soc 1988;110;1083-1087.

523 Gaudé J, L'Haridon P, Hamond C, Marchand R, Laurent Y: Composé à structure apatite. I. Structure de l'oxynitrure $Sm_{10}Si_6N_2O_{24}$. Bull Soc Fr Minér Crist 1975;98:214-217.

524 Hamon C, Marchand R, Maunaye M, Gaudé J, Guyader J: Composés à structure apatite III. - Les oxynitrures $Ln_{10}Si_6O_{24}N_2$ et quelques séries analogues $(Ln,M)_{10}Si_6O_{24}X_2$ (X = N, O, F), obtenues par substitution. Rev Chim Minér 1975;12:259-267.

525 Schwarz H: Apatite des Typs $Pb_6K_4(X^VO_4)_4(X^{VI}O_4)_2$ (X^V = P, As; X^{VI} = S, Se). Z Anorg Allg Chem 1967;356:29-35.

526 Apella MC, Baran EJ: Röntgenographische und IR-Spektroskopische Untersuchung der Substitution von Phosphat- durch Sulfat-Ionen im Fluorapatit-Gitter. Z Naturf 1979;34b:1124-1127.

527 Apella MC, Baran EJ: The infrared spectra of some sulfate apatites. Spectrosc Lett 1979;12:1-6.

528 Baud G, Besse JP, Capestan M, Sueur G, Chevalier R: Étude comparative d'apatites contenant l'ion $(ReO_5)^{3-}$. Structure des fluoro et carbonatoapatites. Ann Chim Sci Mater 1980;5:575-583.

529 Baran EJ, Baud G, Besse J-P: Vibrational spectra of some rhenium-apatites containing ReO_5-groups. Spectrochim Acta 1983;39A:383-386.

530 Besse J-P, Baud G, Levasseur G, Chevalier R: Structure cristalline de $Ba_5(ReO_5)_3Cl$: une nouvelle apatite contenant l'ion $(ReO_5)^{3-}$. Acta Cryst 1979;B35:1756-1759.

531 McConnell D: The substitution of SiO_4- and SO_4-groups for PO_4- groups in the apatite structure; ellestadite, the end-member. Am Miner 1937:22:977-986.

532 Roy DM, Eysel W, Dinger D: Hydrothermal synthesis of various carbonate containing calcium hydroxyapatites. Mater Res Bull 1974;9:35-40.

533 Vignoles M, Bonel G, Holcomb DW, Young RA: Influence of preparation conditions on the composition of type B carbonated hydroxyapatite and on the localization of the carbonate ions. Calcif Tissue Int 1988;43:33-40.

534 Dowker SEP, Elliott JC: Infrared study of trapped carbon dioxide in thermally treated apatites. J Solid State Chem 1983;47:164-173.

535 Mann AW, Turner AG: Excess calcium fluoride in fluorapatite. Aust J Chem 1972;25:2701-2703.

536 Berak J, Tomczak-Hudyma I: Phase equilibria in the system $Ca_3(PO_4)_2$—CaF_2. Roczniki Chem Ann Soc Chim Pol 1972;46:2157-2164.

537 Bonel G, Heughebaert J-C, Heughebaert M, Lacout JL, Lebugle A: Apatitic calcium orthophosphates and related compounds for biomaterials preparation. Ann NY Acad Sci 1988;523:115-130.

538 Vignoles M, Bonel G: Sur la localisation des ions fluorure dans les carbonate-apatites de type B. Compt Rend Acad Sci (Paris) Ser C 1978;287:321-324.

539 Nacken R: Ueber die Bildung des Apatits. I. Centralbl Mineral Geol Pal 1912;545-559.

540 Farr TD, Tarbutton G, Lewis HT: System CaO-P_2O_5-HF-H_2O: equilibrium at 25 and 50°. J Phys Chem 1962;66:318-321.

541 Johnson PD: Some optical properties of powder and crystal halophosphate phosphors. J Electrochem Soc 1961;108:159-162.

542 Lapraz D, Baumer A: Thermoluminescence properties of synthetic and natural fluorapatite, $Ca_5(PO_4)_3F$. Phys Stat Sol 1983;80:353-366.

543 Johnson PD: Identification of activator sites in halophosphate phosphors; in Kallmann HP, Spruch GM (eds): Luminescence of Organic and Inorganic Materials. New York, John Wiley, 1962, pp 563-575.

544 Roufosse A, Harvill ML, Gilliam OR, Kostiner E: The hydrothermal growth of chlorapatite. J Cryst Growth 1973;19:211-212.

545 Mengeot M, Harvill ML, Gilliam OR: Hydrothermal growth of calcium hydroxyapatite single crystals. J Cryst Growth 1973;19:199-203.

546 Baumer A, Caruba R, Bizouard H, Peckett A: Chlorapatite de synthèse: substitution et inclusions de Mn, Ce, U et Th en traces. Can Miner 1983;21:567-573.

547 Schulze H, Weinstock N, Müller A, Vandrish G: Raman intensities and force constants of PO_4^{3-}, SO_4^{2-}, ClO_4^-, SeO_4^{2-}, and BrO_4^-. Spectrochim Acta 1973,29A,1705-1709.

548 Baddiel CB, Berry EE: Spectra structure correlations in hydroxy and fluorapatite. Spectrochim Acta 1966;22:1407-1416.

549 Fowler BO: Infrared spectra of apatites; in Brown WE, Young RA (eds): Proceedings of International Symposium on Structural Properties of Hydroxyapatite and Related Compounds. Gaithersburg, 1968, Ch 7. (unpublished, copy of chapter available from BO Fowler).

550 Levitt SR, Condrate RA: Force constant analysis of molecular groups in apatite solids. J Phys Chem Solids 1973;34:1109-1118.

551 Levitt SR, Blakeslee KC, Condrate RA: Infrared spectra and laser-Raman spectra of several apatites. Soc Roy Sci Liège, Mém Coll 1970;20:121-141.

552 Klee WE: The vibrational spectra of the phosphate ions in fluorapatite. Z Krist 1970;131:95-102.

553 Adams DM, Gardner IR: Single-crystal vibrational spectra of apatite, vanadinite, and mimetite: J Chem Soc (Dalton) 1974;1505-1509.

554 Kravitz LC, Kingsley JD, Elkin EL: Raman and infrared studies of coupled PO_4^{-3} vibrations. J Chem Phys 1968;49:4600-4610.

555 Kingsley JD, Mahan GD, Kravitz LC: Background polarization correction to the Davydov splittings of phosphate vibrations. J Chem Phys 1968;49:4610-4617.

556 O'Shea DC, Bartlett ML, Young RA: Compositional analysis of apatites with laser-Raman spectroscopy: (OH,F,Cl)apatites. Arch Oral Biol 1974;19:995-1006.

557 Fowler BO: Infrared studies of apatites. I. Vibrational assignments for calcium, strontium, and barium hydroxyapatite utilizing isotopic substitution. Inorg Chem 1974;13:194-207.

558 Klein E, LeGeros JP, Trautz OR, LeGeros RZ: Polarized infrared reflectance of single crystals of apatites. Dev Appl Spectrosc 1970;7B:13-22.

559 Fowler BO: Polarized Raman spectra of apatites. Abstract No 68 J Dent Res (Special Issue B) 1977;56:B68.

560 Fowler BO: I. Polarised Raman spectra of apatites II. Raman bands of carbonate ions in human tooth enamel. Mineralized Tissue Communications 1977;3 no 68.

561 Boyer LL, Fleury PA: Determination of interatomic interactions in $Ca_{10}(PO_4)_6F_2$ (fluorapatite) from structural and lattice-dynamical data. Phys Rev B 1974;9:2693-2700.

562 Devarajan V, Klee WE: A potential model for fluorapatite. Phys Chem Miner 1981;7:35-42.

563 Kravitz LC, Kingsley JD, Mahan GD: Raman scattering and infrared reflectance studies of single crystal apatites; in Brown WE, Young RA (eds): Proceedings of International Symposium on Structural Properties of Hydroxyapatite and Related Compounds, Gaithersburg, 1968, Ch 8. (unpublished)

564 Fowler BO: Raman and infrared vibrational spectra of apatites. Abstract No 129 J Dent Res (Special Issue A) 1975;54:77.

565 Fowler BO: Raman and infrared vibrational spectra of apatites. Mineralized Tissue Research Communications 1975;1 no 129.

566 Condrate RA, Cornilsen BC, Dykes E: The Raman spectrum of bromapatite. Appl Spectrosc 1975;29:526-527.

567 Yoon HY, Newnham RE: Elastic properties of fluorapatite. Am Miner 1969;54:1193-1197.

568 Gardner TN, Elliott JC, Sklar Z, Briggs GAD: Acoustic microscope study of the elastic properties of fluor- and hydroxyapatite, tooth enamel and bone. J Biomech 1992;25:1265-1277.

569 Egan EP, Wakefield ZT, Elmore KL: Thermodynamic properties of fluorapatite, 15 to 1600 °K. J Am Chem Soc 1951;73:5581-5582.

570 Tacker RC, Stormer JC: A thermodynamic model for apatite solid solutions, applicable to high-temperature geologic problems. Am Miner 1989;74:877-888.

571 Phakey PP, Leonard JR: Dislocations and fault surfaces in natural apatite. J Appl Cryst 1970;3:38-44.

572 Hopkins RH: The origin of decorated dislocation arrays in fluorapatite. J Cryst Growth 1969;6:91-96.

573 Lugt W van der, Caspers WJ: Nuclear magnetic resonance line shape of fluorine in apatite. Physica 1964;30:1658-1666.

574 Hartog H den, Welch DO, Royce BSH: The diffusion of calcium, phosphorous, and OD⁻ ions in fluorapatite. Phys Stat Sol (b) 1972;53:201-212.

575 Royce BSH: The defect structure and ionic transport properties of calcium apatite. J Physique, Colloque C9, Suppl No 1973;34;(C9)327-(C9)332.

576 Dykes E: Diffusion anisotropy in apatite. Nature Phys Sci 1971;233:12-13.

577 Vakhrameev AM, Sergeev NA: NMR study of the mobility of the fluoride ions and hydroxyl groups in apatites. J Struct Chem 1978;19:555-561. (Translation of Zh Strukt Khim 1978;19:640-647)

578 Watson EB, Harrison TM, Ryerson FJ: Diffusion of Sm, Sr, and Pb in fluorapatite. Geochim Cosmochim Acta 1985;49:1813-1823.

579 Rösick U, Zimen KE: Diffusion von ^{45}Ca, ^{85}Sr und ^{32}P in Hydroxylapatit. Biophysik 1973;9:120-131.

580 Tse C, Welch DO, Royce BSH: Calculation of anion migration energy in the calcium apatites. Bull Am Phys Soc 1972;17:256. (abstract)

581 Tse C, Welch DO, Royce BSH: The migration of F⁻, OH⁻ and O^{2-} ions in apatites. Calcif Tissue Res 1973;13:47-52.

582 Wagner G, Haute P van den: Fission-Track Dating. Dordrecht, Kluwer Academic Publishers, 1992.

583 Fleischer RL, Price PB, Walker RM: Nuclear Tracks in Solids: Principles and Applications. Berkeley, University of California Press, 1975, p. 161.

584 Petford N, Miller JA: Three-dimensional imaging of fission tracks using confocal scanning laser microscopy. Am Miner 1992;77:529-533.

585 Petford N, Miller JA: SLM confocal microscopy: an improved way of viewing fission tracks. J Geol Soc (London) 1990;147:217-218.

586 Petford N, Miller JA: The study of fission track and other crystalline defects using confocal scanning laser microscopy. J Microscopy 1993;170:201-212.

587 Green PF, Duddy IR, Gleadow AJW, Tingate PR, Laslett GM: Fission- track annealing in apatite: track length measurements and the form of the Arrhenius plot. Nucl Tracks 1985;10:323-328.

588 Laslett GM, Green PF, Duddy IR, Gleadow AJW: Thermal annealing of fission tracks in apatite: 2. A quantitative analysis. Chem Geol (Isotope Geosci Sect) 1987;65:1-13.

589 Swank RK: Color centers in X-irradiated halophosphate crystals. Phys Rev A 1964;135:A266-A275.

590 Powers JM, Craig RG: Wear of fluorapatite single crystals: I. A method for quantitative evaluation of wear. J Dent Res 1972;51:167-176.

591 Powers JM, Craig RG: Wear of fluorapatite single crystals: II. Frictional behavior. J Dent Res 1972;51:605-610.

592 Powers JM, Craig RG: Wear of fluorapatite single crystals: III. Classification of surface failure. J Dent Res 1972;51:611-618.

593 Powers JM, Ludema KC, Craig RG: Wear of fluorapatite single crystals: IV. Influence of sliding direction on frictional behavior and surface failure. J Dent Res 1973;52:1019-1025.

594 Powers JM, Ludema KC, Craig RG: Wear of fluorapatite single crystals: V. Influence of environment on frictional behavior and surface failure. J Dent Res 1973;52:1026-1031.

595 Powers JM, Ludema KC, Craig RG: Wear of fluorapatite single crystals: VI. Influence of multiple-pass sliding on surface failure. J Dent Res 1973;52:1032-1040.

596 Powers JM, Craig RG, Ludema KC: The surface failure of fluorapatite single crystals. Wear 1973;23:209-224.

597 Yamashita K, Kanazawa T: Hydroxyapatite; in Kanazawa T (ed): Inorganic Phosphate Materials, Materials Science Monograph 52. Amsterdam, Elsevier, 1989, Ch 2, pp 15-54.

598 Mitchell L, Faust GT, Hendricks SB, Reynolds DS: The mineralogy and genesis of hydroxylapatite. Am Miner 1943;28:356-371.

599 Kay MI, Young RA, Posner AS: Crystal structure of hydroxyapatite. Nature 1964;204:1050-1052.

600 Dykes E, Elliott JC: The occurrence of chloride ions in the apatite lattice of Holly Springs hydroxyapatite and dental enamel. Calcif Tissue Res 1971;7:241-248.

601 Burr C, Jakob J, Parker RJ, Strunz H: Über Hydroxylapatit von der Kemmleten bei Hospenthal (Kt. Uri). Schweiz Min Petr Mitt 1935;15:327-339.

602 Bowman RS, Piaseckey LJ: Calcium phosphate catalysts and method of production. US Patent 1964, No 3,149,082.

603 Monma H: Catalytic behavior of calcium phosphates for decomposition of 2-propanol and ethanol. J Catalysis 1982:75:200-203.

604 John M, Schmidt J: High-resolution hydroxyapatite chromatography of proteins. Anal Biochem 1984;141:466-471.

605 Gorbunoff MJ: The interaction of proteins with hydroxyapatite. I. Role of protein charge and structure. Anal Biochem 1984;136:425-432.

606 Gorbunoff MJ: The interaction of proteins with hydroxyapatite. II. Role of acidic and basic groups. Anal Biochem 1984;136:433-439.

607 Gorbunoff MJ, Timasheff SN: The interaction of proteins with hydroxyapatite. III. Mechanism. Anal Biochem 1984;136:440-445.

608 Elliott JC: Monoclinic space group of hydroxyapatite. Nature, Physical Science 1971;230:72.

609 Elliott JC, Mackie PE, Young RA: Monoclinic hydroxyapatite. Science 1973;180:1055-1057.

610 Posner AS, Perloff A, Diorio AF: Refinement of the hydroxyapatite structure. Acta Cryst 1958;11:308-309.

611 Elliott JC: Further studies of the structure of dental enamel and carbonate apatites using polarized infrared spectroscopy. J Dent Res 1962;41:1251. (abstract)

612 Mengeot M, Bartram RH, Gilliam OR: Paramagnetic holelike defect in irradiated calcium hydroxyapatite single crystals. Phys Rev B 1975;11:4110-4124.

613 Elliott JC, Young RA: Conversion of single crystals of chlorapatite into single crystals of hydroxyapatite. Nature 1967;214:904-6.

614 McLean JD, Nelson DGA: High-resolution n-beam lattice images of hydroxyapatite. Micron 1982;13:409-413.

615 Sudarsanan K, Young RA: Structure of strontium hydroxide phosphate, $Sr_5(PO_4)_3OH$. Acta Cryst 1972;B28:3668-3670.

616 Van Rees HB, Mengeot M, Kostiner E: Monoclinic-hexagonal transition in hydroxyapatite and deuterohydroxyapatite single crystals. Mater Res Bull 1973;8:1307-1310.

617 Hitmi N, LaCabanne C, Young RA: OH⁻ dipole reorientability in hydroxyapatites: effect of tunnel size. J Phys Chem Solids 1986;47:533-546.

618 Hitmi N, Lamure-Plaino E, Lamure A, LaCabanne C, Young RA: Reorientable electric dipoles and cooperative phenomena in human tooth enamel. Calcif Tissue Int 1986;38:252-261.

619 Hitmi N, LaCabanne C, Young RA: OH⁻ reorientability in hydroxyapatites: effect of F⁻ and Cl⁻. J Phys Chem Solids 1988;49:541-550.

620 Young RA, Holcomb DW: Variability of hydroxyapatite preparations. Calcif Tissue Int 1982;34 Suppl 2:S17-S32.

621 Seuter AMJH: Existence region of calcium hydroxylapatite and the equilibrium with coexisting phases at elevated temperatures; in Anderson JS, Roberts MW, Stone FS (eds): Reactivity of Solids. London, Chapman and Hall, 1972, pp 806-812.

622 Christensen JH, Reed RB: Density of aqueous solutions of phosphoric acid. Ind Eng Chem 1955;47:1277-1280.

623 Meyer JL, Fowler BO: Lattice defects in nonstoichiometric calcium hydroxyapatite. A chemical approach. Inorg Chem 1982;21:3029-3035.

624 Doi Y, Moriwaki Y, Aoba T, Okazaki M, Takahashi J, Joshin K: Carbonate apatite from aqueous and non-aqueous media studied by ESR, IR, and X-ray diffraction: effect of NH_4^+ ions on crystallographic parameters. J Dent Res 1982;61:429-434.

625 Vignoles M, Bonel G, Young RA: Occurrence of nitrogenous species in precipitated B-type carbonated hydroxyapatites. Calcif Tissue Int 1987;40:64-70.

626 Hayek E, Newesely H: Pentacalcium monohydroxyorthophosphate (hydroxylapatite). Inorganic Syntheses 1963;7:63-65.

627 Hayek E, Stadlmann W: Darstellung von reinem Hydroxylapatit für Adsorptionszwecke. Angew Chem 1955;67:327.

628 Fowler BO: Infrared studies of apatites. II. Preparation of normal and isotopically substituted calcium, strontium, and barium hydroxyapatite and spectra-structure-composition correlations. Inorg Chem 1974;13:207-214.

629 Avnimelech Y, Moreno EC, Brown WE: Solubility and surface properties of finely divided hydroxyapatite. J Res Natn Bur Stand A Phys Chem 1973;77A:149-155.

References

630 Bell LC, Posner AM, Quirk JP: The point of zero charge of hydroxyapatite and fluorapatite in aqueous solutions. J Coll Interface Sci 1973;42:250-261.

631 Mika H, Bell LC, Kruger BJ: The role of surface reactions in the dissolution of stoichiometric hydroxyapatite. Arch Oral Biol 1976;21:697-701.

632 Bell LC, Mika H: The pH dependence of the surface concentrations of calcium and phosphorus on hydroxyapatite in aqueous solutions. J Soil Sci 1979;30:247-258.

633 Palmer JW, Rosenstiel TL: Process of preparing hydroxylapatite. US Patent 1989, No 4,849,193.

634 Moreno EC, Gregory TM, Brown WE: Preparation and solubility of hydroxyapatite. J Res Natn Bur Stand 1968;72A:773-782.

635 Aoba T, Moreno EC: Preparation of hydroxyapatite crystals and their behavior as seeds for crystal growth. J Dent Res 1984;63:874-880.

636 Moreno EC, Kresak M, Zahradnik RT: Physicochemical aspects of fluoride-apatite systems relevant to the study of dental caries. Caries Res 1977;11(Suppl 1):142-160.

637 Arends J, Christoffersen J, Christoffersen MR, Eckert H, Fowler BO, Heughebaert JC, Nancollas GH, Yesinowski JP, Zawacki SJ: A calcium hydroxyapatite precipitated from an aqueous solution. An international multimethod analysis. J Cryst Growth 1987;84:515-532.

638 Silverman SR, Fuyat RK, Weiser JD: Quantitative determination of calcite associated with carbonate-bearing apatites. Am Miner 1952;37:211-222.

639 Mayer I, Wahnon S, Cohen S: Preparation of hydroxyapatites via the MSO_4 sulphates (M= Ca, Sr, Pb and Eu). Mater Res Bull 1979;14:1479-1483.

640 Jarcho M, Bolen CH, Thomas MB, Bobick J, Kay JF, Doremus RH: Hydroxylapatite synthesis and characterization in dense polycrystalline form. J Mater Sci 1976;11:2027-2035.

641 Irvine GD: Synthetic bone ash. Brit Pat No 1,589,915. 1981.

642 Kijima T, Tsutsumi M: Preparation and thermal properties of dense polycrystalline oxyhydroxyapatite. J Am Ceram Soc 1979;62:455-460.

643 Rowles SL: Studies on non-stoichiometric apatites; in Fearnhead RW, Stack MV (eds): Tooth Enamel, Proc Int Symp, London, 1964. Bristol, John Wright & Sons, 1965, pp 23-25 and pp 56-57.

644 Mortier A, Lemaitre J, Rodrique L, Rouxhet PG: Synthesis and thermal behavior of well-crystallized calcium-deficient phosphate apatite. J Solid State Chem 1989;78:215-219.

645 Heughebaert JC, Zawacki SJ, Nancollas GH: The growth of nonstoichiometric apatite from aqueous solution at 37 °C. I. Methodology and growth at pH 7.4. J Colloid Interface Sci 1990;135:20-32.

646 Zawacki SJ, Heughebaert JC, Nancollas GH: The growth of nonstoichiometric apatite from aqueous solution at 37 °C. II. Effects of pH upon the precipitated phase. J Colloid Interface Sci 1990;135:33-44.

647 Lorah JR, Tartar HV, Wood L: A basic phosphate of calcium and of strontium and the adsorption of calcium hydroxide by basic calcium phosphate and by tricalcium phosphate. J Am Chem Soc 1929;51:1097-1106.

648 Carlström D: X-ray crystallographic studies on apatites and calcified tissues. Acta Radiol Suppl 121, 1955.

649 Rogers AF: Dahllite (Podolite) from Tonopah, Nevada: Vœlckerite, a new basic calcium phosphate; remarks on the chemical composition of apatite and phosphate rock. Am J Sci Ser 4 1912;33:475-482.

650 Voelcker JA: Die chemische Zusammensetzung des Apatits nach eigenen vollständigen Analysen. Berichte Chem Ges 1883;16:2460-2464.

651 Lagergren C, Carlström D: Crystallographic studies of calcium and strontium hydroxyapatites. Acta Chem Scand 1957;11:545-550.

652 Verbeeck RMH, Heijligers HJM, Driessens FCM, Schaeken HG: Effect of dehydration of calcium hydroxylapatite on its cell parameters. Z Anorg Allg Chem 1980;466:76-80.

653 Trombe J-C, Montel G: Sur la préparation de l'oxyapatite phosphocalcique. Compt Rend Acad Sci (Paris) Ser C 1971;273,462-465.

654 Kukura M, Bell LC, Posner AM, Quirk JP: Radioisotope determination of the surface concentrations of calcium and phosphorus on hydroxyapatite in aqueous solutions. J Phys Chem 1972;76:900-904.

655 Trombe JC, Montel G: Some features of the incorporation of oxygen in different oxidation states in the apatite lattice. I. On the existence of calcium and strontium oxyapatites. J Inorg Nucl Chem 1978;40:15-21.

656 Wallaeys R: Étude d'une apatite carbonatée obtenue par synthèse dans l'état solide; in: Silicon, Sulphur, Phosphorus, IUPAC Colloquium, Weinheim, Verlag Chemie, 1954, pp 183-190.

657 Taïtaï A, Lacout JL: Hydroxylation and fluorination of europium containing oxyapatites. J Phys Chem Solids 1987;48:629-633.

658 Yamashita K, Owada H, Nakagawa H, Umegaki T, Kanazawa T: Trivalent-cation-substituted calcium oxyhydroxyapatite. J Am Ceram Soc 1986;69:590-594.

659 Trombe JC, Montel G: Sur le spectre d'absorption infrarouge des apatites dont les tunnels contiennent des ions bivalent et des lacunes. Comp Rend Acad Sci (Paris) Ser C 1973;276:1271-1274.

660 Gourdon J-C, Rey C, Chachaty C, Trombe J-C, Pescia J: Résonance paramagnétique eléctronique d'une apatite oxygénée de synthèse. Comp Rend Acad Sci (Paris) Ser B 1973;276:559-562.

661 Dugas J, Rey C: Electron spin resonance characteristics of superoxide ions in some oxygenated apatites. J Phys Chem 1977;81:1417-1419.

662 Montrejaud M, Rey C, Trombe JC, Montel G: Sur l'aptitude du réseau apatitique à fixer des molécules d'oxygène; in: Physico-Chimie et Cristallographie des Apatites d'Intérêt Biologique, Colloques Internationaux du Centre National de la Recherche Scientifique No 230 Paris, 1973, CNRS, 1975, pp 481-486.

663   Collin RL: Strontium-calcium hydroxyapatite solid solutions precipitated from basic, aqueous solutions. J Am Chem Soc 1960;82:5067-5069.

664   Bigi A, Foresti E, Marchetti F, Ripamonti A, Roveri N: Barium calcium hydroxyapatite solid solutions. J Chem Soc Dalton Trans 1984;1091-1093.

665   Suzuki T, Ishigaki K: Removal of toxic Pb^{2+} ions by synthetic hydroxyapatites. Chem Eng Commun 1985;34:143-151.

666   Okazaki M, Takahashi J, Kimura H: F^- uptake inhibition by excess phosphate during fluoridated apatite formation. Caries Res 1985;19:342-347.

667   Okazaki M: Heterogeneous synthesis of fluoridated hydroxyapatites. Biomaterials 1993 (in press).

668   LeGeros RZ: The unit-cell dimensions of human enamel apatite: effect of chloride incorporation. Arch Oral Biol 1974;20:63-71.

669   Rey C, Trombe JC, Montel G: Sur la fixation de la glycine dans le réseau des phosphates à structure d'apatite. J Chem Res 1978;188:2401-2416.

670   Bacquet G, Vo Quang Truong, Vignoles M, Bonel G: EPR detection of acetate ions trapping in B-type carbonated fluorapatites. J Solid State Chem 1981;39:148-153.

671   Khattech I, Jemal M: Sur la décomposition thermique d'apatites precipitées en presence d'ions acetate. Ann Chim (Paris) 1985;10:653-656.

672   Jemal M, Khattech I: Simultaneous thermogravimetry and gas chromatography during decomposition of carbonate apatites. Thermochim Acta 1989;152:65-76.

673   Hayek E, Lechleitner J, Böhler W: Hydrothermalsynthese von Hydroxylapatit. Angew Chem 1955;67:326.

674   Hayek von E, Böhler W, Lechleitner J, Petter H: Hydrothermalsynthese von Calcium-Apatiten. Z Anorg Allg Chem 1958;295:241-246.

675   Blakeslee KC, Condrate RA: Vibrational spectra of hydrothermally prepared hydroxyapatite. J Am Ceram Soc 1971;54:559-563.

676   Perloff A, Posner AS: Preparation of pure hydroxyapatite crystals. Science 1956;124:583-584.

677   Lapraz D, Baumer A, Iacconi P: On the thermoluminescent properties of hydroxyapatite $Ca_5(PO_4)_3OH$. Phys Stat Sol 1979;54:605-613.

678   Simpson DR: Carbonate in hydroxylapatite. Science 1965;147:501-502.

679   Jullmann H, Mosebach R: Zur Synthese, Licht- und Doppelbrechung des Hydroxylapatits. Z Naturf 1966;21:493-494.

680   Baumer A, Argiolas R: Synthèses hydrothermals et déterminations RX d'apatites chlorée, fluorée ou hydroxylée. Neues Jb Miner Mh 1981;344-348.

681   Arends J, Schuthof J, Linden WH van der, Bennema P, Berg PJ van den: Preparation of pure hydroxyapatite single crystals by hydrothermal recrystallization. J Cryst Growth 1979;46:213-220.

682   Jongebloed WL, Berg PJ van den, Arends J: The dissolution of single crystals of hydroxyapatite in citric acid and lactic acids. Calc Tiss Res 1974;15:1-9.

683   Ioku K, Yoshimura M, Sōmiya S: Hydrothermal synthesis of ultrafine hydroxyapatite single crystals. J Jpn Chem Soc 1988;1565-1570. (in Japanese)

684 Eysel W, Roy DM: Hydrothermal flux growth of hydroxyapatites by temperature oscillation. J Cryst Growth 1973;20:245-250.

685 Engel G: Hydrothermalsynthese von Bleihydroxylapatiten $Pb_5(XO_4)_3OH$ mit X = P, As, V. Naturwissenschaften 1970;57:355.

686 Riboud PV: Dosage de l'eau dans l'hydroxyapatite à haute température. Compt Rend Acad Sci (Paris) Ser C 1969;269:691-694.

687 Vatassery GT, Armstrong WD, Singer L: Determination of hydroxyl content of calcified tissue mineral. Calcif Tissue Res 1970;5:183-188.

688 Meyer JL: Hydroxyl content of solution-precipitated calcium phosphates. Calcif Tissue Int 1979;27:153-160.

689 Fowler BO, Hailer AW, Meyer JL: Infrared and Raman studies of hydroxide quantities in apatitic calcium phosphates. Abstract No 391 J Dent Res (Special Issue A) 1980;59:365.

690 Baumer A, Ganteaume M, Klee WE: Determination of OH ions in hydroxyfluorapatites by infrared spectroscopy. Bull Minér 1985;108:145-152.

691 Nishino M, Yamashita S, Aoba T, Okazaki M, Moriwaki Y: The laser-Raman spectroscopic studies on human enamel and precipitated carbonate-containing apatites. J Dent Res 1981;60:751-755.

692 Cho G, Yesinowski JP: Multiple-quantum NMR dynamics in the quasi-one-dimensional distribution of protons in hydroxyapatite. Chem Phys Lett 1993;205:1-5.

693 Gee A, Deitz VR: Pyrophosphate formation upon ignition of precipitated basic calcium phosphates. J Am Chem Soc 1955;77:2961-2965.

694 Quinaux N: Influence des composes alcalino-terreux sur la formation de groupes pyrophosphates par chauffage du phosphate tricalcique hydrate. Bull Soc Chim Biol (Paris) 1964;46:561-578.

695 Siew C, Gruninger SE, Chow LC, Brown WE: Procedure for the study of acidic calcium phosphate precursor phases in enamel mineral formation. Calcif Tissue Int 1992;50:144-148.

696 Winand L: Séparation de mélanges ortho- pyrophosphate sur échangeur anionique. J Chromatography 1962;7:400-404.

697 Arends J, Davidson C L: HPO_4^{2-} content in enamel and artificial carious lesions. Calcif Tissue Res 1975;18:65-79.

698 Joris SJ, Amberg CH: The nature of deficiency in nonstoichiometric hydroxyapatite. II. Spectroscopic studies of calcium and strontium hydroxyapatite. J Phy Chem 1971;75:3172-3178.

699 Rey C, Collins B, Goehl T, Dickens IR, Glimcher MJ: The carbonate environment in bone mineral: a resolution-enhanced Fourier transform infrared spectroscopy study. Calcif Tissue Int 1989;45:157-164.

700 Miquel JL, Facchini L, Legrand AP, Rey C, Lemaitre J: Solid state NMR to study calcium phosphate ceramics. Colloids and Surfaces 1990;45:427-433.

701 Featherstone JDB, Pearson S, LeGeros RZ: An infrared method for quantification of carbonate in carbonated apatites. Caries Res 1984;18:63-66.

702   Scheib RM, Thrasher RD, Lehr JR: Chemical composition determination of francolite apatites by Fourier transform infrared (FTIR) spectroscopy. Proc Soc Photo-optical Instrum Engineers 1981;289:289-291.

703   Elliott JC, Holcomb DW, Young RW: Infrared determination of the degree of substitution of hydroxyl by carbonate ions in human dental enamel. Calcif Tissue Int 1985;37:372-375.

704   Rey C, Renugopalakrishnan V, Shimizu M, Collins B, Glimcher MJ: A resolution-enhanced Fourier transform infrared spectroscopic study of the environment of the CO_3^{2-} ion in the mineral phase of enamel during its formation and maturation. Calcif Tissue Int 1991;49:259-268.

705   Guillaume M: Applications du dosage par voie diffractométrique. Bull Soc Roy Sci, Liège 1961;30:340-366.

706   Balmain N, Legros R, Bonel G: X-ray diffraction of calcined bone tissue: a reliable method for the determination of bone Ca/P molar ratio. Calcif Tissue Int 1982;34:S93-S98.

707   Bett JAS, Christner LG, Hall WK: Studies of the hydrogen held by solids. XII. Hydroxyapatite catalysts. J Am Chem Soc 1967;89:5535-5541.

708   Barton SS, Harrison BH: Surface properties of hydroxyapatites I. Enthalpy of immersion. J Colloid Interface Sci 1976;55:409-414.

709   Driessens FCM, Verbeeck RMH, Heijligers HJM: Some physical properties of Na- and CO_3-containing apatites synthesized at high temperatures. Inorganica Chim Acta 1983;80:19-23.

710   Posner AS, Perloff A: Apatites deficient in divalent cations. J Res Natn Bur Stand 1957;58:279-286.

711   Posner AS, Stutman JM, Lippincott ER: Hydrogen-bonding in calcium-deficient hydroxyapatite. Nature 1960;188:486-487.

712   Stutman JM, Posner AS, Lippincott ER. Hydrogen bonding in the calcium phosphates. Nature 1962;193:368-369.

713   Winand L: Étude physico-chimique du phosphate tricalcique hydraté et de l'hydroxylapatite. Ann Chim (Paris) 13th Series 1961;6:951-967.

714   Kühl G von, Nebergall WH: Hydrogenphosphat- und Carbonatapatite. Z Anorg Allg Chem 1963;324:313-320.

715   Berry EE: The structure and composition of some calcium-deficient apatites. Bull Soc Chim Fr (Spec No) 1968;1765-1770.

716   Berry EE: The structure and composition of some calcium-deficient apatites II. J Inorg Nucl Chem 1967;29:1585-1590.

717   Delay A, Friedli C, Lerch P: Comportement thermique et composition des phosphates calciques de Ca/P molaire supérieur à 3/2. Bull Soc Chim Fr (special. no) 1974:839-844.

718   Trautz OR: X-ray diffraction of biological and synthetic apatites. Ann NY Acad Sci 1955;60:696-712.

719   LeGeros RZ, Bonel G, Legros R: Types of "H_2O" in human enamel and in precipitated apatites. Calcif Tissue Res 1978;26:111-118.

720 Young RA, Holcomb DW: Role of acid phosphate in hydroxyapatite lattice expansion. Calcif Tissue Int 1984;36:60-63.

721 Holcomb DW, Young RA: Thermal decomposition of human tooth enamel. Calcif Tissue Int 1980;31:189-201.

722 Labarthe J-C, Bonel G, Montel G: Sur la structure et les propriétés des apatites carbonatées de type B phospho-calciques. Ann Chim (Paris) 14th Series 1973;8:289-301.

723 Blumenthal NC: Mechanisms of inhibition of calcification. Clin Orthop Rel Res 1989;247:279-289.

724 Kemenade van MJJM, Bruyn de PL: A kinetic study of precipitation from supersaturated calcium phosphate solutions. J Colloid Interface Sci 1987;118:564-585.

725 Boskey AL, Posner AS: Formation of hydroxyapatite at low supersaturation. J Phys Chem 1976;80:40-45.

726 Moreno EC, Zahradnik RT, Glazman A, Hwu R: Precipitation of hydroxyapatite from dilute solutions upon seeding. Calcif Tissue Res 1977;24:47-57.

727 Moreno EC, Varughese K: Crystal growth of calcium apatites from dilute solutions. J Cryst Growth 1981;53:20-30.

728 Koutsoukos P, Amjad Z, Tomson MB, Nancollas GH: Crystallization of calcium phosphates. A constant composition study. J Am Chem Soc 1980;102:1553-1557.

729 Koutsoukos PG, Nancollas GH: The morphology of hydroxyapatite crystals grown in aqueous solution at 37°C. J Cryst Growth 1981;55:369-375.

730 Koutsoukos PG, Nancollas GH: Influence of strontium ion on the crystallization of hydroxylapatite from aqueous solution. J Phys Chem 1981;85:2403-2408.

731 Koutsoukos PG, Nancollas GH: The effect of lithium on the precipitation of hydroxyapatite from aqueous solution. Colloids and Surfaces 1986;17:361-370.

732 Wilson JWL, Werness PG, Smith LH: Inhibitors of crystal growth of hydroxyapatite: a constant composition approach. J Urol 1985;134:1255-1258.

733 Koutsoukos PG, Nancollas GH: Crystal growth of calcium phosphates—epitaxial considerations. J Cryst Growth 1981;53:10-19.

734 Ebrahimpour A, Zhang J, Nancollas GH: Dual constant composition method and its application to studies of phase transformation and crystallization of mixed phases. J Cryst Growth 1991;113:83-91.

735 Newesely H: Changes in crystal types of low solubility calcium phosphates in the presence of accompaning ions. Arch Oral Biol (Spec Suppl) 1961;6:174-180.

736 Meyer JL, Nancollas GH: The effect of stannous and fluoride ions on the rate of crystal growth of hydroxyapatite. J Dent Res 1972;51:1443-1450.

737 Eanes ED, Meyer JL: The influence of fluoride on apatite formation from unstable supersaturated solutions at pH 7.4. J Dent Res 1978;57:617-624.

738 Hoek van den WGM, Feenstra TP, Bruyn de PL: Influence of fluoride on the formation of calcium phosphates in moderately supersaturated solutions. J Phys Chem 1980;84:3312-3317.

739 Saleeb FZ, Bruyn PL de: Surface properties of alkaline earth apatites. J Electroanal Chem 1972;37:99-118.

740 Chander S, Fuerstenau DW: Interfacial properties and equilibria in the apatite-aqueous solution system. J Colloid Interface Sci 1979;70:506-516.

741 Suzuki T, Hatsushika T, Hayakawa Y: Surface characteristics of various synthetic apatites; in Eon C (ed): Proc 2nd Int Congr Phosphorus Compounds, Boston. IMPHOS, Paris, 1980, pp 165-174.

742 Brown WE, Chow LC, Mathew M: Thermodynamics of hydroxyapatite surfaces. Croatica Chem Acta 1983;56:779-787.

743 Brown WE, Chow LC: Surface equilibria of sparingly soluble crystals. Colloids and Surfaces 1983;7:67-80.

744 Chander S, Fuerstenau DW: Solubility and interfacial properties of hydroxyapatite: a review; in Misra DN (ed): Adsorption on and Surface Chemistry of Hydroxyapatite. New York, Plenum Press, 1984, pp 29-49.

745 Somasundaran P, Wang YHC: Surface chemical characteristics and absorption properties of apatite; in Misra DN (ed): Adsorption on and Surface Chemistry of Hydroxyapatite. New York, Plenum Press, 1984, pp 129-149.

746 Bell LC, Mika H, Kruger BJ: Synthetic hydroxyapatite-solubility product and stoichiometry of dissolution. Arch Oral Biol 1978;23:329-336.

747 Verbeeck RMH, Steyaert H, Thun HP, Verbeek F: Solubility of synthetic calcium hydroxyapatites. J Chem Soc Faraday I 1980;76:209-219.

748 Verbeeck RMH, Thun HP, Driessens FCM: Effect of dehydration of hydroxyapatite on its solubility behavior. Z Physik Chem NF 1980;119:79-84.

749 Aoba T, Moreno EC: Changes in the solubility of enamel mineral at various stages of porcine amelogenesis. Calcif Tissue Int 1992;50:266-272.

750 Moreno EC, Aoba T: Comparative solubility study of human dental enamel, dentin, and hydroxyapatite. Calcif Tissue Int 1991;49:6-13.

751 Jahnke RA: The synthesis and solubility of carbonate fluorapatite. Am J Sci 1984;284:58-78.

752 Moreno EC, Kresak M, Zahradnik RT: Fluoridated hydroxyapatite solubility and caries formation. Nature 1974;247:64-65.

753 Chin KOA, Nancollas GH: Dissolution of fluorapatite. A constant-composition kinetics study. Langmuir 1991;7:2175-2179.

754 Verbeeck RMH, Khan RA, Thun HP, Driessens FCM: Solubility of thermal synthesized fluorhydroxyapatites. Bull Soc Chim Belg 1979;88:751-759.

755 Mishra RK, Chander S, Fuerstenau DW: Effect of ionic surfactants on the electrophoretic mobility of hydroxyapatite. Colloids and Surfaces 1980;1:105-119.

756 Lin J, Raghavan S, Fuerstenau DW: The adsorption of fluoride ions by hydroxyapatite from aqueous solution. Colloids and Surfaces 1981;3:357-370.

757 Benton DP, Bullock JI, Danil de Namor AF, Ingram GS: Calorimetric studies of the interaction between hydroxyapatite and certain anions in aqueous solution. Caries Res 1980;14:110-114.

758 Misra DN (ed): Adsorption on and Surface Chemistry of Hydroxyapatite. New York, Plenum Press, 1984.

759 Misra DN: Isoionic isotope exchange with hydroxylapatite and the dilution effect. J Res Natn Bur Stand 1979;84:395-406.

760 Wilson AD, Prosser HJ, Powis DM: Mechanism of adhesion of polyelectrolyte cements to hydroxyapatite. J Dent Res 1983;62:590-592.

761 Davies AK, Cundall RB, Dandiker Y, Slifkin MA: Photo-oxidation of tetracycline adsorbed on hydroxyapatite in relation to the light-induced staining of teeth. J Dent Res 1985;64:936-939.

762 Moreno EC, Kresak M, Kane JJ, Hay DI: Adsorption of proteins, peptides, and organic acids from binary mixtures onto hydroxylapatite. Langmuir 1987;3:511-519.

763 Moreno EC, Kresak M, Hay DI: Adsorption of salivary proteins onto Ca apatites. Biofouling 1991;4:3-24.

764 Tanabe T, Aoba T, Moreno EC, Fukae M: Effect of fluoride in the apatitic lattice on adsorption of enamel proteins onto calcium apatites. J Dent Res 1988; 67:536-542.

765 Fujisawa R, Kuboki Y: Preferential adsorption of dentin and bone acid proteins on the (100) face of hydroxyapatite crystals. Biochim Biophys Acta 1991;1075:56-60.

766 Cowan MM, Taylor KG, Doyle RJ: Energetics of the initial phase of adhesion of *Streptococcus sanguis* to hydroxlyapatite. J Bacteriol 1987;169:2995-3000.

767 Chander S, Fuerstenau DW: An XPS study of the fluoride uptake by hydroxyapatite. Colloids and Surfaces 1985;13:137-144.

768 Yesinowski JP, Mobley MJ: ^{19}F MAS-NMR of fluoridated hydroxyapatite surfaces. J Am Chem Soc 1983;105:6191-6193.

769 Kreinbrink AT, Sazavsky CD, Pyrz JW, Nelson DGA, Honkonen RS: Fast-magic-angle-spinning ^{19}F NMR of inorganic fluorides and fluoridated apatitic surfaces. J Magnetic Resonance 1990;88:267-276.

770 White DJ, Bowman WD, Faller RV, Mobley MJ, Wolfgang RA, Yesinowski JP: ^{19}F MAS-NMR and solution chemical characterization of the reactions of fluoride with hydroxyapatite and powdered enamel. Acta Odont Scand 1988;46:375-389.

771 Borneman-Starinkevitch ID: On some isomorphic substitutions in apatite. Dokl Acad Sci SSSR 1938;19:253-255.

772 Borneman-Starinkevitch ID: On isomorphic substitutions in apatite. Dokl Acad Nauk SSSR 1939;22:113-115.

773 Arends J, Nelson DGA, Dijkman AG, Jongebloed WL: Effect of various fluorides on enamel structure and chemistry; in Guggenheim B (ed) Cariology Today, Int Congr, Zürich, 1983. Basel, Karger, 1984, pp 245-258.

774 Tsuda H, Arends J: Detection and quantification of calcium fluoride using micro-Raman spectroscopy. Caries Res 1993;27:249-257.

775 Christoffersen MR, Christoffersen J, Arends J: Kinetics of dissolution of calcium hydroxyapatite VII. The effect of fluoride ions. J Cryst Growth 1984;67:107-114.

776 Chander S, Chiao CC, Fuerstenau DW: Transformation of calcium fluoride for caries prevention. J Dent Res 1982;61:403-407.

777 Jordan TH, Wei SHY, Bromberger SH, King JC: $Sn_3F_3PO_4$: the product of the reaction between stannous fluoride and hydroxyapatite. Arch Oral Biol 1971;16:241-246.

778 Berndt AF: Reaction of stannous fluoride with hydroxyapatite: the crystal structure of $Sn_3PO_4F_3$. J Dent Res 1972;51:53-57.

779 Chander S, Fuerstenau DW: On the dissolution and interfacial properties of hydroxyapatite. Colloids and Surfaces 1982;4:101-120.

780 Jordan TH, Schroeder LW, Dickens B, Brown WE: Crystal structure of stannous hydroxide phosphate, a reaction product of stannous fluoride and apatite. Inorg Chem 1976;15:1810-1814.

781 Ingram GS: The reaction of monofluorophosphate with apatite. Caries Res 1972;6:1-15.

782 Ingram GS: Reaction between apatite and monofluorophosphate: modification by fluoride and condensed phosphate. Caries Res 1977;11:30-38.

783 Christoffersen J: Dissolution of calcium hydroxyapatite. Calcif Tissue Int 1981;33:557-560.

784 Christoffersen J, Christoffersen MR: Kinetics of dissolution of calcium hydroxyapatite. Faraday Discuss Chem Soc 1984;77:235-242.

785 Gramain Ph, Thomann JM, Gumpper M, Voegel JC: Dissolution kinetics of human enamel powder. I. Stirring effects and surface calcium accumulation. J Colloid Interface Sci 1989;128:370-381.

786 Napper DH, Smythe BM: The dissolution kinetics of hydroxyapatite in the presence of kink poisons. J Dent Res 1966;45:1775-1783.

787 Wu M-S, Higuchi WI, Fox JL, Friedman M: Kinetics and mechanisms of hydroxyapatite crystal dissolution in weak acid buffer using the rotating disk method. J Dent Res 1976;55:496-505.

788 Fawzi MB, Fox JL, Dedhiya MG, Higuchi WI, Hefferren JJ: A possible second site for hydroxyapatite dissolution in acidic media. J Colloid Interface Sci 1978;67:304-311.

789 Fox JL, Higuchi WI, Fawzi MB, Wu M-S: A new two-site model for hydroxyapatite dissolution in acidic media. J Colloid Interface Sci 1978;67:312-330.

790 Patel MV, Fox JL, Higuchi WI: Physical model for non-steady-state dissolution of dental enamel. J Dent Res 1987;66:1418-1424.

791 Patel MV, Fox JL, Higuchi WI: Effect of acid type on kinetics and mechanism of dental enamel demineralization. J Dent Res 1987;66:1425-1430.

792 Nelson DGA, Barry JC, Shields CP, Glena R, Featherstone JDB: Crystal morphology, composition and dissolution behavior of carbonated apatites prepared at controlled pH and temperature. J Colloid Interface Sci 1989;130:467-479.

793 Nelson DGA, Featherstone JDB, Duncan JF, Cutress TW: Effect of carbonate and fluoride on the dissolution behaviour of synthetic apatites. Caries Res 1983;17:200-211.

794 Higuchi WI, Cesar EY, Cho PW, Fox JL: Powder suspension method for critically re-examining the two-site model for hydroxyapatite dissolution kinetics. J Pharm Sci 1984;73:146-153.

795 Margolis HC, Moreno EC: Kinetics of hydroxyapatite dissolution in acetic, lactic, and phosphoric acid solutions. Calcif Tissue Int 1992;50:137-143.

796 Okazaki M, Takahashi J, Kimura H, Aoba T: Crystallinity, solubility, and dissolution rate behavior of fluoridated CO_3 apatites. J Biomed Mater Res 1982;16:851-860.

797 Christoffersen J, Christoffersen MR, Kjærgaard N: The kinetics of dissolution of calcium hydroxyapatite in water at constant pH. J Cryst Growth 1978;43:501-511.

798 Christoffersen J, Christoffersen MR: Kinetics of dissolution of calcium hydroxyapatite II. Dissolution in non-stoichiometric solutions at constant pH. J Cryst Growth 1979;47:671-679.

799 Christoffersen J: Kinetics of dissolution of calcium hydroxyapatite III. Nucleation-controlled dissolution of a polydisperse sample of crystals. J Cryst Growth 1980;49:29-44.

800 Thomann JM, Voegel JC, Gumpper M, Gramain Ph: Dissolution kinetics of human enamel powder II. A model based on the formation of a self-inhibiting surface layer. J Colloid Interface Sci 1989;132:403-412.

801 Thomann JM, Voegel JC, Gramain Ph: Kinetics of dissolution of calcium hydroxyapatite powder. III: pH and sample conditioning effects. Calcif Tissue Int 1990;46:121-129.

802 Budz JA, LoRe M, Nancollas GH: Hydroxyapatite and carbonated apatite as models for the dissolution behavior of human enamel. Adv Dent Res 1987;1:314-321.

803 Budz JA, Nancollas GH: The mechanism of dissolution of hydroxyapatite and carbonated apatite in acidic solution. J Cryst Growth 1988;91:490-496.

804 Daculsi G, Kerebel B, Kerebel LM: Mechanisms of acid dissolution of biological and synthetic apatite crystals at the lattice pattern level. Caries Res 1979;13:277-289.

805 Mayer, I, Voegel JC, Brès EF, Frank RM: The release of carbonate during the dissolution of synthetic apatites and dental enamel. J Cryst Growth 1988;87:129-136.

806 Brès EF, Barry JC, Hutchison JL: A structural basis for the carious dissolution of the apatite crystals of human tooth enamel. Ultramicroscopy 1984;12:367-372.

807 Daculsi G, LeGeros RZ, Mitre D: Crystal dissolution of biological and ceramic apatites. Calcif Tissue Int 1989;45:95-103.

808 Christoffersen J, Christoffersen MR: Kinetics of dissolution of calcium hydroxyapatite V. The acidity constant for the hydrogen phosphate surface complex. J Cryst Growth 1982;57:21-26.

809   Reynolds EC, Riley PF, Storey E: Phosphoprotein inhibition of hydroxyapatite dissolution. Calcif Tissue Int 1982;34:S52-S56.

810   Christoffersen J, Christoffersen MR, Christensen SB, Nancollas GH: Kinetics of dissolution of calcium hydroxyapatite VI. The effects of adsorption of methylene diphosphonate, stannous ions and partly-peptized collagen. J Cryst Growth 1983;62,254-264.

811   Christoffersen MR, Christoffersen J: The inhibitory effects of ATP, ADP, and AMP on the rate of dissolution of calcium hydroxyapatite. Calcif Tissue Int 1984;36:659-661.

812   Christoffersen MR, Thyregod HC, Christoffersen J: Effects of aluminium(III), chromium(III), and iron(III) on the rate of dissolution of calcium hydroxyapatite crystals in the absence and presence of the chelating agent desferrioxamine. Calcif Tissue Int 1987;41:27-30.

813   Arends J, Jongebloed WL: Apatite single crystals. Formation, dissolution and influence of CO_3^{2-} ions. Rec Trav Chim Pays-Bas 1981;100:3-9.

814   Anderson P: Real-time X-ray Absorption Studies and Their Interpretation via Numerical Solutions of Diffusion and Reaction Equations of Model Systems for Dental Caries. PhD Thesis, University of London, 1988.

815   Anderson P, Elliott JC: Coupled diffusion as basis for subsurface demineralisation in dental caries. Caries Res 1987;21:522-525.

816   Leaist DG: Subsurface dissolution and precipitation during leaching of porous ionic solids. J Colloid Interface Sci 1987;118:262-269.

817   Anderson P, Elliott JC: Subsurface demineralization in dental enamel and other permeable solids during acid dissolution. J Dent Res 1992;71:1473-1481.

818   Nelson DGA, Williamson BE: Low-temperature laser Raman spectroscopy of synthetic carbonated apatites and dental enamel. Aust J Chem 1982;35:715-727.

819   Nelson DGA, Featherstone JDB: Preparation, analysis, and characterization of carbonated apatites. Calcif Tissue Int 1982;34:S69-S81.

820   Kislovsky LD, Knobovets RG: On the sensitivity of infrared spectra of apatite monocrystals to isomorphous substitutions. Dokl Akad Nauk SSSR 1968;179:1432-1435. (in Russian)

821   Knubovetz RG, Kislovsky LD: Study of anion substitution in apatite using infra- red spectroscopy; in Sobolev VS (ed): Physics of Apatite (Spectroscopic Investigation of Apatite). Academy of Sciences of the USSR, Siberian Branch, Trans Inst Geol Geophys Issue 50. Novosibirsk, Nauka, 1975, pp 63-88. (in Russian)

822   Iqbal Z, Tomaselli VP, Fahrenfeld O, Möller KD, Ruszala FA, Kostiner E: Polarized Raman scattering and low frequency infrared study of hydroxyapatite. J Phys Chem Solids 1977;38:923-927.

823   Engel G, Klee WE: Infrared spectra of the hydroxyl ions in various apatites. J Solid State Chem 1972;5:28-34.

824   Cant NW, Bett JAS, Wilson GR, Hall WK: The vibrational spectrum of hydroxyl groups in hydroxyapatites. Spectrochim Acta 1971;27A:425-439.

825 Reisner I, Klee WE: Temperature dependence of the ν(OH) bands of hydroxyapatite. Spectrochim Acta 1982;38A:899-902.

826 Elliott JC: The determination of the fluorine substituting for hydroxyl ions using infrared spectroscopy. Paper at Brit Div Int Assoc Dental Res, Edinburgh, 1964. J Dent Res 1964;43:959. (abstract)

827 Fowler BO: Infrared spectra of hydroxy-fluor-apatite. Abstract No 247 in Program and Abstracts of papers presented to the 45th General Meeting of the International Association for Dental Research, Washington, March, 1967, p 98.

828 Young RA, Lugt W van der, Elliott JC: Mechanism for fluorine inhibition of diffusion in hydroxyapatite. Nature 1969;223:729-730.

829 Klee WE: OH-Ionen in natürlichen Fluorapatiten. Neues Jb Miner Mh 1974;127-143.

830 Amberg CH, Luk HC, Wagstaff KP: The fluoridation of nonstoichiometric calcium hydroxyapatite, an infrared study. Can J Chem 1974;52:4001-4006.

831 Freund F, Knobel RM: Distribution of fluorine in hydroxyapatite studied by infrared spectroscopy. J Chem Soc Dalton Trans 1977:1136-1140.

832 Menzel B, Amberg CH: An infrared study of the hydroxyl groups in a nonstoichiometric calcium hydroxyapatite with and without fluoridation. J Colloid Interface Sci 1972;38:256-264.

833 Maiti GC, Freund F: Incorporation of chlorine into hydroxy-apatite. J Inorg Nucl Chem 1981;43:2633-2637.

834 Winand L, Duyckaerts G: Étude infrarouge de phosphates de calcium de la famille de l'hydroxylapatite. Bull Soc Chim Belg 1962;71:142-150.

835 Aue WP, Roufosse AH, Glimcher MJ, Griffin RG: Solid-state phosphorus-31 nuclear magnetic resonance studies of synthetic solid phases of calcium phosphate: potential models of bone mineral. Biochemistry 1984;23:6110-6114.

836 Yesinowski JP: High resolution NMR spectroscopy of solids and surface-adsorbed species in colloidal suspension: ^{31}P NMR spectra of hydroxyapatite and diphosphonates. J Am Chem Soc 1981;103:6266-6267.

837 Bonar LC, Shimizu M, Roberts JE, Griffin RG, Glimcher MJ: Structural and composition studies on the mineral of newly formed dental enamel: A chemical, X-ray diffraction, and ^{31}P and proton nuclear magnetic resonance study. J Bone Miner Res 1991;6:1167-1176.

838 Roufosse AH, Aue WP, Roberts JE, Glimcher MJ, Griffin RG: Investigation of the mineral phases of bone by solid-state phosphorus-31 magic angle sample spinning nuclear magnetic resonance. Biochemistry 1984;23:6115-6120.

839 Lugt W van der, Knottnerus DIM, Perdok WG: Nuclear magnetic resonance investigation of fluoride ions in hydroxyapatite. Acta Cryst 1971;B27:1509-1516.

840 Vakhrameev AM, Gabuda SP, Knubovets RG: ^{1}H and ^{19}F NMR in apatites of the type $Ca_5(PO_4)_3[F_{1-x}(OH)_x]$. J Struct Chem 1978;19:256-261 (English Trans of Zh Strukt Khim 1978;19:298-304).

841 Knubovetz RG, Gabuda SP: The study into isomorphous substitution of F by OH groups in apatite as found by NMR; in Sobolev VS (ed): Physics of Apatite (Spectroscopic Investigation of Apatite). Academy of Sciences of the USSR,

Siberian Branch, Trans Inst Geol Geophys Issue 50. Novosibirsk, Nauka, 1975, pp 100-112. (in Russian)

842 Yesinowski JP, Wolfgang RA, Mobley MJ: New NMR methods for the study of hydroxyapatite surfaces; in Misra DN (ed): Adsorption on and Surface Chemistry of Hydroxyapatite. New York, Plenum Press 1984, pp 151-175.

843 Code RF, Gelman N, Armstrong RL, Hallsworth RS, Lemaire C, Cheng P-T: Field dependence of ^{19}F NMR in rat bone powders. Phys Med Biol 1990;35:1271-1286.

844 Gelman N, Code RF: NMR spin-echo study of ^{19}F environments in rat bone mineral. J Magnetic Resonance 1992;96:290-301.

845 Chapman MR, Miller AG, Stoebe TG: Thermoluminescence in hydroxyapatite. Med Phys 1979;6:494-499.

846 Lapraz D, Baumer A: Chloroapatite, $Ca_5(PO_4)_3Cl$: Thermoluminescent properties. Phys Stat Sol 1981;68:309-319.

847 Davies JE: Surface dependent emission of low energy electrons (exoemission) from apatite samples; in Misra DN (ed): Adsorption on and Surface Chemistry of Hydroxyapatite. New York, Plenum Press, 1984, pp 71-95.

848 Katz JL, Ukraincik K: On the anisotropic elastic properties of hydroxyapatite. J Biomech 1971;4:221-227.

849 Egan EP, Wakefield ZT, Elmore KL: High-temperature heat content of hydroxyapatite. J Am Chem Soc 1950;72:2418-2421.

850 Perdok WG, Christoffersen J, Arends J: The thermal lattice expansion of calcium hydroxyapatite. J Cryst Growth 1987;80:149-154.

851 Takahashi T, Tanase S, Yamamoto O: Electrical conductivity of some hydroxyapatites. Electrochim Acta 1978;23:369-373.

852 Maiti GC, Freund F: Influence of fluorine substitution on proton conductivity of hydroxyapatite. J Chem Soc Dalton Trans 1981;949-955.

853 Ash R, Barrer RM, Coughlan B: Sorption and flow of gases in compacts of synthetic hydroxyapatite. J Colloid Interface Sci 1972;38:61-74.

854 McConnell D: A structural investigation of the isomorphism of the apatite group. Am Miner 1938;23:1-19.

855 Henry TH: On francolite, a supposed new mineral. Phil Mag 1850;36:134-135.

856 Brögger WC, Bäckström H: Über den Dahllit, ein neues Mineral von Ödegården, Bamle, Norwegen. Neues Jb Miner Geol Paläont 1890;223-224. (abstract)

857 Bonel G, Montel G: Sur une nouvelle apatite carbonatée synthétique. Comp Rend Acad Sci (Paris) 1964:258;923-926.

858 Solly RH: Francolite, a variety of apatite from Levant Mine, St Just, Cornwall. Miner Mag 1886;7:57-58.

859 Hendricks SB, Jefferson ME, Mosley VM: The crystal structure of some natural and synthetic apatite-like substances. Z Krist 1932;81:352-369.

860 Gruner JW, McConnell D: The problem of the carbonate-apatites. Z Krist 1937;97A:208-215.

861 Belov NV: On some isomorphic substitutions in the apatite group. Dokl Acad Sci SSSR 1939;22:89-92.

862 Hendricks SB, Hill WL: The inorganic constitution of bone. Science 1942;96:255-257.

863 McConnell D: The problem of the carbonate apatites. IV. Structural substitutions involving CO₃ and OH. Bull Soc Fr Minér Crist 1952;75:428-445.

864 Borneman-Starinkevitch ID, Belov NV: On carbonate apatites. Dokl Acad Nauk SSSR 1953;90:89-92. (in Russian)

865 Trautz OR: Crystallographic studies of calcium carbonate phosphates. Ann NY Acad Sci 1960;85:145-160.

866 Elliott JC: The interpretation of the infra-red absorption spectra of some carbonate-containing apatites; in Fearnhead RW, Stack MV (eds): Tooth Enamel, Proc Int Symp, London, 1964. Bristol, John Wright & Sons, 1965, pp 20-22 and pp 50-57.

867 Elliott JC: Some observations on the crystal chemistry of carbonate-containing apatites; in Hardwick JL, Dustin J-P, Held HR (eds): Proceedings of the 9th ORCA Conference, Paris. Arch Oral Biol Spec Suppl, Oxford, Pergamon Press, 1963, pp 277-282.

868 Thewlis J, Glock GE, Murray MM: Chemical and X-ray analysis of dental, mineral and synthetic apatites. Trans Faraday Soc 1939;35:358-363.

869 Maslennikov BM, Kavitskaya FA: On the phosphatic substance of phosphorites. Dokl Acad Nauk SSSR 1956;109:990-992. (in Russian)

870 Gassmann T. Über die künstliche Darstellung des Hauptbestandteiles der Knochen und der Zähne. Hoppe-Seyl Z 1928;178:62-67.

871 Elliott JC: Interpretation of carbonate bands in infrared spectrum of dental enamel. J Dent Res 1963;42:1081. (abstract)

872 LeGeros RZ, Trautz OR, LeGeros JP, Klein E: Carbonate substitution in the apatite structure. Bull Soc Chim Fr (Special No) 1968:1712-1718.

873 Klement R, Trömel G: Hydroxylapatit, der Hauptbestandteil der anorganischen Knochen- und Zahnsubstanz. Hoppe-Seyl Z 1932;213:263-269.

874 Pobequin T: Distinction analytique des carbonates de calcium au moyen de méthodes d'étude physiques. Chim Analyt; 1954;8:203-210.

875 Herman H, Dallemagne MJ: The main mineral constituent of bone and teeth. Arch Oral Biol 1961;5:137-144.

876 McConnell D: Recent advances in the investigation of the crystal chemistry of dental enamel. Arch Oral Biol 1960;3:28-34.

877 McConnell D: The crystal structure of bone mineral. Clin Orthop Rel Res 1962;23:253-268.

878 Klement R, Hüter F, Köhrer K: Bildet sich Carbonatapatit in wässrigen Systemen? Z Elektrochem 1942;48:334-336.

879 Romo LA: Synthesis of carbonato-apatite. J Am Chem Soc 1954;76:3924-3925.

880 Ames LL: The genesis of carbonate apatites. Economic Geol 1959;54:829-841.

881 Neuman WF, Toribara TY, Mulryan BJ: The surface chemistry of bone. IX. Carbonate:phosphate exchange. J Am Chem Soc 1956;78:4263-4266.

882 Akhavan Niaki AN: Contribution à l'étude des substitutions dans les apatites. Ann Chim (Paris) 1961;51-79.

883   Mohseni-Koutchesfehani S: Contribution à l'étude des apatites barytiques. Ann Chim (Paris) 1961;463-479.

884   Elliott JC: Synthetic and biological carbonate-containing apatites; in Brown WE, Young RA (eds): Proceedings of International Symposium on Structural Properties of Hydroxyapatite and Related Compounds, Gaithersburg, 1968, Ch 11. (unpublished, copy of chapter available from JC Elliott).

885   Trautz OR, Zapanta RR: Synthetic carbonate apatite. J Dent Res 1960;39:664. (abstract)

886   Bonel G, Montel G: Étude comparée des apatites carbonatées obtenues par différentes méthodes de synthèse; in Schwab G-M (ed): Reactivity of Solids. Amsterdam, Elsevier, 1965, pp 667-675.

887   Eitel W: Über Karbonatphosphate der Apatitgruppe. Schr Königsberger Gelehrt Ges Naturw Kl 1924;1:159-177.

888   Newesely H: Kristallchemische und mikromorphologische Untersuchungen an Carbonat-Apatiten. Monatsh Chemie 1963;94:270-280.

889   Ito A, Aoki H, Akao M, Miura N, Otsuka R, Tsutsumi S: Structure of borate groups in boron-containing apatite. J Ceram Soc Jpn Inter Ed 1988;96:695-697.

890   Smith JP, Lehr JR: An X-ray investigation of carbonate apatites. J Agr Food Chem 1966;14:342-349.

891   McClellan GH: Mineralogy of carbonate fluorapatites. J Geol Soc (London) 1980;137:675-681.

892   Labarthe J-C, Bonel G, Montel G: Sur la localisation des ions carbonate dans le réseau des apatites calciques. Compt Rend Acad Sci (Paris) 1971;273:349-351.

893   Bacquet G, Vo Quang Truong, Vignoles M, Bonel G: ESR of the F^+ centre in B-type carbonated hydroxyapatite. Phys Stat Sol 1981;68:K71-K74.

894   Beshah K, Rey C, Glimcher MJ, Schimizu M, Griffin RG: Solid state carbon-13 and proton NMR studies of carbonate-containing calcium phosphates and enamel. J Solid State Chem 1990;84:71-81.

895   Binder G, Troll G: Coupled anion substitution in natural carbon-bearing apatites. Contr Miner Petrol 1989;101:394-401.

896   Trombe J-C, Bonel G, Montel G: Sur les apatites carbonatées préparées à haute température. Bull Soc Chim Fr (special No) 1968;1708-1711.

897   Driessens FCM, Verbeeck RMH, Kiekens P: Mechanism of substitution in carbonated apatites. Z Anorg Allg Chemie 1983;504:195-200.

898   Deans T: Francolite from sedimentary ironstones of the Coal Measures. Miner Mag 1938;25:135-139.

899   McConnell D, Gruner JW: The problem of the carbonate-apatites. III. Carbonate-apatite from Magnet Cove, Arkansas. Am Miner 1940;25:157-167.

900   Hutton CO, Seelye FT: Francolite, a carbonate-apatite from Milburn, Otago. Trans Roy Soc New Zealand 1942;72:191-198.

901   Pupke F: Die Optischen Anomalien bei Apatit. Doctoral Thesis, Rheinischen Friedrich-Wilhelms-Universita"t, Bonn, 1908.

902   McConnell D: The crystal chemistry of carbonate apatites and their relationship to the composition of calcified tissue. J Dent Res 1952;31:53-63.

903   Aoki H, Akao M, Miura N, Ito A, Shimizukawa Y, Nakamura S, Otsuka R: Crystal structure of carbonate-bearing hydroxyapatite from Durango, Mexico. Tokyo Ika Shika Daigaku Iyo Kizai Kenkyusho Hokoku (Reports of the Institute for Medical & Dental Engineering) 1985;19:15-20. (in Japanese)

904   Perdikatsis B: X-ray powder diffraction study of francolite by the Rietveld method. Mater Sci Forum 1991;79-82:809-814.

905   Baur WH: The geometry of polyhedral distortions. Predictive relationships for the phosphate group. Acta Cryst 1974;B30:1195-1215.

906   White WB: The carbonate minerals; in Farmer VC (ed): The Infrared Spectra of Minerals. London, The Mineralogical Society, 1974, Ch 12, pp 227-284.

907   Fraser RDB: Infra-red spectra; in Alexander P (ed): A Laboratory Manual of Analytical Methods of Protein Chemistry, Vol 2. Oxford, Pergamon Press, 1960, pp 285-351.

908   Okazaki M: F$^-$-CO$_3$$^{2-}$ interaction in IR spectra of fluoridated CO$_3$-apatites. Calcif Tissue Int 1983;35:78-81.

909   Roux P, Bonel G: Étude par diffraction des rayons X et par spectrométrie d'absorption infrarouge des apatite carbonatées de type A phospho-calcique et arsénio-calcique «haute pression». Bull Minér 1978;101:448-452.

910   Nadal M, Legros J-P, Bonel G, Montel G: Mise en évidence d'un phénomène d'order-désordre dans le réseau des carbonate-apatites strontique. Compt Rend Acad Sci (Paris) 1971;272:45-48.

911   El Feki H, Rey C, Vignoles M: Carbonate ions in apatites: infrared investigations in the ν_4 CO$_3$ domain. Calcif Tissue Int 1991;49:269-274.

912   Massuyes M, Trombe J-C, Bonel G, Montel G: Étude par spectrométrie d'absorption dans l'infrarouge, de l'ion carbonate dans quelques calciques préparées à haute température. Comp Rend Acad Sci (Paris) 1969;268:941-944.

913   Young RA, Bartlett ML, Spooner S, Mackie PE, Bonel G: Reversible high temperature exchange of carbonate and hydroxyl ions in tooth enamel and synthetic hydroxyapatite. J Biol Phys 1981;9:1-26.

914   Hitmi N, LaCabanne C, Bonel G, Roux P, Young RA: Dipole co-operative motions in an A-type carbonated apatite, Sr$_{10}$(AsO$_4$)$_6$CO$_3$. J Phys Chem Solids 1986;47:507-515.

915   Bonel G, Montel G: Sur l'introduction des ions CO$_3$$^{2-}$ dans le réseau des apatites calciques. Compt Rend Acad Sci (Paris) 1966;263:1010-1013.

916   Trombe J-C, Bonel G, Montel G: Influence de la chaux sur la formation d'apatites carbonatées à haute température. Compt Rend Acad Sci (Paris) Ser C 1967;265:1113-1116.

917   Bonel G, Labarthe JC, Vignoles C: Contribution à l'étude structurale des apatites carbonatées de type B; in: Physico-Chimie et Cristallographie des Apatites d'Intérêt Biologique, Colloques Internationaux du Centre National de la Recherche Scientifique No 230 Paris, 1973. Paris, CNRS, 1975, pp 117-125.

918   Roux P, Bonel G: Évolution structural sous haute pression des apatites carbonatées de type B. Ann Chim (Paris) 1980;5:397-405.

919 Vignoles C, Trombe J-C, Bonel G, Montel G: Sur la décomposition thermique des apatites carbonatées préparées en milieu sodé. Compt Rend Acad Sci (Paris) Ser C 1975;280:275-277.

920 Ellies LG, Nelson DGA, Featherstone JDB: Crystallographic structure and surface morphology of sintered carbonated apatites. J Biomed Mater Res 1988;22;541-553.

921 Simpson DR: Effect of pH and solution concentration on the composition of carbonate apatite. Am Miner 1967;52:896-902.

922 Doi Y, Moriwaki Y, Aoba T, Takahashi J, Joshin K: ESR and IR studies of carbonate-containing hydroxyapatites. Calcif Tissue Int 1982;34:178-181.

923 Neuman WF, Mulryan BJ: Synthetic hydroxyapatite crystals. III. The carbonate system. Calcif Tissue Res 1967;1:94-104.

924 Bachra BN, Trautz OR, Simon SL: Precipitation of calcium carbonates and phosphates. I. Spontaneous precipitation of calcium carbonates and phosphates under physiological conditions. Arch Biochem Biophys 1963;103:124-138.

925 Bachra BN, Trautz OR, Simon SL: Precipitation of calcium carbonates and phosphates. II. A precipitation diagram for the system calcium-carbonate-phosphate and the heterogeneous nucleation of solids in the metastability region; in Hardwick JL, Held HR, König KG (eds); Advances in Fluorine Research and Dental Caries Prevention, vol 3. Oxford, Pergamon Press, 1965, pp 101-118.

926 Bachra BN, Trautz OR, Simon SL: Precipitation of calcium carbonates and phosphates. III. The effect of magnesium and fluoride ions on the spontaneous precipitation of calcium carbonates and phosphates. Arch Oral Biol 1965;10:731-738.

927 Blumenthal NC, Betts F, Posner AS: Effect of carbonate and biological macromolecules on formation and properties of hydroxyapatite. Calcif Tissue Res 1975;18:81-90.

928 Elliott JC: Infrared spectrum of the carbonate ion in carbonate-containing apatites. J Dent Res 1961;40:1284. (abstract)

929 Emerson WH, Fischer EE: The infra-red absorption spectra of carbonate in calcified tissues. Arch Oral Biol 1962;7:671-683.

930 Rey C, Renugopalakrishnan V, Collins B, Glimcher MJ: Fourier transform infrared spectroscopic study of the carbonate ions in bone mineral during aging. Calcif Tissue Int 1991;49:251-258.

931 Zapanta-LeGeros R: Effect of carbonate on the lattice parameters of apatite. Nature 1965;206:403-404.

932 Simpson DR: The nature of alkali carbonate apatites. Am Miner 1964;49:363-376.

933 Bacquet G, Vo Quang Truong, Bonel, G, Vignoles M: Résonance paramagnétique électronique du centre F⁺ dans les fluorapatites carbonatées de type B. J Solid State Chem 1980;33:189-195.

934 Clark JS, Turner RC: Reactions between solid calcium carbonate and orthophosphate solutions. Can J Chem 1955;33:665-671.

935 Nathan Y, Lucas J: Synthèse de l'apatite à partir du gypse; application au problème de la formation des apatites carbonatées par précipitation directe. Chem Geol 1972;9:99-112.

936 Mayer I, Featherstone JDB, Nagler R, Noejovich M, Deutsch D, Gedalia I: The thermal decomposition of Mg-containing carbonate apatites. J Solid State Chem 1985;56:230-235.

937 Khattech I, Jemal M: Étude de la décomposition thermique de fluorapatites carbonatées. Thermochim Acta 1985;95:119-128.

938 Khattech I, Jemal M: Décomposition thermique de fluorapatite carbonatées de type B "inverse". Thermochim Acta 1987;118:267-275.

939 Callens FJ, Verbeeck RMH, Naessens DE, Matthys PFA, Boesman ER: The effect of carbonate content and drying temperature on the ESR-spectrum near g = 2 of carbonated calciumapatites synthesized from aqueous media. Calcif Tissue Int 1991;48:249-259.

940 Winand L: Étude physico-chimique de divers carbonatapatites. Bull Soc Chim Fr (special no) 1968:1718-1721.

941 Grøn P, Spinelli M, Trautz O, Brudevold F: The effect of carbonate on the solubility of hydroxylapatite. Arch Oral Biol 1963;8:251-263.

942 Thorpe JF, Whiteley MA: Cyanates. In Thorpe's Dictionary of Applied Chemistry (4th ed) Vol. III. London, Longmans, Green and Co, 1939, pp 463-512.

943 Harries JE, Hasnain SS, Shah JS: EXAFS study of structural disorder in carbonate-containing hydroxyapatite. Calcif Tissue Int 1987;41:346-350.

944 Nelson DGA, Wood GJ, Barry JC, Featherstone JDB: The structure of (100) defects in carbonated apatite crystallites: a high resolution electron microscope study. Ultramicroscopy 1986;19:253-266.

945 Okazaki M, Moriwaki Y, Aoba T, Doi Y, Takahashi J: Solubility behavior of CO_3 apatites in relation to crystallinity. Caries Res 1981;15:477-483.

946 Okazaki M, Moriwaki Y, Aoba T, Doi Y, Takahashi J, Kimura H: Crystallinity changes of CO_3-apatites in solutions at physiological pH. Caries Res 1982;16:308-314.

947 Nelson DGA, Featherstone JDB, Duncan JF, Cutress TW: Paracrystalline disorder of biological and synthetic carbonate substituted apatites. J Dent Res 1982;61:1274-1281.

948 Cappellen P van, Berner RA: A mathematical model for the early diagenesis of phosphorus and fluorine in marine sediments: apatite precipitation. Am J Sci 1988;288:289-333.

949 Weatherell JA, Robinson C: The inorganic composition of teeth; in Zipkin I (ed): Biological Mineralization. New York, John Wiley 1973, Ch 3, pp 43-74.

950 Dallemagne MJ, Richelle LJ: Inorganic chemistry of bone; in Zipkin I (ed): Biological Mineralization. New York, John Wiley, 1973, pp 23-42.

951 Pellegrino ED, Biltz RM: Mineralization in the chick embryo I. Monohydrogen phosphate and carbonate relationships during maturation of the bone crystal complex. Calcif Tissue Res 1972;10:128-135.

952 Lorcher K, Newesely H: Calcium carbonate (calcite) as a separate phase besides calcium phosphate apatite in medullary bone of laying hens. Calcif Tissue Res 1969;3:358-362.

953 Wix P, Mohamedally SM: The significance of age-dependent fluoride accumulation in bone in relation to daily intake of fluoride. Fluoride 1980;13:100-104.

954 Baud CA, Moghissi-Buchs M: Étude par diffraction des rayons X de la fixation *in vivo* du strontium dans la substance minérale osseuse. Compt Rend Acad Sci (Paris) 1965;260:5390-5391.

955 Pellegrino ED, Biltz RM: The composition of human bone in uremia. Medicine 1965;44:397-418.

956 Mann S, Webb J, Williams RJP (eds): Biomineralization, Chemical and Biochemical Perspectives. Weinheim, VCH Verlagsgesellschaft, 1989.

957 Brown WE, Eidelman N, Tomazic B: Octacalcium phosphate as a precursor in biomineral formation. Adv Dent Res 1987;1:306-313.

958 Jacquet J, Very JM, Flack HD: The 2θ determination of diffraction peaks from 'poor' powder samples. Application to biological apatite. J Appl Cryst 1980;13:380-384.

959 Young RA, Mackie PE: Crystallography of human tooth enamel: initial structure refinement. Mater Res Bull 1980;15:17-29.

960 Trautz OR: Crystalline organization of dental mineral; in Miles AEW (ed): Structural and Chemical Organization of Teeth. New York, Academic Press, 1967, pp 189 and 193.

961 Frazier PD: X-ray diffraction analysis of human enamel containing different amounts of fluoride. Arch Oral Biol 1967;12:35-42.

962 Iijima M, Kamemizu H, Wakamatsu N, Goto T, Moriwaki Y: Thermal decomposition of *Lingula* shell apatite. Calcif Tissue Int 1991;49:128-133.

963 Glas J-E, Omnell K-Å, Studies on the ultrastructure of dental enamel. 1. Size and shape of the apatite crystallites as deduced from X-ray diffraction data. J Ultrastruct Res 1960;3:334-344.

964 Posner AS, Eanes ED, Harper RA, Zipkin I: X-ray diffraction analysis of the effect of fluoride on human bone apatite. Arch Oral Biol 1963;8:549-570.

965 Eanes ED, Zipkin I, Harper RA, Posner AS: Small-angle X-ray diffraction analysis of the effect of fluoride on human bone apatite. Arch Oral Biol 1965;10:161-173.

966 Grynpas MD, Simmons ED, Pritzker KPH, Hancock RV, Harrison JE: Is fluoridated bone different from non-fluoridated bone?; in Yousuf Ali S (ed): Cell Mediated Calcification and Matrix Vesicles. Amsterdam, Elsevier, 1986, pp 409-414.

967 Fowler BO, Kuroda S: Changes in heated and in laser-irradiated human tooth enamel and their probable effects on solubility. Calcif Tissue Int 1986;38:197-208.

968 Mayer I, Schneider S, Sydney-Zax M, Deutsch D: Thermal decomposition of developing enamel. Calcif Tissue Int 1990;46:254-257.

969 Poyart CF, Bursaux E, Fréminet A: The bone CO_2 compartment: evidence for a bicarbonate pool. Respiration Physiology 1975;25:89-99.

970 Legros R, Godinot C, Torres L, Mathieu J, Bonel G: Sur la stabilité thermique des carbonates du tissu osseux. J Biol Bucalle 1982;10:3-9.

971 Gao XJ, Elliott JC, Anderson P: Scanning microradiographic study of the kinetics of subsurface demineralization in tooth sections under constant-composition and small constant-volume conditions. J Dent Res 1993;72:923-930.

972 Gao XJ, Elliott JC, Anderson P, Davis GR: Scanning microradiographic and microtomographic studies of remineralisation of subsurface enamel lesions. J Chem Soc Faraday Trans 1993;89:2907-2912.

973 Posner AS: The structure of bone apatite surfaces. J Biomed Mater Res 1985;19:241-250.

974 Patel PR, Brown WE: Thermodynamic solubility product of human tooth enamel: powdered sample. J Dent Res 1975;54:728-736.

975 Moreno EC, Zahradnik RT: Chemistry of enamel subsurface demineralization in vitro. J Dent Res 1974;53:226-235.

976 Moreno EC, Aoba T: Formation and solubility of carbonated tooth minerals; in S Suga, H Nakahara (eds): Mechanisms and Phylogeny of Mineralization in Biological Systems. Tokyo, Springer-Verlag, 1991, Ch 10, pp 179-186.

977 Brasseur H, Dallemagne ML, Melon J: Chemical nature of salts from bones and teeth and of tricalcium phosphate precipitates. Nature (London) 1946;157:453.

978 Little MF: Studies on the inorganic carbon dioxide component of human enamel. II. The effect of acid on CO_2. J Dent Res 1961;40:903-914.

979 Hallsworth AS, Weatherell JA, Robinson C: Loss of carbonate during the first stages of enamel caries. Caries Res 1973;7:345-348.

980 Aoba T, Okazaki M, Takahashi J, Moriwaki Y: X-ray diffraction study on remineralization using synthetic hydroxyapatite pellets. Caries Res 1978;12:223-230.

981 Aoba T, Moriwaki Y, Doi Y, Okazaki M, Takahashi J, Yagi T: The intact surface layer in natural enamel caries and acid-dissolved hydroxyapatite pellets: an X-ray diffraction study. J Oral Path 1981;10:32-39.

982 Gao XJ: Kinetics of Human Teeth De- and Remineralisation: Studies of Real Time Processes with Scanning Microradiography and Other X-ray Techniques. PhD Thesis, University of London, 1992.

983 Featherstone JDB, Duncan JF, Cutress TW: Crystallographic changes in human tooth enamel during *in-vitro* caries simulation. Arch Oral Biol 1978;23:405-413.

984 Hallsworth AS, Robinson C, Weatherell JA: Mineral and magnesium distribution within the approximal carious lesion of dental enamel. Caries Res 1972;6:156-168.

985 Wei SHY, Forbes WC: X-ray diffraction analysis of the reaction between intact and powdered enamel and several fluoride solutions. J Dent Res 1968;47:471-477.

986   Nelson DGA, Jongebloed WL, Arends J: Crystallographic structure of enamel surfaces treated with topical fluoride agents: TEM and XRD considerations. J Dent Res 1984;63:6-12.

987   Tsuda H, Jongebloed WL, Stokroos I, Arends J: Combined Raman and SEM study on CaF_2 formed on/in enamel by APF treatments. Caries Res 1993;27:445-454.

988   Larsen MJ, Jensen SJ: An X-ray diffraction and solubility study of equilibration of human enamel-powder suspensions in fluoride-containing buffer. Arch Oral Biol 1985;30:471-475.

989   Chow LC, Guo MK, Hsieh CC, Hong YC: Reaction of powdered human enamel and fluoride solutions with and without intermediate $CaHPO_4.2H_2O$ formation. J Dent Res 1980;59:1447-1452.

990   Dawes C, Cate ten JM (eds): Proceedings of a Joint IADR/ORCA International Symposium on Fluorides: Mechanisms of Action and Recommendations for Use, March, 1989, J Dent Res 1990;69:505-836.

991   Friedman M, Kalderon Y, Breyer L, Gedalia I: Inhibition of enamel dissolution rates by sodium fluoride and amine fluorides in the presence or absence of strontium ions. Pharm Acta Helv 1986;61:30-32.

992   Grynpas MD, Cheng P-T: Fluoride reduces the rate of dissolution of bone. Bone Mineral 1988;5:1-9.

993   Funduk N, Kydon DW, Schreiner LJ, Peemoeller H, Miljović L, Pintar MM: Composition and relaxation of the proton magnetization of human enamel and its contribution to the tooth NMR image. Magnetic Resonance Med 1984;1:66-75.

994   Code RF, Armstrong RL, Hallsworth RS, Lemaire C, Cheng P-T: Concentration dependence of fluorine impurity spin-lattice relaxation rate in bone mineral. Phys Med Biol 1992;37:211-221.

995   Herzfeld J, Roufosse A, Haberkorn RA, Griffen RG, Glimcher MJ: Magic angle spinning in inhomogeneously broadened biological systems. Phil Trans Roy Soc B 1980;289:459-469.

996   Ackerman JL, Rayleigh DP, Glimcher MJ: Phosphorus-31 magnetic resonance imaging of hydroxyapatite: A model for bone imaging. Magnetic Resonance Med 1992;25:1-11.

997   Daculsi G, Menanteau J, Kerebel LM, Mitre D: Length and shape of enamel crystals. Calcif Tissue Int 1984;36:550-555.

998   Jackson SA, Cartwright AG, Lewis D: The morphology of bone mineral crystals. Calcif Tissue Res 1978;25:217-222.

999   Swancar JR, Scott DB, Simmelink JW, Smith TJ: The morphology of enamel crystals; in Fearnhead RW and Stack MV (eds): Tooth Enamel II. Bristol, John Wright, 1969, pp 233-239.

1000 Jongebloed WL, Molenaar I, Arends J: Morphology and size distribution of sound and acid-treated enamel crystallites. Calcif Tissue Res 1975;19:109-123.

1001 Voegel JC, Frank RM: Stages in the dissolution of human enamel crystals in dental caries. Calcif Tissue Res 1977;24:19-27.

1002 Brès EF, Voegel J-C, Frank RM: High resolution electron microscopy of human enamel crystals. J Microscopy 1990;160:183-201.

1003 Nakahara H, Kakei M: The central dark line in developing enamel crystallite: an electron microscopic study. Bull Josai Dent Univ 1983;12:1-7.

1004 Iijima M, Tohda H, Suzuki H, Yanagisawa T, Moriwaki Y: Effects of F⁻ on apatite-octacalcium phosphate intergrowth and crystal morphology in a model system of tooth enamel formation. Calcif Tissue Int 1992;50:357-361.

1005 Cuisinier FJG, Steuer P, Senger B, Voegel JC, Frank RM: Human amelogenesis I: High resolution electron microscopy study of ribbon-like crystals. Calcif Tissue Int 1992;51:259-268.

1006 Bres EF, Steuer P, Voegel J-C, Frank RM, Cuisinier FJG: Observation of the loss of the hydroxyapatite sixfold symmetry in a human fetal tooth enamel crystal. J Microscopy 1993;170:147-154.

1007 Brès EF, Cherns D, Vincent R, Morniroli J-P: Space-group determination of human tooth-enamel crystals. Acta Cryst 1993;B49:56-62.

1008 Theuns HM, Shellis RP, Groeneveld A, van Dijk JWE, Poole DFG: Relationships between birefringence and mineral content in artificial caries lesions of enamel. Caries Res 1993;27:9-14.

1009 Raemdonck W van, Ducheyne P, Meester P De: Calcium phosphate ceramics; in Ducheyne P, Hastings GW (eds): Metal and Ceramic Biomaterials, Vol II, Strength and Surface. Boca Raton, CRC Press, 1984, pp 143-166.

1010 Heughebaert J-C: Biocéramiques constituées de phosphates de calcium. Silicates Industriels 1988;3-4:37-41.

1011 Ducheyne P, Lemons JE (eds): Bioceramics: Material Characteristics Versus *In Vivo* Behavior. Ann NY Acad Sci, Vol 523, 1988.

1012 Yamamuro T, Hench LL, Wilson J (eds): CRC Handbook of Bioactive Ceramics, Vol I, Bioactive Glasses and Glass-Ceramics. Boca Raton, CRC Press, 1990.

1013 Yamamuro T, Hench LL, Wilson J (eds): CRC Handbook of Bioactive Ceramics, Vol II, Calcium Phosphate and Hydroxylapatite Ceramics. Boca Raton, CRC Press, 1990.

1014 Groot K de: Medical applications of calciumphosphate bioceramics. J Ceram Soc Jpn 99;943-953:1991.

1015 Aoki H: Science and Medical Applications of Hydroxyapatite. Tokyo, Japanese Association of Apatite Science, 1991.

1016 Hench LL, Wilson J (eds): An Introduction to Bioceramics. Singapore, World Scientific, 1993.

1017 Oonishi H, Aoki H, Sawai K (eds): Bioceramics (Proc 1st Internat Bioceramics Symp, Tokyo), Ishiyaku EuroAmerica Inc, 1989.

1018 Heimke G (ed): Bioceramics 2 (Proc 2nd Internat Bioceramics Symp on Ceramics in Medicine, Heidelberg 1989). Cologne, Deutsche Keramische Gesellschaft eV, 1990.

1019 Hulbert JE, Hulbert SF (eds): Bioceramics 3 (Proc 3rd Internat Symp on Bioceramics in Medicine, Terre Haute, 1990). Terre Haute, Rose-Hulman Institute of Technology, 1992.

1020 Bonfield W, Hastings GW, Tanner KE (eds): Bioceramics 4 (Proc 4th Internat Symp on Ceramics in Medicine, London 1991). Oxford, Butterworth-Heinemann, 1991.

1021 Yamamuro T, Kobubo T, Nakamura T (eds): Bioceramics 5 (Proc 5th Internat Symp on Bioceramics in Medicine, Kyoto, 1992). Kyoto, Kobunshi Kankokai, 1992.

1022 Ducheyne P, Christiansen D (eds): Bioceramics 6 (Proc of the 6th Internat Symp on Bioceramics in Medicine, Philadelphia 1993). Oxford, Butterworth-Heinemann, 1993.

1023 Groot K de, Klein CPAT, Wolke JGC, Blieck-Hogervorst de JMA: Chemistry of calcium phosphate bioceramics, in Yamamuro T, Hench LL, Wilson J (eds): CRC Handbook of Bioactive Ceramics, Vol II, Calcium Phosphate and Hydroxylapatite Ceramics. Boca Raton, CRC Press, 1990, pp 3-16.

1024 Jarcho M: Hydroxylapatite Ceramic. US Patent 1978, No 4,097,935.

1025 Barralet J, Best S, Bonfield W: Preparation and sintering of carbonate-substituted apatites, in Ducheyne P, Christiansen D (eds): Bioceramics 6 (Proc of the 6th Internat Symp on Bioceramics in Medicine, Philadelphia 1993). Oxford, Butterworth-Heinemann, 1993, pp 179-184.

1026 Ito K, Ooi Y: Osteogenic activity of synthetic hydroxylapatite with controlled texture—on the relationship of osteogenic quantity with sintering temperature and pore size, in Yamamuro T, Hench LL, Wilson J (eds): CRC Handbook of Bioactive Ceramics, Vol II, Calcium Phosphate and Hydroxylapatite Ceramics. Boca Raton, CRC Press, 1990, pp 39-44.

1027 Klein CPAT, Patka P, Hollander W den: A comparison between hydroxylapatite and β-whitlockite macroporous ceramics implanted in dog femurs, in Yamamuro T, Hench LL, Wilson J (eds): CRC Handbook of Bioactive Ceramics, Vol II, Calcium Phosphate and Hydroxylapatite Ceramics. Boca Raton, CRC Press, 1990, pp 53-60.

1028 Peelen JGJ, Rejda BV, De Groot K: Preparation and properties of sintered hydroxylapatite. Ceramurgia Intern 1978;4:71-74.

1029 Ducheyne P, McGuckin JF: Composite bioactive ceramic-metal materials, in Yamamuro T, Hench LL, Wilson J (eds): CRC Handbook of Bioactive Ceramics, Vol II, Calcium Phosphate and Hydroxylapatite Ceramics. Boca Raton, CRC Press, 1990, pp 175-186.

1030 Pilliar RM, Filiaggi MJ: New calcium phosphate coating methods, in Ducheyne P, Christiansen D (eds): Bioceramics 6 (Proc of the 6th Internat Symp on Bioceramics in Medicine, Philadelphia 1993). Oxford, Butterworth-Heinemann, 1993, pp 165-171.

1031 Kay JF: Bioactive surface coatings for hard tissue biomaterials, in Yamamuro T, Hench LL, Wilson J (eds): CRC Handbook of Bioactive Ceramics, Vol II, Calcium Phosphate and Hydroxylapatite Ceramics. Boca Raton, CRC Press, 1990, pp 111-122.

1032 Groot de K, Klein CPAT, Wolke JGC, Blieck-Hogervorst de JMA: Plasma-sprayed coatings of calcium phosphate, in Yamamuro T, Hench LL, Wilson J (eds): CRC Handbook of Bioactive Ceramics, Vol II, Calcium Phosphate and Hydroxylapatite Ceramics. Boca Raton, CRC Press, 1990, pp 133-142.

1033 Ducheyne P, Cuckler J, Radin S, Nazar E: Plasma sprayed calcium phosphate ceramic linings on porous metal coatings for bone ingrowth, in Yamamuro T, Hench LL, Wilson J (eds): CRC Handbook of Bioactive Ceramics, Vol II, Calcium Phosphate and Hydroxylapatite Ceramics. Boca Raton, CRC Press, 1990, pp 123-131.

1034 Hench LL: Bioactive glasses and glass-ceramics: a perspective, in Yamamuro T, Hench LL, Wilson J (eds): CRC Handbook of Bioactive Ceramics, Vol I, Bioactive Glasses and Glass-Ceramics. Boca Raton, CRC Press, 1990, pp 7-23.

1035 Hench LL, Andersson Ö: Bioactive glasses; in Hench LL, Wilson J (eds): An Introduction to Bioceramics. Singapore, World Scientific, 1993, pp 41-62.

1036 Yoshii S, Yamamuro T, Kitsugi T, Nakamura T, Kokubo T, Oka M, Shibuya T, Takagi M: Bone-bonding capability and mechanical properties of modified A-W glass-ceramic (animal studies); in Yamamuro T, Hench LL, Wilson J (eds): CRC Handbook of Bioactive Ceramics, Vol I, Bioactive Glasses and Glass-Ceramics. Boca Raton, CRC Press, 1990, pp 51-63.

1037 Kitsugi T, Yamamuro T, Yoshii S, Kokubo T, Takagi M, Shibuya T: The influence of substituting B_2O_3 for CaF_2 on the bonding behavior of A-W glass-ceramic to bone tissue, in Yamamuro T, Hench LL, Wilson J (eds): CRC Handbook of Bioactive Ceramics, Vol I, Bioactive Glasses and Glass-Ceramics. Boca Raton, CRC Press, 1990, pp 65-71.

1038 Doyle C: Bioactive composites in orthopedics, in Yamamuro T, Hench LL, Wilson J (eds): CRC Handbook of Bioactive Ceramics, Vol II, Calcium Phosphate and Hydroxylapatite Ceramics. Boca Raton, CRC Press, 1990, pp 195-207.

1039 Ostrowski K, Dziedzic-Gocławska A, Stachowicz W, Michalik J: Accuracy, sensitivity, and specificity of electron spin resonance analysis of mineral constituents of irradiated tissues. Ann NY Acad Sci 1974;238:186-200.

1040 Piper WW, Kravitz LC, Swank RK: Axial symmetric paramagnetic color centers in fluorapatite. Phys Rev A 1965;138:A1802-A1814.

1041 Roufosse A, Stapelbroek M, Bartram RH, Gilliam OR: Oxygen-associated holelike centers in calcium chlorapatite. Phys Rev B 1974;9:855-862.

1042 Callens FJ, Verbeeck RMH, Naessens DE, Matthys PFA, Boesman ER: Effect of carbonate content on the ESR spectrum near g = 2 of carbonated calciumapatites synthesized from aqueous media. Calcif Tissue Int 1989;44:114-124.

1043 Roufosse A, Richelle LJ, Gilliam OR: Electron spin resonance of organic free radicals in dental enamel and other calcified tissues. Arch Oral Biol 1976;21:227-232.

1044 Geoffroy M, Tochon-Danguy HJ: Long-lived radicals in irradiated apatites of biological interest: an e.s.r. study of apatite samples treated with $^{13}CO_2$. Int J Radiat Biol 1985;48:621-633.

1045 Geoffroy M, Tochon-Danguy HJ: ESR identification of radiation damage in synthetic apatites: a study of ^{13}C-hyperfine coupling. Calcif Tissue Int 1982;34:S99-S102.

1046 Doi Y, Aoba T, Okazaki M, Takahashi J, Moriwaki Y: ^{13}C Enriched carbonate apatites studied by ESR: comparison with human tooth enamel apatites. Calcif Tissue Int 1981;33:81-82.

1047 Doi Y, Moriwaki Y, Aoba T, Okazaki M, Takahashi J: Carbonate-derivative centers in X-ray irradiated carbonate-containing apatites. J Osaka Univ Dent Sch 1981;21:75-85.

1048 Gilinskaya LG, Shcherbakova MYa, Zanin YuN: Carbon in the structure of apatite according to electron paramagnetic resonance data. Soviet Physics - Crystallography 1971;15:1016-1019.

1049 Cevc P, Schara M, Ravnik Č: Electron paramagnetic resonance study of irradiated tooth enamel. Radiation Res 1972;51:581-589.

1050 Aoba T, Doi Y, Yagi T, Okazaki M, Takahashi J, Moriwaki Y: Electron spin resonance study of sound and carious enamel. Calcif Tissue Int 1982;34:S88-S92.

1051 Bacquet G, Vo Quang Truong, Vignoles M, Trombe JC, Bonel G: ESR of CO_2^- in X-irradiated tooth enamel and A-type carbonated apatite. Calcif Tissue Int 1981;33:105-109.

1052 Willigen H van, Roufosse AH, Glimcher MJ: Proton and Phosphorus ENDOR on paramagnetic centers in X-irradiated human tooth enamel. Calcif Tissue Int 1980;31:70. (abstract)

1053 Callens FJ, Verbeeck RMH, Matthys PFA, Martens LC, Boesman ER: The contribution of CO_3^{3-} and CO_2^- to the ESR spectrum near g = 2 of powdered human tooth enamel. Calcif Tissue Int 1987;41:124-129.

1054 Callens FJ, Verbeeck RMH, Matthys PFA, Martens LC, Boesman ER, Driessens FCM: The ESR spectrum near g = 2 of carbonated calciumapatites synthesized at high temperatures. Bull Soc Chim Belg 1986;95:589-596.

1055 Callens FJ, Verbeeck RMH, Matthys PFA, Martens LC, Boesman ER: ^{13}C-hyperfine couplings of carbonate radicals in carbonated calciumapatites synthesized at high temperature. Bull Soc Chim Belg 1987;96:165-171.

1056 Callens FJ, Verbeeck RMH, Naessens DE, Matthys PFA, Boesman ER: ESR study of ^{13}C-enriched carbonated calciumapatites precipitated from aqueous solutions. Calcif Tissue Int 1993;52:386-391.

1057 Grün R, Stringer CB: Electron spin resonance dating and the evolution of modern humans. Archaeometry 1991;33:153-199.

1058 Grün R: Die ESR-Altersbestimmungsmethode, Berlin, Springer-Verlag, 1989.

1059 Sakthivel A, Young RA: User's Guide to Programs DBWS-9006 and DBWS-9006PC for Rietveld Analysis of X-ray and Neutron Powder Diffraction Patterns. Atlanta, Georgia Institute of Technology, 1993.

INDEX

$\overline{\alpha}$-tricalcium phosphate
  melting point 47
  structure 37, 41
  thermal stability 34
α-tricalcium phosphate
  $\overline{\alpha}$-TCP, transition to 34, 47
  β-TCP, comparison with 37
  ACP, from heating 61
  apatite, relation with 36
  density 42
  glaserite, relation with 36
  hydrolysis to DCPA, DCPD, OCP or
    Ca-def OHAp 47
  — to double salts of OCP and
    dicarboxylic acids 22
  IR spectrum 50
  lattice parameters 35
  OHAp, from heating 128, 279, 297
  optical properties 43
  plasma spraying, formation in 297
  preparation 44
  single crystals, growth of 45
  structure 35
  XRD pattern 309
β-calcium pyrophosphate
  α-calcium pyrophosphate, transition
    to 61
β-tricalcium phosphate
  α-TCP, transition to 47, 49
  β-TCa,MgP, structure 40
  ACP, from heating 61
  apatite, relation with 71
  BCaPs, from heating 146
  bioceramics of 295
  biological apatites, from heating
    279, 280
  copper in 41
  density 43
  high pressure phase transition 49

  hydrolysis in H_2O_2 to form O-rich
    Aps 132
  IR spectrum 50
  lattice parameters 37
  —, effect of Mg 45
  NMR spectrum 50
  OCP, from heating 20
  optical properties 43
  plasma spraying 297
  polymorph, high pressure 49
  porous ceramics 296
  preparation 43
  resorption in bone 289
  single crystals, growth of 45
  solubility 46
  —, lowering by Mg 46
  structure 37
  —, β-TCa,MgP 40
  XRD pattern 310

Abbreviations
  apatites 3
  CaPs 2
  general 9
Acetate in apatite lattice 136, 254
Acid phosphate
  ACP, in 59, 61, 144
  biological apatites, in 260, 261,
    270, 277
  —, NMR studies of 289
  Ca-def OHAps, effect on lattice
    parameters 151
  —, in 126, 150
  —, IR spectra 150, 180
  —, NMR studies of 183
  CO_3Aps, in 229
  determined by IR 144

— by NMR 141, 145
— by oxalate precipitation 143
enamel mineral, IR spectra 270
—, IR determination in 277
OCP, IR bands in 20
OHAps, determination in 141
whitlockite, in 35, 42
Ammonium
    CO₃Aps precipitated, in 235, 242, 253
    MCPA, in 11
Amorphous calcium phosphate(s)
    acid phosphate in 59, 61, 144
    biological mineral, in 266
    biomineralisation, in 262
    CaPs, in precipitation of 155
    CO₃-containing 54, 230, 239
    composition of 54
    ESR of X-irradiated 61
    EXAFS spectrum 60
    hydrolysis in solid state 55, 149, 153, 184
    — in water 57
    —, kinetics of 58
    IR spectrum 58
    Mg-containing 54, 59, 61
    NMR spectrum 59, 184, 256
    occurrence 53
    OHAp, IR of hydrolysis to 59
    preparation 53
    pyrophosphate yield, on heating 142
    solubility, effect of Mg on 58
    stability in solution 57
    structure 60
    thermal decomposition 59, 61, 142, 239
Antimony
    apatite, location in 91
    ionic radius 84
Apatite(s)
    α-TCP, relation with 36
    β-TCP, relation with 71

abbreviations for 3
acetate in 136, 254
amino-2-ethylphosphate in 136
aragonite, relation with 71
Ca, substitutions for 82
Ca(1) versus Ca(2) substitution 82, 94
colour centres in 109
densities 8
diffusion in 105
dislocations in 105, 168, 290
epitaxial relations with CaPs 74
etching of 107
fission tracks in 107
glaserite, relation with 71
glycine in 136
hardness 104
hexad axis, diagram 67
—, substitutions on 80
higher symmetry structures 74
laser host, use as 109
lattice parameters 8
literature reviews 7
mineralogy 7
name, origin of 5
Nowotny phases, relation with 74
OCP, relation with 13
phosphate, substitutions for 94
refractive indices 8
samuelsonite, relation with 71
silico-carnotite, relation with 71
structure, description 64
TetCP, relation with 51
thermodynamic constants 5, 104, 187
Apatites, biological
    CO₃ IR bands, assignment 230, 270
    CO₃, preferential loss on dissolution 194, 285
    composition 259
    dating of fossil enamel 300
    dissolution 168, 290

FAp in shark enameloid 63
francolite, comparison with 270
lattice parameters 263, 265
lower symmetry suggested 293
Mg, effect on *c*-axis 265
morphology 153, 262, 265, 289
NCO⁻ and NCN²⁻, from heating 253, 278
NMR spectra 266, 287
optical properties 294
surface chemistry 282
thermal decomposition 253, 261, 275
XRD line broadening 265
Aragonite
apatite, relation with 71
lattice parameters 71
oriented CO₃Ap from 198, 248
structure 71
Argon in ashed biological apatites 80
Arsenate
A-AsO₄CO₃Ap, high pressure modification 215
—, lattice parameters 214
—, preparation 213
apatite, substitution for PO₄ 94
CdAsO₄BrAp 81

Barium
A-BaCO₃Ap, lattice parameters 214
—, preparation 194
apatite, preference for Ca(2) site 85
B-BaCO₃FAp 224
Ba,CaOHAps, IR band shifts 181
BaOHAp, heats of immersion 159
—, IR assignments 181
—, lattice parameter changes on heating 129
—, lattice parameters 134
BaReO₅OAp, lattice parameters 81
Ca,BaOHAps, synthesis and lattice parameters 134

Eu,BaFAp 90
ionic radius 84
O-rich Aps 133
Sr,BaFAps 85
Barium vanadate, structure 37
Beryllium, Pb,NaBeF₄FAp 84
Bioactive glasses and glass-ceramics 297
Bioceramic coatings 297
Biomaterials
45S5 Bioglass® 297
A-W bioactive glass ceramic 297
CaPs in 295
composites 298
Biomineralisation 261, 288, 293
Bone mineral
CO₃ loss on drying 281
composition 259
IR spectrum 275
NMR spectrum 288
Boron
BO₂⁻, position in apatite 253
CaPO₄,BO₃BO₂Ap, growth of single crystals 198
—, structure 81, 198, 207
Sr,NaBO₂Ap, structure 81, 84
Bromapatite
Br, location of 219
lattice parameters 81
Raman spectrum 104
steam, reaction with 125
structures of various 81
Bromide
ionic radius 84
MCPA, in 11
Brushite
see also, Dicalcium phosphate dihydrate
discovery 23
occurrence 23

Cadmium
apatites, various 81, 90
CdOHAp, structure 115
ionic radius 84
O-rich Aps 133
Calcite
CO₃Ap from 194, 246, 248, 296
in bone 261, 267
Calcium
diffusion in FAp 105
— in OHAp 106
ionic radius 84
pyrophosphates 1
Calcium fluoride
enamel and F⁻, from 286
NMR spectrum 162, 186
OHAp and F⁻, from reaction in
solution 161
solubility 161
Calcium phosphates
entropies 6
free energies of formation 6
lattice parameters 8, 305
nucleation and crystal growth 2
solubility isotherms 4
— products, calculated 6
—, temperature dependence 9, 13,
138, 157
speciation diagram 5
specific heats 6
standard enthalpies 6
Calcium-deficient hydroxyapatite,
see Hydroxyapatite, Ca-deficient
Calcium-rich "hydroxyapatite" 127, 181
Calcium/phosphorus ratio from β-TCP
on heating BCaPs 146
Caracolite 84
Carbamate, role in thermal
decomposition of CO₃Aps 250,
254
Carbon dioxide
biological apatites, in 80

CO₃Aps precipitated, loss on drying
242, 247, 281
CO₃FAps precipitated, loss on
heating 250
enamel, loss on heating 276, 279
formation during thermal
decomposition of CO₃Aps 251
IR when trapped in CO₃Ap 252
Carbon monoxide from heating CO₃Aps
254, 276
Carbonate
ACP, effect on thermal stability 61
—, in 54
CO₃Aps, effect on rate of
dissolution 290
—, effect on solubility product 257,
282
—, preferential loss on dissolution
194, 285
distortion, effect on IR and Raman
spectra 208
francolite, effect on thermal stability
212, 244
HPO₄, interference with IR
estimation 145
IR and Raman assignments 207
labile, in precipitated CO₃Aps 233,
274, 281
loss in dental caries 285
OCP, effect on hydrolysis 18
OCP/OHAp intercrystalline, in 154
OH replacement in precipitated
CO₃Aps 230, 237
pyrophosphate yield reduction 142
whitlockite, in 42
Carbonate apatite
A- and B-type nomenclature 191
aragonite, from 198, 248
calcite, from 248
CO₃ exchange in solution 242
— location, change on heating 250,
277
— on apatite surface 194, 281, 285

— on hexad axis, position 221

— on hexad axis, space limitation 193, 195, 219

— replacing OH, discovery 194

crystal size, effect of temperature of precipitation 296

DCPA, from hydrolysis of 33

dissolution, rate of 166, 258

ESR of X-irradiated 298

F⁻, reaction with, in solution 162

high pressure forms 215, 226, 228, 245

hydrolysis to OHAp 259

lattice images in EM 293

— parameters, precipitated 238

literature reviewed 229

phase diagram 197

polarised IR study of CO₃ replacing PO₄ 210

realisation of two types of CO₃ substitution 195

single crystals, analogues 198

—, growth of 196

structure, early work 192

summary of structural work 301

thermal decomposition 228, 248

Carbonate apatite, A-type

CaF₂, reaction with on heating 223

CO₃ position in lattice 221

— replacing OH, IR demonstration 195

— tilted in hexad axis 195, 217

— transition moment directions 216

density 223

electric dipole relaxations 222

high pressure modification 215

IR and Raman assignments 215

IR band shifts with CO₃ content 216

lattice parameters of various 214

limited substitution in aqueous preparations 230

monoclinic to hexagonal phase transition 221

NH₃ gas, cyanamide apatite when heated with 254

NMR spectroscopy 222

NO, nitrated apatite when heated with 81, 214

O₂, peroxyapatites when heated with 131

OAp, from heating 130

PO₄ IR bands 218

polarised IR spectrum 216

preparation 213

Rietveld analysis 219

structure 218

thermal decomposition 130

Carbonate apatite, AB-type

see also Carbonate apatite, B-type

A- to B-type ratio, by IR 146, 272

— in precipitated 237

AB-CO₃F,OHAp, solid state synthesis 224

density 225, 227

ESR of X-irradiated 298

formula, high temperature 225

high pressure modification 228

IR and Raman assignments in precipitated 230, 271

IR changes with unit cell volume 224

IR CO₃, complexity in precipitated 231, 271

lattice parameters, precipitated 238

Na-containing, high temperature 227

preparation by heating ACa,CO₃P 239

Raman spectrum 228

thermal decomposition 226, 250

Carbonate apatite, B-type
　see also Carbonate apatite, AB-type
　　and Francolite
　B-CO$_3$ClAps　224
　B-CO$_3$FAp, F in PO$_4$ site　193, 199,
　　203, 245, 274
　—, structural formula　223
　—, synthesis　223
　CO$_3$ environment in precipitated
　　236, 255
　CO$_3$ replacing PO$_4$, EXAFS evidence
　　for　255
　—, polarised IR evidence for　210
　—, XRD evidence for　206
　EXAFS of precipitated　254
　francolite, comparison with　245
　high pressure modification　226
　high temperature compared with
　　precipitated　237
　HPO$_4$, in precipitated　229
　IR assignments for F,CO$_3$ group　273
　IR changes with F content　211, 245
　IR spectrum, precipitated　239
　lattice parameters, precipitated　235,
　　238
　Na-containing, high temperature　224
　—, precipitated　239
　NMR of precipitated　256
　OH IR intensity, precipitated　236
　preparation, precipitation　235
　—, solid state　225
　solubility product of B-CO$_3$F,OHAps
　　257
　structural formula, high temperature
　　Na-containing　223, 227
　—, precipitated　234, 236, 240, 243,
　　245
　synthetic, first systematic study　195
　thermal decomposition　244, 250
　—, Na containing　228
Carbonatites　202
Caries crystals　43

Casein
　DCPD, effect on growth　27
　OCP, effect on growth　27
Central dark line, EM studies of
　　CO$_3$Aps　290
Chlorapatite
　CaCl$_2$-deficient　79, 96, 98
　—, structure　79
　Cl,OHAp, perturbation of OH IR
　　178
　density　79, 98
　dielectric measurements　76
　doped　89, 99
　electric dipole relaxations　78
　factor-group analysis　103
　fluorescent spectrum　79
　ion size limits for Ca replacement
　　83
　IR and Raman spectra　103
　lattice parameters　75, 79, 96, 98
　melting point　98
　Mn-doped　89
　monoclinic structure, loss of　75, 80,
　　88, 96
　monoclinic to hexagonal phase
　　transition　75
　OH,ClAp, preparation　135
　optical properties　75, 99, 109
　preparation, aqueous　97
　—, solid state　96
　single crystals, growth of　98
　steam, reaction with　124, 138
　structure, Ca,SrClAp　86
　—, CaCl$_2$-def　79
　—, Co,CaClAp　88
　—, F,ClAps　80
　—, monoclinic　75
　thermal decomposition　64, 96
　thermoluminescence　186
　UV absorption　109
　XRD pattern, monoclinic　75

Chloride
    Cl,FAp, position in 80
    ClAp, position in 75
    Co,CaClAp, position in 88
    enamel, position in 268
    francolite analogue 224
    francolite, analyses of 202
    Holly Springs, in 111
    ionic radius 84
    MCPA, in 11
    OH IR, perturbation in Cl,OHAp 178
    —, perturbation in enamel 268
    OHAp, effect on growth 156
Chlorspodiosite, preparation 96
Chromium
    apatite, in 91, 94
    CaCrO₄OHAp, structure 116
Citrate, effect on growth of OHAp 156
Cobalt
    ClAp, in 88
    FAp, in 88
    ionic radius 84
Constant composition method
    Ca-def OHAps, preparation with 127
    crystal dissolution studies 166
    OHAp, rate of dissolution 164
    system described 155
Copper in β-TCP 41
Corals, use for biomaterials 296
Crystal growth
    constant composition 155
    similarity with dissolution 164
Cyanamide
    formation in heated CO₃Aps 253
    location in apatite 254
Cyanamide apatite, preparation and lattice constants 254
Cyanate
    enamel, from heating 278
    formation in heated CO₃Aps 253

location in apatite 253

Dahllite
    a-axis reduction with CO₃, comparison with francolite 211
    composition 203
    discovery 191
    francolite, differentiation from 191
    IR spectrum, comparison with francolite 211
    lattice parameters 203
    Ödegården, IR spectrum 211
    —, structural formula 205
    optical properties 204
    type specimen 191
Density measurement 147
Dental caries 145, 169, 261, 294, 299
    dissolution of crystals in 290
    F⁻ and 286
    loss of CO₃ and Mg 285
    model systems 168
    subsurface demineralisation in 282
    whitlockite in 43
Dental enamel
    see also Apatites, biological
    Cl position in 268
    CO₃ location, change on heating 277
    composition 259
    dissolution, rate of 166, 167
    F⁻, reaction with 162, 286
    IR and Raman assignments 267
    IR determination of A- to B-type ratio 272
    laser effects on 279
    lattice parameters 263
    —, changes on heating 152, 278
    polarised IR spectrum 267
    Raman spectrum 274
    Rietveld analysis of structure 265
    solubility product of mineral 282
    thermal decomposition 276

Dentine, composition 259
Deuteration
    Ca-def OHAp 180
    dental enamel 267, 287
    FAp single crystals 105
    OCP 21
    OHAp powders 173, 182
    — single crystals 137
    —, kinetics 147
Dicalcium phosphate anhydrous
    CO_3Ap, aqueous formation from 229
    DCPD, aqueous conversion from 28
    —, from heating 29
    density 33
    entropy 34
    heat capacity 34
    hydrolysis 33
    IR and Raman spectra 34
    lattice parameters 31
    NMR spectrum 34
    occurrence 30
    OCP, from heating 19
    optical properties 32
    phase transition, thermal 31
    preparation 32
    single crystals, growth of 32
    solubility 33
    structure 31
    TetCP, reaction with 50
    thermal decomposition 33
    XRD pattern 309
Dicalcium phosphate dihydrate
    AB-CO_3Ap, from reaction with $CaCO_3$ at high temperature 225
    ACP, from hydrolysis of 57
    biomineralisation, in 262, 266, 288
    CO_3Aps, during precipitation 247
    crystal growth 26, 156
    DCPA, aqueous conversion to 28
    density 26
    dissolution 27, 156

enamel F^- uptake, pretreatment for 286
entropy 29
gypsum, relation with 24
heat capacity 29
hydrolysis to OCP 28
— to OHAp 28, 118, 126
IR and Raman spectra 30
lattice parameters 24
NMR spectrum 30
occurrence 23
OCP, singular point with 28
OHAp, singular point with 28
optical properties 26
preparation 24
single crystals, growth of 25
solubility 27
structure 24
TetCP, reaction with in bone cement 50
thermal decomposition 29
water content measured 139
— loss from 29
XRD pattern 308
Dichroic ratio, definition 208
Diffusion in apatites 105, 147
Dioxyapatites 97
"Direct" carbonate apatites 229
Dissolution of crystals, rate of 164
Durango FAp 63, 104, 107

Ellestadite 94
Epitaxy
    FAp and CaF_2 163
    OHAp and CaPs 74
ESR
    CO_3Aps, X-irradiated 298
    dating of fossil enamel 300
    doped FAp 70, 89, 91
    O-rich Aps 132, 133
    O_3^- in biological apatites 80
    OHAp, monoclinic 113, 186

X-irradiated ACP 61
Europium
    A-Sr,EuCO$_3$Ap, lattice parameters
      214
    apatite, in 90
    Eu,CaOAp 131
    EuAsO$_4$OHAp 134
    ionic radius 84
    sulphoapatite 94
EXAFS
    ACP 60
    CO$_3$Aps, precipitated 254
    OHAp 60
Exoemission from apatites 186

Factor-group analysis 100
Fission-track chronothermometry 107
Fluorapatite
    bioactive glass, in 297
    CaF$_2$, epitaxy with 163
    —, from, in solution 163
    CaF$_2$-deficient 80, 89, 95, 97
    density 63, 97, 104
    diffusion in 105
    dissolution, rate of 166
    distortion from hexagonal 83
    doped, growth of single crystals 98
    Durango 63, 104, 107
    elastic constants 104
    electric dipole relaxations 117
    entropy 104
    F,OHAp, from F$^-$ reaction with
      enamel 286
    —, from F$^-$ reaction with OHAp 161
    —, preparation 135
    —, solubility 158
    F$^-$, with excess 95
    factor-group analysis 100
    Fe-substituted 88
    fluorescent spectrum 79
    francolite, from heating 192
    heat capacity 104

    — content 104
    heats of immersion 159
    ion size limits for Ca replacement
      83
    IR and Raman spectra 100
    lattice parameters 63, 64, 97, 98
    melting point 64, 95, 97
    MgFAp, growth of single crystals
      86
    Mn-doped 89
    Nd-substituted, structure 91
    NMR of single crystals 105
    NMR spectra of OH,FAp 185
    OCP, from 18
    OHAp, high temperature conversion
      to 64
    optical properties 63, 95, 97, 98
    phase transition, low temperature
      70
    preparation, aqueous 97
    —, solid state 95
    Sb-substituted, Rietveld analysis 92
    single crystals, growth of 97
    solubility product 158
    structure 64
    —, Nd-substituted 91
    thermal conductivity 104
    — decomposition 64
    — expansion 104
    thermoluminescence 186
    water adsorption isotherms 160
    wear properties 110
    XRD pattern 311
Fluoride
    ACP, effect on formation 54
    —, effect on hydrolysis 56, 157
    BCaPs, effect on precipitation 156
    biological apatites, in 64, 261, 265
    —, NMR of 288
    Cl,FAp, position in 80
    CO$_3$Aps, effect on IR 211, 224,
      231, 245

—, effect on rate of dissolution 258, 287

DCPD, effect on growth 27

—, effect on hydrolysis 28

dental caries, in 286

enamel, reaction with 163, 286

F,OHAps, protein adsorption on 160

F-H distances in F,OHAp 184

francolite, analyses of 202

—, excess in 199, 202, 206, 211, 212

—, excess in synthetic 245, 274

—, loss on heating 212

Holly Springs OHAp, position in 113

ionic radius 69, 84

NMR, F reaction with OHAp 186

—, F,OHAp 184

OCP, effect on growth 16, 293

—, effect on hydrolysis 18

OH column reversal in OHAp 112, 176

OH IR in enamel, perturbed by 268

— in F,OHAps, perturbed by 175

OHAp, adsorption on 161

—, effect on dissolution 168

—, effect on precipitation 156

—, reaction with 160

Raman study of F⁻ and OHAp reaction 163, 286

Fluoroberyllate 84

Fluorophosphate

Ca-def OHAp, reaction with 163

enamel, reaction with 286

Francolite

a-axis reduction as CO₃ increases 193, 201, 202

birefringence increase with CO₃ 193

Cl analyses 202

CO₂ loss on heating 212

— retention on heating 252

CO₃ content of biaxial sectors 204

CO₃ orientation, from birefringence 204

— from IR 208

CO₃ replacing PO₄, first suggestion 193

—, polarised IR evidence for 210

—, XRD evidence for 206

CO₃, direction of transition moments 208

—, position in lattice 211

CO₃/PO₄ ratio from IR 145

composition of ideal end-member 201

compositional correlations 202

dahllite, IR spectrum comparison 211

discovery 191

Durango, single crystal XRD study 205

enamel, comparison with 270

Epirus, Rietveld analysis 206

F analyses 202

— excess 199, 202

— in PO₄ site 193, 199, 203

— loss on heating 212

FAp, from heating 192

formula, Epirus 206

—, general 201, 202, 246

—, Magnet Cove 205

H₂O loss on heating 212

IR spectra, polarised 209

—, comparison with dahllite 211

lattice parameters 200, 202

Magnet Cove, XRD single crystal study 205

Mg in 87

monoclinic symmetry suggested 204

OH analyses 202

— in PO₄ site 193

— IR polarised spectrum 212

optical properties 193, 204

—, biaxial 204
—, correlated with composition 203
solubility in sea water 258
structural changes on heating 206
synthesis, first 195, 223
thermal decomposition 206, 212
type specimen 191
unit cell contents correlated with
    a-axis 202
XRD, Rietveld 206
—, single crystal 204

Germanium, GeO₄ in apatite 94
Glaserite
    relation with apatite 71
    structure 71
Glycine incorporation in PbOHAp 136
Gypsum, relation with DCPD 24

Halide ions, deficiency and oversized in
    apatite 81
Halophosphate phosphors 89, 91
Halophosphates 63
Hilgenstockite 50
Holly Springs hydroxyapatite
    OH IR 175
    optical properties and lattice
      parameters 111
Hydrogen
    OHAp, determination of 147
    —, position in 113
Hydrogen phosphate, *see* Acid
    phosphate
Hydroxyapatite
    see also Hydroxyapatite, Ca-deficient
    ACP, from hydrolysis of 57, 58
    adsorption on 148, 159
    amino-2-ethylphosphate in 136
    Ba,CaOHAps, IR spectra 181
    biomaterials use 295
    Ca-rich 127
    CaF₂, reaction with 139

CaO, reaction with 52
Cl,OHAp, IR spectrum 178
—, preparation 135
—, Raman spectrum 178
CO₂, reaction with 213
CO₃ uptake in solution 230
crystal growth 154, 156, 293
—, effect of Cl 156
—, effect of F 156
DCPD, from hydrolysis of 28
—, singular point with 28
deuterated, IR 170
deuterium exchange kinetics 147
diffusion in 106
dissolution, inhibitors 168
—, rate of 164, 166
—, single crystals 168
—, subsurface 168
elastic constants 187
electric dipole relaxations 117
electrical conductivity 187
epitaxial relations with CaPs 74
EXAFS spectrum 60, 188
F,OHAp, OH assignments in IR
    178
—, preparation 135
—, Raman spectrum 178
—, solubility 158
F⁻ reduction of dissolution 168
—, reaction with 160
—, reaction with, NMR studies 186
F-H distances in F,OHAp 177, 184
H-content from D-exchange 147
heats of immersion 159
Holly Springs 111
Hospenthal 111
IR and Raman assignments 171
lattice images in EM 115, 290
lattice modes 172
lattice parameters 111, 118
—, best estimate 121
—, monoclinic 112

Mg, effect on precipitation 87

monoclinic structure, loss of 115

— to hexagonal phase transition 116

—, IR and Raman spectra 172

Na in single crystals 137

neutron diffraction 112

NH_3 gas, reaction with 175

NMR changes with temperature 185

— images 289

— of F reaction with OHAp 186

— spectrum 182

—, F,OHAp powders 185

—, F,OHAp single crystals 184

O_2, formation of peroxyapatites with 131

OCP, from heating 19

—, intercrystalline mixtures with 17, 153, 293

—, relation with 13, 71

OH columns, coherence from NMR 182

— ions, in phase transition 117

— ions, ordering 112

OH IR bands, additional bands in various 174

—, high temperature 174

—, intensity loss after heating 129

—, perturbation by F⁻ and Cl⁻ 175

—, surface 175

optical properties 111

— in phase transition 116, 117

—, monoclinic 137

oriented, from aragonite 198

—, from OCP 16, 125

plasma spraying 297

polycrystalline blocks 295

porous 248

— ceramics 296

preparation, from $CaSO_4$ 124

—, large scale 125

—, miscellaneous 124

—, monoclinic, from ClAp 121, 124

—, monoclinic, solid state 121

—, s-OHAp discussed 118

—, s-OHAp, aqueous 119, 122, 124

—, solid state 119, 121, 124

Raman spectrum in phase transition 117

single crystals, growth of 137

SnF_2, reaction with 163

solubility 157

—, F,OHAp 158

Sr,CaOHAps, IR spectra 181

stoichiometry of 118

structure 67, 112

— of various 116

—, monoclinic 113

surface charges 158

— chemistry 159, 282

— measurement, active 147

thermal conductivity 187

— decomposition 128, 188

— decomposition, IR study 179

— diffusivity 187

— expansion 187

thermodynamic constants 187

thermoluminescence 186

water adsorption isotherms 160

—, constitutional, loss 128

XRD pattern, hexagonal 311

—, monoclinic 113

zeta potentials 159

Hydroxyapatite, Ca-deficient

see also Hydroxyapatite

α-TCP, from hydrolysis of 47

ACP, from solid-state hydrolysis 55, 149

Ca/P from β-TCP on heating 146

crystal morphology 153

IR spectrum 180

lattice parameters 126, 151

—, effect of H_2O 151

—, effect of HPO_4 151

—, reversible change in steam 151
models for lattice substitutions 149
monofluorophosphate, reaction with 163
NMR spectrum 151, 182
OCP, intercrystalline mixtures with 151, 153
OH IR, additional bands at higher temperatures 173
OH librational intensity 181
PO$_4$ IR bands 180
preparation 118, 125
— from solid ACP 149
—, constant composition 127, 151
—, ion release from EDTA 126
—, well-crystallised 125
structure 148, 153
thermal decomposition 126, 152, 188
water loss on heating 152
Hydroxyl
    Ca-def OHAps, content of 150
    CO$_3$Aps, IR intensity reduced 235
    —, librational IR intensity reduced 268
    ionic size 69, 84
    OHAp, IR and Raman intensity in 178
    —, IR of surface 175
    —, measurement in 139
    orientation in enamel apatite 267
Hydroxylapatite 111
    see Hydroxyapatite

"Inverse" carbonate apatites 229
Iodide
    apatite, in 81
    ionic radius 84
Iodoapatite 81
Ion pairs 2, 4, 27, 33
IR spectroscopy
    CO$_3$ determination 145

CO$_3$/PO$_4$ ratio in francolites 145
HPO$_4$ determination 144
OH determination 140
Iron
    apatite, in 88
    ionic radius 84
    whitlockite, effect on precipitation 44
    —, in 42

Lanthanides, *see* rare-earths
Lanthanum in apatite 93, 97
Laser(s)
    apatites used as 109
    enamel, effect on 279
Lattice modes, apatite 172, 179
Lead
    Ca,PbOHAp 134
    diffusion in FAp 106
    glycine in PbOHAp 136
    ionic radius 84
    ns-PbOHAp, structure 149
    Pb,NaAp, structure 84
    PbAsO$_4$OHAp 134
    —, IR OH bands 181
    —, single crystals 138
    PbOAp, lattice parameters 82
    —, thermal stability 134
    PbOHAp, preparation 134
    —, single crystals 138
    —, thermal decomposition 129
    PbVO$_4$OHAp 134
    —, IR OH bands 181
    —, single crystals 138
Lithium
    apatite, in 90
    CO$_3$Aps precipitated, in 242
    OHAp, effect on growth 156

Magnesium
    β-TCa,MgP, effect on solubility 46
    —, effect on thermal stability 49

—, structure 40

ACP, effect on formation 54

—, effect on hydrolysis 56

—, effect on solubility 58

—, effect on thermal stability 61

apatites, substitution in 86, 200

— biological, effect on *c*-axis 265, 281

CaPs, effect on precipitation 58

CO₃Aps, preferential loss on dissolution 285

DCPD, effect on growth 27

—, effect on hydrolysis 29

dentine, in 87

francolite, correlated with composition 87, 202

—, in 87

ionic radius 84

loss in dental caries 285

MgFAp, growth of single crystals 86

OCP, effect on growth 15

—, effect on hydrolysis 18

OHAp, effect on growth 154, 156

—, effect on precipitation 87

pyrophosphate yield reduction 142

whitlockite, effect on lattice parameters 45

—, effect on precipitation 44

—, effect on solubility 46

—, presence in 35

Manganese

apatite, in 89, 94

—, position in lattice 89

ESR of Mn in ClAp and FAp 79

ionic radius 84

whitlockite, effect on precipitation 44

—, in 42

Martinite 35, 42

optical properties 43

Mimetite 6

Monetite

see also, Dicalcium phosphate anhydrous

discovery 30

Monocalcium phosphate anhydrous

lattice parameters 10

optical properties 11

preparation 11

solubility 9

structure 10

substituted 11

thermal decomposition 11

XRD pattern 307

Monocalcium phosphate monohydrate

lattice parameters 10

optical properties 11

preparation 11

solubility 9

structure 10

thermal decomposition 11

XRD pattern 307

Monofluorophosphate, reaction with Ca-def OHAp 163

Nasonite, apatite structure of 71

Neodymium in apatite 91

Nitrated apatites 81, 214

Nitrogen

apatite, silicon oxynitrides 91, 94

CO₃Aps, loss on heating 250, 276

cyanamide apatite 254

NCO⁻ and NCN²⁻ in heated CO₃Aps 253, 278

NO₃⁻, NO₂⁻ and NO₂²⁻ in apatite 81, 214

NMR spectroscopy

β-TCP 50

A-CO₃Ap 222

ACP 58, 184, 256

—, solid state hydrolysis of 56

apatites, biological 266, 287

B-CO₃Aps, precipitated 256

Ca-def OHAp 151, 182
DCPA 34
DCPD 30
F,OHAp single crystals 184
F⁻ reaction with OHAp 162
FAp single crystals 105
H₂O determination 141
HPO₄ determination 141, 145
— in OHAp 183
OCP 21
OH determination 141
OH,FAp, changes with temperature 185
—, correlation with IR 177
OHAp 182
Sb-doped FAp 92
Nowotny phases, relation with apatite 74

Octacalcium phosphate
ACP, from hydrolysis of 57
amorphous 55
biomineralisation, in 262, 266
Ca-def OHAp, absence in 151
CaPs, in precipitation of 155
CO₃Ap, during precipitation 247
collapsed 17, 19
crystal growth of 15, 156
DCPD, from hydrolysis of 28
—, singular point with 28
density 15
dicarboxylic acids, double salts with 22
history 12
hydrolysis 17
— in presence of CO₃ 241
— to oriented OHAp 125
IR and Raman spectra 20
lattice parameters 12
NMR spectrum 21
OHAp, intercrystalline mixtures with 17, 153, 293

—, relation with 13, 71
optical properties 13
oriented growth on ion-selective membranes 16
polymorphs 21
preparation 13
solubility product 16
structure 12
thermal decomposition 18
XRD pattern 308
Oxalate precipitation
HPO₄ determination 140, 143
OH determination in BCaPs 140
Oxyapatite
biomaterials use 295
formation in heated enamel 279
hardness 104
IR spectrum 179
laser hosts 109
lattice parameters 129, 130
O,OHAp solubility 157
OH ion, possible need for 127
OHAp, rehydroxylation to 129
porous ceramics 296
preparation from A-CO₃Ap 130
— from OHAp 128
rare-earth 86, 90, 93, 131
—, single crystals 91
single crystals, growth of 91, 97
thermal decomposition to α-TCP and TetCP 129
Oxygen
apatites, biological ashed, in 80
—, O-rich 132
—, peroxide in 131
—, superoxide in 133

Peptides, adsorption on OHAp 159
Peroxyapatites 131
Phase diagrams
Ca(OH)₂-CaCO₃-Ca₃(PO₄)₂-H₂O 197

$Ca(OH)_2$-H_3PO_4-H_2O 4
$CaCl_2$-ClAp   104
CaF_2-FAp   104
CaO-P_2O_5 47
CaO-P_2O_5-H_2O 1, 10
   quaternary   2
SrO-P_2O_5 47
Phosphate(s)
   alkaline earth   1
   enamel, IR dichroism   269
   FAp, IR and Raman assignments
      102
   francolite, IR dichroism   212
   ionic size   69
   IR with divalent ion on apatite hexad
      axis   179
   OHAp, IR bands in   172
Phosphonates, effect on growth of
   OHAp   154
Phosphorite   7, 229
      see also   Rock-phosphate   and
         Francolite
   CO_3/PO_4 ratio by IR   145
   lattice parameters, relation with unit
      cell contents   199
   refractive index correlated with CO_3
      203
Phosphors   89, 91
Phosphorus
   diffusion in FAp   105
   — in OHAp   106
Podolite   191
Polyphosphate(s)
   alkaline earth   1
   OCP, from heating   18
   OHAp, effect on growth   154
Potassium
   CO_3Aps precipitated, in   241, 247
   ionic radius   84
   MCPA, in   11
   PbKAp, structure   84

Proteins
   OHAp, adsorption on   159
   —, effect on dissolution   168
Pyromorphite   6
   normal coordinate analysis   100
Pyrophosphate(s)
   ACP, effect on formation   54
   —, effect on hydrolysis   56
   —, effect on thermal stability   61
   —, from heating   61
   apatites biological, from heating
      194, 260, 261, 277
   Ca-def OHAp, from heating   126,
      142, 188
   calcium   1
   CO_3Aps precipitated, from heating
      229
   DCPA, from heating   29, 34
   OCP, effect on growth   15
   —, effect on hydrolysis   18
   —, from heating   19
   OHAp, effect on growth   156

Quercerite   191

Rare-earths
   apatite, growth of single crystals   97
   —, substitutions in   90
   diffusion in FAp   106
   Durango FAp, in   63
   oxyapatites   86, 93
Rhenium
   A-BaReO$_5$CO$_3$Ap,   single   crystals
      198
   —, structure   219
   apatite, substitutions in   94
   BaReO$_5$OAp, lattice parameters   81
Rietveld analysis   9
   A-CO$_3$Ap   219
   apatite hexad axis content   150
   dental enamel   264

Epirus francolite 206
Mn position in apatite 90
Pb apatites 85
Sb position in apatite 92
Rock-phosphate 7, 9, 23, 30, 192, 203
see also Phosphorite and Francolite
thermal decomposition 212
Rubidium in precipitated CO_3Aps 242

Samarium in apatite 91
Samuelsonite, relation with apatite 71
Selenium, SeO_4 in apatite 94
Silicate
in apatites 86, 90, 93, 94, 97
oxyapatites, laser host 109
Silicon, oxynitrides in apatite 91, 94
Silver in apatite 90
Site-group approximation 99
Sodium
AB-CO_3Aps high temperature, in 227
Ca,NaOHAp 137
CO_3Aps precipitated, in 239, 247
CO_3Aps, CO_3 content influenced by 227, 240, 241
francolite, in 199, 203
ionic radius 84
PbNaAp, structure 84
Solubility isotherms
CaPs 4
—, acidic region 10
Solubility product
β-TCP 46
B-CO_3F,OHAps 258
CaF_2 161
CaPs, calculation of 3
DCPA 33
DCPD 28
enamel apatite 282
FAp 158
MCPM 9

OCP 16
OH,FAp 158
OHAp 157
Speciation diagram, CaPs 5
Staffelite 191
Stanfieldite 41
Strontium
A-Sr,EuCO_3Ap, lattice parameters 214
A-SrAsO$_4$$CO_3$Ap, lattice parameters 214
A-SrCO_3Ap, IR spectrum 218
—, lattice parameters 214
—, preparation 194
apatite, preference for Ca(2) site 85
B-SrCO_3FAp 224
bone apatite, in 261, 266
Ca,SrOHAps 85, 134
—, IR band shifts 181
DCPD, effect on growth 27
diffusion in FAp 106
— in OHAp 106
F$^-$ synergist effect for enamel solubility reduction 287
ionic radius 84
O-rich Aps 133
peroxyapatites 131
Sr,NaBO_2Ap, structure 81, 84
SrAsO$_4$OHAp 134
SrFAp, growth of single crystals 97
SrOAp, possible new form 130
SrOHAp, heats of immersion 159
—, IR assignments 181
—, lattice parameter changes on heating 129
—, lattice parameters 115, 134
—, preparation 134
—, structure 115
—, surface charges 159
SrVO$_4$OHAp 134
Sulphoapatite, europium 94

Sulphur
  calcium phosphate-sulphate hydrate 24
  europium sulphoapatite 94
  PbNaSO₄ClAp, structure 84
  S²⁻ in apatite 90
  SO₄ in apatite hexad axis 82
  — replacing PO₄ in apatite 94
Superoxide ion, location in O-rich Aps 133
Superphosphate 9
Surface adsorption on Ca-def OHAps 148

Tetracalcium phosphate
  apatite, relation with 51
  DCPD or DCPA, room temperature reaction with 50
  density 51
  discovery 50
  HCl gas, reaction with 96
  isostructures 51
  lattice parameters 51
  occurrence 50
  OHAp, from heating 128, 279
  —, hydrolysis to 52
  optical properties 52
  plasma spraying, formation in 297
  preparation 52
  single crystals, growth of 52
  structure 51
  water, high temperature reaction with 52
  XRD pattern 310
Tetracycline, adsorption on OHAp 159
Thermal decomposition
  ACP 61
  apatites, biological 253, 261, 275
  B-CO₃Aps, precipitated 239
  B-CO₃FAps, precipitated 244
  Ca-def OHAp 188
  ClAp 64, 96

CO₃Aps, Na-containing 228
—, precipitated 228, 247, 250, 253, 275
—, summary 248
DCPA 33
DCPD 29
FAp 64
francolite 206, 212
MCPA 11
MCPD 11
OCP 18
OHAp 128, 188
PbOAp 134
Thermoluminescence
  apatites 109
  bone 186
  FAp 186
  OHAp 186
Thomas slag 50
Tin
  OHAp, effect of Sn²⁺ on dissolution 168
  —, reaction with SnF₂ 163
Tricalcium phosphates
  see also ᾱ-, α-, β-tricalcium phosphate
  A-CO₃Ap, from CaCO₃ heated with 213
  CaCl₂, reaction with 96
  CaF₂, reaction with 95
  IR spectra 50
  nomenclature 35
  polymorphs 34, 47, 49

Unit cell contents, determination of 147
Uranium in apatite 107

Vacancies
  apatite Ca sites 79, 84, 90
  apatite hexad axis 79-81, 84, 98, 109, 128, 179, 218

B-CO₃Aps precipitated, in 236
Ca-def OHAp, in 153, 183
OHAp, OH columns 182
—, on heating 128
Vanadate
  A-VO₄CO₃Ap, lattice parameters
    214
  apatite, substitution for PO₄ 94
  CdVO₄BrAp 81
  CdVO₄IAp 81
  VO₄OHAp, structure 115
Vanadinite 6
Vœlckerite 127

Water
  apatite biological, loss on heating
    276, 280
  apatite, location in 80, 153, 212,
    236, 241
  Ca-def OHAp, effect on lattice
    parameters 151
  —, loss on heating 152
  —, reversible thermal loss 151
  CO₃Aps precipitated, loss on heating
    247
  DCPD, loss from 29
  enamel, NMR study of 287
  francolite, analyses of 202
  —, loss on heating 212
  IR absorption in enamel mineral 269
  OCP, loss from 19
  —, loss on heating 20
  OHAp, adsorption on 160

—, determination of 141
—, loss on heating 128
whitlockite, loss on heating 50
Whitlockite
  acid phosphate in 35, 42
  CO₃-containing 42
  DCPA, from 33
  DCPD, from 29
  density 43
  discovery 35
  divalent ions, effect on precipitation
    44
  Fe in 42
  IR spectrum 50
  lattice parameters 42
  —, effect of Mg on 45
  Mn in 42
  nomenclature 35
  occurrence 35
  optical properties 43
  preparation 44
  solubility 46
  structure 37, 42
  — of Mn-containing 42
  thermal decomposition 50
Wollastonite 297

Yttrium
  in apatite 97
  in francolite 205

Zeta potentials, OHAp 159